Thermodynamics, Solubility and Environmental Issues

Thermodynamics, Solubility and Environmental Issues

Edited by

Trevor M. Letcher

Emeritus Professor,
University of KwaZulu-Natal,
Durban, South Africa

Amsterdam • Boston • Heidelberg • London • New York • Oxford
Paris • San Diego • San Francisco • Singapore • Sydney • Tokyo

Elsevier
Radarweg 29, PO Box 211, 1000 AE Amsterdam, The Netherlands
Linacre House, Jordan Hill, Oxford OX2 8DP, UK

First edition 2007

Library of Congress Cataloging-in-Publication Data
A catalog record for this book is available from the Library of Congress

British Library Cataloguing in Publication Data

Letcher, T. M. (Trevor M.)
Thermodynamics, solubility and environmental issues
1. Pollutants – Environmental aspects 2. Solubility
3. Thermodynamics
4. Environmental chemistry
I. Title
577.2′7
ISBN-13: 978-0-444-52707-3

ISBN-13: 9780444527073

For information on all Elsevier publications
visit our website at books.elsevier.com

Printed and bound by CPI Group (UK) Ltd, Croydon, CR0 4YY

Transferred to Digital Print 2011

Preface

Environmental problems are becoming an important aspect of our lives as industries grow apace with populations throughout the world. Solubility is one of the most basic and important of thermodynamic properties, and a property that underlies most environmental issues. This book is a collection of twenty-five chapters of fundamental principles and recent research work, coming from disparate disciplines but all related to solubility and environmental pollution. Links between these chapters, we believe, could lead to new ways of solving new and also old environmental issues. Underlying this philosophy is our inherent belief that a book is an important vehicle for the dissemination of knowledge.

Our book "Thermodynamics, Solubility and Environmental Issues" had its origins in committee meetings of the International Association of Chemical Thermodynamics and in discussions with members of the International Union of Pure and Applied Chemistry (IUPAC) subcommittee on Solubility, in particular Professor Glenn Hefter of the University of Monash, Perth, Dr Justin Salminen of the University of California, Berkeley and Professor Heinz Gamsjäeger of the University of Leoben, Austria. It is a project produced under the auspices of the IUPAC. In true IUPAC style, the authors, important names in their respective fields, come from many countries around the world, including: Australia, Austria, Canada, Brazil, Finland, France, Greece, Italy, Japan, Poland, Portugal, South Africa, Spain, United Kingdom and the United States of America.

The book highlights environmental issues in areas such as mining, polymer manufacture and applications, radioactive wastes, industries in general, agro-chemicals, soil pollution and biology, together with the basic theory and recent developments in the modelling of environmental pollutants – all of which are linked to the basic property of solubility. It includes chapters on:

- Environmental remediation
- CO_2 in natural systems and in polymer synthesis
- Supercritical fluids in reducing pollution
- Corrosion problems
- Plasticizers and environmental pollution
- Surfactants
- Radioactive waste disposal problems
- Ionic liquids in separation processes
- Biodegradable polymers
- Pesticide contamination in humans and in the environment
- Pollution in a mining context
- Green chemistry
- Environmental pollution in soils
- Heavy metal and groundwater issues
- Biological uptake of aluminium
- Environmental problems related to gasoline and its additives
- Ionic liquids in the environment
- Body fluids and solubility
- Biosurfactants and the environment
- Problems related to polymer production
- Basic theory and modelling, linking thermodynamics and environmental pollution

I wish to record my special thanks to Professor Glenn Hefter, Professor Heinz Gamsjäeger and Dr Justin Salminen, who were part of the task team, to the 49 authors and to Joan Anuels of Elsevier, our publishers. They have all helped in producing, what we believe will be a useful and informative book on the importance and applications of solubility and thermodynamics, in understanding and in reducing chemical pollution in our environment.

Trevor M. Letcher
Stratton-on-the-Fosse, Somerset
2 September 2006

Foreword

Thermodynamics, Solubility and Environmental Issues – this book project was started in 2005; it consists of 25 chapters which highlight the importance of solubility and thermodynamic properties to environmental issues.

The opening chapter "An introduction to modelling of pollutants in the environment" by Trevor M. Letcher demonstrates convincingly that equilibrium concepts and simple models lead to realistic predictions of, for example, the concentration of a polychlorinated biphenyl in the fishes of the St. Lawrence River. Relative solubilities expressed by octanol–water and air–water partition coefficients play a crucial role for estimating the distribution of chemicals in the environment. This is pointed out in the introductory chapter as well as in others such as "Estimation of volatilization of organic chemicals from soil" by Epaminondas Voutsas.

Solubility phenomena between solid and aqueous phases are treated in the chapters "Leaching from cementitious materials used in radioactive waste disposal sites" by Kosuke Yokozeki, "An evaluation of solubility limits on maximum uranium concentrations in groundwater" by Teruki Iwatsuki and Randolf C. Arthur, and "The solubility of hydroxyaluminosilicates and the biological availability of aluminium" by Christopher Exley.

Supercritical fluids can be applied to remove polluting materials from the environment. Theory and practice of this technology is of increasing interest at the present time. In "Supercritical fluids and reductions in environmental pollution" by Koji Yamanaka and Hitoshi Ohtaki focus their attention to start with the thermodynamics and structure of supercritical fluids and then describe the supercritical water oxidation process, the extraction of pollutant from soils with supercritical carbon dioxide and other supercritical fluids, and recycle of used plastic bottles with supercritical methanol. Andrew I. Cooper et al. report on "Supercritical carbon dioxide as a green solvent for polymer synthesis". Carbon dioxide is an attractive solvent alternative for the synthesis of polymers; it is non-toxic, non-flammable, inexpensive, and readily available in high purity. These authors have also developed methods for producing CO_2-soluble hydrocarbon polymers or "CO_2-philes" for solubilization, emulsification, and related applications.

Prashant Reddy and Trevor M. Letcher outline the possibilities to use ionic liquids for industrial separation processes in their chapter "Phase equilibrium studies on ionic liquid systems for industrial separation processes of complex organic mixtures". Ionic liquids are a very promising class of solvents which, after extensive research and development, will eventually be applied for the separation of industrially relevant organic mixtures.

Regarding the chapters not yet mentioned suffice it to say that this book altogether elucidates the interplay of solubility phenomena, thermodynamic concepts, and environmental problems. I congratulate the editor and the authors on this remarkable achievement in such a comparatively short time.

Heinz Gamsjäger
Chairman of the IUPAC Analytical Chemistry Division (V),
Subcommittee on Solubility and Equilibrium Data
Professor Emeritus, Montanuniversitiät, Leoben, Austria

List of Contributors

Part I: Basic Theory and Modelling

1. *An Introduction to Modelling of Pollutants in the Environment* by Trevor M. Letcher. Email: Trevor@letcher.eclipse.co.uk

Professor Trevor Letcher, School of Chemistry, University of KwaZulu-Natal, Durban 4041, South Africa. Phone and Fax +44-1761-232311.

2. *Modeling the Solubility in Water of Environmentally Important Organic Compounds* by Ernesto Estrada, Eduardo J. Delgado and Yamil Simón-Manso. Email: estrada66@yahoo.com

Dr Ernesto Estrada, Edificio CACTUS, University of Santiago de Compostela, 15782 Santiago de Compostela, Spain. Phone +34 981 563100, Fax 547077.

3. *Modeling of Contaminant Leaching* by Maria Diaz and Defne Apul. Email: defne.apul@utoledo.edu

Professor Defne Apul, Department of Civil Engineering, University of Toledo, 2801 W Bancroft St, Mail Stop 307, Toledo, OH 43606, USA. Phone +1-419-5308132.

Part II: Industry and Mining

4. *Supercritical Fluids and Reductions in Environmental Pollution* by Koji Yamanaka and Hitoshi Ohtaki†. Email: yamana-k@organo.co.jp

Dr Koji Yamanaka, Organo Co. of Japan, 1-4-9 Kawagishi Toda, Saitama 335-0015, Japan. Phone +81-48-446-1881, Fax +81-48-446-1966.

†Deceased

5. *Phase Equilibrium Studies on Ionic Liquid Systems for Industrial Separation Processes of Complex Organic Mixtures* by Prashant Reddy and Trevor M. Letcher. Email: Trevor@letcher.eclipse.co.uk

Professor Trevor Letcher, School of Chemistry, University of KwaZula-Natal, Durban 4041, South Africa. Phone and Fax +44-1761-232311.

6. *Environmental and Solubility Issues Related to Novel Corrosion Control* by William J. van Ooij and P. Puomi. Email: vanooijwj@email.uc.edu

Professor William van Ooij, Department of Chemical and Materials Engineering, 497 Rhodes Hall, University of Cincinnati, Cincinnati, OH 45221-0012, USA. Phone +1 513 5563194, Fax +1 513 5563773.

7. *The Behavior of Iron and Aluminum in Acid Mine Drainage: Speciation, Mineralogy, and Environmental Significance* by Javier S. España. Email: j.sanchez@igme.es

Professor Javier Sánchez España, Mineral Resources and Geology Division, Geological Survey of Spain, Rios Rosas 23, 28003 Madrid, Spain. Phone +34-913-495740, Fax +34-913-495834.

Part III: Radioactive Wastes

8. *An Evaluation of Solubility Limits on Maximum Uranium Concentrations in Groundwater* by Teruki Iwatsuki and Randolf C. Arthur. Email: iwatsuki.teruki@jaea.go.jp

Dr Teruki Iwatsuki, Mizunami Underground Research Laboratory, 1-64 Yamanouchi, Akeyo-Cho, Mizunami-Shi, Gifu 509-6132 JNC, Japan. Phone +81 572 662244, Fax +81 572 662245.

9. *Leaching from Cementitious Materials Used in Radioactive Waste Disposal Sites* by Kosuke Yokozeki. Email: yokozeki@kajima.com

Dr Kosuke Yokozeki, Kajima Technical Research Institute, 19-1-2 Tobitakyu, Chofu-shi, Tokyo 182-0036, Japan. Phone +81-424-897816, Fax +81-424-897078.

Part IV: Air, Water, Soil and Remediation

10. *Solubility of Carbon Dioxide in Natural Systems* by Justin Salminen, Petri Kobylin and Anne Ojala. Email: justin@newman.cchem.berkeley.edu

Dr Justin Salminen, Department of Chemical Engineering, University of California, Berkeley, CA 94720-1462, USA. Phone 1(510) 642 1972, Fax 1(510) 642 4778.

11. *Estimation of Volatilization of Organic Chemicals from Soil* by Epaminondas Voutsas. Email: evoutsas@chemeng.ntua.gr

Professor Epaminondas C. Voutsas, School of Chemical Engineering, National Technical University of Athens, 9 Heroon Polytechniou Street, Zographou Campus, GR-15780 Athens, Greece. Phone +30 210-7723971, Fax +30 210-7723155.

12. *Solubility and the Phytoextraction of Arsenic from Soil by Two Different Fern Species* by Valquiria Campos. Email: vcampos@usp.br

Dr Valquiria Campos, Polytechnic School, Chemical Engineering Department, University of São Paulo, Rua Marie Nader Calfat, 351 apto 71 Evoluti, Morumbi, São Paulo, SP, Brazil 05713-520. Phone +55 30914663, Fax +55 30313020.

13. *Environmental Issues of Gasoline Additives – Aqueous Solubility and Spills* by John Bergendahl. Email: jberg@wpi.edu

Professor John Bergendahl, Department of Civil and Environmental Engineering, Worcester Polytechnic Institute, Worcester, MA, USA. Phone +1 508-8315772, Fax +1 508-8315808.

14. *Ecotoxicity of Ionic Liquids in an Aquatic Environment* by Daniela Pieraccini, Cinzia Chiappe, Luigi Intorre and Carlo Pretti. Email: dpieraccini@tiscali.it

Dr Daniela Pieraccini, Dipartimento di Chimica Bioorganica e Biofarmacia, University di Pisa, Via Bonanno 33, 56126 Pisa, Italy. Phone +39 050-2219700, Fax +39 050-2219660.

15. *Rhamnolipid Biosurfactants: Solubility and Environmental Issues* by Catherine N. Mulligan. Email: mulligan@civil.concordia.ca

Professor Catherine Mulligan, Department of Building, Civil and Environmental Engineering, 1455 de Maisonneuve Boulevard West, Concordia University, Montreal H3G 1M8, Canada.

16. *Sorption, Lipophilicity and Partitioning Phenomena of Ionic Liquids in Environmental Systems* by Piotr Stepnowski. Email: sox@chem.univ.gda.pl

Professor Piotr Stepnowski, Faculty of Chemistry, University of Gdansk, ul. Sobieskiego 18/19, PL80-952 Gdańsk, Poland. Phone +48 58 5235448, Fax +48 58 5235577.

17. *The Solubility of Hydroxyaluminosilicates and the Biological Availability of Aluminium* by Christopher Exley. Email: c.exley@chem.keele.ac.uk

Professor Christopher Exley, Birchall Centre for Inorganic Chemistry and Materials Science, Lennard-Jones Labs, Keele University, Staffordshire ST5 5BG, UK. Phone +44 1782 584080, Fax +44 1782 712378.

18. Apatite *Group Minerals: Solubility and Environmental Remediation* by M. Clara F. Magalhães and Peter A. Williams. Email: mclara@dq.ua.pt

Professor Clara Magalhães, Departamento de Quimica, Universidade de Aveiro, P-3810 Aveiro, Portugal. Phone +351 234-370200, Fax +351 234-370084.

Part V: Polymer Related Issues

19. *Solubility of Gases and Vapors in Polylactide Polymers* by Rafael Auras. Email: aurasraf@msu.edu

Dr Rafael Auras, School of Packaging, Michigan State University, East Lansing, MI 48824-1223, USA, Phone +1 517-4323254, Fax +1 517-3538999.

20. *Biodegradable Material Obtained from Renewable Resource: Plasticized Sodium Caseinate Films* by Jean-Luc Audic, Florence Fourcade and Bernard Chaufer. Email: jean-luc.audic@univ-rennes1.fr

Dr Jean-Luc Audic, Laboratoire Rennais de Chimie et d'Ingenierie, ENSCR – Universite Rennes 1, 35700 Rennes, France. Phone +33-223-235760, Fax +33 223 235765.

21. *Supercritical Carbon Dioxide as a Green Solvent for Polymer Synthesis* by Colin D. Wood, Bien Tan, Haifei Zhang and Andrew I. Cooper. Email: aicooper@liverpool.ac.uk

Professor Andrew I. Cooper, Donnan and Robert Robinson Laboratories, Department of Chemistry, University of Liverpool, Liverpool L69 3BX, UK.

22. *Solubility of Plasticizers, Polymers and Environmental Pollution* by Ewa Białecka-Florjańczyk and Zbigniew Florjańczyk. Email: evala@ch.pw.edu.pl

Professor Zbigniew Florjańczyk, Faculty of Chemistry, Warsaw University of Technology, ul. Noakowskiego 3, PL-00 664 Warsaw, Poland. Phone +48 22 2347303, Fax +48 22 2347271.

Part VI: Pesticides and Pollution Exposure in Humans

23. *Solubility Issues in Environmental Pollution* by Alberto Acre and Ana Soto. Email: eqaaarce@usc.es

Professor Alberto Arce, Department of Chemical Engineering, University of Santiago de Compostela, E-15782 Santiago de Compostela, Spain. Tel. +34 981 563100 ext 16790, Fax +34 981 528050.

24. *Hazard Identification and Human Exposure to Pesticides* by Antonia Garrido Frenich, F.J. Egea González, A. Marín Juan and J.L. Martínez Vidal. Email: agarrido@ual.es

Dr Antonia Garrido Frenich, Department of Hydrogeology and Analytical Chemistry, University of Almeria, 04071 Almeria, Spain. Phone +34 950015985, Fax +34 950015483.

25. *Solubility and Body Fluids* by Erich Königsberger and Lan-Chi Königsberger. Email: koenigsb@murdoch.edu.au

Dr Erich Königsberger, School of Chemical and Mathematical Sciences, Murdoch University, Murdoch, WA 6150, Australia.

Table of Contents

PART III. RADIOACTIVE WASTES

PART IV. AIR, WATER, SOIL AND REMEDIATION

PART V. POLYMER RELATED ISSUES

PART VI. PESTICIDES AND POLLUTION EXPOSURE IN HUMANS

PART I

BASIC THEORY AND MODELLING

Thermodynamics, Solubility and Environmental Issues
T.M. Letcher (editor)

Chapter 1

An Introduction to Modelling of Pollutants in the Environment

Trevor M. Letcher

School of Chemistry, University of KwaZulu-Natal, Durban, 4041, South Africa

1. INTRODUCTION

This chapter serves as an introduction to our book. We will consider where a pollutant (for example, a fertilizer or herbicide), once introduced into the environment, will end up. When any chemical is released into the air or water, or sprayed on the ground, it will ultimately appear in all parts of the environment which includes the upper and lower atmosphere, lakes and oceans and the soil, and in all animal and vegetable matter, including our bodies. We will use simple models [1] for estimating the amount of a chemical distributed in various parts of the environment, commonly called environmental compartments, and throughout the food chain. We will show the importance of solubility data in these calculations and predictions. Furthermore, our approach will be underpinned by basic thermodynamic principles.

Even at equilibrium, a chemical will be present in quite different concentrations in the different environmental compartments. For example, if a volatile chemical is dissolved in water, its molar concentration in the air above the water, at equilibrium, is generally quite different from its concentration in the water [1, 2].

2. PARTITION COEFFICIENTS

The modelling of the partitioning of a chemical (pollutant) between water and lipid material (body fat) is usually done through a knowledge of the octanol–water partition coefficient, defined as:

$$K_{\mathrm{OW},i} = \frac{c_i^{\mathrm{O}}}{c_i^{\mathrm{W}}} \qquad (1)$$

where c_i^{O} and c_i^{W} are the equilibrium concentrations of species i in the octanol rich and the water layers, respectively, of a mixture of octanol and water (partially miscible at 298.15 K), as depicted in Fig. 1.

n-Octanol, a simple compound, has been chosen as the chemical solvent because it behaves in a similar way to lipid material (a complex material), which includes the fat in our bodies and waxes in plants. In many cases it is not the best model for lipids but it has now been widely accepted. There are so much K_{OW} data available in the literature that it is almost impossible to change now. Correction factors are often applied to K_{OW} when the properties of a particular lipid material are known.

Hydrophobic compounds have large K_{OW} values. The insecticide DDT has a K_{OW} value of $\sim 10^6$ and phenol, a hydrophilic compound, has a K_{OW} of 29 at 298.15 K. Thus, under chemical equilibrium conditions, the concentration of DDT in lipid material will be 10^6 greater than it is in water. This lipid material could indeed be the fat in fish.

Another partition coefficient that is useful in understanding the large difference in concentration of a chemical species i in air (A) and water (W) is the air–water partition coefficient, K_{AW}:

$$K_{\mathrm{AW},i} = \frac{c_i^{\mathrm{A}}}{c_i^{\mathrm{W}}} \tag{2}$$

The air–water partition coefficient is related to the Henry's law constant for sparingly soluble species. Let us consider a species i at equilibrium in a system of air and water. Equating partial pressures of i, p_i, in the air and the water

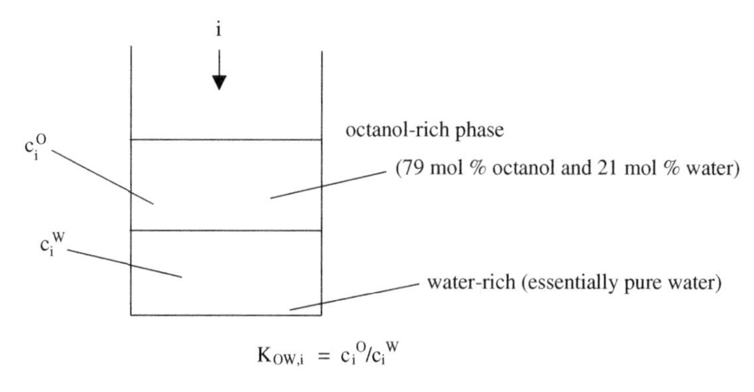

Fig. 1. *n*-Octanol–water partitioning at 298.15 K.

(or equating fugacities or chemical potentials) and using the ideal gas law equation and Henry's law (see Fig. 2):

$$p_i V = n_i RT \tag{3}$$

where n_i is the amount (moles) of i in the air. Applying Raoult's law at low solute concentrations (Henry's law) with one correction factor (the activity coefficient, γ_i^∞):

$$p_i = x_i^{\mathrm{W}} \gamma_i^\infty p_i^{\mathrm{vap}} \tag{4}$$

where x_i^{W} is the mole fraction of i in the water. Rearranging and equating we obtain:

$$p_i = \frac{n_i}{V} RT = c_i^{\mathrm{A}} RT \tag{5}$$

$$p_i = x_i^{\mathrm{W}} \gamma_i^\infty p_i^{\mathrm{vap}} = x_i^{\mathrm{W}} H_i = c_i^{\mathrm{W}} H_i' \tag{6}$$

where H_i and H_i' are the Henry's law constants related to mole fractions and concentrations, respectively. Hence, $K_{\mathrm{AW},i} = c_i^{\mathrm{A}}/c_i^{\mathrm{W}} = H_i'/RT$.

At saturation or the solubility limit, c_i^{W} is the solubility in water, c_i^{WS}, and

$$c_i^{\mathrm{A}} = \frac{p_i^{\mathrm{vap}}}{RT}$$

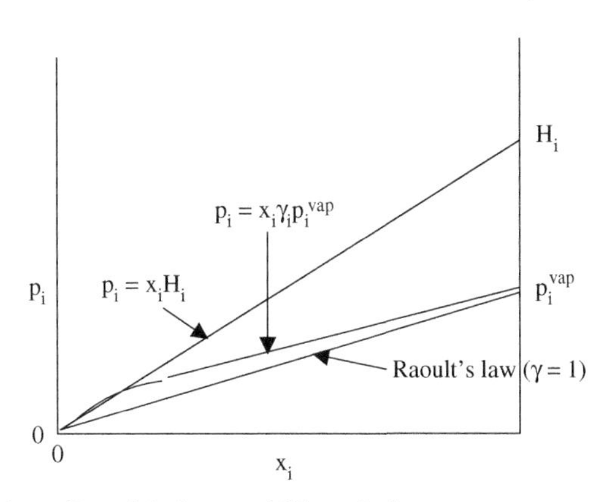

Fig. 2. Ideal behaviour, Raoult's law and Henry's law.

thus

$$K_{\mathrm{AW},i} = \frac{p_i^{\mathrm{vap}}}{c_i^{\mathrm{WS}} RT}$$

This only applies if the solubility is low, which is the case for many pollutants. By way of example [1, 2], the K_{AW} for *n*-hexane is 74 and for DDT it is $\sim 1 \times 10^{-3}$. Thus, if a species *i* is in chemical equilibrium in different phases there can be orders of magnitude difference in the concentration of *i* in the different phases (see Fig. 3).

The Henry's law constant, K_{AW}, and the activity coefficient, γ, can be estimated for a specific chemical from its measured solubility in water. This represents saturation conditions. At saturation, for a very sparingly soluble compound, a pure chemical phase can exist in equilibrium with the saturated solution; thus, Eq. (4) becomes:

$$p_i = p_i^{\mathrm{vap}} = x_i^{\mathrm{W}} \gamma_i^{\infty} p_i^{\mathrm{vap}} \tag{7}$$

and thus, $x_i^{\mathrm{W}} \gamma_i^{\infty} = 1.0$ and $x_i^{\mathrm{W}} \gamma_i^{\infty} = 1/\gamma_i^{\infty}$.

The mole fraction solubility is thus the reciprocal of the activity coefficient. Substances that are sparingly soluble in water have large activity coefficients. They also tend to have large Henry's law constants and air–water partition coefficients, because, as Eq. (6) shows:

$$H_i = \gamma_i^{\infty} p_i^{\mathrm{vap}}$$

The very low vapour pressure of DDT is largely offset by its large activity coefficient. Solubility thus plays a key role in determining the partitioning of chemicals from water to air, lipids and other phases such as soil and sediments.

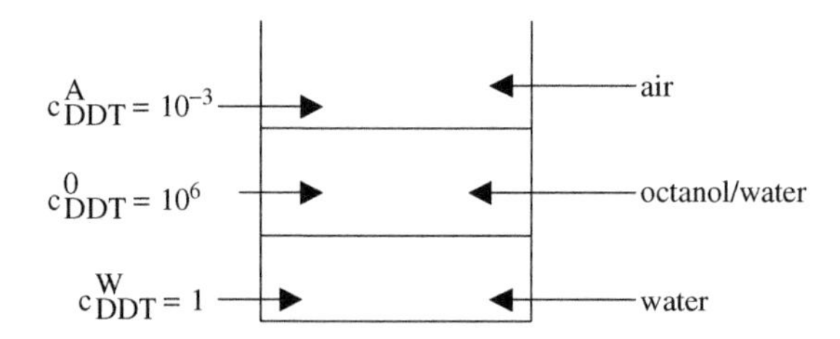

Fig. 3. The partitioning of DDT between water in octanol–water–air mixture based on Eqs. (1 and 2).

3. MODEL ENVIRONMENTS

In order to predict the concentrations of a chemical species in the various regions of a given environment, it is necessary to know the size of each region. These regions or compartments can include the atmosphere (air), water (rivers and lakes), soil (usually to a depth of ~15 cm), river and lake sediments, fish (aquatic biota) and plant material (terrestrial biota). In this chapter we will focus on air, water, soil and aquatic biota. Other compartments such as aerosols, concrete surfaces, etc., can also be considered. The size of each depends on the particular circumstance. By way of example, the sizes of four compartments, including a lake and fish, are given in Table 1 together with relevant properties, needed for the modelling process [1, 2]. These values will be used throughout this chapter.

4. EQUILIBRIUM PARTITION

The criterion for phase equilibrium of a chemical, i, between two phases is equal chemical potential or fugacity, i.e.

$$\mu_i' = \mu_i'' \quad \text{and} \quad f_i' = f_i'' \tag{8}$$

or in the case of ideal vapours:

$$p_i' = p_i'' \tag{9}$$

The equilibrium condition for a species i, partitioning between two liquid phases, e.g. water (W) and chemical (c), is:

$$x_i^{\mathrm{W}} \gamma_i^{\mathrm{W}} p_i^{\mathrm{vap}} = x_i^{\mathrm{c}} \gamma_i^{\mathrm{c}} p_i^{\mathrm{vap}} \tag{10}$$

Table 1
Properties of a model environment

Environmental region	Volume, V (m^3)	f	Density, d (kg m^{-3})
Air	1×10^9		1.2
Water	1×10^6		1000
Soil	15×10^4	0.02	1500
Biota	10	0.05	1000

Note: Here f is the mass fraction of organic material in the particular region, soil or biota.

This would apply to the octanol–water phases and

$$x_i^{\mathrm{W}}\gamma_i^{\mathrm{W}}p_i^{\mathrm{vap}} = x_i^0\gamma_i^0 p_i^{\mathrm{vap}} \tag{11}$$

For very hydrophobic species, x_i^{W} is very small, $\gamma_i^{\mathrm{W}} \approx \gamma_i^{\infty}$, $x_i^0 \approx 1$ and $\gamma_i^0 \approx 1$, so $x_i^{\mathrm{W}} \approx 1/\gamma_i^{\infty}$. As a result, the insecticide, DDT, has a γ_i^{∞} value of $\sim 10^6$. Substituting these values and assuming thermodynamic equilibrium, we see that there can be a difference of many orders of magnitude between the concentration of a particular chemical species in one phase and the concentration of the same species in another phase (see Fig. 3).

We can use these concepts to calculate the relative concentrations of a chemical in equilibrium with air and water. For example, let us consider 1,1,1-trichloroethylene (T) – a common industrial pollutant in the environment defined in Table 1 [1, 2].

At 25°C, its vapour pressure, $p_{\mathrm{T}}^{\mathrm{vap}}$, is 12800 Pa and its solubility in water, $c_{\mathrm{T}}^{\mathrm{W}}$, is 730 g m^{-3}. Because it melts at -32°C we know that it is a liquid at 25°C. Its molar mass (MW) is equal to 133.4 g mol^{-1}, so $x_{\mathrm{T}} = (c_{\mathrm{T}}^{\mathrm{W}}/\mathrm{MW})/(10^6/18) = 0.99 \times 10^{-4}$, where the molar volume of water is 18/10^6 mol m^{-3}, and hence, from Eq. (11), $\gamma_{\mathrm{T}}^{\infty} \approx 10100$, $H_{\mathrm{T}}\gamma_{\mathrm{T}}^{\infty}\, p_{\mathrm{T}}^{\mathrm{vap}} = 1.29 \times 10^8$ Pa (mole fraction)$^{-1}$ and $H_{\mathrm{T}}' = (H_{\mathrm{T}} \times 18)/10^6 = 2322$ Pa m^3 mol^{-1}; hence, $K_{\mathrm{AW,T}} = H_{\mathrm{T}}'/RT = 9.4 \times 10^{-1}$.

The concentration of trichloroethylene in air thus will be 0.94 times its concentration in water, if equilibrium conditions are met.

The value of K_{OW} for 1,1,1-trichloroethylene is 295. Using this value we can quantify the concentration of the pollutant in the fish (biota) from a knowledge of K_{BW} for the pollutant, as the two terms are related by:

$$K_{\mathrm{BW},i}\frac{c_i^{\mathrm{B}}}{c_i^{\mathrm{W}}} = f_{\mathrm{B}}K_{\mathrm{OW},i} \tag{12}$$

where f_{B} is the mass fraction of lipid material (assumed to be equivalent in solvent properties to octanol) in the biota and c_i^{B} the equilibrium concentration of i in the biota. Table 1 indicates that in fish, the f_{B} value is 0.05. If the concentration in water, c_i^{W}, for 1,1,1-trichloroethylene, is 1 ppm (1 mg L^{-1} or 1 gm^{-3}), then c_i^{B} is $\sim$15 ppm. Hence, we can say that a fish, swimming in water contaminated with this particular trichloroethylene at a concentration of 1 ppm, will be contaminated with the pollutant at a concentration of 15 ppm. The group, $f_{\mathrm{B}}K_{\mathrm{OW},i}$, is usually known as the bioconcentration factor.

Following on from Eq. (12) the more hydrophobic the polluting species, the greater will be c_i^{B}. Let us consider the concentration of one type

of hydrophobic pollutant – a polychlorinated biphenyl (one of the infamous PCBs), namely 2,2′,4,4′-tetrachlorobiphenyl in a river such as the St. Lawrence River, and estimate its concentration in the fish in the river [1, 2]. An average value given for PCBs in this river has been reported as 0.3 ppb (0.003 g m^{-3}). Let us assume that the only PCB in the river is 2,2′,4,4′-tetrachlorobiphenyl. For this PCB, log $K_{OW} = 5.90$.
Hence,

$$c_i^{B} = f_{B} K_{OW,i} c_i^{W} = (0.05 \times 7.9 \times 10^{5} \times 0.3)\,\text{ppb} = 11.9 \times 10^{3}\,\text{ppb or g m}^{-3}.$$

The predicted equilibrium concentration of PCB in fish is thus ~40000 times the concentration of the PCB in the water, an amazing magnification!

The experimental value of c_i^{B} for PCBs in the St. Lawrence River is 7×10^{3} ppb which is within an order of magnitude of our prediction based on a very simple model.

A similar treatment can be made for soil contamination. Taking into account that soil contamination (c_i^{S}) is measured in mg kg^{-1} and water is usually mg L^{-1} we have for soil (S):

$$K'_{SW} = \frac{(c_i^{S})'}{c_i^{W}} = \frac{K_{SW}}{d_{S}} = \frac{f_{S} K_{OW}}{d_{S}} \tag{13}$$

where d_S refers to the soil density.

We conclude that the substances of low solubility and high activity coefficients in water will tend to bioconcentrate into fish and any other animal and reach concentrations that are many orders of magnitude higher than the concentrations in water. These are usually referred to as hydrophobic or "water hating".

5. ENVIRONMENTAL DISTRIBUTION

With the above information we can describe the partitioning of a species between the different regions of the environment as described in Table 1. The total mass M_i of a pollutant is:

$$\begin{aligned} M_i = {} & \text{mass } i \text{ in air} + \text{mass } i \text{ in water} \\ & + \text{mass } i \text{ in aquatic biota} + \text{mass } i \text{ in soil} \end{aligned}$$

$$\begin{aligned} M_i &= c_i^{A} V_{A} + c_i^{W} V_{W} + c_i^{B} V_{B} + c_i^{S} V_{S} \\ &= K_{AW,i} c_i^{W} V_{A} + c_i^{W} V_{W} + K_{BW,i} c_i^{W} V_{B} + K_{SW,i} c_i^{W} V_{S} \end{aligned}$$

and hence

$$c_i^{\mathrm{W}} = \frac{M_i}{K_{\mathrm{AW},i}V_{\mathrm{A}} + V_{\mathrm{W}} + K_{\mathrm{BW},i}V_{\mathrm{B}} + K_{\mathrm{SW},i}V_{\mathrm{S}}} \tag{14}$$

Let us consider the distribution, in an environment defined in Table 1, of 1.00 kg of each of the three polluting chemicals: trichloroethylene (T), hexane (H) and DDT (D). The necessary data are [1]:

Pollutant	K_{AW}	$\log K_{\mathrm{OW}}$
Trichloroethylene (T)	0.048	2.29
Hexane (H)	74	4.11
DDT (D)	1×10^{-3}	6.19

The density of air at 25°C and 1 atm is 1185 g m^{-3}. Using the above data, together with the data of Table 1 and Eq. (14), we calculate that $c_{\mathrm{T}}^{\mathrm{W}} = 0.020\,\mathrm{mg\,m^{-3}}$. This is well below the solubility in water of trichloroethylene, which is $1100\,\mathrm{g\,m^{-3}}$.

$$c_{\mathrm{T}}^{\mathrm{A}} = K_{\mathrm{AW}}c_{\mathrm{T}}^{\mathrm{W}} = 0.96\times10^{-3}\,\mathrm{mg\,m^{-3}}$$

$$\text{Mass of T in air} = c_{\mathrm{T}}^{\mathrm{A}}V_{\mathrm{A}} = 0.96\,\mathrm{kg}$$

$$c_{\mathrm{T}}^{\mathrm{S}} = K_{\mathrm{SW,T}}c_{\mathrm{T}}^{\mathrm{W}} = f_{\mathrm{S}}K_{\mathrm{OW,T}}c_{\mathrm{T}}^{\mathrm{W}} = 0.04\,\mathrm{mg\,m^{-3}}$$

$$c_{\mathrm{T}}^{\mathrm{B}} = K_{\mathrm{BW,T}}c_{\mathrm{T}}^{\mathrm{W}} = f_{\mathrm{B}}K_{\mathrm{OW,T}}c_{\mathrm{T}}^{\mathrm{W}} = 0.195\,\mathrm{mg\,m^{-3}}$$

The results given in Table 2 require careful analysis. Most of the trichloroethylene and the hexane are to be found in the air but the highest concentrations are found in the biota, i.e. fish. For the DDT, however, most reside in the soil but again, the highest concentrations are to be found in the fish. The calculations performed here suggest that the biota or the soil (which have the highest chemical concentrations) would be the best to sample if one wished to monitor chemicals in the environment.

We have shown that given only K_{OW} and K_{AW} (or the Henry's law constant) for a pollutant, its concentration in various environmental regions can be predicted. This model is a simple one and assumes that equilibrium takes place at every compartment interface. It ignores flows in and out of the environment, the rate of diffusional processes and other time dependant effects. In the next section we will consider these effects.

Table 2
Results of analysis

Region	Concentration (c_T) (mg m^{-3})	ppb (by mass) (T)	Percentage of total (T)
Trichloroethylene			
Air	0.96×10^{-3}	0.810	97.0
Water	0.020	0.020	3.0
Soil	0.040	0.026	0.0
Biota	0.195	0.195	0.0
Hexane	(c_H)	(H)	(H)
Air	1×10^{-3}	0.843	100.0
Water	13.5×10^{-6}	13.5×10^{-6}	0.001
Soil	1.74×10^{-3}	1.16×10^{-3}	0.003
Biota	8.40×10^{-3}	8.40×10^{-3}	0.000
DDT	(c_D)	(D)	(D)
Air	4.25×10^{-6}	3.586×10^{-3}	0.005
Water	4.25×10^{-3}	4.25×10^{-3}	0.005
Soil	65	43	99.8
Biota	329	329	0.003

Note: The concentrations, c_i^W, were calculated from Eq. (14).

6. ENVIRONMENTAL DISTRIBUTION USING A FLOW MODEL [1]

This model takes into account both the flow of a chemical into and out of the environmental region and the degradation, and formation of the chemical within the region. The model like the previous one assumes that there is phase equilibrium between the environmental regions. The model is based on the mass balance:

Rate of change of i in the environment	=	Rate at which i enters the environment	−	Rate at which i leaves the environment

$$\frac{d}{dt}(c_i^A V_A + c_i^W V_W + c_i^S V_S + c_i^B V_B)$$
$$= \frac{dM_i}{dt} = \left(\frac{dQ_A}{dt} c_{i,IN}^A - \frac{dQ_A}{dt} c_i^A + \frac{dQ_W}{dt} c_{i,IN}^W - \frac{dQ_W}{dt} c_i^W \right)$$
$$- (k_A c_i^A V_A - k_W c_i^W V_W - k_B c_i^B i V_B - (k_{Sc})_i^S V_S) \tag{15}$$

The dQ_A/dt term is the air flow rate and the dQ_W/dt term the water flow velocity. These terms would have units of $m^3 s^{-1}$. The k terms are degradation first order values ($dc/dt = -kc$) [3] specific to each environment region.

The differential equation can be simplified as:

$$\alpha \frac{dc_i^W}{dt} = \beta - \gamma c_i^W \tag{16}$$

where

$$\alpha = K_{AW,i} V_A + V_W + K_{BW,i} V_B + K_{SW,i} V_S \tag{17}$$

$$\beta = \frac{dQ_A}{dt} c_{i,IN}^A + \frac{dQ_W}{dt} c_{i,IN}^W \tag{18}$$

and

$$\gamma = \left(\frac{dQ_A}{dt} + V_A k_A \right) K_{AW,i} + \left(\frac{dQ_W}{dt} + V_W k_W \right) + V_B k_B K_{BW,i} + V_S k_S K_{SW,i} \tag{19}$$

The solution of the equation is:

$$c_i^W(t) = \frac{\beta}{\gamma}\left[1 - \exp\left(\frac{-\gamma t}{\alpha}\right)\right] \tag{20}$$

One way of simplifying this equation is to assume a steady state condition ($t = \infty$) which gives:

$$c_i^W(t) = \frac{\beta}{\gamma} \tag{21}$$

Using the above theory, let us apply it to the real case of DDT pollution.

Although DDT has been out of production for almost two decades, its concentration in water in a particular fresh water lake was found to be 1.00 $mg\,m^{-3}$. That is 1 part per 10^9, i.e. 1 ppb. This is below its solubility in water value which is 3.1 $mg\ m^{-3}$. In this problem we will use the data below: (a) to calculate the DDT concentrations in the air, biota and soil; (b) assuming that the surrounding environment has the same DDT (no flow in or out), to calculate the time that will be required for the DDT concentration to fall by 50%.

For DDT, $K_{AW} = 1 \times 10^{-3}$ and log $K_{OW} = 6.19$ (i.e. $K_{OW} = 1.549 \times 10^6$). In our answer we will assume the model environment given in Table 1 and the first order rate constant for DDT degradation:

by photolysis in water is 5.3×10^{-7} h^{-1};
by hydrolysis in water is 3.6×10^{-6} h^{-1};
by biodegradation in soil is 5.42×10^{-6} h^{-1}.

Assuming that the DDT has reached chemical equilibrium (after all, the time scale is about two decades!) then as

$$K_{AW,D} = \frac{c_D^A}{c_D^W}$$

then

$$c_D^A = c_D^W K_{AW,D} = 1.00 \times 1 \times 10^{-3} = 1.00 \times 10^{-3} \text{ mg m}^{-3}$$

and similarly,

$$c_D^S = c_D^W K_{SW,D} = 1.00 \times 0.01 \times 1.549 \times 10^6 = 15.49 \times 10^3 \text{ mg m}^{-3}$$

$$c_D^B = c_D^W K_{BW,D} = 1.00 \times 0.05 \times 1.549 \times 10^6 = 77.4 \times 10^3 \text{ mg m}^{-3}$$

Again, the highest concentrations of DDT are found in the fish and the soil is also contaminated. Note that the DDT in the fish is 77000 times the concentration of the DDT in the water.

Of interest, the total mass of DDT in this environment (1 km $\times$ 1 km $\times$ 1 km) (see Table 1) is:

$$M = c_D^W V_W + c_D^A V_A + c_D^S V_S + c_D^B V_B$$

$$M(\text{mg}^{-1}) = (1.00 \times 10^6) + (1.00 \times 10^{-3} \times 10^9) + (15.5 \times 10^3 \times 1.5 \times 10^4) + (77.4 \times 10^3 \times 10)$$

$$M(\text{kg}^{-1}) = 1.00 + 1.00 + 232 + 0.8$$

$$M = 234.8 \text{ kg}$$

As we found before, most of the DDT resides in the soil – in this case it is 232 kg out of a total of 234.8 kg.

Since there is no net inflow of DDT and there is no production of DDT, the terms involving dQ/dt and dM/dt are zero and we have:

$$\alpha' \frac{dc_D^W}{dt} = -\gamma' c_D^W$$

where

$$\begin{aligned}\alpha' &= K_{AW,D}V_A + V_W + K_{BW,D}V_B + S_{SW,D}V_S \\ &= (1\times10^{-3}\times10^9) + 10^6 + (0.05\times1.549\times10^6\times10) \\ &\quad + (0.01\times1.549\times10^6\times1.5\times10^4) \\ &= 234.8\times10^6\ \text{m}^3\end{aligned}$$

$$\begin{aligned}\gamma' &= V_W K_W + V_S K_{SW,D} K_S \\ &= 1\times10^6(5.3\times10^{-7} + 3.6\times10^{-6}) + 15\times10^4(0.01\times1.549\times10^6) \\ &\quad \times5.42\times10^{-6} \\ &= (4.13 + 1259.3)\,\text{m}^3\,\text{h}^{-1} = 1263.4\,\text{m}^3\,\text{h}^{-1}\end{aligned}$$

Getting back to our differential equation

$$\frac{dc_D^W}{dt} = -\frac{\gamma'}{\alpha'} c_D^W = 5.365\times10^{-6}(\text{h}^{-1})c_D^W$$

This first order equation [3]:

$$\frac{dc}{dt} = -kc$$

has

$$k = 5.365\times10^{-6}\,\text{h}^{-1}$$

and hence the half life can be calculated:

$$t_{1/2} = \ln\frac{2}{k} = \ln\frac{2}{5.365\times10^{-6}} = 1.29\times10^5\,\text{h} = 14.7\,\text{years}$$

Thus, starting with a concentration of 1.00 mg m^{-3}, the concentration will drop to half this level, i.e. 0.50 mg m^{-3}, in 14.7 years. This shows how persistent DDT is in the environment.

7. ACCUMULATION OF CHEMICALS IN THE FOOD CHAIN

Experimental analysis of chemicals in animals has shown that in some cases the concentration increases with increasing position of the animal in the food chain. Fish eating eagles, for example, have been found to have a much greater concentration of some pesticides than do the fish that the eagles feed upon. This process is known as biomagnification and appears to be a function of the magnitude of the K_{OW} of the chemical species involved. It is found that biomagnification occurs for chemical species with log K_{OW} values greater than 5.0.

Values of log K_{OW} for various insecticides are given in Table 3 [1].

The answer lies in mass transfer effects. A fish loses much of its pollutant in the respired water passing through the gills. If K_{OW} is very large, the concentration in that water will be very low; hence, the flux of chemical is constrained to a low level. Elimination is slow; thus, there is a build up or biomagnification of the chemical that is adsorbed from the food. There may also be increased resistance to transfer through membranes if K_{OW} is very large, i.e. 10^8. This biomagnification effect is much smaller (reported values range from 3 to 30) than the enormous effects (40000 and 77000 times) found under equilibrium conditions of the partitioning of a pesticide between water and fish (see Sections 4 and 6).

Many of the concepts and terms introduced in this chapter will be used in chapters in this book, and especially in Chapters 3, 13, 14 and 24.

ACKNOWLEDGEMENT

I wish to thank Professor Donald MacKay of Trent University, Canada, for many helpful suggestions and to professor Stan Sandler of the University of Delaware, USA, who introduced me to the modelling of pollutants.

Table 3
Values of log K_{OW} for some insecticides

	log K_{OW}
Malathion	2.9
Lindane	3.85
Dieldrin	5.5
DDT	6.19
Mirex	6.9

REFERENCES

[1] D. Mackay, Multimedia Environmental Models, 2nd edn., CRC Press, Lewis Publishers, New York, 2001 (ISBN 1-56670-542-8).
[2] S.I. Sandler, in "Chemical Thermodynamics – Chemistry for the 21st Century" (T.M. Letcher, ed.), Blackwell Science, Oxford, 1999 (Chapter 2).
[3] P. Atkins, Physical Chemistry, 6th edn., p. 767, Oxford University Press, Oxford, 1998.

Thermodynamics, Solubility and Environmental Issues
T.M. Letcher (editor)

Chapter 2

Modeling the Solubility in Water of Environmentally Important Organic Compounds

Ernesto Estrada[a], Eduardo J. Delgado[b] and Yamil Simón-Manso[c]

[a]Complex Systems Research Group, X-Rays Unit, RIAIDT, Edificio CACTUS, University of Santiago de Compostela, 15782 Santiago de Compostela, Spain
[b]Theoretical and Computational Chemistry Group (QTC), Faculty of Chemical Sciences, Casilla 160-C, Universidad de Concepción, Concepción, Chile
[c]INEST Group, PM-USA and Center for Theoretical and Computational Nanosciences, National Institute of Standards & Technology (NIST), 100 Bureau Drive, Stop 8380, Gaithersburg, MD 20899-8380, USA

1. INTRODUCTION

Solubility in water is one of the most important physico-chemical properties of a substance and a knowledge of solubility plays an important part in the environmental fate and transport of xenobiotics (man-made chemicals) in the environment. Substances, which are readily soluble in water, will dissolve freely in water if accidentally spilled and will tend to remain in aqueous solution until degraded. On the other hand, sparingly soluble substances dissolve more slowly and, when in solution, have stronger tendency to partition out of aqueous solution into other phases. They tend to have large air–water partition coefficients or Henry's law constants, and they tend to partition more into soils, sediments and biota. As a result, it is common to correlate partition coefficients from water to these media with solubility in water.

From a qualitative point of view, solubility can be viewed as the maximum concentration that an aqueous solution will tolerate just before the onset of the phase separation. From a thermodynamic point of view, solubility is the concentration of solute A required to reach the following equilibrium:

$A(\mathrm{p}) \leftrightarrow A(\mathrm{aq.})$

where p refers to a specific aggregation state of the solute, namely, solid, liquid or gas state. The condition of thermodynamic equilibrium at T and P constants requires the equality of the chemical potential of the solute A in both phases, namely: $\mu_A(T, P, x_A) = \mu_{A(\mathrm{p})}(T, P)$, where x_A is the solute molar fraction in the saturated solution, in other words it is the solubility of the solute in the molar fraction scale.

Although, the experimental determination of solubility is not difficult, there are some justifications to develop models that allow predicting it. This is especially important in environmental studies where the compounds of interest are toxic, carcinogenic or undesirable for other reasons. In addition, the ability to predict this property is important for designing novel pharmaceutical products and agrochemicals whose solubility can be predicted before carrying out the synthesis. Design of novel compounds may, in this way, be guided by the results of calculations.

Accordingly, very extensive studies have been carried out on solubility resulting in diverse theories of solute–solvent interactions that form the basis of the knowledge for the understanding of solubility. These theories are based on concepts ranging from quantitative analysis to statistical and quantum mechanics.

Molecular sciences look for explanations of macroscopic properties, e.g., solubility, from the microscopic properties of matter. Statistical mechanics is one of such disciplines, which links those two pictures through the probabilistic treatment of particle ensembles. The application of Kirkwood's continuum solvent approach to nondissociating fluids resulted in a variety of simulation techniques. Applications of such techniques to study phase equilibria have been reported widely in literature [1–10]. Although some simple hydrocarbons can nowadays be reasonably well described by molecular modeling (molecular dynamics and Monte Carlo simulations), water and especially water mixtures, still represent challenges for such simulations techniques despite 30 years of active parameterization of appropriate force-fields. This is due to the extremely strong and complicated electrostatic and hydrogen-bond interactions.

Quantum-chemical methods, with the explicit inclusion of solvation free energy into the framework of the MO SCF method, have been developed to the point that they are useful tools for predicting thermodynamic properties and phase behavior of some substances to an accuracy useful in engineering calculations [11–20]. Among these methods, the SM_x solvation models of Cramer and Truhlar, and the Conductor-like Screening Model for Real Solvents, COSMO-RS, of Klamt, are presumably the most widely used methods to study phase equilibria. For instance, the aqueous solubility of 150 drug-like compounds and 107 pesticides were calculated using COSMO-RS method [21].

The above-mentioned methods based on either statistical mechanics or quantum mechanics allow the prediction of rather accurate values of solubility, but these methods are time-consuming and can hardly be applied to the solubility modeling of large biomolecules or to the large-scale modeling of many hundreds of small molecules [22]. Several less sophisticated, but also much less time-consuming methods based on quantitative structure–property relationships (QSPR) methodology have been developed recently for the prediction of solubility. This family of methods is divided into two groups: methods based on experimentally determined descriptors and methods based solely on molecular structure.

2. QUANTUM CHEMISTRY METHODS

Despite the huge progress in calculating free energies of solvation with dielectric continuum models (DCMs) [17, 23–25], such as COSMO [17] and SM5.42R [25], there has been little work to predict solubilities using them. In DCMs, the solute is considered to be located in a cavity of a dielectric continuum. The solvent polarization is included in the calculation as a boundary condition. Then, the Schrödinger equation is solved self-consistently.

COSMO-RS [21] is a two-step methodology for the prediction of equilibrium properties such as vapor pressure, heat of vaporization, activity and solvent partition coefficients, phase diagrams and solubility. In the first step, COSMO is used to simulate a virtual conductor environment for the molecule and evaluate the screening charge density, σ, on the surface of the molecule. In a second step, the statistical thermodynamics of the molecular interactions is used to quantify the interaction energy of the pair-wise interacting surface segments (σ,σ'). Three major interactions between surface segments are taken into account using appropriate functional forms. The Coulomb interaction between the screening charges σ and σ' on surface pairs, so-called the misfit term, the hydrogen bonding interaction and the van der Waals interactions [20, 26–28].

COSMO-RS is able to calculate the chemical potential (the partial Gibbs free energy) of a compound, either pure or in a mixture from the probability distribution of σ. The solubility of a compound, X, can be calculated from the differences between the chemical potentials of X in solution and pure [21, 26]. COSMOS-RS not only predicts reasonable solubility values but also the correct temperature dependence of solubility, with deviations from experiment below 0.3log(x) [26]. However, some "lack of thermodynamic consistency" has been reported [27].

Truhlar et al. [28] predict the aqueous solubilities of 75 liquid solutes and 15 solid solutes by utilizing a relation between solubility, free energy of solvation and solute vapor pressure. The method is based on SM5.42R solvation model and the classical thermodynamic theory of solutions. In the SM5.42R solvation model the free energy of solvation is written as, $\Delta G_S^0 = \Delta E_E + G_P + G_{\mathrm{CDS}}$, where ΔE_E is the change of electronic energy due to the embedding of the solute into the dielectric environment, G_P the electronic polarization energy, i.e., the mutual polarization of the solute and the solvent, G_{CDS} a semiempirical term that takes into account the nonbulk contributions, i.e., inner solvation-shell effects. It is a parametric function of several solvent descriptors.

Based on the theory of solutions of classical thermodynamics, the standard-state free energy for the solvation processes $A(\mathrm{p}) \leftrightarrow A(\mathrm{aq.})$, at temperature T, can be expressed as:

$$\Delta G_S^0(\mathrm{sol}) = RT \ln \frac{P_A^\bullet}{P^\circ} - RT \ln M_A^{\mathrm{sol}} \tag{1}$$

where M_A^{sol} is the solubility of A (in molarity units), $P_A^\bullet$ and P_A° are respectively the equilibrium vapor pressure of A over pure A and the pressure of an ideal gas at 1 molar concentration and temperature of 298 K. This equation is obtained under the assumption of infinite dilution, i.e., the activity coefficients are equal to 1. Truhlar et al. used a training set of 75 liquid solutes and 15 solid solutes. This set can be considered relatively small for comparison with other solubility models, however their results are very promising with a mean-unsigned error in the logarithm of 0.33–0.88.

In general, the quantum methods need relatively few parameters, are able to handle exotic molecules, transition states and do not make assumptions such as group transferability or additive properties. However, at the present time, they are not very accurate and need at least one time-consuming quantum calculation which limits their use for extremely large pools of molecules.

3. EXPERIMENT-BASED QSPR MODELING

Hansch et al. [29] showed that solubility and octanol–water partition coefficient, K_{OW}, are well correlated for liquid solvents. Numerous further studies have explored and refined this relationship for different classes of compounds, e.g., halogenated hydrocarbons [30, 31], polycyclic aromatics hydrocarbons [32].

The General Solubility Equation of Yalkowsky includes two experimental parameters: the melting point of the solute and its octanol–water partition coefficient. The aqueous solubility of 150 physiologically active compounds has been estimated using this equation [33].

In the solvatochromic or linear solvation energy method, developed by Kamlet et al. [34] and Taft et al. [35], the solubility is predicted from molar volume, melting point and two parameters which express dipolarity/polarizability and hydrogen bond basicity. It has been applied to predict the solubility of very diverse type of compounds.

The mobile order theory (MOT) approach, developed by Huyskens [36], has been widely used by Ruelle et al. to predict the solubility of a diverse set of chemicals of environmental relevance [37, 38]. Paasivirta et al. [39] using this approach estimated the solubility of 73 persistent organic pollutants as a function of temperature.

4. STRUCTURE-BASED QSPR MODELING

These models exploit two different paradigms [22]. One relies on the concept of the structural additivity of properties. According to this hypothesis, any property in the form of a continuous smooth function can be expanded into a linear function in some predefined structural features such as atoms, bonds and chemical functional groups. This approach – also known as fragment-based or group contribution scheme – has been actively pursued by the groups of Klopman and Zhu [40] and Wendoloski et al. [41]. Klopman et al. derived contribution coefficients for many groups and successfully correlated them with the aqueous solubilities of 1168 organic compounds. The UNIFAC (universal quasi chemical method (UNIQUAC) functional group activity coefficient) method, an extension of the UNIQUAC, has been tested by different authors [42–44] to ascertain its applicability to water solubility.

The other approach involves the derivation of various molecular characteristics (descriptors) solely from molecular structure. Depending on the level of consideration, one can use constitutional, topological, geometric, electrostatic or quantum-chemical descriptors.

Jurs et al. have published several articles on the correlation of aqueous solubility with molecular structure using both multilinear regression (MLR) and artificial neural networks (ANN). Successful nine-parameter regression models are reported for three sets of compounds (hydrocarbons, halogenated hydrocarbons and ethers) and a fourth model is also reported for the combined set of all compounds [45]. Later the same authors reported MLR and ANN models for the prediction of aqueous solubility for a diverse set of heteroatom-containing organic compounds [46]. A set of nine descriptors was found that effectively linked the aqueous solubility to each structure. It is also reported that PCBs have larger errors associated with them than any other class of compounds in the dataset. In an attempt to improve the overall error of the model, the set of nine descriptors was used to build ANN model. The root of mean squared (rms) error of the PCBs alone was 0.51 log units, which was significantly higher than the overall errors in the model. Later, Jurs and Mitchell [47] also reported an ANN model for the prediction of solubility of 332 organic compounds. The model involving nine descriptors has an rms error of 0.395 log units, and the squared correlation coefficient between the experimental and calculated values is 0.97. A large dataset of diverse organic compounds was used by Huuskonen [48] who reported a QSPR model for predicting aqueous solubility of 1297 organic compounds using topological and electrotopological indices.

Delgado reports a QSPR model [49] for the prediction of log(1/S) for a set of 50 chlorinated hydrocarbons including chlorinated benzenes, dibenzo-*p*-dioxins and PCBs. The model involves only two molecular descriptors, one geometry-dependent descriptor and one charged partial surface area (CPSA) descriptor. The geometric descriptor is the area of the shadow of the molecule projected on a plane defined by the X and Y axes (XY shadow); the CPSA descriptor is the surface weighted atomic partial negative surface area (WNSA-3). The model has a squared correlation coefficient of 0.97 and standard deviation of 0.45 log units.

On the other hand, Gao et al. [50], using principal component regression analysis, derived a QSPR model for the prediction of solubility of a set of 930 diverse compounds, including pharmaceuticals, pollutants, nutrients, herbicides and pesticides. The model, which involves 24 molecular descriptors, predicts the log(S) with a squared correlation coefficient of 0.92 and rms error of 0.53 log units.

Delgado et al. [51] reported QSPR models for the solubility of herbicides stressing the importance of considering the aggregation state of the solute. It is found that the phase of the solutes plays a fundamental role in the development of QSPR model from both a statistical and a physical point of view. From a statistical point of view, it is observed that the predictive performance of the models drops drastically when the QSPR model, obtained for a given phase, is used to predict the solubility of the same set of compounds in another phase. From a physical point of view, when the compounds considered in the training set are in different phases, the physical interpretation of the descriptors involved in the model is obscured because the descriptors which appear in the correlation equation are a sort of average encoding the different physical mechanisms existing for the different phases in the solubility phenomenon.

The above-mentioned models use different types of molecular descriptors to quantitatively predict solubility using multiple linear regression and artificial neural network. These molecular descriptors encode topological, geometric, electronic and quantum-chemical molecular features responsible for the observed solubility. Even though these models predict solubility relatively well, the number of descriptors involved in the models is rather high. It is highly desirable to have models with fewer descriptors which allow their application in a more straight forward way, and, on the other hand, to preserve the principle of parsimony. In this direction Estrada [52–55] have defined the quantum-connectivity indices by using a combination of topological invariants, such as interatomic connectivity, and quantum-chemical information, such as atomic charges and bond orders. These indices also contain important three-dimensional information, incorporated by the quantum-chemical parameters used in their definition. Therefore, these indices are richer in chemical information than traditional molecular descriptors since they encode, at the same time, both topological and quantum-chemical features of molecules. They have been successfully applied to the prediction of aqueous solubility of environmentally important compounds [56].

5. THE QUANTUM-CONNECTIVITY INDICES

"Classical" connectivity indices are based on the concept of vertex degree, which is the number of edges incident to a vertex [57, 58]. The degree of a vertex designated by k is represented by δ_k. In the simple graph representation of acetone, 2-methylpropene and 2-methylpropane there is one vertex with degree 3 and three vertices with degree 1. However, when pseudo-graphs are considered it is necessary to use the concept of valence degree δ_k^{v} introduced by Kier and Hall [58]. In this case the valence degree for a vertex k is defined as a count of electrons σ, π and n orbitals (excluding hydrogen atoms). It is equivalent to consider the number of multiple edges and loops incident to a vertex in the pseudo-graph (loops are doubly incident to a vertex). In this case the oxygen atoms of the acetone has $\delta_k^{\mathrm{v}} = 6$, which differentiates it from the CH_2 group in 2-methylpropene, $\delta_k^v = 2$ and from CH_3 of 2-methylpropane, $\delta_k^v = 1$. However, with this molecular representation the C_{sp^2} atom of the carbonyl group of acetone is not differentiated from the C_{sp^2} of the ethylenic group, i.e., they have $\delta_k^v = 4$. In addition to this pseudograph representation, there is no differentiation of geometric isomers as it does not take into consideration the three-dimensional molecular structure.

We have overcome these difficulties by considering weighted pseudographs in the context of quantum-chemical molecular orbital approaches [52–55]. Instead of using simple entire numbers for counting the number of multiple bonds and loops in the pseudo-graph we employ quantum-chemical parameters for weighting the edges and vertices of the graph, respectively. As a measure of the bond multiplicity we use the bond order, which is defined as the sum of the products of the corresponding atomic orbital coefficients over all the occupied molecular spin-orbitals. On the other hand, we weight a vertex of the graph by means of the charge density of the corresponding atoms in the molecule, Q_i, which is defined as the number of valence electrons, Z_i, minus the atomic charge, q_i: $Q_i = Z_i - q_i$.

Then, we introduce several new definitions of vertex degree in the context of quantum-connectivity. The first is defined as the sum of bond orders of all bonds that are incident with the corresponding vertex [52]:

$$\delta_i(\rho) = \sum_i \rho_{ij} \tag{2}$$

A second vertex degree is defined as the number of edges (including loops) which are incident with the corresponding vertex minus the atomic charge assigned to it. In other words, it is the charge density minus the number of hydrogen atoms bonded to the corresponding vertex [53]:

$$\delta_i(q) = Q_i - h_i \tag{3}$$

A correction for hydrogen atoms is introduced in this scheme according to the following formula [53]:

$$\delta_i^c(q) = Q_i - \sum_j Q_{hj} \tag{4}$$

where Q_{hj} is the atomic charge density of the jth hydrogen atom bonded to the atom i. For elements beyond the second row of the periodic table of elements the values of $\delta_i(q)$ and $\delta_i^c(q)$ are calculated in a similar way as for the valence connectivity [57] index by dividing the right part of expressions (3) and (4) by $(Z_i - Z_i^v - 1)$, where Z_i and Z_i^v are the total number of electrons and the number of valence electrons in the *i*th atom.

In a similar way that vertex degree is defined, an edge degree is also known in graph theory. It corresponds to the number of edges that are adjacent to the corresponding edge. The following relationship exists between vertex and edge degrees:

$$\delta(e_k) = \delta_i + \delta_j - 2 \tag{5}$$

where $\delta(e_k)$ is the degree of the edge *k* in G, which is incident with vertices *i* and *j*. Thus, using this expression we have extended the vertex degrees previously defined to edge degrees.

Quantum-connectivity indices are finally calculated by an expression analogous to that of the connectivity indices but using weighted vertex degrees instead of simple vertex degrees [52, 53]:

$$^{h}\Omega_t(w) = \sum_{s=1} \left[\delta_i(w)\delta_j(w)\cdots\delta_{h+1}(w) \right]_s^{-0.5} \tag{6}$$

where w represents the weighting scheme used, e.g., bond orders, charge density or charge density corrected for hydrogen atoms. The product is over the $h+1$ vertex degrees in the subgraph having h edges, and the summation is carried out over all subgraphs of type t in the molecule. The different types of subgraphs studied in the molecular connectivity scheme are: path, clusters, path-clusters and rings, which are designed as p, C, pC and Rg, respectively according to their original definitions [58].

The bond quantum-connectivity indices based on weighted molecular graphs, are calculated in a similar way to that of their topological analogues [59, 60]. They are defined as follows [54]:

$$^{h}\varepsilon_t(w) = \sum_{p=1} \left[\delta(e_i)(w)\delta(e_j)(w)\cdots\delta(e_h)(w) \right]_p^{-0.5} \tag{7}$$

6. MODELING SOLUBILITY WITH QUANTUM-CONNECTIVITY

We have used quantum-connectivity indices to model the solubility of a set of organic compounds of environmental relevance. This dataset is formed by 53 compounds, including 30 pesticides chemicals and other chemicals which can be found as contaminants in the environment, such as polycyclic aromatic compounds (PAHs) and chlorobiphenyls. The solubility of these compounds is expressed as ln(S) and covers a wide range of solubility values, from the very low solubility of Benzo[a]pyrene ln(S) = −11.96 to the very soluble pesticide Amitrole, ln(S) = 8.11. Using quantum-connectivity indices we have developed the following quantitative model to predict the solubility of these compounds [56]:

$$\begin{aligned} &\ln(S) = 10.487 - 2.575[{}^{2}\Omega_{p}^{C}(q)] - 112.61[{}^{6}\varepsilon_{\mathrm{Rg}}(\rho)] \\ &N = 53, R^{2} = 0.9257, s = 1.007, F = 311.26 \end{aligned} \tag{8}$$

where R is the correlation coefficient, s is the standard deviation and F is the Fisher ratio of the regression model.

The first index is the quantum-connectivity index of path order two, based on charge density weighted graphs. The second is a bond quantum-connectivity index for rings of order six, based on bond order weighted graphs. We tested the predictability of this model using an external prediction series of 30 compounds. In Fig. 1, it can be seen that the model obtained by using quantum-connectivity indices show good predictability for the external prediction dataset of compounds, which are not directly related to those in the training set. This is, of course, an important characteristic of quantitative models toward their usability in practical problems of predicting properties for new chemicals.

The mean effects, i.e., the coefficient of the variable in the QSAR/QSPR model multiplied by the mean value of the variable in the dataset, for ${}^{2}\Omega_{\mathrm{p}}^{\mathrm{C}}(q)$ and ${}^{6}\varepsilon_{\mathrm{Rg}}(\rho)$ in the model are 8.817 and 2.810, respectively. This indicates that the first descriptor plays the most important role in predicting water solubility for this dataset of compounds. This descriptor, ${}^{2}\Omega_{\mathrm{p}}^{\mathrm{C}}(q)$, is defined on the basis of paths of order two, i.e., a sequence of three consecutive atoms: A-B-C. The number of this fragment increases rapidly with the number of substituents on a specific site. Thus, the quantum-connectivity index ${}^{2}\Omega_{\mathrm{p}}^{\mathrm{C}}(q)$ controls the influence of the number of substitutions at different sites in a molecule. For instance, 1,1,2-trichloroethane is more soluble in water, ln(S) = 3.54, than 1,1,2,2-tetrachloroethane, ln(S) = 2.82. It could be though that this is due to the increase in the number of chlorine atoms in the second molecule with respect to the first one. However, if we consider 1,1,1-trichloroethane we can see that it is the least soluble of the three compounds, ln(S) = 2.27, despite it has a chlorine atom less than 1,1,2,2-tetrachloroethane.

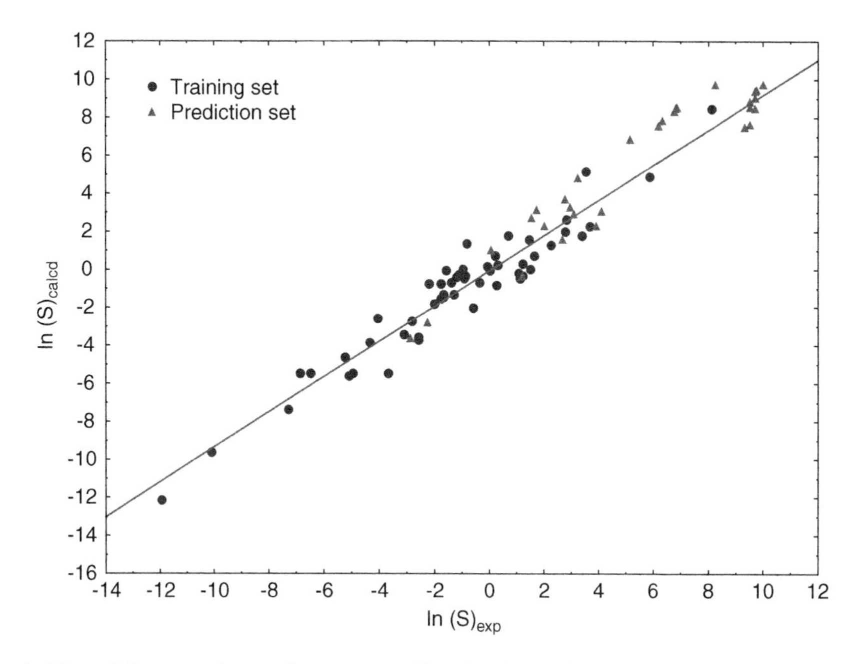

Fig. 1. Plot of the experimental versus predicted values of aqueous solubility of environmentally relevant organic compounds according to the model developed using quantum-connectivity indices.

The differences arise, however, because 1,1,1-trichloroethane has more A-B-C fragments due to the fact that the three substitutions are at the same carbon atom. Consequently, it has the largest value of ${}^{2}\Omega_{\mathrm{p}}^{\mathrm{C}}(q) = 3.560$, while 1,1,2,2-tetrachloroethane has ${}^{2}\Omega_{\mathrm{p}}^{\mathrm{C}}(q) = 3.054$ and 1,1,2-trichloroethane has ${}^{2}\Omega_{\mathrm{p}}^{\mathrm{C}}(q) = 2.068$.

The other quantum-connectivity index ${}^{6}\varepsilon_{\mathrm{Rg}}(\rho)$ accounts for the influence of bond order weighted cyclic fragments of six bonds, i.e., six-atoms rings. The higher the number of such rings in the molecule, the higher the value of this descriptor. The bond order weighting scheme used, also ensures a differentiation among six-membered rings in different chemical environments. The compound in our dataset with the higher value of ${}^{6}\varepsilon_{\mathrm{Rg}}(\rho)$ is benzo[a]pyrene followed by benz[a]anthracene. The first has five condensed benzene rings and the second has four. They have the lowest solubilities of all the compounds studied here. All these rings are clearly hydrophobic units, which will decrease water solubility in any molecular framework in which they are present. In Fig. 2 it can be observed that the general trend is that the rings with more propensity to be in contact with the solvent have higher

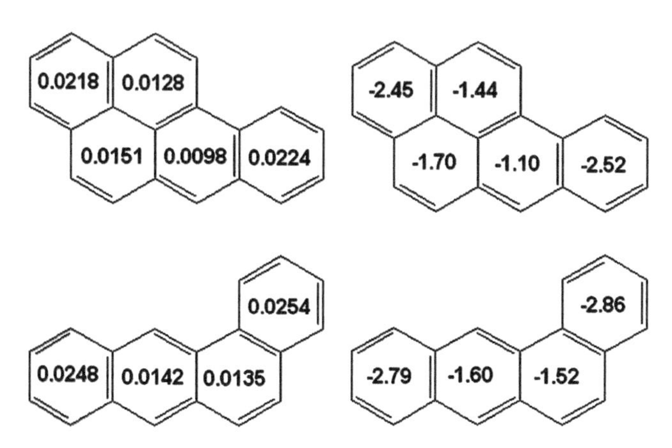

Fig. 2. Contribution of the aromatic rings of two polycylic aromatic hydrocarbons (PAHs) to the quantum-connectivity index ${}^{6}\varepsilon_{\mathrm{Rg}}(\rho)$ (left) and to aqueous solubility expressed as ln(S) (right). The compound at the top is benzo[a]pyrene and the one at the bottom is benz[a]anthracene.

contribution to ${}^{6}\varepsilon_{\mathrm{Rg}}(\rho)$ compared to those having less propensity to be in contact with water. In consequence the first contributes more significantly to decrease water solubility of these compounds. In other words, those rings which are more exposed to solvent, such as terminal rings in benz[a]anthracene, have more chance to be in contact with water, having more influence in water solubility than those with less contacts, such as internal rings.

On the left side of Fig. 2 we give the contribution of each aromatic ring to the quantum-connectivity index ${}^{6}\varepsilon_{\mathrm{Rg}}(\rho)$. The sum of the values of these contributions gives the value of the index ${}^{6}\varepsilon_{\mathrm{Rg}}(\rho)$ for the respective compound. On the right side of this figure we give the contribution of each aromatic ring to the water solubility expressed as ln(S). The sum of these values plus the contribution of the other quantum-connectivity index and the intercept of the model gives the solubility predicted for the respective compound. For instance, for benzo[a]pyrene, the sum of these contributions is -9.21, which after summing the contributions of the other index and intercept gives -12.00, in agreement with the experimental value for this compound which is -11.96.

7. CONCLUDING REMARKS

Today there is a large arsenal of methods and tools for predicting the aqueous solubility of organic compounds which can impact the environment in different ways. These methods range from empirical and semiempirical QSPR approaches to more sophisticated quantum-chemical approaches. The election of one or another method depends very much on the characteristics of the problem we are trying to solve. In making this election we have to consider that "models are to be used, not believed" [61] in such a way that the elected method solves our problem in the best possible way using the minimum amount of resources. In this sense we have illustrated here the use of the quantum-connectivity indices in modeling aqueous solubility of a series of environmentally important organic compounds. These molecular descriptors condense quantum-chemical, geometrical and topological information into single indices. Thus, they are appropriated for describing quantitatively the aqueous solubility of organic compounds using QSPRs methodologies.

ACKNOWLEDGMENTS

E.E. thanks the program "Ramón y Cajal", Spain for partial financial support. Partial financial support from FONDESYT Project 7020464 for international cooperation is also acknowledged.

REFERENCES

[1] M.P. Allen and D.J. Tildesley, Computer Simulation of Liquids, Oxford University Press, Oxford, 1987.
[2] D. Frenkel and B. Smit, Understanding Molecular Simulation, Academic Press, San Diego, CA, 1996.
[3] J.I. Siepmann, Adv. Chem. Phys., 105 (1999) 443.
[4] E.M. Duffy, W.L. Jorgensen, J. Am. Chem. Soc., 122 (2000) 2878.
[5] J.J. Potoff, J.R. Errington and A.Z. Panagiotopoulos, Mol. Phys., 97 (1999) 1073.
[6] J.M. Stubbs, B. Chin, J.J. Potoff and J.I. Siepmann, Fluid Phase Equilib., 183–184 (2001) 301.
[7] J.R. Errington and A.Z. Panagiotopoulos, J. Chem. Phys., 111 (1999) 9731.
[8] W.L. Jorgensen and J. Tirado-Rives, Perspect. Drug Discov. Des., 3 (1995) 123.
[9] P. Kollman, Chem. Rev., 93 (1993) 2395.
[10] W.L. Jorgensen, in "Encyclopedia of Computational Chemistry" (P.v.R. Schleyer, ed.), p. 1061, Vol. 2, Wiley, New York, 1998.
[11] J. Tomasi and M. Persico, Chem. Rev., 94 (1994) 2027.
[12] J. Li, T. Zhu, G.D. Hawkins, P. Winget, D.A. Liotard, C.J. Cramer and D.G. Truhlar, Theor. Chem. Acc., 103 (1999) 9.
[13] G.D. Hawkins, D.A. Liotard, C.J. Cramer and D.G. Truhlar, J. Org. Chem., 63 (1998) 4305.
[14] P. Winget, C.J. Cramer and D.G. Truhlar, Environ. Sci. Technol., 34 (2000) 4733.
[15] S.I. Sandler, J. Chem. Thermodyn., 31 (1999) 3.

[16] S.I. Sandler, Fluid Phase Equilib., 210 (2003) 147–160.
[17] A. Klamt and G. Schüürmann, J. Chem. Soc. Perkin Trans., 2 (1993) 799.
[18] A. Klamt, in "Encyclopedia of Computational Chemistry" (P.v.R. Schleyer, ed.), Vol. 2, Wiley, New York, 1998.
[19] A. Klamt and F. Eckert, Fluid Phase Equilib., 172 (2000) 43.
[20] A. Klamt, J. Phys. Chem., 99 (1995) 2224.
[21] A. Klamt, F. Eckert, M. Horning, M.E. Beck and T. Bürger, J. Comput. Chem., 23 (2002) 275.
[22] A.R. Katritzky, A.A. Oliferenko, P.V. Oliferenko, R. Petrukhin, D.B. Tatham, U. Maran, A. Lomaka and W.E. Acree, Jr., J. Chem. Inf. Comput. Sci., 43 (2003) 1794.
[23] C.J. Cramer and D.G. Truhlar, Chem. Rev., 99 (1999) 2161.
[24] J. Tomasi, B. Menucci and R. Cammi, Chem. Rev., 105 (2005) 2999.
[25] T. Zhu, J. Li, G.D. Hawkins, C.J. Cramer and D.G. Truhlar, J. Chem. Phys., 109 (1998) 9117.
[26] A. Klamt and F. Eckert, AIChE J., 48 (2002) 369.
[27] S.T. Lin and S.I. Sandler, Ind. Eng. Chem. Res., 41 (2002) 899.
[28] J.D. Thompson, C.J. Cramer and D.G. Truhlar, J. Chem. Phys., 119 (2003) 1661.
[29] C. Hansch, J.E. Quinlan and G.L. Lawrence, J. Org. Chem., 33 (1968) 347.
[30] Y.B. Tewari, M.M. Miller, S.P. Wasik and D.E. Martire, J. Solution Chem., 11 (1982) 435.
[31] C.T. Chiou, V.H. Freed, D.W. Schmedding and R.L. Kohnert, Environ. Sci. Technol., 11 (1977) 475.
[32] S.H. Yalkowsky and S.C. Valvani, J. Chem. Eng. Data, 24 (1979) 127.
[33] Y. Ran and S.H. Yalkowsky, J. Chem. Inf. Comput. Sci., 41 (2001) 354.
[34] M.J. Kamlet, R.M. Doherty, P.W. Carr, D. Mackay, M.H. Abraham and R.W. Taft, Environ. Sci. Technol., 22 (1988) 503.
[35] R.W. Taft, J.L.M. Abboud, M.J. Kamlet and M.H. Abraham, J. Solution Chem., 14 (1985) 153.
[36] P.L. Huyskens, J. Mol. Struct., 270 (1992) 197.
[37] P. Ruelle and U.W. Kesselring, Chemosphere, 34 (1997) 275.
[38] J.S. Chickos, G. Nichols and P. Ruelle, J. Chem. Inf. Comput. Sci., 42 (2002) 368.
[39] J. Paasivirta, S. Sinkkonen, P. Mikkelson, T. Rantio and F. Wania, Chemosphere, 39 (1999) 811.
[40] G. Klopman and H. Zhu, J. Chem. Inf. Comput. Sci., 41 (2001) 439.
[41] V.N. Viswanadhan, A.K. Ghose and J.J. Wendoloski, Perspect. Drug. Discov., 19 (2000) 85.
[42] S. Banerjee, Environ. Sci. Technol., 19 (1985) 369.
[43] S. Banerjee and P.H. Howard, Environ. Sci. Technol., 22 (1988) 839.
[44] W.B. Arbuckle, Environ. Sci. Technol., 20 (1986) 1060.
[45] T.M. Nelson and P.C. Jurs, J. Chem. Inf. Comput. Sci., 34 (1994) 601.
[46] J.M. Sutter and P.C. Jurs, J. Chem. Inf. Comput. Sci., 36 (1996) 100.
[47] B.E. Mitchell and P.C. Jurs, J. Chem. Inf. Comput. Sci., 38 (1998) 489.
[48] J. Huuskonen, J. Chem. Inf. Comput. Sci., 40 (2000) 773.
[49] E.J. Delgado, Fluid Phase Equilib., 199 (2002) 101.
[50] H. Gao, V. Shanmugasundaram and P. Lee, Pharm. Res., 19 (2002) 497.

[51] E.J. Delgado, J.B. Alderete, A.R. Matamala and G.A. Jaña, J. Chem. Inf. Comput. Sci., 44 (2004) 958.
[52] E. Estrada and L.A. Montero, Mol. Eng., 2 (1992) 363.
[53] E. Estrada, J. Chem. Inf. Comput. Sci., 35 (1995) 708.
[54] E. Estrada, J. Chem. Inf. Comput. Sci., 36 (1996) 837.
[55] E. Estrada and E. Molina, J. Chem. Inf. Comput. Sci., 41 (2001) 791.
[56] E. Estrada, E.J. Delgado, J.B. Alderete and G.A. Jaña, J. Comput. Chem., 25 (2004) 1787.
[57] M. Randić, J. Am. Chem. Soc., 97 (1975) 6609.
[58] L.B. Kier and L.H. Hall, Molecular Connectivity in Chemistry and Drug Research, Academic Press, Letchworth, Herts, UK, 1976.
[59] E. Estrada, J. Chem. Inf. Comput. Sci., 35 (1995) 31.
[60] E. Estrada, N. Guevara and I. Gutman, J. Chem. Inf. Comput. Sci., 38 (1998) 428.
[61] H. Theil, Principles of Econometrics, p. vi, Wiley, New York, 1971.

Thermodynamics, Solubility and Environmental Issues
T.M. Letcher (editor)

Chapter 3

Modeling of Contaminant Leaching

Maria Diaz and Defne Apul

Department of Chemical Engineering, Department of Civil Engineering, University of Toledo, Toledo, OH, USA

1. OVERVIEW OF SIGNIFICANCE

An understanding of leaching of contaminants from waste matrices has traditionally been important for managing landfills and contaminated soils. More recently, leaching of contaminants has also become important for proper management of byproduct materials (e.g. coal ashes, foundary sand, steel slag) that can be reused in various applications (e.g. road construction). Leaching and transport of pollutants in the subsurface are often studied in the context of the hydrogeology and geochemistry of the system, associated modeling of the release and transport of pollutants, the risk posed by the pollutants to the human health and the interpretation of these scientific results in the context of environmental regulations [1]. To accurately estimate the risk an accurate characterization of the source term is necessary since if leaching of contaminants is not properly assessed the remainder of the analysis is also flawed.

The extent and rate of leaching of contaminants depends on two major phenomena: (1) the hydrology and therefore the hydraulic properties (e.g. soil moisture retention curve, both saturated and unsaturated hydraulic conductivity) of the matrix, and (2) the geochemistry of the matrix. To model transport, the advective–dispersion equation is used which is based on Darcy's law. In unsaturated conditions, the advective velocity necessary for the advective–dispersion equation is estimated from Richards equation. Appropriate equations necessary for modeling the geochemistry of leaching are various but they are often based on equilibrium mass action laws. There are various degrees of sophistication in contaminant leaching modeling. While more detailed models translate into more accurate predictions, they also require a deeper understanding of the system and more computational effort, resulting in higher costs. A compromise needs to be reached between accuracy, practicality and cost.

In this chapter, we present various equilibrium-based geochemical modeling approaches that can be used for the analysis of leaching of various species from contaminated media. We focus on two important processes that can chemically limit the concentration of the contaminant released or leached: dissolution/precipitation and adsorption. We progress from a simpler approach based on aqueous speciation of chemicals to more complicated approaches that in addition require dissolution/precipitation and/or adsorption calculations. Examples for applications of the geochemical modeling approaches discussed in this chapter are provided in Table 1.

2. GEOCHEMICAL MODELING

Geochemical modeling is usually approached in two different ways: time dependent or equilibrium way. Therefore, we can find chemical kinetic models where reaction rates data are needed and time is a primary factor, or chemical equilibrium models, where reactions are assumed to take place in a very short period of time and the system is in equilibrium. Chemical equilibrium is the most common approach to geochemical modeling and it is the focus of this chapter. Within equilibrium modeling, forward models refer to the case where the starting water chemistry is defined and an attempt is made to model water evolution by dissolution and precipitation of mineral phases and gases. The output of such a program has the following form [13]:

$$\text{Initial water} + \text{Reacting phases} = \text{Predicted water} + \text{Product phases}$$

2.1. Dissolution/Precipitation

The first step in forward modeling is the calculation of the equilibrium distribution of mass among complexes and redox couples. The goal in this step is to analyze the presence of possible minerals in the system that may control the release and concentration of certain contaminants. In this step, solid precipitation is suppressed, allowing the calculation of saturation indices (SI) of the water sample with respect to different mineral phases.

$$\text{SI} = \log(\text{IAP}) - \log K_{\text{sp}} \quad (1)$$

where log(IAP) is the logarithm of the ion activity product calculated according to the stoichiometry of each mineral and K_{sp} the solubility product constant of that mineral. The activity is defined as a surrogate of concentration

Table 1
Examples of geochemical modeling applications

Approach	Data	Program	Application	Reference
Geochemical speciation	Leaching experiments at various pH values	MINTEQA2	Study of the weathering mechanisms of municipal solid waste incinerator bottom ash	[2]
Geochemical speciation	SEM/EDX and XRD mineralogical composition of the unleached material	SOLTEQ-B (a modified version of MINTEQA2 that includes Berner's model of CSH dissolution)	Prediction of leaching from cement-stabilized waste	[3]
Geochemical speciation	Batch experiments	ORCHESTRA	Leaching of Cr(VI) from chromite ore processing residue	[4]
Dissolution/precipitation	Leaching experiments	EQ3/6 Rel.7.2a package	Leaching from cemented waste forms	[5]
Modeling geochemical speciation and dissolution/precipitation	pH-stat leaching experiments	MINTEQA2	Identification of controlling processes for pH-stat leaching behavior of contaminants	[6]
Dissolution/precipitation and adsorption	Column experiments	PHREEQC, HGC and MINTEQA2	Study of contaminant movement in alluvium	[7]
Geochemical speciation combined with ion exchange	Sorption edges and sorption isotherms	MINTEQ	Study of sorption of divalent metals in calcite	[8]

Table 1 (Continued)

Approach	Data	Program	Application	Reference
Surface complexation modeling	Potentiometric titration	FITEQL, MICROQL	Heavy metal adsorption on natural sediment	[9]
Multi-surface adsorption modeling	Selective extractions for site concentration and total metal concentration	ORCHESTRA	Prediction of leachate concentration from contaminated soils	[10]
Multi-surface geochemical modeling	Selective extractions for site concentration and total metal concentration	ECOSAT	Prediction of leaching from sandy soil and evaluation of the contribution of different surfaces to metal binding	[11]
Multi-surface geochemical modeling		WHAM and SCAMP	Solid–solution partitioning of metals in different environmental systems	[12]

to take into account the interaction of ions with other ions that can mask them. The activity can be expressed as:

$$A = \gamma[C] \tag{2}$$

where γ is the activity coefficient (usually less than 1) and C the molar concentration of the substance (mol/L). The activity coefficient is a function of the ionic strength. For dilute solutions ($I \leq 0.1$) the theoretical Debye–Hückel equation can be used to estimate the activity coefficient [14]:

$$\log\gamma = -\frac{BZ^2\sqrt{I}}{1 + D\alpha\sqrt{I}} \tag{3}$$

where Z is the charge, α the effective hydrated radius of the ion for which γ is being calculated and I the ionic strength. B and D are functions of the density of water, temperature and dielectric constant; at 25°C, $B = 0.5092$ and $D = 0.3283$. For intermediate ionic strengths ($0.1 \leq I \leq 0.7$) positive terms can be added to the Debye–Hückel equation to generate an extended version such as the empirical Davies equation [14]:

$$\log\gamma = -BZ^2\left(\frac{\sqrt{I}}{1 + \sqrt{I}} - 0.3I\right) \tag{4}$$

Suppressing solid precipitation will allow the identification of potential solubility controlling minerals. When SI = 0 the water is at thermodynamic equilibrium with respect to the mineral. When SI > 0, water is supersaturated with respect to the mineral and this mineral should precipitate. On the other hand, if SI < 0, water is under-saturated with respect to the mineral and this mineral should dissolve. Minerals with an SI value close to 0 (usually $-1 \leq \mathrm{SI} \leq 1$) are considered to be present in the system. Once the possible minerals present have been identified the leachate composition in equilibrium with the selected minerals is calculated.

The main assumption in this step is that equilibrium has been reached; however, there is no way to determine the certainty of this assumption. The time to reach equilibrium depends on the system matrix and on each element. The system boundary, which will determine the residence time, will also affect the validity of the equilibrium assumption. Also, an important limitation of this step is the uncertainty about the values of thermodynamic constants for the mineral phase because pure minerals are more of the exception than the rule [13].

The geochemical modeling of mineral dissolution/precipitation has been widely used in applications of diverse nature. Meima and Comans [2] used forward modeling to study the leaching process from municipal incinerator bottom ash. They used the speciation code MINTEQA2 with a few modifications to the stability constant database. They were able to identify the mineral phases that controlled the leaching of elements such as Al. In a different work from Meima et al. [15] the same modeling approach was used to study the effect of carbonation processes in municipal solid waste bottom ash in the leaching of copper and molybdenum. They found that concentrations of Mo were very close to equilibrium concentrations with powellite ($CaMoO_4$) while for Cu they could not identify a specific mineral that would control the leaching process suggesting that other processes may be having an effect. Van Herreweghe et al. [6] used the geochemical modeling approach to identify chemical associations of heavy metals in contaminated soils and controlling factors of the leaching process. Geelhoed et al. [4] tried to identify the process controlling the leaching of Cr(VI) from chromite ore processing residue (COPR). Chemical speciation calculations were performed within the ORCHESTRA modeling framework resulting in an accurate prediction of Cr(VI) concentration and other elements present in solution in the pH range 10–12. Ettler et al. [16] also included speciation calculations in their study of heavy metal liberation from metallurgical slags. They used the models PHREEQC and EQ3NR to calculate the speciation of Pb, Zn and As from Pb–Zn metallurgical slags under different leaching conditions. Li et al. [17] used geochemical modeling to study heavy metal speciation and leaching behavior in cement based solidified/stabilized waste materials. They used MINTEQA2 model; their results showed that Cu and Zn oxides were very important during the leaching process. Mijno et al. [3] also studied compositional changes in cement-stabilized waste during leaching tests. They used the geochemical equilibrium model SOLTEQ-B (a modification of MINTEQA2) to incorporate the solubility behavior of calcium silicate hydrate. They successfully predicted leachate concentrations of Ca, Si, S and to a certain extent Al, Pb and Zn. Carlsson et al. [18] used speciation calculations to study infiltrating water composition in the vadose zone of a remediated tailings' impoundment.

2.2. Adsorption/Desorption Processes

At low concentrations, a substance can be unsaturated with respect to its minerals in which case the dissolution/precipitation of minerals would not be the mechanism that controls the concentration of the substance. In such cases, adsorption/desorption processes may be more relevant. Adsorption/

desorption processes refer to conditions where matter accumulates in the interface between a solid and an aqueous phase. According to the EPA [19] these are most likely the processes controlling contaminant movement when chemical equilibrium exists. In order to characterize these processes adsorption isotherm models have been proposed. Adsorption isotherms can be classified into high affinity, Langmuir, constant partition and sigmoidal-shape isotherm classes ([20] and references therein).

2.2.1. Adsorption Isotherm Models

2.2.1.1. Freundlich isotherms This isotherm provides a relationship between sorbed concentration and dissolved concentration as follows:

$$\bar{C} = K_{\mathrm{f}} C^{a} \tag{5}$$

where $\bar{C}$ is the sorbed concentration (M/M), C the dissolved concentration (M/L^3), and K_{f} and a must be determined for each chemical in each porous medium (M^a/L^{3a}).

Moradi et al. [21] used nonlinear Freundlich isotherm to account for the sorption of cadmium in a sewage-amended soil. Köhne et al. [22] used the nonlinear Freundlich isotherm to incorporate sorption in their multi-process herbicide transport analysis in structured soils. At low concentration the constant a becomes unity and a linear relation is obtained:

$$\bar{C} = K_{\mathrm{d}} C \tag{6}$$

In this case the slope of the isotherm (K_{d}) is called the distribution coefficient. K_{d} values are determined for specific materials under specific conditions, so the extent of their applicability is very limited. This approach describes the sorption process as if there was no limit on the amount of sorbing sites available although this might not be a problem if working at low concentrations. Also, this approach neglects the competitive effect from other dissolved species [23, 24].

Steefel et al. ([23] and references therein) noted that the K_{d} approach does not account for pH, competitive ion effects or oxidation–reduction reactions. As a consequence, K_{d} values may vary by orders of magnitude from one set of conditions to another. Chen [25] also highlighted these limitations by comparing numerical modeling results of contaminant transport using a multi-component coupled reactive mass transport model and a K_{d} based transport model. The conclusion from this work was that K_{d} values vary with location and time and this variation could not be accounted for in the model.

Table 2
Descriptive statistics and distribution coefficients (K_d) data set for soils [19]

	Cadmium K_d (mL/g)	Cesium K_d (mL/g)	Strontium K_d (mL/g)	Thorium K_d (mL/g)
Mean	226.7	651	355	54000
Standard deviation	586.6	1423	1458	123465
Minimum	0.50	7.1	1.6	100
Maximum	4360	7610	10200	500000
No. of samples	174	57	63	17

Several attempts have been made in order to correlate the K_d values to physico-chemical properties of soils; however, not much success has been accomplished. Carlon et al. [26] reported a correlation between pH and soil–water distribution coefficient (K_d) for Pb: $\log K_d = 1.99 + 0.42\text{pH}$. The EPA [19] collected K_d values for cadmium, cesium, chromium, lead, plutonium, radon, strontium, thorium, tritium and uranium in soils. The variability in K_d values can be many orders of magnitude as shown in Table 2.

Many organic contaminants show high affinity for solid organic matter. For the sorption of organic compounds into organic matter the distribution coefficient can be written as:

$$K_d = K_{oc} f_{oc} \tag{7}$$

where K_{oc} is the partition coefficient of the solute in an organic carbon medium and f_{oc} the mass fraction of organic carbon. The K_{oc} value is often estimated through empirical relationships with the octanol–water partition coefficient or the solubility for that compound; for example, for chlorinated hydrocarbons K_{oc} can be estimated through $\log K_{oc} = -0.557 \log S + 4.277$, where S is the solubility in μmol/L ([27] and references therein).

The K_d approach is very commonly used in transport models because mathematically it is relatively easy to incorporate. Goyette and Lewis [28] highlighted the utility of K_d values in screening level ground water contaminant transport models of inorganic ions with the caution that experimental conditions such as pH, electrolyte composition and soil type are similar to those being modeled. Viotti et al. [29] used K_d values to model phenol transport in an unsaturated soil. Schroeder and Aziz [30] used this approach to account for PCBs sorption into dredged materials. Buczko et al. [31] used the Freundlich approach to model chromium transport in unsaturated zone.

Seuntjens et al. [32] also used Freundlich isotherms cadmium sorption onto soil in their study of the aging effect on Cd transport in sandy soils. Huang et al. [33] used K_d approach in their transport model of fly ash landfills. Casey et al. [34] used K_d values to model transport of hormones in agricultural soils. Berkvist and Jarvis [35] used K_d values to model cadmium transport in sludge-amended soils. Bahaminyakamwe et al. [36] used K_d values to account for sorption in their study of copper mobility in sewage sludge-amended soils. Piggott and Cawlfield [37] used K_d values to incorporate sorption in their one dimensional transport model. They used a probabilistic approach to study the sensitivity of the model to the different parameters. The probabilistic approach allows accounting for the limitations of the K_d approach discussed earlier. Bou-Zeid and El-Fadel [38] also included partitioning coefficient in order to account for the sorption process in a landfill facility. A sensitivity analysis of their model suggested that the partitioning coefficient was among the most important parameters of the model and should be selected carefully.

2.2.1.2. Langmuir isotherms This isotherm has the following form:

$$\bar{C} = \frac{K_l \bar{S} C}{1 + K_l C} \tag{8}$$

where K_l is the Langmuir constant (M^{-1}/L^3), $\bar{C}$ the sorbed concentration (M/M), C the dissolved concentration (M/L^3) and $\bar{S}$ the maximum sorption capacity (M/M).

Manderscheid et al. [39] determined Langmuir parameters for SO_4^{2-} sorption in soils from two forested catchments in Germany. These parameters were included in the chemical equilibrium model, Model of Acidification of Groundwater in Catchments (MAGIC), in order to study the effect of isotherm variability on the prediction of SO_4^{2-} fluxes with seepage. Langmuir isotherms are not commonly used in transport models because of the computational burden they introduce due to their nonlinearity and also because many researchers, perhaps unjustifiably, often report just the K_d.

2.2.2. Ion Exchange

The previous sorption models are based on experimental data and come in very handy when the system is very complex and not a lot of knowledge on the chemistry is available. However, chemical principles provide means for calculating sorption isotherms and a variety of ion exchange models are available. An example of one kind of ion exchange process is described by the following equation:

$$mC_1^n + n\bar{C}_2 \leftrightarrow m\bar{C}_1 + nC_2^m$$

where n is the valence for ion 1, m the valence for ion 2, C_i the concentration of the ionic species i in solution phase (M/L^3) and $\bar{C}_i$ the concentration of the ionic species i in solid phase (M/M).

The equilibrium constant for that reaction can be written as:

$$K_{eq} = \frac{[\bar{C}_1]^m [C_2]^n}{[C_1]^m [\bar{C}_2]^n} \tag{9}$$

Zachara et al. [8] described sorption of divalent metals (Ba, Sr, Cd, Mn, Zn, Co and Mo) on calcite with a model that included aqueous speciation and $Me^{2+} - Ca^{2+}$ exchange on cation specific surface sites. Engesgaard and Traberg [40] included ion exchange in the modeling of contaminant transport at a waste residue deposit. They found that ion exchange was the dominant process, with Na^+, K^+ and NH_4 from the leachate exchanging with an initial soil population of Ca^{2+} and Mg^{2+}.

2.2.3. Surface Complexation Models (SCMs)

As the system becomes more complex the applicability of adsorption isotherms and ion exchange models becomes inadequate. Surface complexation models offer a more universal description of the sorption process by taking into account important variables affecting sorption processes such as pH, ionic strength and aqueous speciation. This chapter provides an overview of the SCMs; more detailed explanations of the derivations of these models can be found in geochemistry and environmental geochemistry textbooks such as those by Langmuir [14] and Benjamin [41]. Assumptions made in these models are:

- Sorption takes place at sites having specific coordinative properties which will react with sorbing solutes to form surface complexes similar to aqueous complex formation.
- Mass law equations can be applied to reactions at these sites.
- Electric double layer (EDL) theory can account for the effect of surface charge on sorption. Surface charge (σ) and electric potential

(ψ) are consequences of chemical reactions involving the surface functional groups.
- The apparent binding constants determined for the mass law adsorption equations are empirical parameters related to thermodynamic constants via activity coefficients of the surface species.
- Three frequently used SCMs can be found based on the EDL structure used: constant capacitance model (CCM), double layer model (DLM) and triple layer model (TLM). For all three models the electrostatic variables are fitting parameters and the available complexation reaction constants are model dependent.

2.2.3.1. Modeling single sorbent systems SCMs have originally been developed for specific minerals but have been applied to complicated matrices (i.e., soils, sediments). Martin-Garin et al. [42] used a CCM SCM and a cation exchange model in order to reproduce cadmium uptake by calcite in a stirred flow through reactor. They were able to account for the effect of variable solution composition on Cd^{2+} adsorption on the calcite–aqueous solution interface. Both models were able to predict the experimental results for the most part. However, the CCM was able to account for the effect of pH and p_{CO_2} on Cd^{2+} binding to calcite surfaces. Pokrovsky and Schott [43] incorporated a CCM to describe brucite's dissolution kinetics. Dzombak and Morrel [44] conducted extensive studies in order to describe the DLM for specific binding of metal cations and oxy-anions at the hydrous ferric oxide/aqueous interface (DMDLM). Dzombak and Morrel's model has been widely used in applications of diverse nature. Brown et al. [7] used DMDLM to incorporate adsorption into a reactive transport model of a creek basin. Dijkstra et al. [45] included the DMDLM to account for sorption onto iron hydrous oxides in municipal solid waste incinerator bottom ash. Karthikeyan and Elliot [46] combined the DMDLM for hydrous oxides of iron and aluminum in order to simulate the adsorption of Cu over a range of pH and surface loading conditions. Csoban and Joo [47] selected the DMDLM to model sorption of Cr(III) onto silica and aluminum oxide.

The TLM offers a more comprehensive description of the solid–solution interface ([48] and references therein). For this reason Smith [48] determined TLM parameters to describe Pb, Cd and Zn adsorption in a recycled iron bearing material that is treated as a hydrous oxide in aqueous solution. Sarkar et al. [49] applied the TLM to represent the adsorption of Hg by quartz and gibbsite. Villalobos et al. [50] also used the TLM to model carbonate sorption onto goethite in order to study variability in goethite surface site density. Lumsdon and Evans [51] determined surface complexation parameters for goethite for the three models.

2.2.3.2. Modeling multiple sorbent systems Two different approaches can be used when modeling complicated matrices or matrices where more than one sorbent is present. The first approach consists of defining the matrix as an ensemble of different mineral phases. It describes the entire process as the sum of the contribution of the process on each surface. Dijkstra et al. [10] used this multi-surface approach to model the leaching process in contaminated soils. They defined three different surfaces: organic matter, iron/aluminum (hydr)oxides and clay. The iron/aluminum (hydr)oxides were modeled using DMDLM [44] mentioned in Section 2.2.3.1. Weng et al. [11] also used a multi-surface approach to study Cu^{2+}, Cd^{2+}, Zn^{2+}, Ni^{2+} and Pb^{2+} activity in sandy soils. They defined organic matter, clay silicate and iron hydroxides as the available surface sites. They were able to predict measured activities over a wide pH range for all metals except Pb^{2+}. They found that organic matter was the most relevant sorptive surface for Cu^{2+}, Cd^{2+}, Zn^{2+} and Ni^{2+} in this soil.

The second approach is to consider the material as a whole and determine model parameters for it. Wen et al. [9] used the three SCMs to describe the sorption process of Cu and Cd on natural aquatic sediment. They concluded that all three models were able to simulate the experimental results very well. Davis et al. [52] used both approaches to model Zn^{2+} adsorption by sediment describing them as the general composite and the component additivity approach. They were able to determine model parameters for the general composite model and successfully predict Zn^{2+} adsorption onto the sediment over a range of chemical conditions. They concluded that the component additivity approach requires a deep characterization of the sorbents present in the matrix.

Wersin ([53] and references therein) used PHREEQC to include a combination of ion exchange and surface complexation processes to describe a bentonite backfill pore water. Vico [54] also used the ion exchange–surface complexation combination to model the sorption process of Cu on sepiolite.

2.2.4. Organic Matter Complexation

Organic matter plays an important role in the binding of pollutants in natural environments. Several models are available to replicate the interaction of contaminants with natural organic matter. Nonideal competitive adsorption (NICA)-Donnan (ND) model is among the most popular ones. This model describes the interaction between cations and humic substances. ND is a combination of the NICA model, which accounts for the adsorption onto a heterogeneous material and the electrostatic interactions between ions and the humic material, and the Donnan model which considers the humic material as an electrically neutral phase having a particular volume throughout which there is a uniform electrostatic potential known as the Donnan potential [55–58]. ND model is described by the following set of equations:

$$\theta_{i,T} = \frac{Q_i}{Q_{\max}} = \frac{(\tilde{K}_i C_{\mathrm{D},i})^{n_i}}{\sum_j (\tilde{K}_j C_{\mathrm{D},j})^{n_j}} \cdot \frac{\left[\sum_j (\tilde{K}_j C_{D,j})^{n_j}\right]^P}{1+\left[\sum_j (\tilde{K}_j C_{D,j})^{n_j}\right]^P} \tag{10}$$

where Q_i is the total amount of component i bound to the humic acid (mol/kg), $Q_{\max}$ the number of sites (mol/kg), $\tilde{K}_j$ the median affinity constant for component j, $C_{\mathrm{D},j}$ the local concentration of j near the binding sites (mol/L), n_j the ion specific nonideality and p the width of the affinity distribution, which is common to all components.

Electrostatic interactions are taken into account by relating the local and bulk concentrations through a Boltzmann factor which is linked to the Donnan potential as shown in Eq. (11):

$$C_{\mathrm{D},j} = C_j \exp\left(\frac{-e\psi_{\mathrm{D}}}{kT}\right) \tag{11}$$

where C_j is the concentration (or activity) of species j in solution (mol/L), e the charge of the electron, ψ_{D} the uniform Donnan potential, k the Boltzmann's constant and T the absolute temperature.

The electroneutrality condition is given by:

$$\frac{Z}{V_{\mathrm{D}}} + \sum_j Z_j (C_{\mathrm{D},j} - C_j) = 0 \tag{12}$$

where Z is the net charge of the humic substance (equiv/kg), V_{D} the volume of water in the Donnan phase (L/kg) and Z_j the charge of j including its sign.

The volume of the Donnan phase is related to the ionic strength by Eq. (13):

$$\log V_{\mathrm{D}} = b(1 - \log I) - 1 \tag{13}$$

where *b* is the coefficient dependent on the type of humic substance and *I* the ionic strength.

Milne et al. [59, 60] compiled an extensive data set for proton and ion binding by humic and fulvic acids. This work led to the NICA-Donnan generic parameters that can be used in the absence of specific data. Generic parameters are reported for 23 metal ions: Al, Am, Ba, Ca, Cd, Cm, Co, Cr^{III}, Cu, Dy, Eu, Fe^{II}, Fe^{III}, Hg, Mg, Mn, Ni, Pb, Sr, Th^{IV}, $U^{VI}O_2$, $V^{III}O$ and Zn. Weng et al. [11] incorporated the NICA-Donnan model with generic parameters to model organic matter from a sandy soil. Dijkstra et al. [10] also used NICA-Donnan with generic parameters to model organic matter from different soils.

3. SUMMARY

Modeling of contaminant leaching is a complex and diverse discipline. From simplified to more complicated models there are several options available. The selection of the appropriate model is based on the needs and most importantly on the resources available such as data, sorbent and sorbate characteristics, and reaction constants, among others. The literature review presented in this chapter includes a wide range of applications for contaminant leaching: from natural systems that have been contaminated to engineered systems where a byproduct needs to be evaluated for potential reuse as opposed to disposal.

REFERENCES

[1] F.M. Dunnivant and E. Anders, A Basic Introduction to Pollutant Fate and Transport, p. 480, Wiley, 2006.
[2] J.A. Meima and R.N.J. Comans, Environ. Sci. Technol., 31 (1997) 1269.
[3] V. Mijno, L.J.J. Catalan, F. Martin and J. Bollinger, J. Colloid Interface Sci., 280 (2004) 465.
[4] J.S. Geelhoed, J.C.L. Meeussen, S. Hillier, D.G. Lumsdon, R.P. Thomas, J.G. Farmer and E. Paterson, Geochim. Cosmochim. Acta, 66 (2002) 3927.
[5] B. Kienzer, P. Vejmelka, H. Herbert, H. Meyer and C. Altenhein-Haese, Nucl. Technol., 129 (2000) 101.
[6] S. Van Herreweghe, R. Swennen, V. Cappuyns and C. Vandecasteele, J. Geochem. Explor., 76 (2002) 113.
[7] J.G. Brown, R.L. Bassett and P.D. Glynn, Appl. Geochem., 15 (2000) 35.
[8] J.M. Zachara, C.E. Cowan and C.T. Resch, Geochim. Cosmochim. Acta, 55 (1991) 1549.
[9] X. Wen, Q. Du and H. Tang, Environ. Sci. Technol., 32 (1998) 870.

[10] J.J. Dijkstra, J.C.L. Meeussen and R.N.J. Comans, Environ. Sci. Technol., 38 (2004) 4390.
[11] L. Weng, E.J.M. Temminghoff and W.H. Van Riemsdijk, Environ. Sci. Technol., 35 (2001) 4436.
[12] S. Lofts and E. Tipping, Environ. Geochem. Health, 21 (1999) 299.
[13] O. Sracek, P. Bhattacharya, G. Jacks, J. Gustafsson and M. von Brömssen, Appl. Geochem., 19 (2004) 169.
[14] D. Langmuir, Aqueous Environmental Geochemistry, 600 pp., Prentice Hall, Upper Saddle River, NJ, 1997.
[15] J.A. Meima, R.D. van der Weijden, T.T. Eighmy and R.N.J. Comans, Appl. Geochem., 17 (2002) 1503.
[16] V. Ettler, M. Mihaljevic, J. Touray and P. Piantone, Bull. Soc. Geol. France, 173(2) (2002) 161.
[17] X.D. Li, C.S. Poon, H. Sun, I.M.C. Lo and D.W. Kirk, J. Hazard. Mater., A82 (2001) 215.
[18] E. Carlsson, B. Öhlander and H. Holmström, Appl. Geochem., 18 (2003) 659.
[19] USEPA, Understanding Variation in Partitioning Coefficient, K_d, Values, USEPA, Washington, DC, 1999.
[20] C. Hinz, Geoderma, 99 (2001) 225.
[21] A. Moradi, K.C. Abbaspour and M. Afyuni, J. Contam. Hydrol., 79 (2005) 187.
[22] J.M. Köhne, S. Köhne and J. Šimůnek, J. Contam. Hydrol., 85 (2006) 1.
[23] C.I. Steefel, D.J. DePaolo and P.C. Lichtner, Earth Planet. Sci. Lett., 240 (2005) 539.
[24] C.M. Bethke and P.V. Brady, Ground Water, 38 (2000) 435.
[25] Z. Chen, Comput. Geosci., 29 (2003) 351.
[26] C. Carlon, M.D. Valle and A. Marcomini, Environ. Pollut., 127 (2004) 109.
[27] M.D. La Grega, P.L. Buckingham and J.C. Evans, Hazardous Waste Management, 1202 pp., McGraw-Hill, Boston, MA, 2001.
[28] M.L. Goyette and B.G. Lewis, J. Environ. Eng., 121(7) (1995) 537.
[29] P. Viotti, M.P. Papini, N. Stracqualursi and C. Gamba, Ecol. Modell., 182 (2005) 131.
[30] P.R. Schroeder and N.M. Aziz, J. Environ. Eng., 125(9) (1999) 835.
[31] U. Buczko, L. Hopp, W. Berger, W. Durner, S. Peiffer and M. Scheithauer, J. Plant Nutr. Soil Sci., 167 (2004) 284.
[32] P. Seuntjens, K. Tirez, J. Šimůnek, M.Th. van Genuchten, C. Cornelis and P. Geuzens, J. Environ. Qual., 30 (2001) 1040.
[33] C. Huang, C. Lu and J. Tzeng, J. Environ., Eng., 124 (8) (1998) 767.
[34] F.X.M. Casey, H. Hakk, J. Šimůnek and G.L. Larsen, Environ. Sci. Technol., 38 (2004) 790–798.
[35] P. Berkvist and N. Jarvis, J. Environ. Qual., 33 (2004) 181.
[36] L. Bahaminyakamwe, J. Šimůnek, J.H. Dane, J.F. Adams and J.W. Odom, Soils Sci., 171(1) (2006) 29.
[37] J.H. Piggott and J.D. Cawlfield, J. Contam. Hydrol., 24 (1996) 97.
[38] E. Bou-Zeid and M. El-Fadel, Waste Manag., 24 (2004) 681.
[39] B. Manderscheid, C. Jungnickel and C. Alewell, Soil Sci., 165(11) (2000) 848.
[40] P. Engesgaard and R. Traberg, Water Resour. Res., 32 (1996) 939.
[41] M. Benjamin, Water Chemistry, 668 pp., McGraw-Hill, Boston, MA, 2002.

[42] A. Martin-Garin, P. Van Cappellen and L. Charlet, Geochim. Cosmochim. Acta, 67 (2003) 2763.
[43] O.S. Pokrovsky and J. Schott, Geochim. Cosmochim. Acta, 68 (2004) 31.
[44] D.A. Dzombak and F.M.M. Morrel, Surface Complexation Modelling: Hydrous Ferric Oxide, Wiley, New York, 1990.
[45] J.J. Dijkstra, H.A. van der Sloot and R.N.J. Comans, Waste Manag., 22 (2002) 531.
[46] K.G. Karthikeyan and H.A. Elliot, J. Colloid Interface Sci., 220 (1999) 88.
[47] K. Csoban and P. Joo, Colloids Surf., 151 (1999) 97.
[48] E.H. Smith, Water Res., 30 (10) (1996) 2424.
[49] D. Sarkar, M.E. Essington and K.C. Misra, Soil Sci. Soc. Am. J., 63 (1999) 1626.
[50] M. Villalobos, M.A. Trotz and J.O. Leckie, J. Colloid Interface Sci., 268 (2003) 273.
[51] D.G. Lumsdon and L.J. Evans, J. Colloid Interface Sci., 164 (1994) 119.
[52] J.A. Davis, J.A. Coston, D.B. Kent and C.C. Fuller, Environ. Sci. Technol., 32 (1998) 2820.
[53] P. Wersin, J. Contam. Hydrol., 61 (2003) 405.
[54] L.I. Vico, Chem. Geol., 198 (2003) 213.
[55] M.F. Bennedetti, W.H. Van Riemsdijk and L.K. Koopal, Environ. Sci. Technol., 30 (1996) 1805.
[56] D.G. Kinniburgh, C.J. Milne, M.F. Benedetti, J.P. Pinheiro, J. Filius, L.K. Koopal and W.H. Van Riemsdijk, Environ. Sci. Technol., 30 (1996) 1687.
[57] L.K. Koopal, W.H. Van Riemsdijk and D. Kinniburgh, Pure Appl. Chem., 73 (2001) 2005.
[58] E. Tipping, Cation Binding by Humic Substances, 434 pp., Cambridge University Press, New York, NY, 2002.
[59] C.J. Milne, D.G. Kinniburgh and E. Tipping, Environ. Sci. Technol., 35 (2001) 2049.
[60] C.J. Milne, D.G. Kinniburgh, W.H. Van Riemsdijk and E. Tipping, Environ. Sci. Technol., 30 (2003) 958.

PART II

INDUSTRY AND MINING

Thermodynamics, Solubility and Environmental Issues
T.M. Letcher (editor)

Chapter 4

Supercritical Fluids and Reductions in Environmental Pollution

Koji Yamanaka and Hitoshi Ohtaki[†]

Organo Co. of Japan, 1-4-9 Kawagishi, Toda, Saitama 335-0015, Japan

1. INTRODUCTION

The development of high temperature–high pressure techniques in recent years has meant that supercritical fluids (SCFs) can now be used in industry. SCFs are currently being used to decompose organic wastes, used in supercritical extraction of materials, preparation of extra-small particles and in high temperature–high pressure steam in power stations and nuclear reactors.

The density, ρ, of a SCF is intermediate between the density of its liquid and gas phases and varies drastically around the critical points of t_c and p_c. Fluids with this wide range of density can be used successfully as solvents.

In supercritical water (SCW), the solubility of inorganic electrolytes decreases enormously, and this change can be applied to the preparation of nano-size particles of inorganic substances.

SCFs have extremely high diffusibility, due to the large thermal mobilities of the molecules, and thus can penetrate into micropores of substances to extract target compounds and can clean surfaces with complicated shapes.

In this chapter, we focus our attention on the thermodynamics and structure of SCFs and the application to the treatment of removing polluting materials from our environment.

2. SUPERCRITICAL FLUIDS

An SCF is defined as a fluid in the state where both the temperature and pressure are higher than the critical points. The critical temperature, t_c, the

[†]Deceased on 5 November 2006.

Table 1
Critical temperature t_c, critical pressure, ρ_c and critical density, ρ_c of some substances

Substance	t_c (°C)	ρ_c (MPa)	ρ_c (g cm^{-3})
CO_2	31	7.38	0.47
C_2H_6	32	4.88	0.20
N_2O	36	7.24	0.45
C_3H_8	97	4.25	0.22
NH_3	132	11.28	0.24
C_6H_{14}	234	2.97	0.23
CH_3OH	239	8.09	0.27
C_2H_5OH	243	6.38	0.28
$C_6H_5CH_3$	318	4.11	0.29
H_2O	274	22.06	0.32

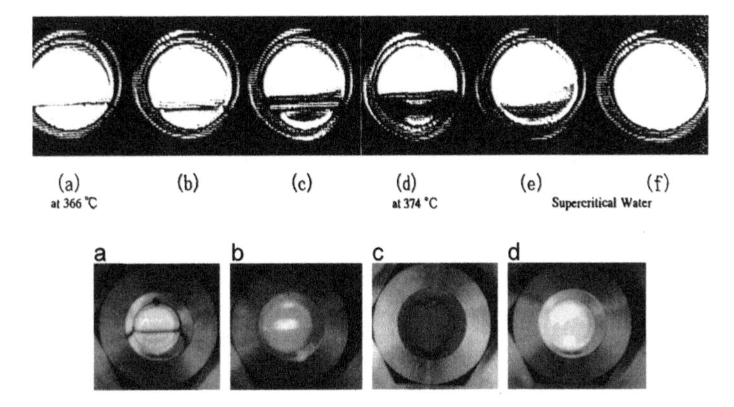

Photograph 1. The phase change of water ($t_c = 374$°C, $p_c = 22.06$ MPa) [2] and methanol ($t_c = 239.4$°C, $p_c = 8.09$MPa) [3] from liquid to supercritical states. (I) Water: (a) at 366°C; (b)–(c) 366°C $< t <$ 374°C; (d) at 374°C; (e)–(f) 374°C $< t$ (supercritical state). (II) Methanol: (a) at ambient temperature; (b) at 238°C (subcritical just below t_c); (c) at 239.4°C (t_c); (d) at 245°C (supercritical state).

critical pressure, p_c and the critical density, ρ_c, of several fluids are listed in Table 1 [1]. The change of the phase of water [2] and methanol [3] are shown in Photograph 1. The values of t_c and p_c are: $t_c = 374$°C and $p_c = 22.06$ MPa for water and $t_c = 239.4$°C and $p_c = 8.09$ MPa for methanol. Under ambient conditions (Photograph 1, Ia and IIa), two phases, i.e., liquid and gas, and their interface are clearly observed. With elevating temperature, the interface diffuses because the densities of the liquid and gas phases approach

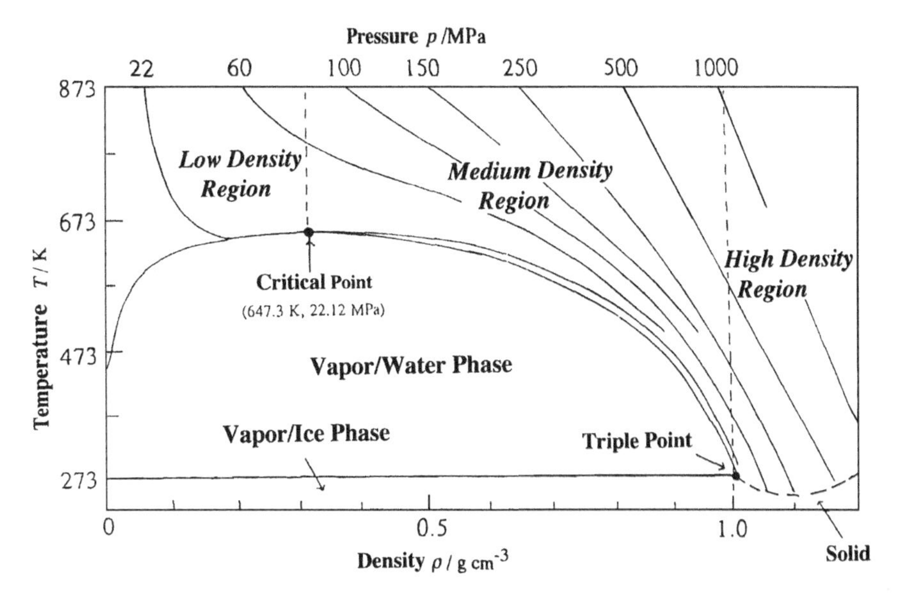

Fig. 1. Phase diagram of water [4, 50, 51].

each other. Under a subcritical condition, fluids become reddish opaque, due to the scattering of light with short wavelengths caused by density fluctuations resulting from the formation of microscopic agglomerates. At the critical point the density of the two phases become identical and the interface disappears (Photograph 1, Ie and IIc), and the fluids turn into one transparent phase (Photograph 1, If and IId) as the agglomerates decompose to much smaller clusters as a result of increased thermal motions.

The phase diagram of water [4, 50, 51] is shown in Fig. 1. SCW is classified into three categories, low-, medium- and high-density regions because of the parameter's importance in relation to thermodynamic and dynamic properties. In the supercritical region many physicochemical properties of water can be represented by the single parameter, density, rather than the parameters, temperature and pressure [50, 51].

In the low-density region, the density of water, ρ, is less than the critical density ($\rho < \rho_c$). The density in the medium density region is an intermediate between ρ_c and $1\,g\,cm^{-3}$ ($\rho_c \leq \rho \leq 1\,g\,cm^{-3}$), while for SCW in the high-density region, it is larger than $1\,g\,cm^{-3}$ ($1\,g\,cm^{-3} < \rho$).

The phase diagram of carbon dioxide is represented in Fig. 2 [5], in which the supercritical region is depicted by the gray area. Above the critical temperature, fluids are usually not liquefied under pressure except for

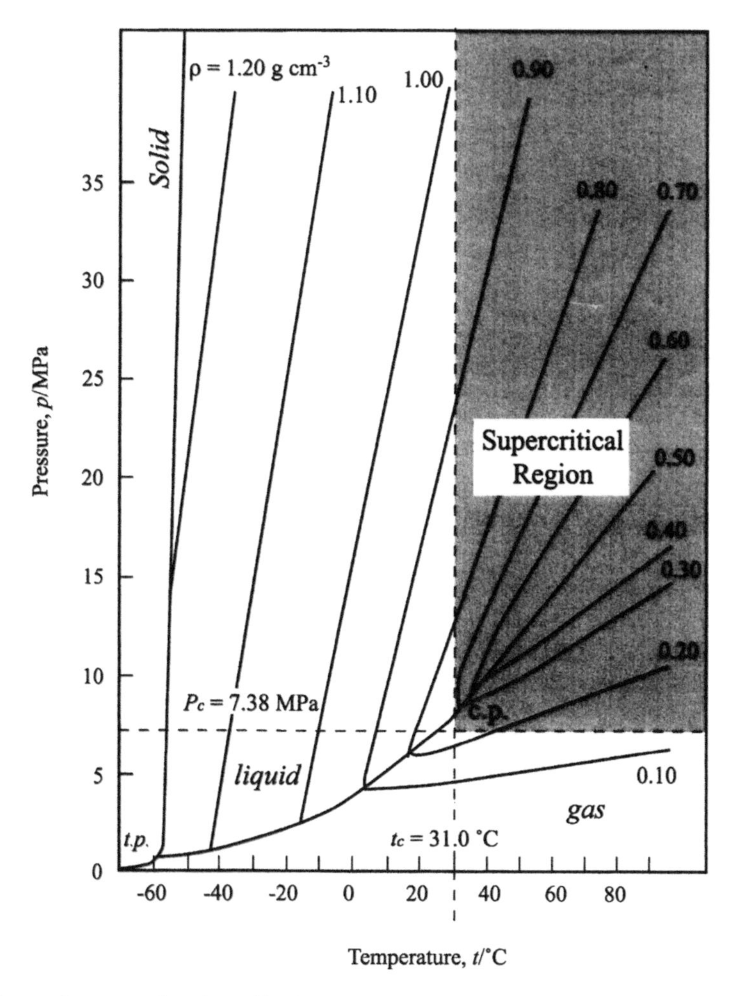

Fig. 2. Phase diagram of carbon dioxide [5].

some substances which are solid under extremely high pressures. Carbon dioxide is one example, which solidifies at a pressure above 570 MPa at t_c.

Under the supercritical conditions the solubility of electrolytes in water dramatically decreases, while solubility of non-electrolytes in water steeply increases. Therefore, SCW can be a good solvent for making inorganic microparticles and also for oxidizing organic wastes by dissolving organic oxidizing substances in water containing the waste. This characteristic property of water can play a key role in the supercritical water oxidation (SCWO) process for removal of hazardous organic wastes and chlorinated hydrocarbons by decomposing them to carbon dioxide, water and inorganic acids [6].

Among various physicochemical properties of SCW, the dielectric constant, ε, and the autoprolysis constant, K_w, play significant roles in the intermolecular interactions of water and ionic solutes.

Various SCFs, other than SCW, have been examined for the treatment of organic wastes and chlorinated hydrocarbons, and some of them have been commercially used. Supercritical methanol is often used, instead of SCW, under the conditions of $t = 240–300°C$ and $p = 8–15$ MPa for selective decomposition or methylation of organic materials. One example is the recycle of polyethylene terephthalate (PET) bottles decomposing to its monomers by supercritical methanol [7].

Supercritical carbon dioxide ($SCCO_2$) is another useful solvent in industry. The critical temperature and pressure of carbon dioxide are $t_c = 30.9°C$ and $p_c = 7.4$ MPa, which can easily be achieved by using relatively simple and inexpensive equipment. From the 1960s, $SCCO_2$ has been used to extract flavors and other compounds. It is also used as the mobile phase in chromatographic separation of various substances. Coffee decaffeinating and hop extraction by $SCCO_2$ (developed in Germany in 1970s) are successful examples of the application of SCFs in industry. A hop extraction industry reached a capacity of 50000 ton per year in 1990 in Germany [5]. $SCCO_2$ is used as an extracting solvent for organic compounds from wastes as a pretreatment, before decomposition or chemical analysis [8]. Further details of the use of SCFs in the reduction of pollutants in our environment will be discussed later.

3. REFERENCES FOR THERMODYNAMIC PROPERTIES OF SUPERCRITICAL FLUIDS

3.1. References for Thermodynamic Properties of Supercritical Water

The largest industrial SCW application is in thermal power stations. Here the temperature and pressure of the SCW are 538–560°C and 24.3–31.0 MPa, respectively. The SCW is used as an energy transmission medium to turbines. Because of the large heat capacity of water, it has been used as an energy carrier since the development of steam engines by James Watt in 1769. Chemical thermodynamics was largely developed for the improvement of the energy efficiency of engines, and thus the thermodynamic properties of water and its vapor have been extensively investigated over a wide range of temperature and pressure.

Numerical values of the thermodynamic properties of water, vapor and SCW were compiled in the International Skeleton Steam Table (IST) published by the International Conference on the Properties of Water and Steam (ICPWS), which was hosted by the International Association for the Properties of Water and Steam (IAPWS) [9]. The latest version of IST was released in 1985 (IST-1985), which covered specific volumes and enthalpies of water, vapor and SCW over the temperature and pressure ranges of 273.15–1073.15 K and 0.1–1,000 MPa, respectively. Data for surface tension [10], dielectric constants (permittivities) [11], ionic products of solvents [12], critical points [13], viscosities [14–16], heat capacities [14, 15] and refractive indices [17] of light and heavy water have been published. IAPWS provides equations by which necessary data can be evaluated over a wide range of temperature and pressure. The most recent publication of formulae appeared in 1995 [18], and the software implementing the formulae is available from National Institute of Standards and Technology (NIST) [19]. Thermodynamic data for industrial use of SCW are also provided by IAPWS-IF97 [20, 21] and the software is distributed through the American Society of Mechanical Engineers (ASME) [22, 23].

Three-dimensional diagrams of t-p-ρ [6, 24] and t-ρ-η (η: viscosity) [4, 25, 26] of water are shown in Fig. 3a (also in Fig. 1) and b, respectively.

In Fig. 3a, it is seen that the density of SCW is sensitive to both temperature and pressure around the critical point. As a result, the solubility of substances in water can easily be controlled by changing the density.

The variation of the viscosity, η, of water with temperature and density is shown in Fig. 3b. The viscosity of water decreases sharply with temperature from 0 to ~200°C, and then, gradually decreases further. However, the density dependence of viscosity is, by contrast, rather simple.

The behavior of the relative dielectric constant of water ($\varepsilon_r = \varepsilon/\varepsilon_0$, where ε_0 denotes the dielectric constant of vacuum) is more or less similar to that of viscosity (Fig. 4) [4, 11, 27]. Due to the breaking of hydrogen bonds of water, the dielectric constant rapidly decreases with temperature. For comparison, the relative dielectric constants of acetonitrile (CH_3CN), methanol (CH_3OH) and liquid ammonia (NH_3) at room temperature are indicated in Fig. 4.

The autoprotolysis constant of water, K_w, changes with temperature, pressure and density. The t-ρ-K_w and t-p-ρ-K_w diagrams of water are shown in

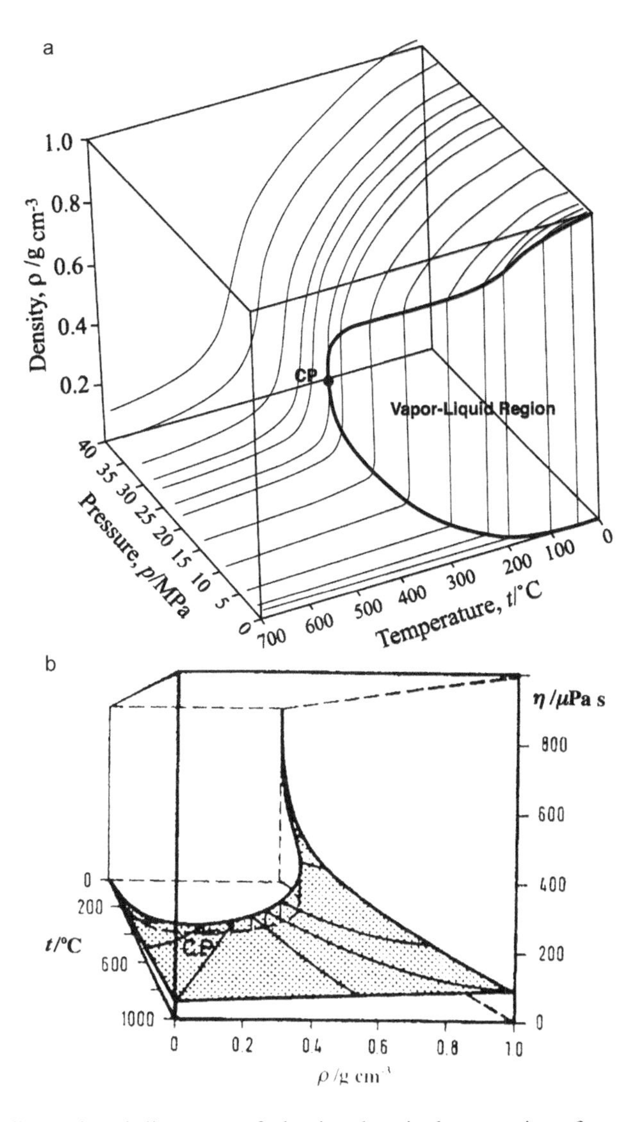

Fig. 3. Three-dimensional diagrams of physicochemical properties of water. (a) t-p-ρ [6, 24]. (b) t-ρ-η [4, 26, 27].

Fig. 5a and b, respectively [4, 12]. From the figures, one could expect that one mole of water might dissociate into $10^{-1.5}$ mol dm^{-3} H^+ and $10^{-1.5}$ mol dm^{-3} OH^- ions under 10^5 atmospheric pressure (10^4 MPa or 10 GPa) at 1000°C. This would mean that under these conditions, the water was both highly acidic and basic at the same time. This condition might be achieved deep underground,

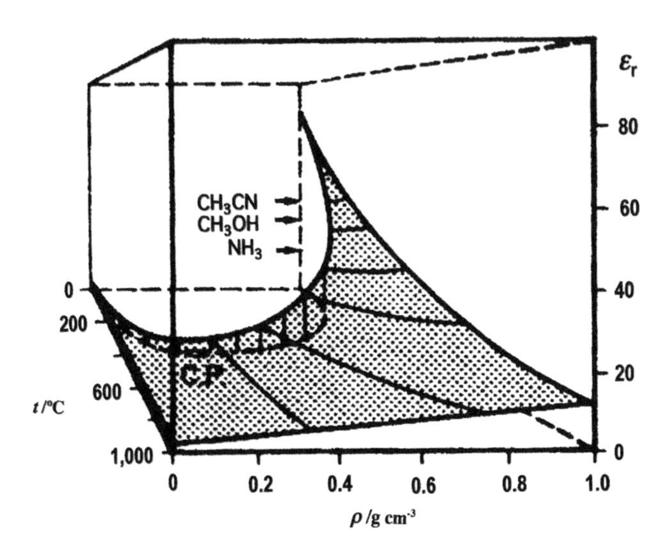

Fig. 4. The three-dimensional t-ρ-ε_r diagram of water [4, 11, 25].

where rocks and stones might be dissolved. However, no evidence has been found for this even under such ultra-supercritical conditions with a very high density of $\rho \sim 1.5\,g\,cm^{-3}$. Fig. 5c shows another plot of the autoprotolysis constant, log K_w, and relative dielectric constant, ε_r, of water against temperature at 25 MPa [11, 12]. With the increase in temperature, the value of log K_w increases slightly, and then sharply decreases at ~400°C. The relative dielectric constant of water also decreases with temperature and a sharp drop in ε_r occurs around 400°C.

3.2. References for Thermodynamic Properties of Supercritical Carbon Dioxide and Other Fluids

The thermodynamic data of carbon dioxide and other fluids are compiled in "International Thermodynamic Tables of the Fluid State" published by the International Union of Pure and Applied Chemistry (IUPAC) [28] in the form of tables and equations of state. Thermodynamic data and equations of state are also provided by "Journal of Physical and Chemical Reference Data" for argon [29], nitrogen [29], oxygen [29], carbon dioxide [30], methane [31], ethane [31], propane [31], butane [31, 32], isobutene [31, 32], ethylene [29] and methanol [33]. "Fluid Phase Equilibria", "Journal of Supercritical Fluids" and "Chemical Engineering Science" are also good sources of thermodynamic and thermochemical data of SCFs. Data for phase diagrams,

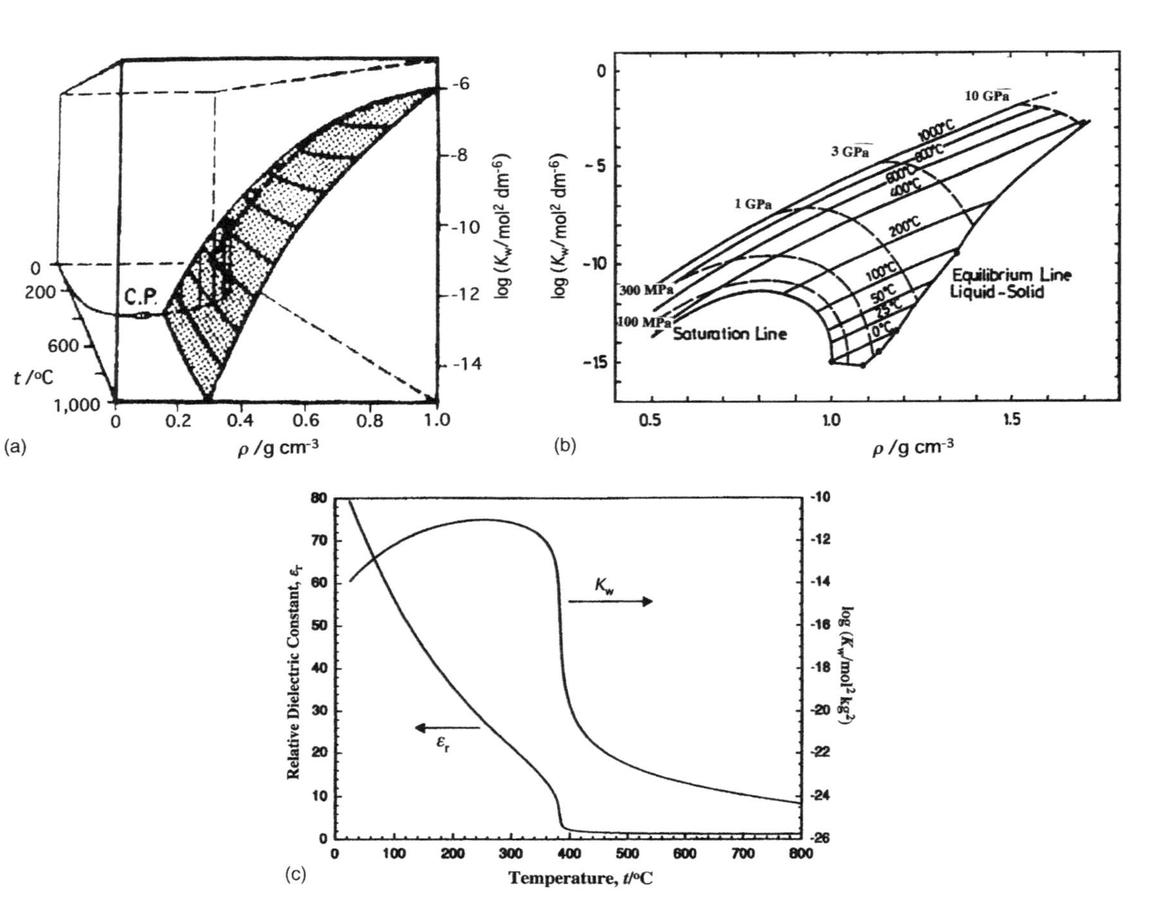

Fig. 5. The diagram of the autoprotolysis constant, K_w, of water with temperature and pressure. (a) The three-dimensional t-ρ-log K_w diagram, (b) the four-dimensional t-p-ρ-log K_w diagram, (c) the variation of log K_w and dielectric constant, ε_r, with temperature under 25 MPa [11, 12, 42].

diffusion coefficients of molecules in pure and mixed fluids are available [34–39]. One can access various thermodynamic databases of pure fluids through the websites of NIST database [19] and "Chemical Web Book" [40]. Data for mixed fluids for engineering purposes can be found in the "Dortmund Data Bank" [41].

4. SOLUBILITY OF ELECTROLYTES AND NON-ELECTROLYTES IN SUPERCRITICAL FLUIDS

4.1. Solubility of Organic and Inorganic Substances in Supercritical Water

Around the critical point, the dissolving power of water for solutes changes dramatically. In SCW, the solubility of inorganic electrolytes decreases sharply, while non-polar hydrophobic organic substances exhibit a high solubility and sometimes become completely miscible. The characteristic change in the dissolving ability of SCW plays a key role in the SCWO technology, where waste organics and oxidants (oxidizing reagent or air) are well mixed with SCW. The changes in ε_r and K_w with temperature and density (Fig. 5) essentially contribute to the SCWO method. Although the autoprotolysis constant, K_w behaves as if SCW is a mixture of a strong acid and a strong base at extremely high temperatures, pressures and density, the value of K_w is $\sim 10^{-23}$ mol^2 dm^{-6} around the critical point, where the density becomes one-third of that of normal water [12].

An example is the solubility of benzene in water at high temperature and pressure [43, 44]. The solubility of benzene in water is ~0.07 wt% at ambient conditions, but is 7–8 wt% at 260°C, and the solubility is practically independent of pressure at any particular temperature. The solubility increases to 18 and 35 wt% at 287 and 295°C, respectively, and at 20–25 MPa, benzene becomes totally miscible at 300°C. Fig. 6 shows the critical solubility curves of binary mixtures of water with H_2, N_2, CO_2, Ar, Xe, CH_4 and benzene [45]. On the left-hand side of the curves the mixtures are separated into two phases and on the right-hand side they exist as one-phase fluids.

In contrast to the case of non-polar organic substances, the solubility of inorganic compounds decreases exponentially in SCW as seen in Fig. 7. Sodium chloride dissolves in water up to 37 wt% at 300°C, but the solubility decreases to only 120 ppm in SCW at 550°C and 25 MPa. The maximum solubility of calcium chloride is 70 wt% in subcritical water, but the solubility goes down to 3 ppm at 500°C and 25 MPa. Most inorganic salts are spar-

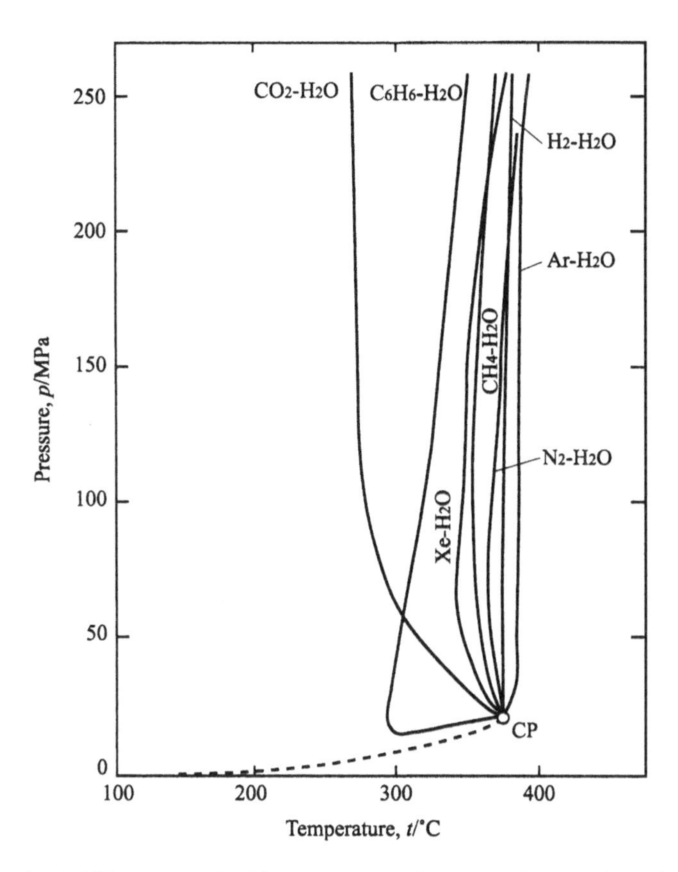

Fig. 6. Critical solubility curves for binary system of water and non-polar substances. CP: critical point of pure water [45].

ingly soluble in water under sub- and supercritical conditions due to insufficient formation of hydration shells around ions because of the low density of the water and high mobility of water molecules.

4.2. Solubility of Organic and Inorganic Substances in Supercritical Carbon Dioxide

The solubility of solid ionic solutes is low in non-polar solvents, while non-polar solutes show higher solubilities than ionic salts. The solubility of non-polar solutes is affected by intermolecular interactions in the solid and solute–solvent interactions and the number density of molecules in the solvents. The vapor pressure of the solid solute can be a good indicator of intermolecular interactions in the solid state. Carbon dioxide is non-polar

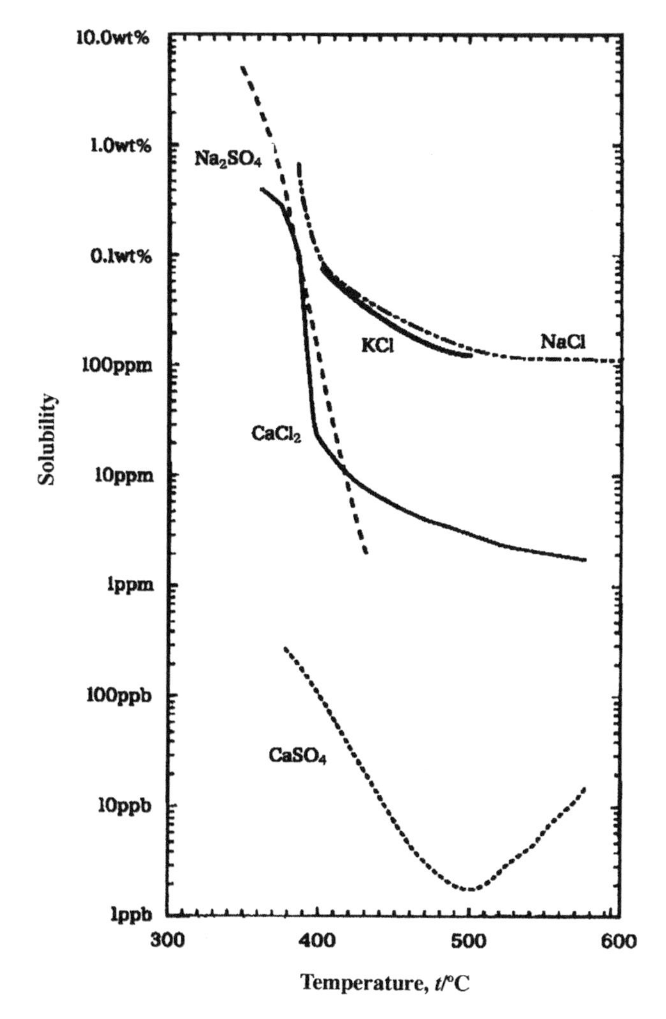

Fig. 7. Solubility of electrolytes in supercritical water at 25 MPa [6].

(dipole moment, $\mu = 0$), but has a quadrupole moment. The solubility of a solute in $SCCO_2$ depends on the molecular weight and the polarity of the solute.

Fig. 8 shows the variation of the mole fraction, y_2, of naphthalene in saturated $SCCO_2$ solutions with temperature and pressure [46]. Subscript 2 denotes the solute, i.e., naphthalene, and the component 1 is of course CO_2. The solubility of naphthalene in CO_2 is strongly affected by pressure around

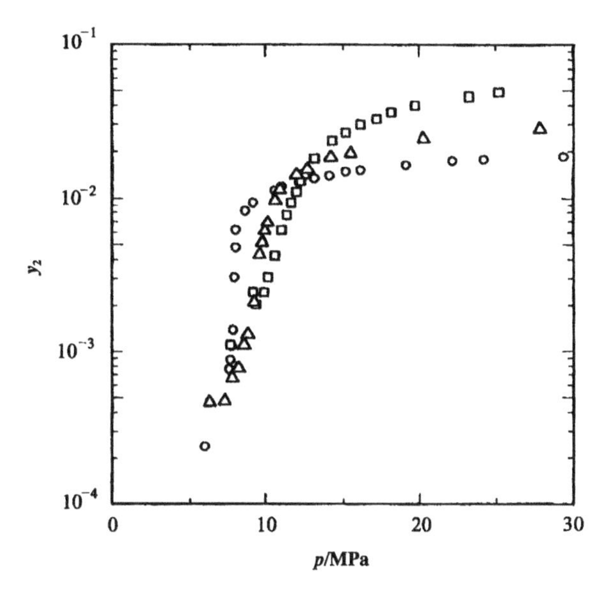

Fig. 8. Solubility of naphthalene in supercritical CO_2. y_2 denotes the mole fraction of naphthalene at saturation [45]. O: 35°C, △: 45°C, and □: 55°C.

the critical point, $p_c = 7.38$ MPa, but at higher pressures the effect is small. The steep increase in the solubility of naphthalene in $SCCO_2$ with pressure around p_c is convenient for extracting naphthalene from a mixture. The target solute is dissolved in $SCCO_2$ from a solid mixture by pressurizing the solution and is precipitated by releasing the pressure.

Fatty acids and alcohols dissolve in $SCCO_2$ and the concentration of fatty acids at saturation increase steeply with pressure around p_c of CO_2 (Fig. 9) [47, 48]. The general trend is that the larger the polarity and molecular size of the solute molecules, the smaller the solute solubility.

$SCCO_2$ is not a good solvent for polar substances when compared with the usual organic liquids. In order to improve the dissolving ability of substances in $SCCO_2$, an entrainer or a co-solvent is added to $SCCO_2$ [49]. In Fig. 10 the effect of ethanol and octane as the entrainer (co-solvent) is shown for the maximum solubility of stearic acid in $SCCO_2$. The mole fraction of stearic acid, y_2, at saturation in $SCCO_2$ increases with the concentration of the entrainers. Ethanol has a larger effect than octane, because polar ethanol can better interact with polar stearic acid than non-polar octane.

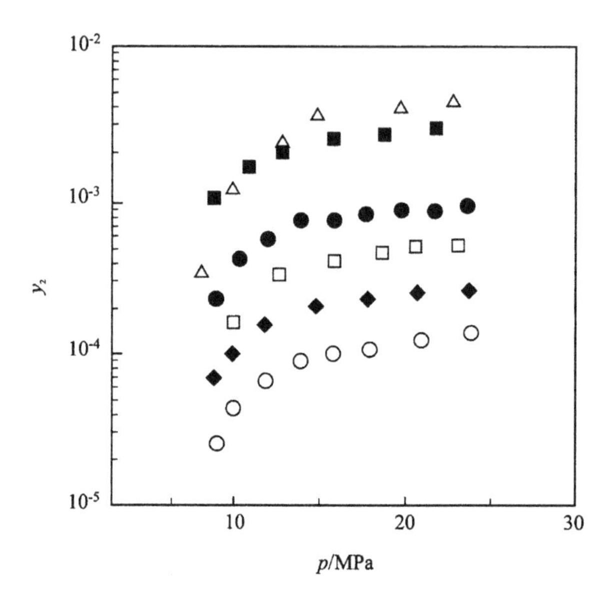

Fig. 9. Solubility of fatty acids and alcohols in supercritical CO_2 at 35°C. y_2 denotes the mole fraction of solutes at saturation [47, 48]. △:myristic acid, □: palmitic acid, ○: stearic acid, ■: cetyl alcohol, ●: stearyl alcohol, ◆: arahydryl alcohol.

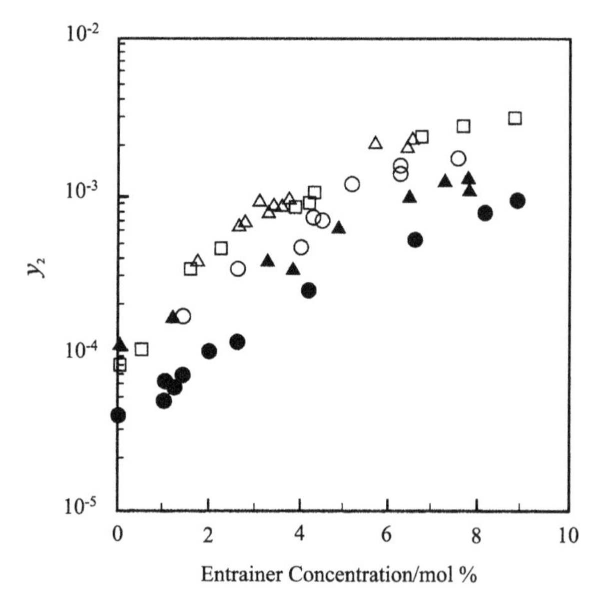

Fig. 10. The entrainer effect of ethanol and octane on the solubility of stearic acid in supercritical CO_2 [49]. ○: 9.9 MPa, □: 14.8 MPa, △: 19.7 MPa for ethanol, and ●: 9.9 MPa, ▲: 19.7 MPa for octane.

5. STRUCTURE OF SUPERCRITICAL WATER

Structural information on SCFs is essentially important not only for the understanding of thermodynamic and thermochemical properties, and the unique reactions of substances in the SCFs, but also for the improvement and extension of industrial applications of SCFs. However, due to technical problems, few studies have been done on the structure of SCFs. The recent progress in analytical methods and equipment has improved the situation. High-speed computers are now widely used to simulate the behavior of molecules in SCFs at high temperatures and pressures, which is sometimes difficult to realize by experiments. Since water is the most attractive substance for chemists, structural investigations of SCW have been carried out by various workers using methods such as X-ray and neutron diffraction techniques, NMR, Raman and infrared spectroscopies. Computer simulations provide useful information for the molecular arrangements and dynamics in SCW [50, 51].

Interatomic spatial correlations are represented by the radial distribution function, $D(r)$, which can be obtained by the X-ray and neutron diffraction methods. From the peak position, the peak area, and the peak width, the interatomic distance, r, the number of atoms within the atom pair, n and the mean-square amplitude of the distance of the atom-pair, σ, respectively, can be estimated.

The X-ray diffraction method at high temperatures and high pressures has been applied to the study on the structure of SCW since the 1980s [52, 53], but the early results were not reliable due to technical difficulties. In 1993 Yamanaka and his coworkers reported X-ray diffraction results for the structure of high temperature and high pressure water at $T = 300\text{–}649$ K and $p = 0.1\text{–}98.1$ MPa by using an imaging plate detector [54]. The radial distribution curves in the form of $D(r) - 4\pi r^2 \rho_0$ at various temperatures and pressures are represented in Fig. 11, where ρ_0 denotes the average electron density in the system. The peaks appearing at $r = 290$, 450 and 670 pm indicate the tetrahedral ice-like hydrogen-bonded structure of liquid water [55].

The peaks at higher r values gradually disappeared with an increase in temperature and pressure, indicating the breaking of the ice-like hydrogen-bonded tetrahedral structure of water. The peak analysis was performed for the first peak around 200–400 pm of the radial distribution function in the form of $D(r)/4\pi r^2 \rho_0$, and the peak was de-convoluted into two peaks, I and II, as seen in Fig. 12. The structure parameters, r, n and half-width at the half-height of the peak, σ, which corresponds to the mean-square amplitude of bonds, are summarized in Table 2.

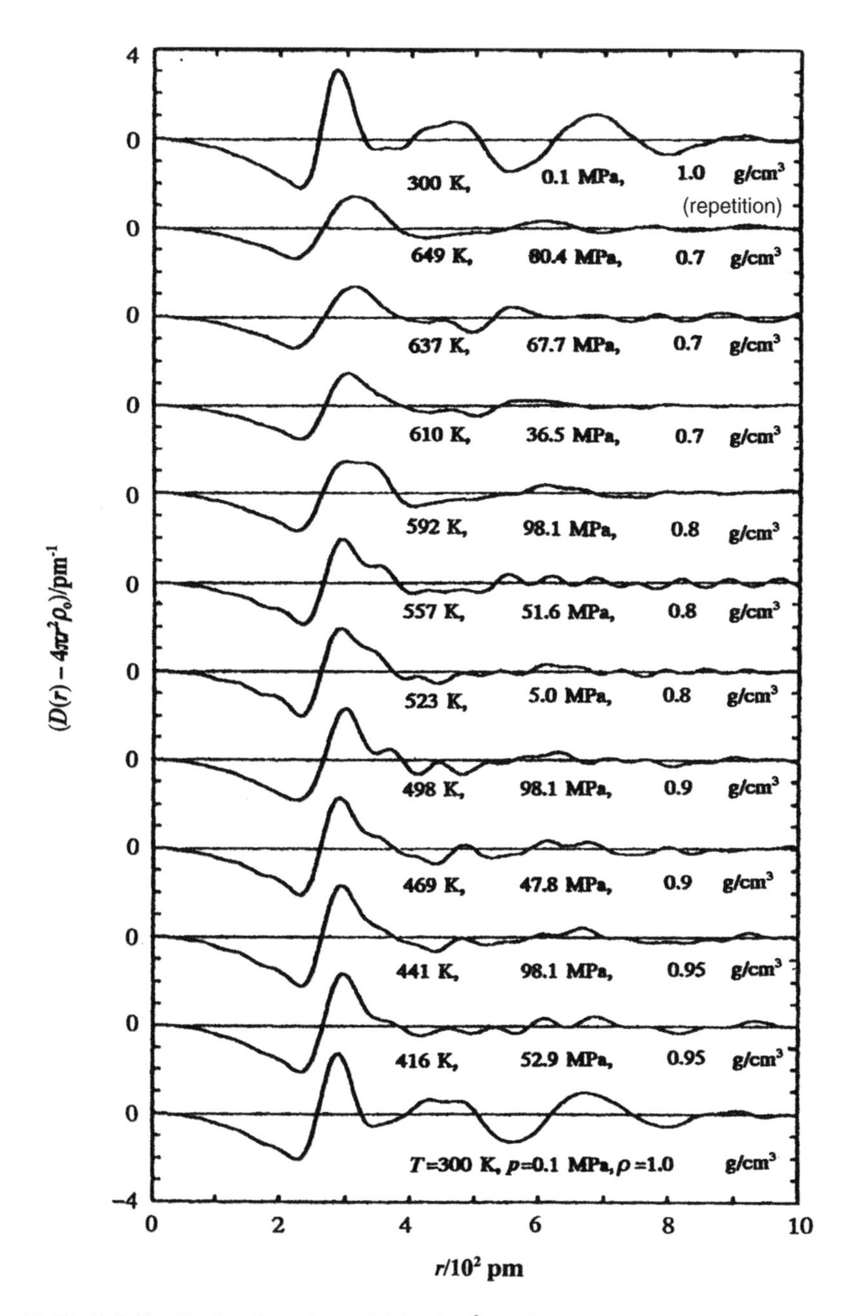

Fig. 11. Radial distribution functions, $D(r)-4\pi r^2\rho_0$ of water at various temperatures and pressures [54].

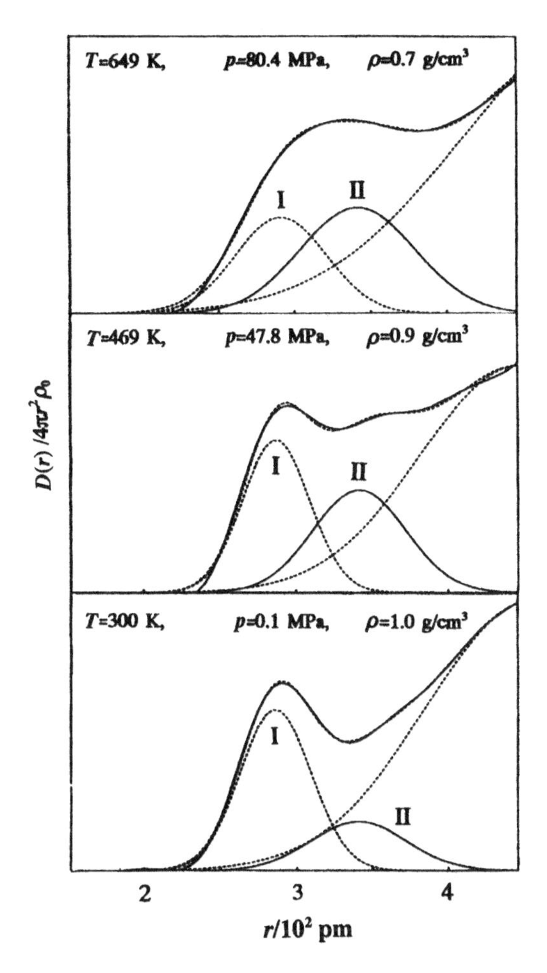

Fig. 12. The peak analysis of radial distribution functions of water, $D(r)/4\pi r^2\rho_0$ at various temperatures and pressures given in Fig. 10. The experimental values are given by solid lines and the calculated peaks I and II with the structural parameters given in Table 2 are indicated by dots [54].

The water–water distance at the nearest neighbor, r_I, increased and the number of water molecules at the nearest neighbor, n_I, decreased with the increase in temperature and pressure. The results indicated that the hydrogen bonds between water molecules were elongated and the hydrogen-bonded water structure was going to be decomposed at elevated temperature and pressure. The σ_I value also increases with t and p, and the result showed that the water–water interaction was weakened when t and p were elevated.

Table 2
The structural data of water molecules at the nearest (peak I) and the second nearest (peak II) neighbors determined by the analysis of radial distribution functions at various temperatures and pressures including the supercritical region

t (°C)	p (MPa)	ρ (g cm^{-3})	Peak I			Peak II		
			n_I	r_I (pm)	σ_I (pm)	n_{II}	r_{II} (pm)	σ_{II} (pm)
300	0.1	1.0	3.1	287	28	1.3	341	37
416	52.9	0.95	2.8	291	27	1.8	341	34
441	98.1	0.95	2.8	290	27	1.9	340	32
469	47.8	0.9	2.5	287	26	2.4	342	36
498	98.1	0.9	2.4	290	27	2.4	342	36
532	5.0	0.8	2.1	287	26	2.2	339	34
557	51.6	0.8	2.1	289	25	2.1	344	35
592	98.1	0.8	2.0	287	30	2.4	343	34
610	36.5	0.7	1.8	293	31	1.9	348	42
637	67.7	0.7	1.7	293	33	2.0	343	42
649	80.4	0.7	1.6	292	35	2.3	342	44

n: the number of water molecules, r: the intermolecular distance, σ: the half-width at the half-height of the peak [54]. Estimated errors in n, r and σ are $\perp$0.1, 1, 1 pm, respectively.

In contrast to the change in n_I, n_{II} increased with the increase in t and p, and r_{II} and s_{II} also increased with t and p.

The values of n, r and σ obtained from peaks I and II are plotted against temperature and density in Fig. 13. In the figure, the sum of $n_I + n_{II}$ is also plotted, which is practically unchanged in spite of the change in the density.

The n_I value was 1.6–1.8 in the SCW at $\rho = 0.7\,g\,cm^{-3}$, and the result indicated that ~40% of hydrogen-bonds in the ice-like water structure ($n = 4$) still remain in the SCW. The σ value increased, which suggested that the H_2O–H_2O bonds in the SCW were largely fluctuating.

An interesting result is given in Fig. 14 for the H_2O–H_2O distance (r) in water collected from the literature plotted against $\rho^{-1/3}$ [50, 51, 56]. Since $\rho^{-1/3}$ corresponds to the statistically averaged intermolecular distance in liquids, the plot represents the comparison of the experimentally obtained water–water distance (r) with the statistically averaged intermolecular distance. In high and medium density water with $\rho = 1.04$–$0.9\,g\,cm^{-3}$ under ambient and subcritical conditions, the water–water distance was proportional to the average distance, and the fact suggested that the thermal expansion of water was mainly caused by the elongation of the water–water hydrogen bond distance. However, in SCW of low density where

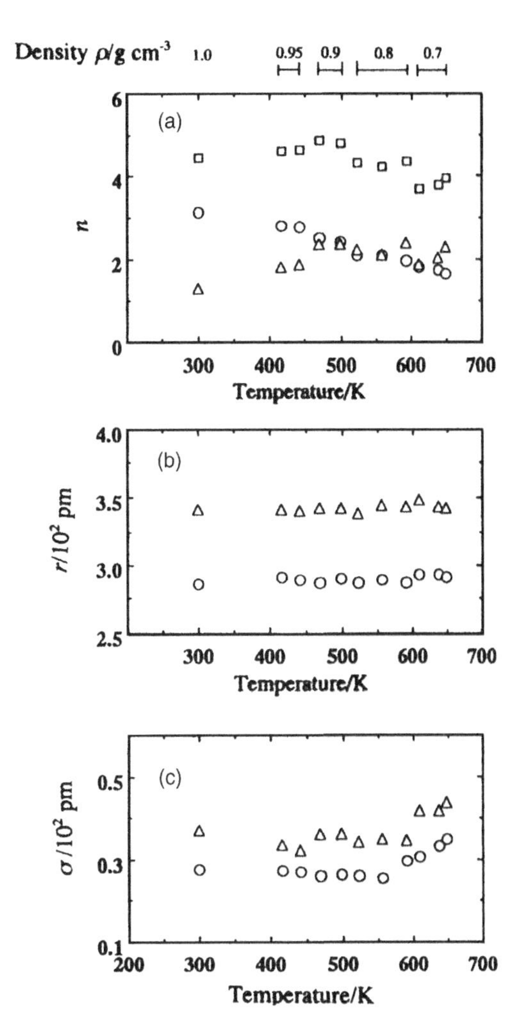

Fig. 13. The structural parameters of water with density and temperature. (a) The number of water molecules at the nearest neighbors n, (b) interatomic distance r, (c) half-width at half-height σ of the peaks I (○) and II (△) and their sum (□) [54].

$\rho^{-1/3} > 1.02\,g^{-1/3}\,cm$, the intermolecular distance, r, was practically unchanged, although the data were scattered to some extent. The result indicated that water molecules were still hydrogen-bonded even in the SCW with a distance of $\sim 293 \pm 3\,pm$.

In the low density SCW, small clusters should be formed in which the H_2O–H_2O hydrogen-bonded distance is kept $\sim 293\,pm$. The clusters may be

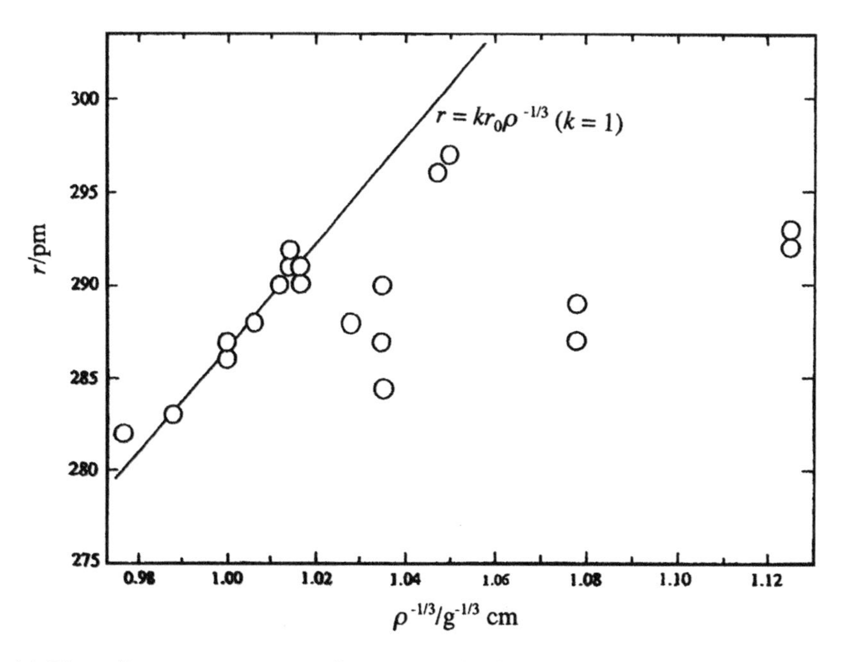

Fig. 14. The adjacent water–water distance r at high temperatures and pressures plotted against $\rho^{-1/3}$. The straight line represents the change in r calculated by assuming a homogeneous expansion of water with $\rho^{-1/3}$, $r = kr_0\rho^{-1/3}$ and $k = 1$ (r_0: O–O distance in ambient water) [56].

dispersed in the SCW, although the co-existence of monomerly dispersed gas-like water molecules together with clusters cannot be denied.

The average number of the nearest water molecules in water, n (n_I in Fig. 13) is plotted against the density of water, ρ in Fig. 15. The plot monotonously decreases with a decrease in density from normal water ($n = 4.4$ under an ambient condition) to SCW ($n = 1.6$–1.8). From the two plots in Figs. 14 and 15, it can be concluded that the clusters formed in SCW are not so large that many of water molecules in the clusters locate around the surface of the clusters, but have smaller numbers of adjacent molecules than those in the bulk. Therefore, water molecules are heterogeneously dispersed in SCW by forming small clusters, in which water molecules are bound by hydrogen-bonds with the intermolecular distance of $r \sim 293$ pm, probably together with monomer dispersed molecules.

The position of hydrogen atoms can be estimated experimentally by means of neutron diffraction and theoretically by computer simulations, and on the basis of these data, the orientation of water molecules in SCW can

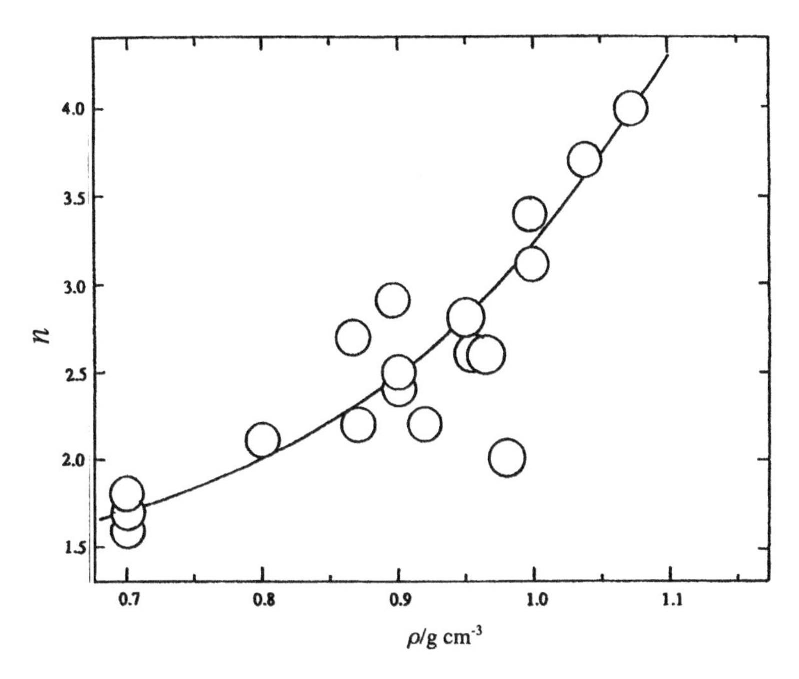

Fig. 15. The number of water molecules at the first hydration shell, n, plotted against ρ [56].

be discussed. Molecular dynamic simulations of SCW were performed by employing TIP4P [57, 58], SPC [59] and SPCE [60] potentials. A neutron diffraction measurement was examined for SCW [61]. The pair correlation functions, $g_{OO}(r)$ of the O–O atom pair found by molecular dynamics simulations and neutron diffraction experiment, together with the $g_{OO}(r)$ function derived from the X-ray diffraction data [54], are compared in Fig. 16.

The shape of the $g_{OO}(r)$ functions obtained from MD simulations and X-ray diffraction data are similar, but that from the neutron diffraction measurement is very different from the others. The result from the neutron diffraction study is today considered to be less reliable.

High temperature–high pressure NMR techniques provide useful information on the intermolecular interactions of SCW. The ^{1}H NMR signal of water observed at 30°C was sharp and shifted toward the low field side with temperature at constant density ($\rho = 0.4\,g\,cm^{-3}$ at high temperatures and pressures; the pressure was changed in an NMR tube with a constant volume) [62, 63]. The signal was broadened with the increase in the temperature as seen in Fig. 17a. The chemical shift σ of water is plotted against temperature in the liquid and gas phases in Fig. 17b.

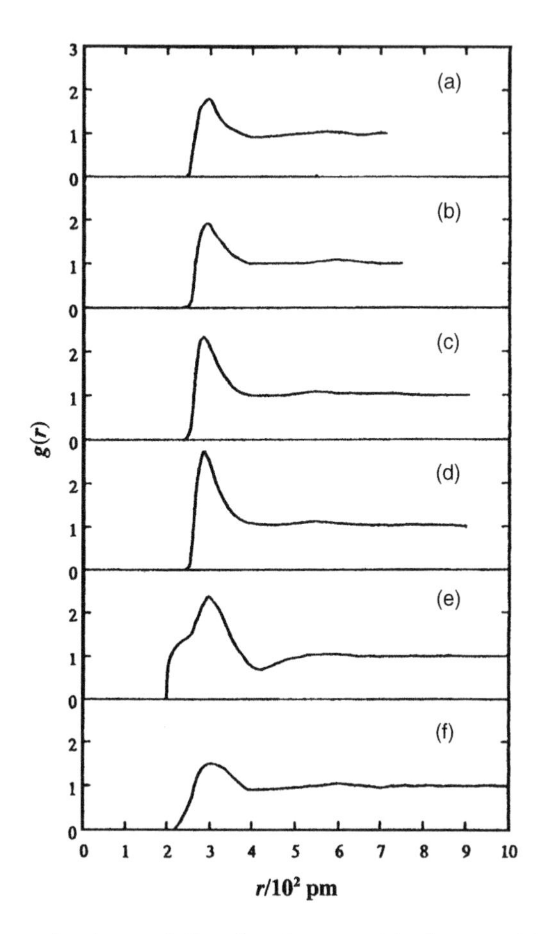

Fig. 16. Comparison of pair correlation functions $g_{OO}(r)$ of supercritical water obtained by various methods.

	Method	Potential function	T/K	ρ (g cm^{-3})	Reference
(a)	MD	TIP4P	725	0.75	57
(b)	MC	TIP4P	673	0.7	58
(c)	MD	SPC	647	0.48	59
(d)	MD	SPCE	652	0.33	60
(e)	ND	–	673	0.66	61
(f)	XD	–	649	0.7	54

MD: molecular dynamics, MC: Monte Carlo, ND: neutron diffraction, XD: X-ray diffraction.

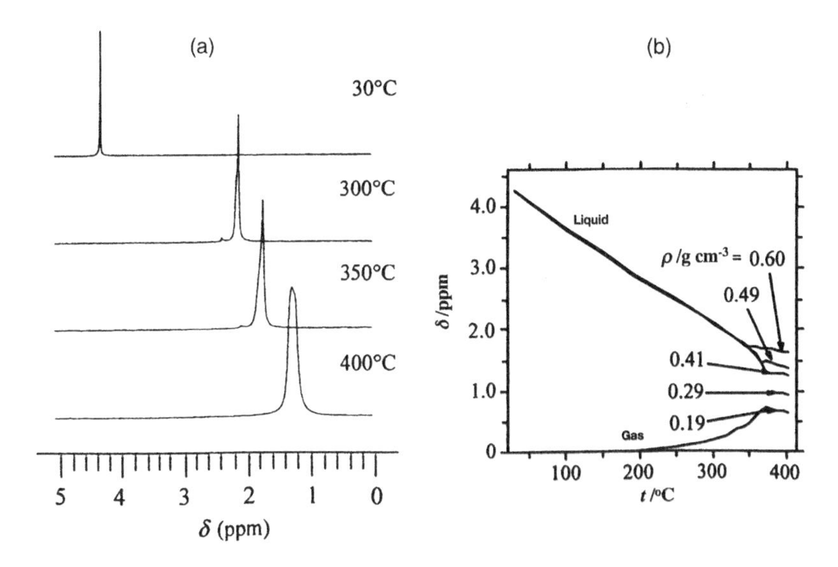

Fig. 17. ^{1}H NMR signals of water. (a) The shape and position of ^{1}H NMR signals at various temperatures. (b) ^{1}H NMR chemical shift δ as a function of temperature and density. The filling factor of water is 0.4, which makes density of supercritical water to be $\rho = 0.4\,g\,cm^{-3}$ [62, 63].

The plot of chemical shift depicted in Fig. 17b shows that the NMR chemical shifts of liquid water move to the high field side on the equilibrium curve with an increase in temperature, while reverse is true for water in the gas phase. A plot of the number of hydrogen-bonded water molecules, n_{HB}, estimated from the NMR chemical shifts at various temperatures and densities is given in Fig. 18. The result shows that hydrogen bonds still exist in SCW even at 400°C and $\rho = 0.2\,g\,cm^{-3}$.

Studies by IR and Raman spectrometry provide important information on the nature of water–water bonds. The O–D stretching vibration band in HDO solution containing 8.5 mol% D_2O in H_2O was measured at 303–673 K and 5–400 MPa and the result is depicted in Fig. 19 [64]. The O–D band was observed at $2520\,cm^{-1}$ under ambient condition. A sharp band appearing at $2719\,cm^{-1}$ in the solution of $\rho = 0.0165\,g\,cm^{-3}$ and $T = 673$ K was ascribed to free HOD molecules. The band disappeared in SCW with $\rho = 0.095\,g\,cm^{-3}$. On the other hand, a new band was observed around $2650\,cm^{-1}$ in water with $\rho = 0.036\,g\,cm^{-3}$ and the intensity of the band increased and shifted toward the low frequency side with increasing density. In water with $\rho = 0.9\,g\,cm^{-1}$ the band appeared at $2600\,cm^{-1}$. The band was ascribed to the O–D frequency of hydrogen-bonded water in SCW.

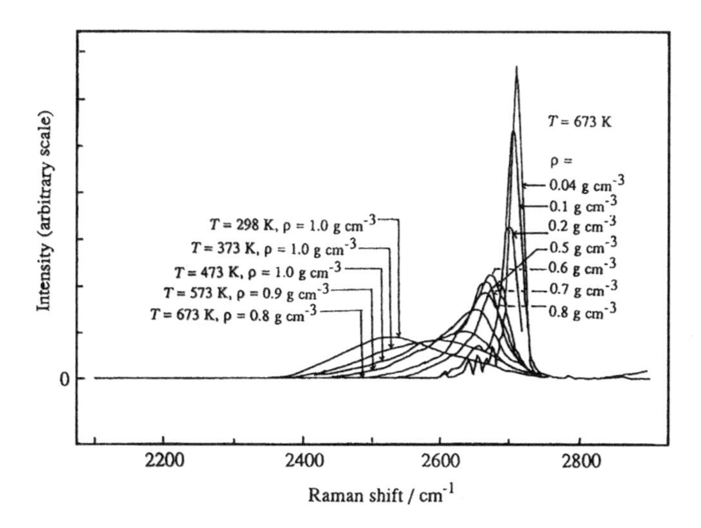

Fig. 18. The average number of hydrogen bonds, n_{HB}, of water estimated by ^{1}H NMR measurements [62, 63].

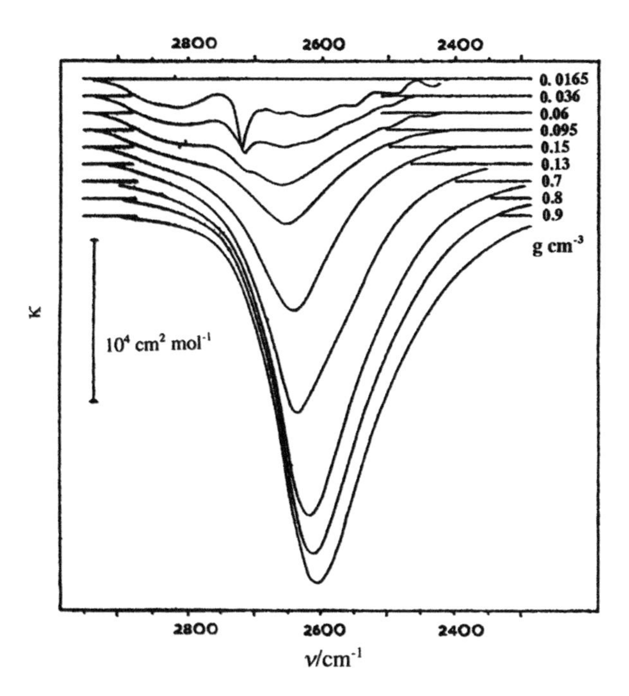

Fig. 19. The frequency of the O–D stretching vibration, ν, in the IR spectra of water observed at 673 K at various densities. κ denotes the molar extinction coefficient in the $cm^2\ mol^{-1}$ unit [64].

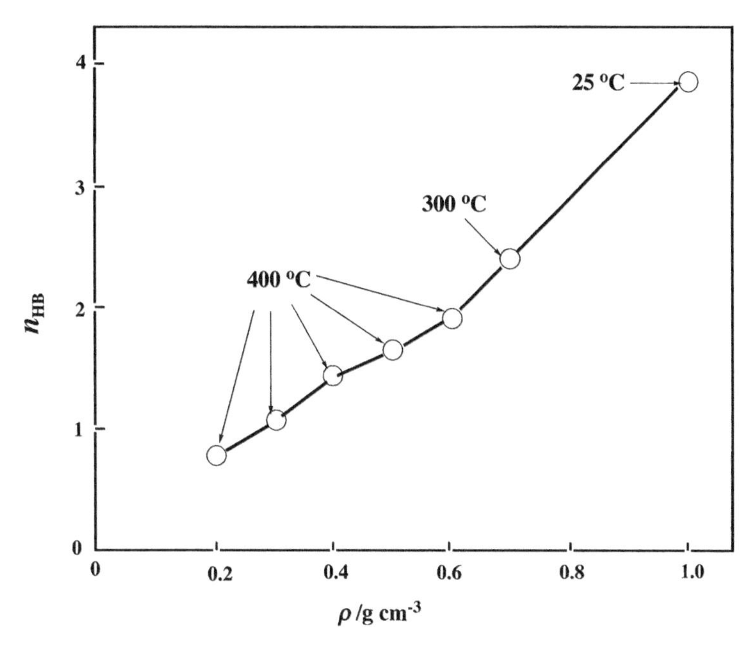

Fig. 20. Isotropic Raman spectra for the D–O stretching frequency, ν, in 9.7 mol% HDO in H_2O at various temperatures and densities [65].

The O–D stretching vibration band was observed by the Raman spectroscopy over the temperature range of 298–673 K within a relatively narrow density range of 0.8–1.0 g cm^{-3} and a wide density range from 0.04 to 0.8 g cm^{-3} at 673 K (Fig. 20) [65]. The band around 2500 cm^{-1} was attributed to the hydrogen bonded O–D stretching vibration, and the band around 2700 cm^{-1} was assigned to the O–D bond in non-hydrogen-bonded water. The band assignment is consistent with the IR measurement [64]. Although the two state model (hydrogen-bonded and free water molecules) may be too simple to explain the IR and Raman spectra, these results showed the existence of hydrogen bonds in relatively dense SCW.

Since SCW consists of heterogeneously dispersed small clusters, oligomers and probably gas-like monomer water molecules, non-polar organic substances can easily be mixed with it. On the other hand, due to the lack of the number density of water molecules, insufficient hydration shells are formed around the inorganic ions and the dissociation of inorganic salts is suppressed.

6. APPLICATION OF SUPERCRITICAL FLUIDS FOR REDUCING POLLUTANTS

6.1. Supercritical Water Oxidation Process

One of the major problems we have today is the disposal of various organic wastes and hazardous compounds. The ideal solution would be to decompose them into innocuous substances. Chemical oxidation processes by ozone and hydrogen peroxide, photo oxidation processes by ultraviolet irradiation, wet oxidation processes by applying high temperature and high pressure (200–300°C and 1.5–10 MPa) and incineration processes at 800–2000°C have been examined at a laboratory scale and some of them have already been scaled up. However for most of these techniques, the efficiency of decomposition of the wastes and polluting materials are not high enough. Although the incineration process could achieve the effective decomposition, the process produces hazardous and toxic by-products such as dioxanes, NO_x and SO_x, which are often discharged into the atmosphere.

In 1982 it was reported that the oxidation process by using SCW is effective in the decomposition of persistent organic wastes, and the decomposition could be greater than 99.99% for most wastes. The SCWO process for the treatment of pollutants, wastes and hazardous compounds attracted much attention in recent years [6, 66–74]. Efficiencies for decomposition of chlorinated organic compounds are listed in Table 3 [72, 73]. The SCWO processes are usually operated at ~600–650°C, which is lower than the temperature of incineration processes.

Organic wastes including chlorine, sulfur and nitrogen atoms are oxidized and decomposed into water, CO_2, N_2, HCl, H_2SO_4 and HNO_3. The acids formed are usually neutralized by a sodium hydroxide solution injected into a reactor. The neutralized acids are precipitated as salts. For the treatment of aqueous solutions containing organic wastes of 1–20 wt%, the SCWO process is more effective in energy and cost than the incineration process. Slurries containing organic wastes such as human metabolic wastes [74], biomass and organic polluted soils are pumped into a reactor. The SCWO process is performed in a fully closed system, and no additional abatement facilities for exhaust gases and discharged water are required.

Major problems of the SCWO process are (1) the removal of precipitates formed in the reactor and scales accumulated in the equipment [75, 76]. They reduce heat efficiencies of the reactor and prevent the stable operation of the process. Metals included in the wastes are often oxidized to metal oxides, which are also precipitated in the reactor. (2) Corrosion of reactors is

Table 3
Destruction efficiencies of chlorinated organic compounds by SCWO [71–73]

Compound	Destruction rate (%)
Carbon tetrachloride	>96.53
Chlorinated dibenzo-1,4-dioxanes[a]	>99.9999
Chloroform	>98.83
2-Chlorophenol	>99.997
2-Chlorotoluence	>99.998
DDT[b]	>99.997
4,4′-Dichlorobiphenyl	99.993
1,2-Dichloroethane	99.99
PCBs[c]	>99.995
1,1,2,2-Tetrachloroethylene	99.99
Trichlorobenzenes[d]	99.99
1,1,1,-Trichloroethane	>99.99997
1,1,2-Trichloroethane	>99.981

[a]Mixtures of chloro-substituted dibenzo-1,4-dioxane.
[b]Dichlorodiphenytrichloroethane.
[c]Polychlorinated biphenyl (or polychlorobiphenyl).
[d]Mixtures of 1,2,3-, 1,2,4- and 1,3,5-trichlorobenzene.

Chloro-substituted dibenzo-1,4-dioxane

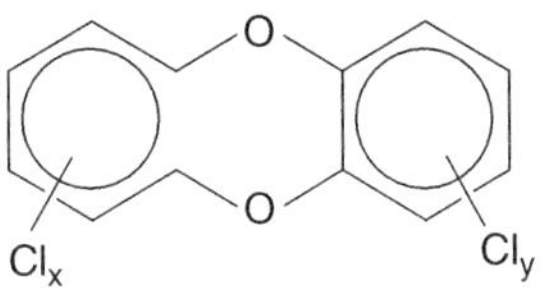

DDT

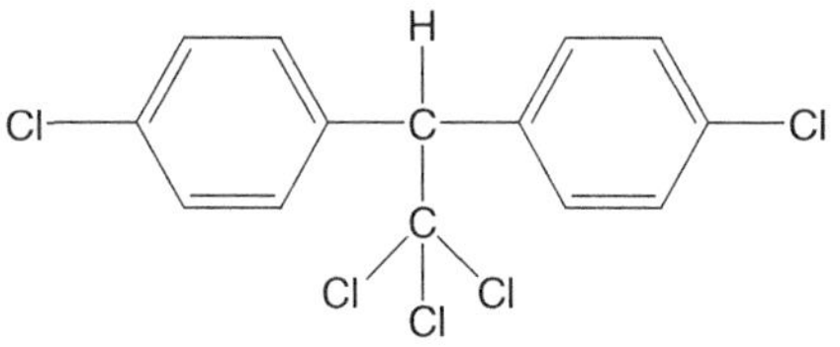

another serious matter in the SCWO process. Oxidation of materials used for reactors is accelerated by the acids produced. As a result, corrosion resistive materials have to be used for the equipment, and this is usually costly.

In order to solve the problem caused by precipitates in reactors, a two-zone technique involving supercritical and subcritical chambers was exploited. A flow-chart of an SCWO process (MODAR process) featuring a two-zone reactor is shown in Fig. 21 [66]. In the upper zone, which is for SCW at ~600°C, the wastes are decomposed and the products are neutralized. In the lower zone (subcritical), in which the temperature of water is ~200°C, the precipitates which formed in the upper zone are dissolved.

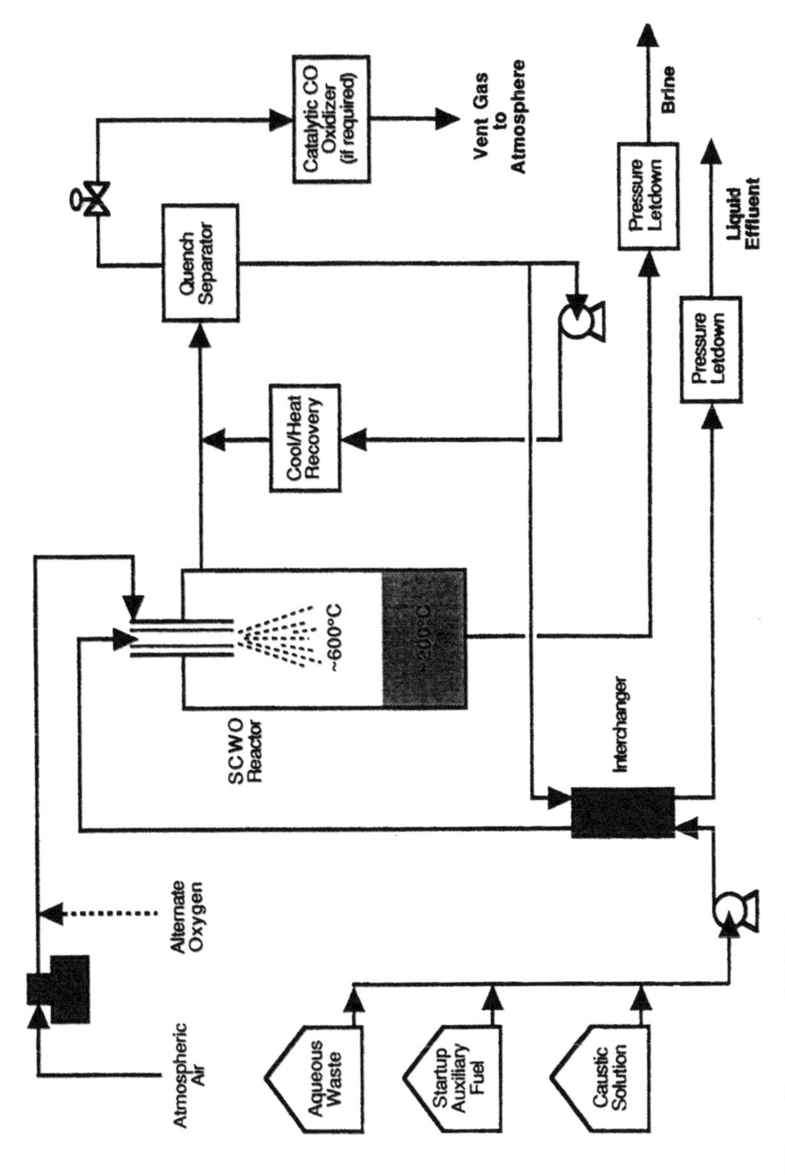

Fig. 21. Flow diagram of SCWO process ("MODAR process") featuring two-zone reactor [66].

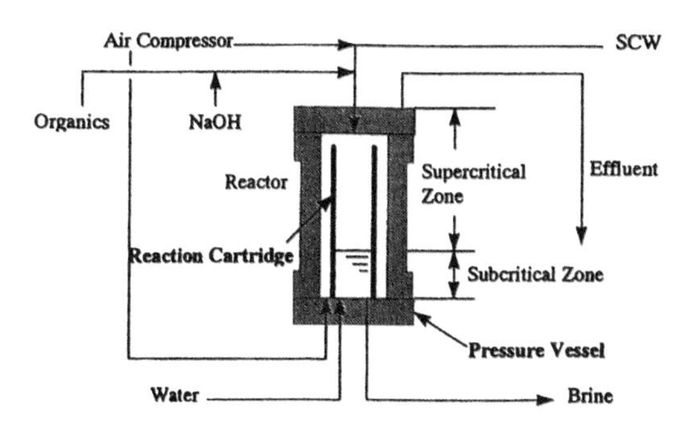

Fig. 22. Schematic presentation of dual shell pressure-balanced vessel for SCWO [77].

Organic wastes, an alkaline solution (used for the neutralization of the acids produced), air as an oxidant and supercritical water, are mixed and preheated, and then injected into the supercritical zone of the reactor. The wastes are oxidized and neutralized in this area. The SCF flows downwards during the reaction, and precipitates formed during the reaction fall down into the subcritical zone and are dissolved in subcritical water. The high temperature fluid turns back and upward due to convection, and then goes out from a vent.

The two-zone type reactor has been improved with the introduction of a dual shell pressure type reactor (Fig. 22) [67, 77]. In the dual shell reactor, pressurized air is introduced between the wall of the pressure vessel and a reaction cartridge. A pilot plant of the dual shell type reactor can treat two tons of aqueous solutions containing 10% organic wastes per day (Photograph 2).

The corrosion problem could be avoided by using highly corrosion resistive material. Nickel alloys like Inconel 625 or Hastelloy C-276 are examples. Even some noble metals are used for the cartridge. Since the thickness of the wall of the cartridge can be reduced in dual shell type reactors, the cost for constructing the SCWO reactor can be drastically reduced.

6.2. Extraction of Pollutant from Soils with Supercritical Carbon Dioxide and Other Supercritical Fluids

The dissolving capacity of $SCCO_2$ depends on the temperature and pressure. When a target material is dissolved in $SCCO_2$ at a high pressure, the material is precipitated when the pressure is released. As a result, $SCCO_2$ can be used for extraction of various materials. Usually no distillation and

Photograph 2. Supercritical water oxidation reactor situated at Organo Corporation, Saitama, Japan. (Partly funded by New Energy and Industrial Technology Development Organization (NEDO), which is an auxiliary organization of the Ministry of Economy, Trade and Industry of Japan).

other additional processes are required. Extraction of caffeine and hop are examples. $SCCO_2$ can also be employed for extracting polluting materials from soil. By extracting contaminants, together with a preconcentration step, it is possible to analyze minute amounts of contaminants by conventional analytical methods. The use of $SCCO_2$ together with an entrainer at the pretreatment step for the contamination analysis in soils, a high recovery rate of contaminants was realized compared with the standard analytical process of SW846 Method 8330 in USA [78]. The recovery of dioxanes from soil was improved from 40 to 90% by adding 2% of methanol to $SCCO_2$ [79]. The addition of 2% methanol to supercritical N_2O could extract dioxanes from soil with the extraction efficiency of greater than 98% [79]. It is possible with a combination of gas chromatography and $SCCO_2$ extraction, to analyze volatile contaminants in soils within 10 min [80, 81]. Successful removals of dichlorodiphenyltrichloroethane (DDT) [82] and chlorinated aromatic compounds such as trichlorophenyl and polychlorinated biphenyls (PCBs) [83] have been reported. Since the 1990s, SCF extraction processes for soil remedy have been scaled up for commercial use.

Supercritical propane has been applied to the treatment of sludge from wastewater treatment facilities in petroleum refineries with a treatment capacity of 300 barrels per day [84].

6.3. Recycle of Used Plastic Bottles with Supercritical Methanol

Supercritical methanol can be used under milder conditions than those used in SCW. Supercritical methanol is used for recycling processes of waste plastics to monomers. PET is decomposed by supercritical methanol to dimethylterephthalate and ethylene glycol. Supercritical methanol can perform the process within 1/10 reaction time with 100% recovery of recycling without the use of catalysts. The method is more advantageous than other methods using liquid methanol and liquid ethylene glycol with catalysts. The decomposition of PET can be made at 330°C and 8.1 MPa [7, 85]. Another advantage of this process is that the evolution of gaseous products is negligible, because the decomposition reactions stop at the formation of monomers and do not proceed to CO_2. The monomers can be used again for producing PET, and the "bottle-to-bottle" recycle is realized.

7. CONCLUDING REMARKS

The application of SCFs to the treatment of wastes and pollutants and to the extraction of contaminants from soil will be extended to other fields in the near future. The cost for the construction of plants will be reduced by the improvement of equipment and materials. The kind of fluids to be used will also be improved and many other technologies will be developed according to world needs. On the other hand, reactivities of SCFs, the nature of intramolecular bonds and intermolecular interactions, which are closely related to the reactivities, are not well elucidated yet. Studies on thermodynamics and thermochemistry and intra- and intermolecular interactions from microscopic points of view of SCFs will help solve some of these problems. Furthermore structural studies of SCFs have rarely been made for compounds other than water. Structural investigations of liquids at high temperature and high pressure are also very attractive and important in relation to life sciences. Further investigations into this area should be encouraged.

REFERENCES

[1] T. Furuya and Y. Arai, Chem. Ind. (KAGAKU KOGYO) (in Japanese), 46 (1995) 14.
[2] N. Asahi, Private Communication.

[3] T. Sako, Chemistry (KAGAKU) (in Japanese), 52 (1997) 31.
[4] E.U. Franck, in "The Physics and Chemistry of Aqueous Ionic Solutions" (M.-C. Bellissent-Funel and G.W. Neilson, eds.), p. 337, D. Reidel Publ. Co., Dordrecht, The Netherlands, 1987.
[5] K. Nagahama, in "Handbook of Supercritical Fluids. Application in Food Industries (in Japanese)" (K. Nagahama and I. Suzuki, eds.), p. 18, Science Forum, Tokyo, Japan, 2002.
[6] J.W. Tester, H.R. Holgate, F.J. Armellini, P.A. Webley, W.R. Killilea, G.T. Hong and H.E. Barner, ACS Symposium Series, Emerging Technologies for Hazardous Waste Management III (1991), and references therein.
[7] T. Sako, T. Sugata, K. Otake, Y. Nakazawa, S. Sato, K. Namiki and M. Tsugumi, J. Chem. Eng. Jpn., 30 (1997) 342.
[8] W. McGovern and P.N. Rice, Pollut. Eng., September (1988) 122.
[9] http://www.iapws.org/
[10] N.B. Vargaftik, B.N. Volkov and L.D. Voljak, J. Phys. Chem. Ref. Data, 12 (1983) 817.
[11] M. Uematsu and E.U. Franck, J. Phys. Chem. Ref. Data, 9 (1980) 1291.
[12] W.L. Marshall and E.U. Franck, J. Phys. Chem. Ref. Data, 10 (1981) 295.
[13] J.M.H. Levelt Sengers, J. Straub, K. Watanabe and P.G. Hill, J. Phys. Chem. Ref. Data, 14 (1985) 193.
[14] N. Matsunaga and A. Nagashima, J. Phys. Chem. Ref. Data, 12 (1983) 933.
[15] J.V. Sengers and J.T.R. Watson, J. Phys. Chem. Ref. Data, 15 (1986) 1291.
[16] A. Nagashima, J. Phys. Chem. Ref. Data, 6 (1977) 1133.
[17] I. Thormählen, J. Straub and U. Grigull, J. Phys. Chem. Ref. Data, 14 (1985) 933.
[18] W. Wagner and A. Pruss, J. Phys. Chem. Ref. Data, 31 (2002) 387.
[19] http://www.nist.gov/srd/nist10.htm
[20] W. Wagner, J.R. Cooper, A. Dittmann, J. Kijima, H.J. Kretzschmar, A. Kruse, R. Mares, K. Oguchi, I. Stocker, O. Sifner, Y. Takaishi, I. Tanishita, J. Trubenbach and T. Willkommen, J. Eng. Gas Turb. Power, 122 (2000) 150.
[21] W.T. Parry, J.C. Bellows, J.S. Gallagher and A.H. Harvey, ASME International Steam Tables for Industrial Use, ASME press, New York, NY, 2000.
[22] http://www.asme.org/research/wsts/steam.html
[23] http://members.asme.org/catalog/ItemView.cfm?ItemNumber=I00420
[24] L. Haar, J.S. Gallagher and G.S. Kell, NBS/NRC Steam Tables, Hemisphere Publishing Corp., New York, 1984.
[25] U. Grigull (ed.), "Properties of Water and Steam in SI-Units", Springer-Verlag, Heidelberg, Germany, 1982.
[26] K.H. Dudziak, and E.U. Franck, Ber. Bunsenges., Phys. Chem., 70 (1966) 1120.
[27] K. Heger, M. Uematsu and E.U. Franck, Ber. Bunsenges., Phys. Chem., 84 (1980) 758.
[28] S. Angus, B. Armstrong and K.M. Reduck, "Carbon Dioxide: International Thermodynamic Tables of the Fluid State-3", Pergamon Press, Oxford, 1976.
[29] B.A. Younglove, J. Phys. Chem. Ref. Data, 11(1) (1982).
[30] R. Span and W. Wagner, J. Phys. Chem. Ref. Data, 25 (1996) 1509.
[31] B.A. Younglove and J.F. Ely, J. Phys. Chem. Ref. Data, 16 (1987) 577.
[32] S.S. Chen, R.C. Wilhoit and B.J. Zwolinsky, J. Phys. Chem. Ref. Data, 4 (1975) 859.

[33] R.D. Goodwin, J. Phys. Chem. Ref. Data, 16 (1987) 799.
[34] R.E. Fornari, P. Alessi and I. Kikic, Fluid Phase Equilib., 57 (1990) 1.
[35] R. Dohrn and G. Brunner, Fluid Phase Equilib., 106 (1995) 213.
[36] M. Christov and R. Dohrn, Fluid Phase Equilib., 202 (2002) 153.
[37] T. Adrian, M. Wendland, H. Hasse and G. Maurer, J. Supercrit. Fluids, 12 (1998) 185.
[38] F.P. Lucien and N.R. Foster, J. Supercrit. Fluids, 17 (2000) 111.
[39] H. Higashi, Y. Iwai and Y. Arai, Chem. Eng. Sci., 56 (2001) 3027.
[40] http://webbook.nist.gov.chemistry/
[41] http://www.ddbs.de
[42] E.U. Franck, S. Rosenzweig and M. Christoforakos, Ber. Bunsenges., Phys. Chem., 94 (1990) 199.
[43] J. Connolly, J. Chem. Eng. Data, 11 (1966) 13.
[44] C.J. Robert and W.B. Key, AIChE J., 5 (1959) 285.
[45] E.U. Franck, Pure Appl. Chem., 98 (1981) 1401.
[46] Y. Tsekhanskaya, Russ. J. Phys. Chem., 38 (1964) 1173.
[47] Y. Iwai, J. Chem. Eng. Data, 36 (1991) 430.
[48] Y. Iwai, J. Chem. Eng. Data, 38 (1993) 506.
[49] Y. Koga, Fluid Phase Equilib., 125 (1996) 115.
[50] H. Ohtaki, T. Radnai and T. Yamaguchi, Chem. Soc. Rev., (1997) 41.
[51] M. Nakahara, T. Yamaguchi and H. Ohtaki, Recent Res. Dev. Phys. Chem., 1 (1997) 17.
[52]. N. Dem'yanets, Zh. Struk. Khim., 24 (1983) 66.
[53] Yu. E. Gorbatyi and Yu. N. Dem'yanets, Zh. Struk. Khim., 24 (1983) 74.
[54] K. Yamanaka, T. Yamaguchi and H. Wakita, J. Chem. Phys., 101 (1994) 9830.
[55] A.H. Narten and H.A. Levy, J. Chem. Phys., 55 (1971) 2263.
[56] T. Radnai and H. Ohtaki, Mol. Phys., 87 (1996) 103.
[57] R.D. Mountain, J. Chem. Phys., 90 (1989) 1866.
[58] A.G. Kalinichev, Z. Naturforsch., 46a (1991) 433.
[59] P.T. Cummings, H.D. Cochran, J.M. Simonson, R.E. Mesmer and S. Karaborni, J. Chem. Phys., 94 (1991) 5606.
[60] Y. Guissani and B. Guillot, J. Chem. Phys., 98 (1993) 8221.
[61] P. Postorino, R.H. Tromp, M.-A. Ricci, A.K. Soper and G.W. Neilson, Nature, 366 (1993) 668.
[62] N. Matubayasi, C. Wakai and M. Nakahara, Phys. Rev. Lett., 78 (1997) 2573, 4309.
[63] N. Matubayasi, C. Wakai and M. Nakahara, J. Chem. Phys., 107 (1997) 9133.
[64] E.U. Franck and K. Roth, Disc. Faraday Soc., 43 (1967) 108.
[65] W. Kohl, H.A. Lindner and E.U. Franck, Ber. Bunsenges., Phys. Chem., 95 (1991) 1586.
[66] J.E. Barner, C.Y. Huang, T. Johnson, G. Jacobs, M.A. Martch and W.R. Killilea, J. Hazardous Mater., 31 (1992) 1.
[67] H.R. Holgate and J.W. Tester, J. Phys. Chem., 98 (1994) 800.
[68] H.R. Holgate and J.W. Tester, J. Phys. Chem., 98 (1994) 810.
[69] S.B. Hawthorne, Yu Yang and D.J. Miller, Anal. Chem., 66 (1994) 2912.
[70] G.E. Bennet and K.P. Johnston, J. Phys. Chem., 98 (1994) 441.
[71] US Patent 4,338,199, July 6 (1982).

[72] T.B. Thomason and M. Modell, Hazard. Wastes, 1 (1984) 453.
[73] M. Modell, in "Supercritical Water Oxidation, Standard Handbook of Hazardous Waste Treatment and Disposal" (H.M. Freeman, ed.), p. 8 and p. 153, McGraw Hill, New York, NY, 1989.
[74] Y. Ohba, K. Nojiri, H. Suzugaki and A. Suzuki, Proceedings of Water Environment Federation Technical Conference, Singapore, March, 1998.
[75] M. Hodes, P.A. Marrone, G.T. Hong, K.A. Smith and J.W. Tester, J. Supercrit. Fluids, 29 (2004) 265.
[76] P.A. Marrone, M. Hodes, K.A. Smith and J.W. Tester, J. Supercrit. Fluids, 29 (2004) 289.
[77] A. Suzuki, T. Oe, N. Anjo, H. Suzugaki and T. Nakamura, Proceedings of 4th International Symposium on Supercritical Fluids, May 11–14, Sendai, Japan, p. 895, 1997.
[78] P.G. Thrne, Gov. Rep. Announce. Index (US), 94 (1994) 463.
[79] F.I. Onuska and K.A. Terry, J. High Res. Chromatogr., 12 (1989) 357.
[80] L. Zaiyou, J. Microcolumn Sep., 4 (1992) 199.
[81] W. Mingin, Anal. Chem., 65 (1993) 2185.
[82] F.C. Knopf, B. Brady and F.R. Groves, CRC Crit. Rev. Environ. Control, 15 (1985) 237.
[83] C.A. Eckert, J.G. Van Alsten and T. Stoicos, Environ. Sci. Technol., 20 (1986) 319.
[84] D.W. Hall, Environ. Prog., 9 (1990) 98.
[85] T. Sako, I. Okajima, T. Sugata, K. Otake, T. Yorita, Y. Takebayashi and C. Kanzawa, Polymer J., 32 (2000) 178.

Thermodynamics, Solubility and Environmental Issues
T.M. Letcher (editor)

Chapter 5

Phase Equilibrium Studies on Ionic Liquid Systems for Industrial Separation Processes of Complex Organic Mixtures

Prashant Reddy[a] and Trevor M. Letcher[b]

[a]Department of Chemical Engineering, University of KwaZulu-Natal, Durban 4041, South Africa
[b]School of Chemistry, University of KwaZulu-Natal, Durban 4041, South Africa

1. INTRODUCTION

The emergence of room temperature ionic liquids (RTILs) or low temperature molten salts as "green designer solvents" [1] with highly unusual, yet desirable, physico-chemical properties has challenged the inherent limitations of conventional molecular solvents and subsequently culminated in a plethora of potential applications [2]. Although the first report of a "useful" ionic liquid[1] dates back to 1914 [3], the synthesis of the first air-stable ionic liquid in 1992 [4] and the commercial availability of ionic liquids as research chemicals in 1999 [5] provided impetus for stimulating research activities. Consequently, investigations on ionic liquids related to their potential applications have become an active hub of research efforts in recent years, as apparent in the marked increase in journal publications [6] and workshops [7] devoted to ionic liquids.

The concentration of research efforts on ionic liquid research is attributed to increasing concerns over the deleterious impact of inefficient and unsafe chemical process technologies on the integrity of the environment

[1]The terms ionic liquid and room temperature ionic liquid will be used interchangeably from this point, unless otherwise stated, where the former would ordinarily refer to an ionic liquid that is in a liquid state under ambient conditions and the latter to an ionic liquid that has a melting point lower than 373.15 K.

and human health [2]. The greatest costs, coupled with solvent and energy usage and effluent generation, are most frequently incurred in separation unit operations [8]. This trend is most evident in the chemical and petrochemical industries, where complex mixtures of paraffins, cyclo-paraffins, olefins, aromatics and oxygenates must be subjected to sufficient number of separation processes or stages to obtain the required product streams. The methods of choice for industrial separation processes have been solvent-intensive or solvent-enhanced separation processes involving extractive distillation, azeotropic distillation [8] and liquid–liquid extraction [9].

It has been estimated that despite the significantly high energy consumption, distillation processes account for 90% of all separation operations in the chemical industry [8]. Common to the above processes is the addition of a selective solvent (usually polar) to alter the distribution of the components between the two respective phases to allow for a more efficient separation process. Traditionally, solvent-intensive industrial separation processes have been plagued by toxicity, flammability, corrosiveness and flammability in the use of solvents such as *N*-methyl-2-pyrrolidone (NMP), tetrahydrothiophene-1,1-dioxide (sulpholane), dimethylsulphoxide (DMSO), ethylene glycols [10], etc.

The feasibility of the use of alternative, cleaner and relatively environmentally benign separation technologies such as supercritical fluid extraction [11] and membrane separation processes [12] has been explored as a possible solution. However, the inherent advantages of distillation and traditional extraction processes over these methods [8] coupled with the high costs associated with plant shutdown and the replacement of well-established technologies currently make these alternative processes quite unattractive. A more pragmatic approach for the development of cleaner separation technologies is the search for viable alternatives for the replacement of traditional solvents in these processes, at the forefront of which is ionic liquids.

The distinguishing feature of ionic liquids is that they are a non-aqueous liquid phase composed entirely of ions, which are bulky, polar and asymmetrical (especially the cation). This allows for a large liquidus range in the face of strong Coulombic forces. This unusual structure endows ionic liquids with equally unusual physical and chemical properties such as negligible vapour pressures under ambient conditions, a large liquidus range, high densities, high decomposition temperatures (up to 473 K) and variable miscibility characteristics in organic, inorganic and polymeric materials [2]. The generalized structure of an ionic liquid is based on the combination of organic cations, which can be categorized as "aromatic" or "onium" and inorganic or organic anions, which are either "conjugate" or "ate" bases [11]. The cations that have been typically employed in ionic liquid synthesis have been stable nitrogen-containing aromatic heterocycles in the form of pyridinium and imidazolium ions, together with quarternary ammonium and phosphonium salts. The substituents on the cations are usually *n*-alkyl chains of variable length but other functional groups such as hexyloxymethyl and methoxy can also be employed. Greater freedom is exercised in the choice of anion to couple or "mate" with the organic cation in the form of the poorly nucleophilic tetrafluoroborate [BF_4], hexafluorophosphate [PF_6], acetate [CH_3CO_2], *p*-toluenesulphonate [$C_7H_7SO_3$], methylsulphate [CH_3SO_4] or bis(trifluoromethylsulphonyl)imide [$(CF_3SO_2)_2N$] ions. Through an intuitive or "trial and error" approach to the variation in the structures of ionic liquids (cations, anions and substituents), often marked variation in the physico-chemical properties such as hydrophobicity, density, conductivity, viscosity, liquidus range and solvation behaviour is observed. This allows for the design of task-specific ionic liquids, whose structure can be "optimized" for a specific purpose, which is highly critical for applications such as the specific extraction of toxic heavy metals from industrially polluted wastewater [13]. It has been estimated that the number of potential ionic liquid structures from possible anion–cation couplings is 10^6 [14], which is an unprecedented occurrence for any class of chemical substances. Consequently, it is this flexibility in the design and function of a solvent which is indeed a great advantage over conventional molecular solvents, which belong to distinct structural classes with chemical and solvent properties determined by the functional groups present.

To extend ionic liquids from the realm of mere theoretical interest in research laboratories to actual implementation in industrial separation processes, conclusive experimental studies, which allow for an effective assessment of the feasibility of ionic liquids in industrial separation processes, are required. For the design engineer trying to size separation equipment, optimizing separation sequences, predicting operating costs and design control schemes [15], reliable knowledge of the phase equilibrium behaviour (solubilities, vapour–liquid equilibria, liquid–liquid equilibria) of the system in the presence of the added solvent is required.

In this chapter, the application of key experimental techniques for the determination of thermodynamic properties of ionic liquid–organic mixture will be discussed as a means for the effective screening of ionic liquids as solvents in industrial separation processes.

2. SOLUBILITY STUDIES ON IONIC LIQUID–ORGANIC MIXTURES AND APPLICATION TO LIQUID–LIQUID EXTRACTION

Solvent extraction or more correctly, liquid–liquid extraction [10] is a major industrial separation process employed in the chemical and petrochemical industries. It is an attractive, less-energy-intensive alternative to solvent-enhanced distillation processes (extractive/azeotropic distillation) for the separation of close-boiling or azeotropic systems. However, in the petrochemical industry, an extractive distillation step in the liquid–liquid extraction sequence is necessary to obtain the desired product in high purity as well as to effect solvent regeneration. Consequently, separation sequences such as the sulpholane process [16] are seen as hybrid processes which allow for the processing of feedstocks of a much wider boiling and composition range than would be possible for either process alone, i.e. for aromatic–aliphatic mixtures, liquid–liquid extraction is suitable for 20–65 wt.% aromatics and extractive distillation is suitable for 65–90 wt.% aromatics [17]. In liquid–liquid extraction, separation is achieved purely as result of selective solvent–solute interactions and liquid phase splitting, dependant on the chemical natures of the components, and extractive distillation combines the selective properties of a solvent in significantly altering the relative volatility of the system to allow for separation in the generated vapour and liquid phases.

2.1. Description of the Liquid–Liquid Extraction Process

The origin of the large-scale industrial implementation of the liquid–liquid extraction process for organic mixtures dates back to the early 1930 s for the separation of aliphatic–aromatic mixtures. The Edeleanu process [10] employed the use of liquid sulphur dioxide for the selective extraction of aromatics from the kerosene fraction of an oil refinery process to produce low aromatic lamp oil. As a result of shifting trends in energy and heating technologies, coupled with the identification of new solvent systems, interest in this process has been superseded by other processes. The benzene, toluene, xylene (BTX) aromatic fraction, obtainable from an aliphatic–aromatic mixture produced from various hydrocarbon processing operations such as reformed petroleum naptha (reformate), pyrolysis gasoline by-products from ethylene crackers (pygas) and coal liquid by-products from coke ovens, i.e. coke oven light oil (COLO) [16], is an invaluable feedstock for the petrochemical industry. Commercial liquid–liquid extraction processes [10, 16, 18] that have been of interest include the Arosolvan process (NMP/water), sulpholane process (2,3,4,5-tetrahydrothiophene-1,1-dioxide or sulpholane), IFP (DMSO), Formex (*N*-formyl morpholine), Udex (glycol/water), Tetra (tetraethylene glycol) and newer processes such as the Carom process (tetraethylene glycol/Carom concentrate).

Further information on the above processes can be obtained from the review articles by Bailes et al. [10, 19], Treybal [20], Hanson [21] and the UOP website [16].

The generalized liquid–liquid extraction process for an aromatic–aliphatic separation is based on the introduction of the upstream hydrocarbon (aromatic–aliphatic) feedstock into the extractor, which is then contacted with a stream of lean solvent on trays in a counter-current fashion. This results in the creation of two immiscible or partially miscible phases, i.e. an organic (aliphatic)-rich phase and a solvent-rich phase, with which the aromatic components then selectively interact with and dissolve into, resulting in favourable separations. The phase containing the desired component (known as the solute) dissolved in the added solvent, i.e. the solvent-rich phase, is known as the extract phase, which is the aromatics-laden stream with a small amount of non-aromatics. The second stream, known as the raffinate phase, is the remaining organic content of the original feedstock, not extracted into the solvent phase, i.e. the aliphatic hydrocarbons together with a residual amount of aromatics. The raffinate phase usually exits the extractor at the top and the denser aromatics-rich solvent phase exits at the bottom. As a result of both streams not being of the desired purity due to the selective nature of the solvent, both streams have to be subjected to work-up processes. The aromatics-rich solvent stream is firstly purged of the non-aromatic content through the use of extractive distillation and/or steam stripping in a stripper section and then sent to a solvent recovery section where the large boiling point difference between the solvent and the aromatics content is exploited to effect the product separation and solvent recovery. In the sulpholane process [16], this is conducted under vacuum to allow for energy savings. The solvent stream is then recycled back into the extractor and the aromatics fraction can then be separated into the respective components through the use of a distillation train. Additional post-purification of the aromatics fraction in the form of clay treatment can be employed to remove trace olefins. The raffinate stream is treated in a water-washing sequence to remove the water-soluble solvent, which can then be recovered to maximize solvent regeneration. The raffinate stream can be used in gasoline formulation or as aliphatic solvents.

2.2. Solvent Choice Criteria for Liquid–Liquid Extraction

Solvent choice criteria for liquid–liquid extraction require consideration of key solvent thermo-physical and chemical properties; some of these are listed below.

2.2.1. Selectivity and Capacity

The favourable partitioning of the desired solute between the raffinate and the extract phases is dependant on the relative affinities (physical interactions) of the solute species for the two phases. This is expressed through a ratio of the distribution coefficients for the separation of an aliphatic–aromatic mixture, partitioning between two phases, known as the selectivity (S) [22], as defined below:

$$S = \frac{[x_{\text{aromatic}}/x_{\text{aliphatic}}]_{\text{extract}}}{[x_{\text{aromatic}}/x_{\text{aliphatic}}]_{\text{raffinate}}} \tag{1}$$

where x refers to a composition, i.e. the component mole fraction. As can be inferred from Eq. (1), a selectivity of much greater than unity is required for an efficient separation.

The capacity of a solvent is defined by the distribution coefficient of the aromatic content in the two phases, i.e.

$$C = \frac{[x_{\text{aromatic}}]_{\text{extract}}}{[x_{\text{aromatic}}]_{\text{raffinate}}} \tag{2}$$

The solvent capacity indicates the amount of the solute phase that can be accommodated in the solvent-rich phase; hence, a large capacity is desirable from the perspective of the economy of the process as it allows for a smaller solvent-to-feed ratio for a given selectivity or extraction efficiency. Unfortunately, the solvent selectivity and capacity are frequently inversely related, as is the case for solvents such as NMP with a high capacity and a low selectivity. Sulpholane has the highest selectivity and capacity of all the commercially significant extraction solvents [16] for aliphatic–aromatic separations. Both capacity and selectivity can vary quite markedly across the feedstock composition range, i.e. as a function of the aromatic concentration. Consequently, these two factors, as a function of feedstock composition, have to be cumulatively assessed to allow for a process that is both economical (large capacity) and efficient (large selectivity). The selectivities and capacities of ionic liquid solvents, as determined in experimental liquid–liquid equilibrium, will be compared to those of commercial solvents in a later discussion.

2.2.2. Boiling Point

A large boiling point difference between the solvent and the mixture components is desirable to facilitate the product separation from the solvent and for solvent regeneration through extractive distillation or stripping. If the volatility of the solvent is extremely small, flash distillation and vacuum distillation are extremely attractive processes, where the latter minimizes energy consumption and solvent decomposition. The higher boiling point also extends the flexibility of the process in terms of the feedstock composition by allowing for the processing of components with higher boiling point ranges. The normal boiling points of NMP and sulpholane are 475 K and 560.5 K, respectively [23]. In the sulpholane process for aromatic–aliphatic mixtures with up to 68% aromatics [16], solvent recovery is achieved with minimal energy input as vacuum distillation is employed. Under ambient conditions, the volatility of ionic liquids is negligible and this would have tremendous consequences for energy savings in solvent recovery.

2.2.3. Thermal Stability

In addition to air, moisture and photochemical stability, the thermal stability is an important aspect of improving the economy of the process. The occurrence of thermally induced polymerization or decomposition reactions results in a loss of solvent recovery potential, specialized facilities for the treatment and post-purification of solvents and product streams and poor flexibility in the optimization of the thermal profile of the process (solvent extraction and extractive distillation steps). *N*-Methyl pyrrolidone has been shown to be chemically and thermally stable in the Arosolvan process. Sulpholane is reported to be stable to 493 K and undergoes some decomposition at 558 K [23]. In the sulpholane process, the influence of oxygen on solvent stability in the form of minor oxidative degradation has been observed under normal operating conditions. Consequently, the exclusion of air in the feed to the extraction unit has been advocated for this process together with the inclusion of a solvent regenerator unit. The latter operates by removing oxidized solvent from a small side-stream of the circulating solvent that is directed towards the solvent regenerator unit [16]. Ionic liquids exhibit excellent thermal stability and lack of sensitivity to oxygen would be advantageous with respect to the processing and recovery of the solvent.

2.2.4. Density and Viscosity

A large difference between the solvent and the raffinate phases (as provided by a solvent with a high density) is desirable to allow for rapid and efficient phase separation in the liquid–liquid extraction. However, a low viscosity solvent is preferred from a theoretical and practical standpoint as it allows for an improved phase separation and mass transfer, together with a decreased load in terms of pumping duties for the solvent. NMP has an intermediate density (1.0259 g cm^{-3} at 298.15 K) [23] and an optimum viscosity (viscosity coefficient = 1.666 cP at 298.15 K) [23]. Sulpholane, on the other hand, has a rather high density (1.2604 g cm^{-3} at 303.15 K) [23] and a high viscosity (viscosity coefficient = 10.286 cP at 303.15 K) [23]. As with NMP and sulpholane, the majority of ionic liquids have densities that are higher than that of water, with values ranging from 0.949 g cm^{-3} to as high as 1.85 g cm^{-3} [24] being reported. Ionic liquid viscosities, as with densities, are equally as variable with values from 18 cP to 33070 cP and are generally quite high.

2.2.5. Extract Products (Aromatics) Range

The product range is a limiting factor in determining the capacity of the solvent as it determines the maximum amount of the solute that can be accommodated in the solvent phase. If this limit is exceeded, phase splitting ceases and the separation process breaks down.

2.2.6. Melting Point and Liquidus Range

The melting point of the solvent of choice should ideally be below ambient temperature so as not to necessitate the introduction of any special requirements such as steam-tracing of equipment and process lines to prevent solidification of the solvent. In this regard, the sulpholane process is burdened by the rather high melting point (301.60 K) [23] of its solvent. A large liquidus range for the solvent also facilitates the storage and the containment of the solvent, together with maximizing solvent recovery and the efficiency of the extraction process. The melting point of RTILs is frequently below 273.15 K and a large liquidus range ($>$200 K) is generally observed for most ionic liquids.

2.2.7. Handling Ability

A solvent which is non-toxic (for human health and environmental concerns), non-flammable and non-corrosive facilitates handling and processing. The latter consideration is particularly important for the economical design of extraction plants, e.g. a plant constructed from carbon steel would suffice for a fairly non-corrosive solvent (e.g. sulpholane). Where corrosion rates are significant, the use of stainless steel as a construction material for the plant is necessitated. The majority of ionic liquids have low corrosion tendencies. Unfortunately, toxicity data on ionic liquids are still rather scarce [29].

2.2.8. *Availability and Cost*

Liquid–liquid extraction is a solvent-intensive process and since solvent recovery is a finite process, the solvent should be available in high purity at moderate cost. Both sulpholane and other commercially significant solvents are obtained at a lower cost than ionic liquids due to a current lack of commercial significance and large-scale preparation of ionic liquids.

2.2.9. *Solvent and Process Flexibility*

The possibility of optimizing or enhancing the solvent's properties for a more effective extraction through fairly simple, inexpensive means is an invaluable option for the process engineer. This has most frequently been achieved through the addition of modifiers or polar mixing components to the solvent to alter the selectivity, capacity, aromatics range and boiling point. An example of this is to be found in the Arosolvan process, where a polar mixing component (water or monoethylene glycol) is employed to improve the capacity of NMP. Process flexibility primarily relates to the ability of the process to maintain high levels of recovery of a pure product in response to variation in the aromatics content and the boiling point range of the feedstock.

2.3. Measurement and Representation of Ternary Liquid–Liquid Equilibria

The acquisition of experimental liquid–liquid equilibria data for solvent–organic mixtures of interest is invaluable for the screening of potential solvents for liquid–liquid extraction. The graphical representation of ternary liquid–liquid equilibria is most suitably represented through the use of a triangular phase diagram (Fig. 1), where the miscibility characteristics of the system as a function of overall composition are shown.

Each vertex represents a pure component and the arrangement shown in Fig. 1 for the solvent, original solvent and solute is most convenient for an assessment of the extraction efficiency of the solvent for the solute. The concentration units are mole fractions such that the sum of the mole fractions of three components A, B and C for each point on the diagram satisfies the following:

$$x_A + x_B + x_C = 1 \quad (3)$$

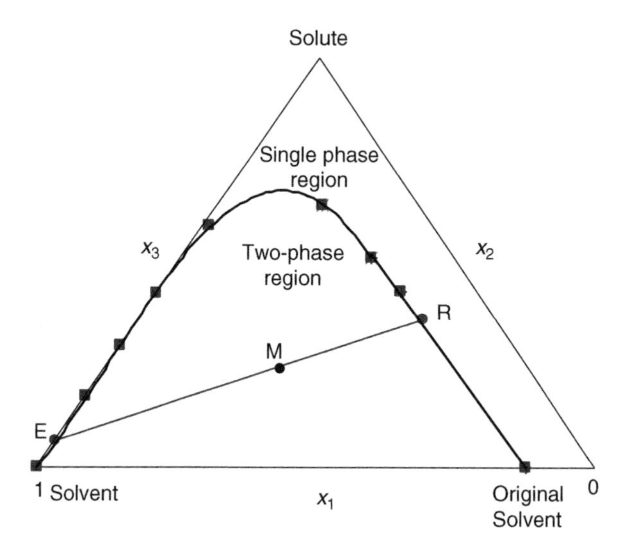

Fig. 1. Representation of ternary liquid–liquid equilibria.

The binodal curve separates the single-phase region from the two-phase region. Any overall system composition (*M*) in that region will spontaneously split into two phases, i.e. the extract phase (E) and the raffinate phase (R). The compositions of the equilibrium conjugate phases lie on the curve on either end of the tie line that passes through the overall system composition. As the tie lines become shorter, the plait point is approached, at which concentration, only one liquid phase exists.

In general, the size of the immiscibility region and the favourable sloping of the tie lines are the key graphical criteria used for liquid–liquid equilibria ternary diagrams in the assessment of the suitability of a solvent for a separation process. A large region under the bimodal curve affords greater flexibility for the extraction process as the integrity of the extraction process, i.e. the existence of two immiscible phases, is maintained for a wider range of overall system compositions, resulting in higher capacities and aromatics ranges. In addition to the basic requirement of the sloping of the tie lines towards the solvent axis, the steeper the tie lines, the larger is the disparity in the relative concentration of the solute in the extract and raffinate, i.e. higher selectivities.

For a discussion of the different types of binodal curves encountered, experimental acquisition of liquid–liquid equilibrium data and the theoretical treatment of liquid–liquid equilibria (correlation and prediction), the review articles by Sorensen et al. [9, 25, 26] can be consulted.

2.4. Current Trends in Liquid–Liquid Extraction Processes and the Screening of Ionic Liquid Solvents

With the advent of the diversification of specialty chemical markets, the development of new technologies and processes, a dire need for greener chemical processes, the availability of renewable chemical feedstocks and energy resources, together with an increased need for ultra-pure chemicals, greater demands have been placed on process engineers for the design of more efficient, economical and flexible separation processes for new and more challenging separations (alkane–alkene, water–alkanol, carboxylic acid–water, alkene–alkanol, etc.). Ionic liquids have been identified as a class of substances with favourable properties to be employed as potential solvents in liquid–liquid extraction processes. As with most other fields of interest, 1,3-dialkylimidazolium ionic liquids have been the focus of attention for researchers investigating the liquid–liquid equilibria of ionic liquid–organic mixtures.

The measurement of partition coefficients or distribution ratios of various classes of organic solutes in an ionic liquid–water biphasic system through the use of the shake-flask or slow-stirring methods has been the subject of numerous investigations [24, 27–31]. In the studies of Huddleston et al. [27] and Carda-Broch et al. [28, 29], the charged state of the organic solute, i.e. pH of the extraction medium, was found to influence the relative magnitude of the partition coefficients. These findings are significant for the design of liquid–liquid extraction processes as it allows for flexibility and an improvement in the selectivity of the extraction through the use of the charged state of a solute as a control variable. However, the use of partition coefficients in the design of separation processes is not always very useful as these are often not reported as a function of composition.

The measurement of the solubility characteristics of ionic liquid–organic solute mixtures, i.e. binary liquid–liquid equilibria, has also been an active area of research [32–42]. In particular, the study of the miscibility characteristics of 1,3-dialkylimidazolium–alcohol systems has been studied by several research groups. In these studies, the effects of alkyl chain length on the cation, the nature of the anion and the alkyl chain length of the alcohol were investigated to elucidate the structure–property correlations responsible for influencing the solubility behaviour (upper critical solution temperature or UCST) of ionic liquid–alcohol mixtures. The study of Marsh et al. [42] was particularly significant as it involved an investigation of the effect of a water impurity on the UCST of a water–alcohol binary system. It was shown that the addition of a 1.7 mass% water impurity results in a 10 K decrease in the UCST, from which Marsh et al. [42] concluded that a water content of lower than 0.02 mass% is required to obtain a UCST within the range of experimental uncertainty of 0.05 K. Although binary liquid–liquid equilibria data do not allow for a direct screening or assessment of the separation efficiency of an ionic liquid for a particular separation problem, they do allow for an isolation of the structure–phase equilibrium (solubility) behaviour of the ionic liquid to allow for a better understanding of solute–solvent interactions in optimizing the structure of an ionic liquid for a specific separation problem.

Studies on ternary liquid–liquid equilibria have centred on specific challenges facing the chemical, petrochemical, pharmaceutical and biochemical industries. A summary of the ternary liquid–liquid equilibria data (selectivities, capacities) for aliphatic–aromatic separations with ionic liquids is presented in Table 1. The aromatic content (expressed as a percentage) is included to provide an indication of the region in which the selectivity and capacity values are determined (since these are a function of the overall composition). Selectivity and capacity values of traditionally or commercially employed solvents have been included for comparative purposes. The favourable characteristics required for solvents suitable for aromatic–aliphatic separations are the following: large selectivity and capacity values, high solubility of the aromatic components in the ionic liquid, poor solubility of the aliphatic components in the ionic liquid and the availability of fairly simple and inexpensive means to recover the ionic liquid from the extract and raffinate phases.

An examination of Table 1 reveals that the ionic liquid solvents, with the exception of [emim][I_3][2] and [bmim][I_3] in the study of Selvan et al. [45], are in general inferior as extraction solvents to sulpholane. However, even the former ionic liquid solvents have to be excluded as potential candidates for liquid–liquid extraction due the corrosive nature of halide-containing ionic liquids [17, 50]. There is in general a high solubility of the aromatic components coupled with a low solubility of the aliphatic components in the ionic liquid, which is quite favourable for aromatic–aliphatic separations. This is as a result of the delocalized π electron clouds associated with aromatic systems producing a strong electrostatic field for greater interaction with the ionic liquid than is possible for a saturated aliphatic molecule [51]. With increasing chain length of the aliphatic component, there is decreased solubility of the alkane in the ionic liquid and in general higher selectivity values. A comparison of the anion effect for [hmim][BF_4] versus [hmim][PF_6] in the work of Letcher and Reddy [47] yields very little difference between the two ionic liquids in an aromatic–aliphatic separation, with the exception of a slighter higher

[2]See Appendix I for an explanation of ionic liquid abbreviations.

Table 1
Selectivities, capacities and solubilities of traditional and ionic liquid solvents in liquid–liquid equilibria investigations of aromatic–aliphatic mixtures

Solvent	Mixture	Temperature (K)	Alkane solubility[a] (MF%)	Aromatic solubility (MF%)	Aromatics content (MF%)	Selectivity	Capacity
Sulpholane [43]	Benzene–heptane	303.15	1.5	–	5.4	22.2	0.46
Sulpholane [43]	Benzene–dodecane	303.15	–	–	18.0	43.6	0.62
Sulpholane [17]	Toluene–heptane	313.15	0.5	–	5.9	30.9	0.31
NMP [44]	Toluene–hexadecane	298.2	1.2	–	10.2	17.6	0.36
[emim][I_3] [45]	Toluene–heptane	318.15	–	70.7	37.0	48.8	2.20
[bmim][I_3] [45]	Toluene–heptane	308.15	–	77.7	39.5	30.1	2.30
[omim][Cl] [46]	Benzene–heptane	298.2	13.1	67.3	15.4	7.7	0.58
[omim][Cl] [46]	Benzene–dodecane	298.2	2.4	67.3	21.6	38.3	0.87
[omim][Cl] [46]	Benzene–hexadecane	298.2	1.9	67.3	15.6	43.3	0.51
[hmim][BF_4] [47]	Benzene–heptane	298.2	9.8	74.6	10.4	8.4	0.81
[hmim][BF_4] [47]	Benzene–dodecane	298.2	3.6	74.6	10.0	12.1	0.49
[hmim][BF_4] [47]	Benzene–hexadecane	298.2	1.2	74.6	6.6	11.2	0.27
[hmim][PF_6] [47]	Benzene–heptane	298.2	8.4	78.4	13.9	11.2	0.92
[hmim][PF_6] [47]	Benzene–dodecane	298.2	5.3	78.4	12.8	8.8	0.50
[hmim][PF_6] [47]	Benzene–hexadecane	298.2	2.2	78.4	7.5	18.5	0.43
[emim][OSO_4] [48]	Benzene–heptane	298.2	14.0	88.0	13.4	2.7	0.49
[emim][OSO_4] [48]	Benzene–hexadecane	298.2	0.56	88.0	24.6	8.0	1.53
[omim] [$MDEGSO_4$] [48]	Benzene–heptane	298.2	10.9	97.0	18.5	1.5	0.32
[omim] [$MDEGSO_4$] [48]	Benzene–hexadecane	298.2	4.7	97.0	19.1	12.4	0.87
[emim][Tf_2N] [49]	Benzene–cyclohexane	297.65	–	–	27.8	17.7	–

[a]Mole fraction percentage.

solubility of the aromatic component in [hmim][PF_6]. As for the [I_3]-based ionic liquids, ionic liquids containing fluorinated anions such as [BF_4] and [PF_6] are now becoming unpopular due to the tendency of these ionic liquids to undergo hydrolysis [42, 52]. This is highly undesirable as it results in the formation of hydrogen fluoride, which is corrosive and toxic. Consequently, the current trends in ionic liquid synthesis and research are in the development and study of halogen-free and hydrolysis-stable ionic liquids [53] such as those investigated by Deenadayalu et al. [48]. As can be seen from Table 1, the selectivities and capacities of these ionic liquids for aromatic–aliphatic separations are low; however, the solubility of the aromatic components in the ionic liquids (especially [omim] [$MDEGSO_4$]) is reasonably high.

Meindersma et al. [17] conducted a comprehensive study on the selection of ionic liquids for aromatic–aliphatic separations by investigating the extraction efficiency of ionic liquids as a function of cation, anion, alkyl chain length, temperature and composition. As in the study of Selvan et al. [45], it was observed that a shorter alkyl group on the cation was desirable for higher selectivities (but lower capacities).

This effect was, however, heavily anion-dependant as the reverse trend was observed for [emim][BF_4] versus [bmim][BF_4]. Interestingly enough, the absence of one alkyl group on the imidazolium cation such as in [Hmim] (where H is hydrogen) results in a lower selectivity. In general, it was observed that ionic liquids based on pyridinium cations such as 4-methyl-*N*-methylpyridinium or [mebupy] in the form of [mebupy][BF_4] and [mebupy] [$MeSO_4$] had higher selectivities and capacities than sulpholane.

Other separations of relevance to the chemical and petrochemical industries, including mixtures such as alkene–alkanol, alkanol–water, alkanol–ether, alkane–alkanol, etc., have also been the subjects of investigations. This is summarized in Table 2. The study of Letcher and Reddy [54] yielded interesting results for the alkene–alkanol separation, where superior extraction efficiency was observed for [hmim][BF_4] relative to [hmim][PF_6]. This could probably be attributed to stronger hydrogen-bonding between the alcohol and the [BF_4] anion as opposed to the [PF_6] anion, an effect that was observed in solubility study of Crosthwaite et al. [40].

The results observed for the alkene–alkanol systems in the study of Letcher et al. [55], i.e. very high selectivities and capacities, are explained in terms of a high preferential affinity of the ionic liquid for the alcohol (hydrogen-bonding, polar interactions) over the aliphatic component. However, as explained above, ionic liquids based on halogen anions are undesirable. The studies of the research group of Arce et al. [56–58] in the selective

Table 2
Selectivities, capacities and solubilities of traditional and ionic liquid solvents in liquid–liquid equilibria investigations of miscellaneous mixtures

Solvent	Mixture	Temperature (K)	Solute content (MF%)	Selectivity	Capacity
[bmim][PF_6] [42]	Water–ethanol[a]	298.15	10.0	–	0.17
[hmim][PF_6] [54]	1-Hexene–ethanol[a]	298.2	21.2	6.5	0.92
[hmim][PF_6] [54]	1-Heptene–ethanol[a]	298.2	35.5	6.0	0.85
[hmim][BF_4] [54]	1-Hexene–ethanol[a]	298.2	32.6	42.7	5.9
[bmim][BF_4] [54]	1-Heptene–ethanol[a]	298.2	47.5	32.6	3.5
[omim][Cl] [55]	Heptane–methanol[a]	298.2	44.7	7800	447
[omim][Cl] [55]	Heptane–ethanol[a]	298.2	37.1	400	53
[omim][Cl] [55]	Dodecane–ethanol[a]	298.2	53.2	1800	76
[omim][Cl] [55]	Hexadecane–methanol[a]	298.2	85.9	730000	859
[omim][Cl] [55]	Hexadecane–ethanol[a]	298.2	67.9	46000	679
[omim][Cl] [56]	*tert*-Amylethylether–ethanol[a]	298.15	22.1	122.2	18.6
[bmim][TfO] [57]	*tert*-Amylethylether–ethanol[a]	298.15	24.7	44.6	4.1
[bmim][TfO] [58]	Ethyl *tert*-butylether–ethanol[a]	298.15	14.4	20.3	3.4

[a]The solute to be extracted in the binary mixture.

extraction of ethanol from ether–ethanol mixtures show that the use of ionic liquid solvents in this type of separation is indeed feasible as a result of favourable selectivity and capacity values. Furthermore, the use of [bmim] [TfO] in the purification of ethyl *tert*-butylether through the efficient extraction of ethanol has been shown to be superior than that of water [58], which is the commonly used solvent for this type of separation.

As can be inferred from Tables 1 and 2, liquid–liquid equilibrium studies have largely been limited to ionic liquid–organic mixtures containing imidazolium-based ionic liquids. Consequently, an accurate assessment is not feasible with such a limited pool of phase equilibrium data. However, in general, the ionic liquid solvents have shown to be inferior to commercial solvents currently employed in liquid–liquid equilibrium processes.

3. THE DETERMINATION OF ACTIVITY COEFFICIENTS AT INFINITE DILUTION FOR THE SELECTION OF ENTRAINERS IN EXTRACTIVE DISTILLATION

Solvent-enhanced distillation processes such as extractive and azeotropic distillation are necessary for the separation of close-boiling or azeotropic mixtures, where the use of liquid–liquid extraction is not feasible (e.g. aliphatic–aromatic mixtures with an aromatics content higher than 65% [16], poor selectivities or capacities for liquid–liquid extraction), and for solvent regeneration in a liquid–liquid extraction plant. The selection of suitable solvents or entrainers for extractive distillation is an integral part of ensuring that an efficient and economical extractive distillation process is achieved. There are currently four broad categories [59] of entrainers that have been proposed in the form of solid ionic salts, high-boiling organic liquids, a combination of ionic salts and organic liquids and ionic liquids. Although the first three categories of potential entrainers have been suitable for incorporation into commercial extractive distillation processes [59], the high selectivities and favourable thermo-physical properties offered by ionic liquids provide an opportunity to lower energy usage and operating costs, and to improve the process efficiency.

Activity coefficients at infinite dilution (γ_{13}^{∞}) of a solute (1) in a solvent phase (3) are invaluable for the development of correlative and predictive thermodynamic models and for the selection of solvents for extractive/ azeotropic distillation, liquid–liquid extraction and solvent-aided crystallization. The experimental determination of γ_{13}^{∞} can be achieved through the use of techniques such as ebulliometry [60], headspace chromatography [61], static methods [62], the dilutor technique [63] and gas–liquid chromatography (glc). The latter technique has the advantage that it is a relatively inexpensive, simple, reliable, well-established and rapid technique for the measurement of γ_{13}^{∞}. However, the limitations associated with the use of the glc technique include the following: (i) the injected solute must be appreciably volatile at the column temperature, (ii) there must be instant equilibration of the solute between the mobile and the stationary phases, (iii) the carrier gas must be inert and insoluble in the stationary phase, (iv) surface effects, in the form of adsorption of the solute onto either the liquid solvent or the inert support material on which it is immobilized, must be non-existent and (v) column bleed must be negligible at the column operating temperature. A more comprehensive review of the glc theory and the experimental technique for γ_{13}^{∞} determination is provided in the works of Letcher [64] and Conder and Young [65]. Ionic liquids are well suited to the use of the glc method for physico-chemical measurements due to their negligible volatilities ensuring that column bleeding does not take place.

3.1. Solvent Choice Criteria for Extractive Distillation

The selection of a suitable solvent for extractive distillation is roughly guided by the same considerations that influence the selection of solvents for liquid–liquid extraction (see Section 2.2). However, the acquisition of experimental phase equilibrium (VLE) data to obtain phase compositions or finite concentration activity coefficients, in the definition of a selectivity factor for an extractive distillation process, is a time-consuming, tedious and expensive exercise.

Through the measurement of γ_{13}^{∞}, selectivity and capacity can be defined for the condition of infinite dilution as follows:

$$S_{12}^{\infty} = \frac{\gamma_1^{\infty}}{\gamma_2^{\infty}} \tag{4}$$

$$C_{12}^{\infty} = \frac{1}{\gamma_2^{\infty}} \tag{5}$$

where components 1 and 2 are to be separated ($\gamma_1^{\infty} > \gamma_2^{\infty}$) by the solvent (3) and component 2 is to be extracted (in the definition of C_{12}^{∞}). The smaller the value of γ_{13}^{∞}, the stronger is the interaction between the solvent and the solute, with values much lower than unity indicating particularly strong affinity of the solvent for the solute.

In terms of the S_{12}^{∞} value, an effective entrainer is capable of different types of physical interactions (polar, dispersive, hydrogen-bonding, etc.) with the two components to be separated to allow for the S_{12}^{∞} value to be much larger than unity to allow for a better separation by altering the relative volatility of the system, where the relative volatility at infinite dilution is:

$$\alpha_{12}^{\infty} = \frac{P_1^{\mathrm{s}}\gamma_1^{\infty}}{P_2^{\mathrm{s}}\gamma_2^{\infty}} \tag{6}$$

In terms of capacity (C_{12}^{∞}), it can be seen that components with small γ_{13}^{∞} values (strong interaction and solubility in the solvent phase) will have a large capacity, which is favourable for an economical process as a lower solvent-to-feed ratio and circulation rate are possible.

The use of the above quantities in the selection of potential solvents for extractive distillation has to be approached with caution as the physico-chemical properties of the system at infinite dilution can differ quite markedly from that at finite concentration [16], where for conventional extractive distillation, the latter represents the technically relevant composition range. Since the condition at infinite dilution is one of maximum non-ideality and the value of the activity coefficient is highly concentration-dependant, selectivity values at finite concentration can be much lower than those at infinite dilution [59].

3.2. Experimental γ_{13}^{∞} Determinations for Organic Solutes in Ionic Liquid Solvents

The current pool of γ_{13}^{∞} data for ionic liquid–organic solute interactions exceeds those obtained by alternative techniques such as headspace chromatography, dilutor and the static technique. As mentioned previously, experimental investigations on ionic liquids from an industrial perspective have largely been guided by separation problems of interest to the chemical and petrochemical industries such as alkane–aromatic, cyclo-alkane–aromatic and alkane–alkene mixtures. In this regard, selectivities at infinite dilution (S_{12}^{∞}) are presented in Table 3 for (i) *n*-hexane–benzene, (ii) cyclohexane–benzene and (iii) *n*-hexane–1-hexene separations in various ionic liquid and commercially significant solvents.

It can be observed from Table 3 that the γ_{13}^{∞} values decrease in the following hierarchy: *n*-hexane > cyclohexane > 1-hexene > benzene. This is of course in accordance with the relative strengths of the ionic liquid–organic solute interactions. Saturated non-polar aliphatic molecules in the form of

Table 3

A summary of the S_{12}^{∞} values for the separation problems: (i) *n*-hexane–benzene, (ii) cyclohexane–benzene and (iii) *n*-hexane–1-hexene

Solvent	T (K)	γ_{13}^{∞} (benzene)	γ_{13}^{∞} (*n*-hexane)	γ_{13}^{∞} (cyclohexane)	γ_{13}^{∞} (1-hexene)	S_{12}^{∞}		
						(i)	(ii)	(iii)
NMP [66]	298.15	1.3	15.4	7.9	9.5	11.8	6.1	1.6
Sulpholane [67]	303.15	2.16	75.0	34.2	–	34.7	2.19	–
[hmim][BF_4] [68]	298.15	0.96	22.1	12.9	10.9	23.0	13.4	2.03
[hmim]PF_6] [69]	298.15	1.03	22.50	12.65	10.42	21.8	12.3	2.16
[mmim][TF_2N][a] [70]	303.15	1.34	39.9	22.7	17.2	29.8	16.9	2.32
[memim][Tf_2N] [71]	313	1.097	25.267	14.859	–	23.0	13.5	–
[emim][TF_2N][a] [70]	303.15	1.19	27.9	15.6	12.7	23.4	13.1	2.20
[bmim][TF_2N][a] [70]	303.15	0.881	14.2	8.64	7.34	16.1	9.81	1.93
[hmim][TF_2N] [72]	298.15	0.674	8.33	5.50	4.74	12.4	8.16	1.76
[omim][TF_2N] [73]	303.15	0.63	5.32	3.64	3.29	8.44	5.78	1.62
[omim][Cl] [74]	298.15	1.99	17.2	10.5	11.1	8.64	5.28	1.55
[emim][ESO_4][a] [70]	303.15	2.73	106	56.7	48.2	38.8	20.8	2.20
[emim][Tos][b] [75]	323.15	2.61	142.52	38.57	44.78	55.8	14.8	3.25
[bmim][OSO_4][b] [75]	323.15	0.95	4.75	2.83	3.38	5.00	2.98	1.41
[moim][$MDEGSO_4$] [76]	298.15	1.40	12.9	7.86	8.24	9.20	5.61	1.40
[epy][TF_2N] [77]	303.15	1.26	33.5	18.5	14.6	26.6	14.7	2.29
[py][$EOESO_4$][a] [77]	303.15	3.81	45.2	36.5	19.7	11.9	9.58	2.29
[mebupy][BF_4] [78]	313	1.639	60.44	29.03	–	36.9	17.7	–
[TdH_3P][$TPfEPF_3$] [79]	308.15	0.20	0.66	0.49	0.53	3.30	2.45	1.25
[$MeBu_3P$][$MeSO_4$] [80]	308.15	1.01	10.28	5.52	6.40	10.2	5.5	1.6

[a]Using the dilutor technique.
[b]Corrected for interfacial adsorption.

n-hexane interact very weakly with the ionic liquid solvent through dispersive forces and consequently have the largest values. Values of γ_{13}^{∞} that are less than unity are frequently obtained for aromatic compounds, indicating a very strong affinity of the ionic liquid solvent for aromatic compounds (as was observed in the high solubilities of the aromatic compounds in the ionic liquid solvents in Table 1). The existence of delocalized or mobile π electron clouds above and below the plane of the aromatic ring creates a very strong electrostatic field available for interaction with the ionic liquid solvent, with simulation studies [51] indicating a great deal of ordering in the ionic liquid cation–aromatic and ionic liquid anion–aromatic interactions.

A wide range of γ_{13}^{∞} and values have been obtained for the organic solutes in the ionic liquid solvents in Table 3 indicating that the (i) cation, (ii) anion and (iii) substituent chain length have a major influence on the nature of the ionic liquid–organic solute interaction. The effect of the anion in the [hmim][BF_4] and [hmim][PF_6] structures has a negligible effect on an aromatic–aliphatic separation (similar S_{12}^{∞} values), which is consistent with the ternary liquid–liquid equilibria studies of Letcher and Reddy [54]. However, for the [Tf_2N] analogue, a much lower selectivity is observed due to a relatively greater interaction of the ionic liquid with the alkane phase. Lei et al. [59] postulate that the interaction between an ionic liquid and a hydrocarbon solute with a mobile electron cloud (alkenes, aromatics) is enhanced favourably through the incorporation of anions which have steric shielding (number and bulk of attached groups) around the anion charge centre. However, since the [BF_4], [PF_6] and [Tf_2N] anions all conform to the above, this cannot be used as a basis for the discrepancy in the S_{12}^{∞} values. It is probably related to the packing effects in liquid phase structure of the ionic liquid, where a bulky anion inhibits the close packing of the cations and anions, allowing for greater dispersive interactions between the alkyl substituents on the cation and the alkane. For [omim][Cl], the lack of steric shielding around the anion contributes to low S_{12}^{∞} values.

The high selectivity of [emim][Tos] is due to the aromatic ring present in the tosylate anion, which allows for greater discrimination in an aromatic–aliphatic separation, favouring the former.

Despite the advantage of sulphate-based anions in ionic liquids over fluorinated or halide-containing anions in terms of "greenness", the former ([bmim][OSO_4] and [moim][$MDEGSO_4$]) have in general not proven to be effective in the separation of aromatic–aliphatic mixtures, with the exception of [emim][ESO_4]. This indicates that a small sulphate-based anion is more effective for the separation of the separation problems presented in

Table 3. For the pyridinium-based ionic liquids, a much stronger interaction is expected between the pyridinium ring of the cation and the aromatic solute [24]. However, it can be observed that the incorrect choice of anion and the alkyl substituents on the pyridinium ring can adversely affect this. Pyridinium ionic liquids based on the [mebupy] cation, such as [mebupy] [BF_4], have proven to be quite effective as entrainers, as observed by Meindersma et al. [17] in liquid–liquid equilibria studies.

Quaternary ammonium or phosphonium ionic liquids have not been considered as serious candidates as potential solvents as the selectivities obtained are generally very low [16]. This is clearly observed in Table 3 for the two phosphonium-based ionic liquids. The relative magnitude of the γ_{13}^{∞} values for [TdH_3P][$TfEPF_3$] are surprising, as values much lower than unity have been obtained for saturated non-polar aliphatic solutes such as hexane. This is uncharacteristic for dispersive interactions.

The work done by Muletet and Jaubert [75] is significant and raises concerns over the accuracy of γ_{13}^{∞} values which have been obtained without a correction for interfacial adsorption at the gas–liquid interface.

Based on the evidence of the research findings presented above, pyridinium-based ionic liquids, in particular, those with alkyl substituents, and aromatic-based anions seem to show the greatest promise in being the most effective for the separation of saturated–unsaturated/aromatic hydrocarbons. Consequently, an ideal candidate for the above separations would be obtained through the coupling of the [mebupy] cation and the [Tos] anion, i.e. [mebupy] [Tos].

4. ASSESSMENT OF THE POTENTIAL OF IONIC LIQUIDS AS SOLVENTS IN SEPARATION PROCESSES

An effective assessment of the use of ionic liquids as solvents in commercial liquid–liquid extraction and extractive distillation processes requires consideration of numerous factors associated with the thermo-physical and chemical properties of ionic liquids as well as the economical and environmental impact of the use of ionic liquids in the chemical and petrochemical industry.

The advantages of the use of ionic liquid solvents over commercially employed solvents such as sulpholane and NMP can be summarized as follows:

(i) Flexibility in the pure component physico-chemical and thermophysical properties of ionic liquids such as viscosity, density, heat capacity, liquidus range, etc., allows for flexibility in the design of separation processes and in the processing of fluid streams. This allows for the phase equilibrium properties (solubility, VLE) of ionic liquid–organic mixtures to be optimized through the judicious variation of the cation, anion and substituents in the ionic liquid structure. This was clearly evident in the studies of Letcher and Reddy [54], Meindersma et al. [17] and Crosthwaite et al. [40].

(ii) The high density of ionic liquid solvents makes them ideal for liquid–liquid extraction processes as this allows for rapid and efficient phase separation.

(iii) The general non-corrosive, non-toxic and non-flammable nature of some ionic liquids eradicates concerns over safety, handling and containment of ionic liquids.

(iv) The high thermal stabilities, boiling points and liquidus range of ionic liquids maximize solvent recovery and allow for the use of novel, less expensive techniques for solvent regeneration (vacuum distillation, flash distillation, pervaporation). Studies by Meindersma et al. [17] on the regeneration of ionic liquids from hydrocarbon mixtures showed that this process had no effect on the integrity of the ionic liquid, even after repeated recycling.

(v) Due to the low solubility of the ionic liquid in the raffinate phase, especially for aromatic–aliphatic separations, recovery of ionic liquid from the raffinate phase is a fairly simple process, where the water solubility of the ionic liquid can be exploited in water-washing stage for solvent recovery.

The disadvantages or impediments associated with the incorporation of ionic liquids in separation technology are the following:

(i) The corrosive nature of ionic liquids containing halogenated anions (e.g. $[I_3]$) as in the study by Selvan et al. [45] and the tendency of fluorinated anions such as $[BF_4]$ and $[PF_6]$ to undergo hydrolysis [42, 53] do not bode well for the use of these types of ionic liquids. It is for these types of ionic liquids that a great deal of experimental data have been accumulated and great promise (favourable selectivities) has been displayed by these types of ionic liquids for the separation of complex organic mixtures.

(ii) The availability of ionic liquids at moderate cost and in high purity is unfortunately not a reality at this stage [53, 81], with costs of ~€300 to €2600 L^{-1}. Consequently, unless an economical commercial scale production of ionic liquids is realized, commercially employed solvents such as sulpholane and NMP will not be displaced by ionic liquids.

(iii) The investigation of ionic liquids in biotechnology applications [82] has revealed that there is considerable anti-microbial activity on the part of ionic liquids. Coupled with the aqueous solubility of ionic liquids, the effect of ionic liquids on aquatic ecosystems has to be assessed.

(iv) Despite claims by researchers that the presence of minute amounts of impurities in ionic liquids does not adversely affect the phase equilibrium properties of ionic liquids, studies by researchers such as Marsh et al. [42] have shown that the effect of a small amount of water (1.7%) shifts the UCST of a [bmim][PF_6]–alcohol mixture 10 K. This clearly indicates a concern over ionic liquid purity.

(v) The high viscosity of ionic liquids is undesirable for the liquid–liquid extraction process and in general the processing of a viscous liquid is undesirable (as high solvent regeneration and circulation rates are difficult to achieve).

5. CONCLUSION

The acquisition of experimental phase equilibrium data for ionic liquid–organic mixtures has displayed that ionic liquids are suitable for the separation of industrially relevant organic mixtures, in particular, for aromatic–aliphatic separation problems. Ionic liquids are indeed a very promising class of solvents that offer numerous advantages over commercially applicable solvents, most notably in flexibility, separation efficiency and economy. However, considerable challenges (biotoxicity, viscosity, solvent production cost and purity) have to be addressed before the full scale industrial implementation of ionic liquids in separation processes can be realized.

APPENDIX I: LIST OF ABBREVIATIONS FOR IONIC LIQUID NOMENCLATURE

Cations [$^+$]

[mmim]	1,3-dimethylimidazolium
[memim]	1,2-dimethyl-3-ethylimidazolium
[emim]	1-ethyl-3-methylimidazolium
[bmim]	1-butyl-3-methylimidazolium
[hmim]	1-methyl-3-hexylimidazolium
[omim]	1-octyl-3-methylimidazolium
[py]	pyridinium
[epy]	1-ethylpyridinium
[mebupy]	1-butyl-4-methylpyridinium
[TdH_3P]	trihexyl (tetradecyl) phosphonium
[$MeBu_3P$]	tributylmethylphosphonium

Anions [$^-$]

[BF_4]	tetrafluoroborate
[Cl]	chloride
[I_3]	triiodide
[PF_6]	hexafluorophosphate
[$MeSO_4$]	methylsulphate
[ESO_4]	ethylsulphate
[OSO_4]	ocytlsulphate
[$EOESO_4$]	ethoxyethylsulphate
[Tos]	*p*-toluenesulphonate
[Tf_2N]	bis (trifluorosulphonyl) imide
[TfO]	trifluoromethanesulphonate
[$MDEGSO_4$]	diethyleneglycolmonomethylethersulphate
[$TPfEPF_3$]	tris (pentafluoroethyl) trifluorophosphate

REFERENCES

[1] M. Freemantle, Chem. Eng. News, 76 (1998) 32–37.
[2] M. Freemantle, Chem. Eng. News, 78 (2000) 37–50.
[3] P. Walden, Bull. Acad. Imper. Sci. (St. Petersburg), (1914) 1800.
[4] J.S. Wilkes and M.J. Zaworotko, J. Chem. Soc., Chem. Commun., (1992) 965.
[5] Solvent Innovation, available from http://www.solvent-innovation.de.
[6] ISI, Essential Science Indicators Special Topics: Ionic Liquids, ISI, 2004, available from http://www.esitopics.com/ionic-liquids/index.html. ISI, 3501 Market St, Philadelphia, PA 19104, USA.
[7] NATO Advanced Research Workshop, Green Industrial Applications of Ionic Liquids, Heraklion, Crete, Greece, 2000.
[8] J. Gmehling, in "The Basis for the Synthesis, Design and Optimization of Thermal Separation Processes" (T.M. Letcher, ed.), pp. 1–13, Chemical Thermodynamics: A Chemistry for the 21st Century, Blackwell Science, Oxford, 1999.
[9] J.M. Sorensen, T. Magnussen, P. Rasmussen and A. Fredenslund, Fluid Phase Equilib., 2(4) (1979) 297–309.
[10] P.J. Bailes, C. Hanson and M.A. Hughes, in "Liquid–Liquid Extraction: Non-Metallic Materials" (L. Ricci, ed.), pp. 232–237, Separation Techniques 1: Liquid–Liquid Systems, McGraw-Hill, New York, NY, 1980.
[11] R.M. Pagni, "Ionic liquids as alternatives to traditional organic and inorganic solvents", presented at NATO Advanced Research Workshop, Green Industrial Applications of Ionic Liquids, Heraklion, Crete, Greece, 2000.
[12] G.W. Meindersma and M. Kuczynski, Fluid Phase Equilib., 113 (1996) 285–292.
[13] A.E. Visser, R.P. Swatloski, W.M. Reichert, R. Mayton, S. Sheff, A. Wierzbicki, J.H. Davis, Jr. and R.D. Rogers, Chem. Commun., (2001) 135–136.

[14] M.J. Earl and K.R. Seddon, Pure Appl. Chem., 72 (2000) 1391–1398.
[15] M.A. Gess, R.P. Danner and M. Nagvekar, Thermodynamic Analysis of Vapour–Liquid Equilibria: Recommended Models and a Standard Database, American Institute of Chemical Engineers, New York, 1991.
[16] UOP, available from http://www.uop.com.
[17] G.W. Meindersma, A.J.G. Podt and A.B. de Haan, Fuel Proc. Technol., 87 (2005) 59–70.
[18] Process Economics Program Report 30: BTX, Aromatics, available from http://www.sriconsulting.com/PEP/Public/Reports/Phase_II/RP030/.
[19] P.J. Bailes, C. Hanson and M.A. Hughes, Chem. Eng., 19 (January) (1976) 86–100.
[20] R.E. Treybal, Liquid Extraction, 2nd edn., McGraw-Hill, New York, 1963.
[21] C. Hanson (ed.), Recent Advances in Liquid–Liquid Extraction, Pergamon Press, Oxford, 1971.
[22] T.M. Letcher and N. Deenadayalu, J. Chem. Eng. Data, 44 (1999) 1178–1182.
[23] J.A. Riddick, W.B. Bunger and T.K. Sakano, Organic Solvents: Physical Properties and Methods of Purification, 4th edn., Wiley, New York, 1986.
[24] C.F. Poole, J. Chromatogr. A, 1037 (2004) 49–82.
[25] J.M. Sorensen, T. Magnussen, P. Rasmussen and A. Fredenslund, Fluid Phase Equilib., 3(1) (1979) 47–82.
[26] T. Magnussen, J.M. Sorensen, P. Rasmussen and A. Fredenslund, Fluid Phase Equilib., 4(1–2) (1979) 151–163.
[27] J.G. Huddleston, H.D. Willauer, R.P. Swatloski, A.E. Visser and R.D. Rogers, J. Chem. Soc. Chem. Commun., (1998) 1765.
[28] S. Carda-Broch, A. Berthod and D.W. Armstrong, Anal. Bioanal. Chem., 375 (2003) 191.
[29] A. Berthod and S. Carda-Broch, J. Liquid Chromatogr. Relat. Technol., 26 (2003) 1493.
[30] M.H. Abraham, A.M. Zissimos, J.G. Huddleston, H.D. Willauer, R.D. Rogers and W.E. Acree, Jr., Ind. Eng. Chem. Res., 42 (2003) 413–418.
[31] J. Liu, Y. Chi, J. Peng, G. Jiang and J.A. Jonsson, J. Chem. Eng. Data, 49 (2004) 1422–1424.
[32] U. Domanska, E. Bogel-Lukasik and R. Bogel-Lukasik, Chem. Eur. J., 9 (2003) 3033–3049.
[33] U. Domanska, E. Bogel-Lukasik and R. Bogel-Lukasik, J. Phys. Chem. B, 107 (2003) 1858–1863.
[34] U. Domanska and E. Bogel-Lukasik, Ind. Eng. Chem. Res., 42 (2003) 6986–6992.
[35] U. Domanska and A. Marciniak, J. Chem. Eng. Data, 48 (451) (2003) 451–456.
[36] U. Domanska and A. Marciniak, J. Phys. Chem. B, 108 (2004) 2376–2382.
[37] U. Domanska and L. Mazurowska, Fluid Phase Equilib., 221 (2004) 73–82.
[38] U. Domanska, Pure Appl. Chem., 77 (2005) 543–557.
[39] U. Domanska, A. Pobudkowska and F. Eckert, J. Chem. Therm., 38(6) (2005) 685–695.
[40] J.M. Crosthwaite, S.N.V.K. Aki, E.J. Maginn and J.F. Brennecke, Fluid Phase Equilib., 228–229 (2005) 303–309.
[41] A. Heintz, J.K. Lehmann and C. Wertz, J. Chem. Eng. Data, 48 (2003) 472–474.

[42] K.N. Marsh, A. Deev, A.C.-T. Wu, E. Tran and A. Klamt, Korean J. Chem. Eng., 19(3) (2003) 67–76.
[43] T.M. Letcher, G.G. Redhi, S.E. Radloff and U. Domanska, J. Chem. Eng. Data, 41 (1996) 634–638.
[44] T.M. Letcher and P.K. Naicker, J. Chem. Eng. Data, 43 (1998) 1034–1038.
[45] M.S. Selvan, M.D. Mckinley, R.H. Dubois and J.L Atwood, J. Chem. Eng. Data, 45 (2000) 841–845.
[46] T.M. Letcher and N. Deenadayalu, J. Chem. Therm., 35 (2003) 67–76.
[47] T.M. Letcher and P. Reddy, J. Chem. Therm., 37 (2005) 415–421.
[48] N. Deenadayalu, K.C. Ngcongo, T.M. Letcher and D. Ramjugernath, J. Chem. Eng. Data, 51(3) (2006) 988–991.
[49] J. Gmehling and M. Krummen, German Patent Application, DE 101 54 052 A1, 10-07-2003 (2003).
[50] J.F. Brennecke and E.J. Maginn, AIChE J., 47(11) (2001) 2384–2389.
[51] C.G. Hanke, A. Johansson, J.B. Harper and R.M. Lynden-Bell, Chem. Phys. Lett., 374 (2003) 85–90.
[52] R.P. Swatloski, J.D. Holbrey and R.D. Rogers, Green Chem., 5(4) (2003) 361–363.
[53] C. Jork, M. Seiler, Y.A. Beste and W. Arlt, J. Chem. Eng. Data, 49 (2004) 852–857.
[54] T.M. Letcher and P. Reddy, Fluid Phase Equilib., 219 (2004) 107–112.
[55] T.M. Letcher, N. Deenadayalu, B. Soko, D. Ramjugernath and P.K. Naicker, J. Chem. Eng. Data, 48(4) (2003) 904–907.
[56] A. Arce, O. Rodriguez and A. Soto, J. Chem. Eng. Data, 49 (2004) 514–517.
[57] A. Arce, O. Rodriguez and A. Soto, Ind. Eng. Chem. Res., 43 (2004) 8323–8327.
[58] A. Arce, H. Rodriguez and A. Soto, Chem. Eng. J., 115 (2006) 219–223.
[59] Z. Lei, W. Arlt and P. Wasserscheid, Fluid Phase Equilib., 241 (2006) 290–299.
[60] D.M. Trampe and C.A. Eckert, J. Chem. Eng. Data, 35 (1990) 156.
[61] A. Hussam and P.W. Carr, Anal. Chem., 57 (1985) 793.
[62] P. Alessi, M. Fermeglia and S.I. Sandler, J. Chem. Eng. Data, 37 (1992) 484.
[63] J.C. Leroi, J.C. Masson, H. Renon, J.C. Fabries and H. Sannier, Ind. Eng. Chem. Proc. Des. Dev., 16 (1977) 139–144.
[64] T.M. Letcher, in "Activity Coefficients at Infinite Dilution from Gas–Liquid Chromatography", (M.L. McGlashan, ed.), pp. 46–70, Chemical Thermodynamics, Vol. II, Specialist Periodical Reports, The Chemical Society, London, 1978.
[65] J.R. Conder and C.L. Young, Physicochemical Measurement by Gas Chromatography, Wiley, Chichester, 1979.
[66] T.M. Letcher and P.G. Whitehead, J. Chem. Thermodyn., 29 (1997) 1261–1268.
[67] T.M. Letcher and W.C. Moollan, J. Chem. Thermodyn., 27 (1995) 867–872.
[68] T.M. Letcher, B. Soko, P. Reddy and N. Deenadayalu, J. Chem. Eng. Data, 48 (2003) 1587–1590.
[69] T.M. Letcher, B. Soko, D. Ramjugernath, N. Deenadayalu, A. Nevines and P.K. Naicker, J. Chem. Eng. Data, 48 (2003) 708–711.
[70] M. Krummen, P. Wasserscheid and J. Gmehling, J. Chem. Eng. Data, 47 (2002) 1411–1417.
[71] A. Heintz, D.V. Kulikov and S.P. Verevkin, J. Chem. Thermodyn., 34 (2002) 1341–1347.

[72] T.M. Letcher, A. Marciniak, M. Marciniak and U. Domanska, J. Chem. Thermodyn., 37 (2005) 1327–1331.
[73] R. Kato and J. Gmehling, J. Chem. Thermodyn., 37 (2006) 603–620.
[74] W. David, T.M. Letcher, D. Ramjugernath and J.D. Raal, J. Chem. Thermodyn., 35 (2003) 1335–1341.
[75] F. Mutelet and J.N. Jaubert, J. Chromatogr. A, 1102 (2006) 256–267.
[76] N. Deenadayalu, S.H. Thango, T.M. Letcher and D. Ramjugernath, J. Chem. Thermodyn., 38(5) (2005) 542–546.
[77] R. Kato and J. Gmehling, Fluid Phase Equilib., 226 (2004) 37–44.
[78] A. Heintz, D.V. Kulikov and S.P. Verevkin, J. Chem. Eng. Data, 46 (2001) 1526–1529.
[79] T.M. Letcher and P. Reddy, Fluid Phase Equilib., 235 (2005) 11–17.
[80] T.M. Letcher and P. Reddy, Fluid Phase Equilib., in press (2007).
[81] A. Soto, A. Arce and M.K. Khoshkbarchi, Sep. Purif. Technol., 44 (2005) 242–246.
[82] A.G. Fadeev and M.M. Meagher, Chem. Commun., (2001) 295.

Thermodynamics, Solubility and Environmental Issues
T.M. Letcher (editor)

Chapter 6

Environmental and Solubility Issues Related to Novel Corrosion Control

W.J. van Ooij and P. Puomi

Department of Chemical and Materials Engineering, University of Cincinnati, Cincinnati, OH 45221-0012, USA

1. INTRODUCTION

Steels, galvanized steels and aluminum alloys, which are used in constructions, vehicles and appliances are usually painted with a protective polymer coating for decorative purposes and for extending the life-cycle of the coated metal. Without the protective coating these metals would corrode due to electrochemical dissolution of the metal. However, the paint on the metal only retards the initiation of corrosion and still, even if metals are painted, the costs of corrosion of metals are surprisingly high. The total direct cost of corrosion in USA is determined to be around $280 billion per year, which is ~3% of the US gross domestic product (GDP) [1]. The onset and rate of corrosion in a metal depends on the type of metal and the environment surrounding the metal. Some metals produce porous corrosion products, through which corrosion occurs fast and if the metal has been painted, the corrosion products may lift and delaminate the paint near scratches and the corrosion spreads. Other metals may corrode slower due to, e.g., dense oxide layers protecting the metal and if the metal eventually starts to corrode it may proceed slowly due to thick and dense corrosion products formed on the metal.

Metals and their alloys can corrode in many ways. The forms of corrosion are numerous including galvanic, filiform, cell, crevice, pitting, intergranular, high temperature, uniform, erosion, hydrogen embrittlement, microbial corrosion and stress corrosion cracking. Any form of wet corrosion involves two electrochemical reactions taking place simultaneously on the metal. In the anodic half reaction metal ions are dissolved and in the cathodic half reaction the electrons produced in the first reaction are consumed by hydrogen ions forming hydrogen or are combined with oxygen to form water [2].

$$Fe \rightarrow Fe^{2+} + 2e^{-} \quad \text{anodic reaction} \tag{1}$$

$$2H^{+} + 2e^{-} \rightarrow H_2 \quad \text{cathodic reaction} \tag{2}$$

$$O_2 + 4H^{+} + 4e^{-} \rightarrow H_2O \quad \text{cathodic reaction} \tag{3}$$

The corrosion reaction between the anode and the cathode proceeds, if there is a solution present enabling the transfer of metal ions of the corrosion reaction. The transfer of electrons between the anodic and cathodic areas occurs in the metal. The reaction can, however, be stopped or inhibited, if either the anode or the cathode is passivated or the solution, enabling the corrosive reaction, is removed [3].

Organic paint systems on metals prevent or retard corrosion of metals mainly by preventing water, electrolytes and oxygen to permeate to the metal/paint interface, i.e., they mainly function as physical barriers on metals. The bonding of the paint to the metal is essential in order to prevent premature delamination of the paint layer from the metal [4, 5]. Thin pretreatment layers are usually deposited onto metals before painting in order to improve the adhesion between the paint and the metal. Before pretreatment and painting the metal is thoroughly cleaned in order to remove oil, dust and other impurities that could adversely affect the adhesion of the paint to the metal. Novel corrosion control methods involve treatments that combine several methods of corrosion inhibition. Formerly chromates were used in the pretreatment and primer layers to inhibit corrosion of painted metals. Nowadays, chromate is being replaced by almost equally active non-chromate inhibitors [6]. This has been accomplished by introducing new types of water-borne polymer paint coatings from which the chromate-free (Cr-free) inhibitors can leach out on-demand as from the conventional chromate-containing (Cr-containing) systems [7–11]. The novel methods have also included the improvement of adhesion between the metal and the primer.

This chapter will discuss the thickness and the chemistry of the layers of a conventional three layer paint system, consisting of a pretreatment, primer and topcoat layer. Description of a new intermediate system for corrosion control will be given along with a future system. First however, the corrosion characteristics of a few industrially important metals will be discussed before describing the novel silane-based pretreatments and primers of the new and future systems.

2. CORROSION OF INDUSTRIALLY IMPORTANT METALS

Steels, galvanized steels and aluminum alloys are industrially important metals that are produced in large quantities. Of these materials the corrosion protection of steel is most challenging, even if iron is more noble than zinc or aluminum [2]. Fig. 1 shows a schematic of the corrosion process on iron along with the possible half-cell reactions depending on the type of environment [12].

The hydrated ferric oxide ($Fe_2O_3 \cdot 3H_2O$) that is formed in the reactions between iron, water and oxygen is the red rust that is usually referred with the generic term, rust. Pretreated and painted cold-rolled steel (CRS) when scratched is particularly susceptible for red rust bleeding especially in acidic corrosive conditions [12]. Therefore, steel is often galvanized to protect the steel with a sacrificial zinc alloy coating, which will corrode instead of steel, if the galvanized steel is cut or scratched [13].

The reactivity of metals in different conditions, can be anticipated to some degree by studying Pourbaix diagrams, which provide information on

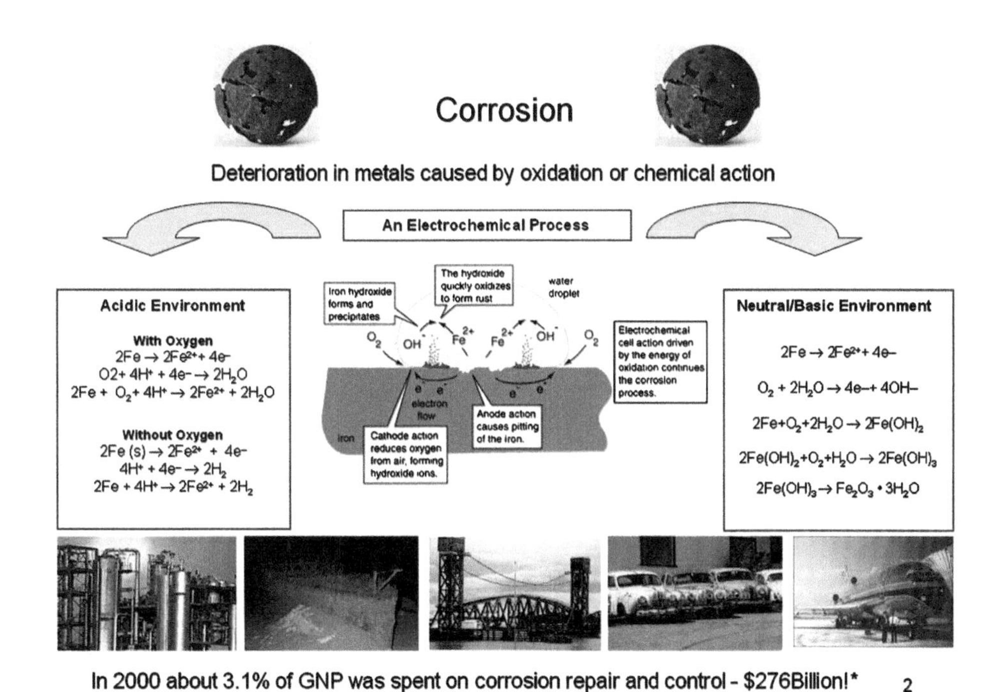

Fig. 1. Chemical corrosion reactions on iron, in aqueous environments [12].

the stability of the oxides/hydroxides on metals. These diagrams illustrate areas of conditions where the metal is passive, corrosive or immune [14]. These regions are, however, only indications, actual corrosion rates cannot be derived from the diagrams. From the Pourbaix diagram of aluminum, shown in Fig. 2a, it can be concluded that aluminum is an amphoteric metal for which the protective oxide film dissolves at low (below pH 4) and high pH (above pH 8–9). Also for aluminum there is a large potential difference (driving force) between the lines representing the cathodic and anodic half-cell reactions. However, aluminum is an excellent example of the fact that the corrosion rate is relatively low due to kinetic limitations, despite the large driving force for the corrosion reactions [2].

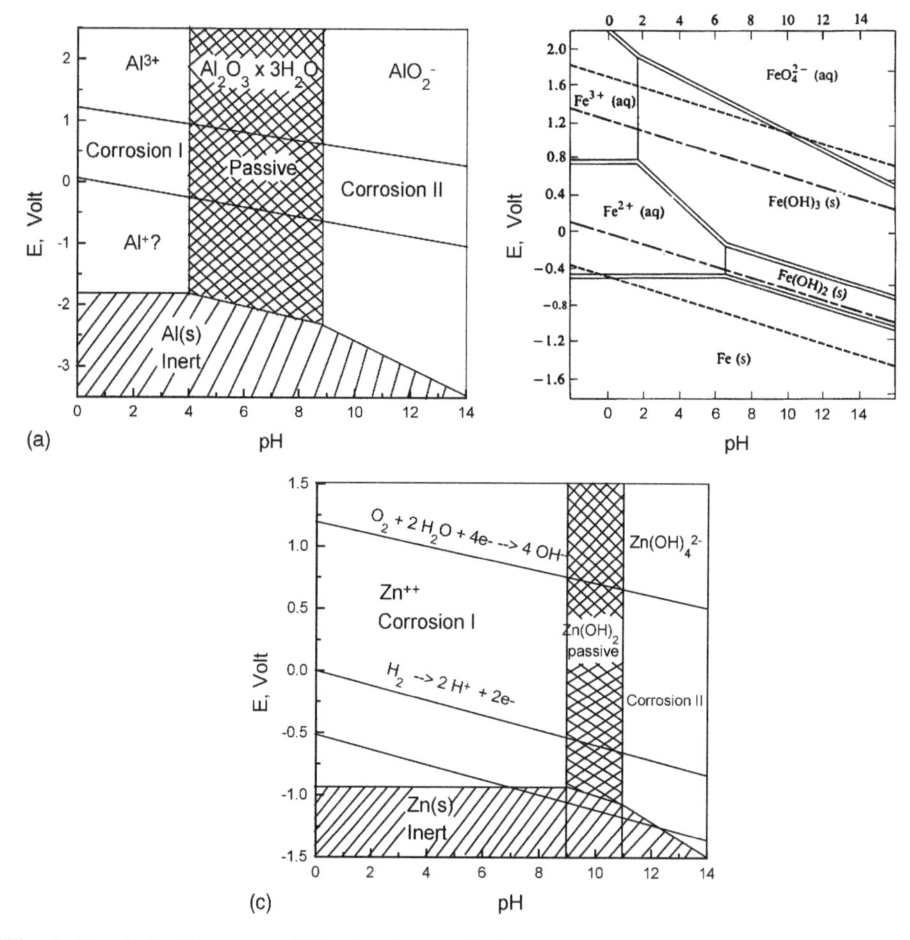

Fig. 2. Pourbaix diagrams of (a) aluminum; (b) iron; and (c) zinc [2, 14].

Iron is similar to aluminum in that a protective oxide forms in nearly neutral solutions. However, for iron the field of oxide stability is substantially greater at elevated pH, and iron is far more resistant to alkaline solutions compared with aluminum. Contributing to the overall resistance of iron, is the generally more noble half-cell electrode potential for the anodic dissolution reactions which lower the driving force for corrosion reactions. It is, however, apparent from Fig. 2b that this resistance disappears in more acidic solutions [2].

For zinc the passive region is even narrower than for aluminum. Zinc has soluble forms of compounds under pH 9 and above pH 11 (Fig. 2c). The figure also shows that in moderately alkaline solutions (pH 9–11) $Zn(OH)_2$ will precipitate, but in a strongly alkaline solution, the solid hydroxide will dissolve as zincate ions, $Zn(OH)_4^{2-}$.

It is well known that aluminum as such is fairly passive, because a very dense and uniform aluminum oxide Al_2O_3 layer is formed onto the metal to protect the metal from corrosion. Highly ductile light weight aluminum alloys that are passed through specific heat treatments can, however, make aluminum susceptible to corrosion. These materials may contain alloying elements such as magnesium and/or copper, which alter and complicate the corrosion behavior of aluminum. Typical forms of corrosion for the alloys are localized and pit corrosion. Due to the dense structure of the aluminum oxide layer, the corrosion rate of aluminum alloys is, however, substantially slower compared with corrosion/dissolution of CRS or HDG steel [15].

Regular HDG steel coatings corrode by dissolving and re-precipitating as zinc-oxide crystals over the surface. Thus, the surface attains an irregular appearance. The corrosion products of zinc are $Zn(OH)_2$, ZnO (dehydration of $Zn(OH)_2$) and basic salts formed as a result of compounding of $Zn(OH)_2$ with $ZnCl_2$ or $ZnCO_3$. The corrosion products are porous and allow free access of corrosive electrolytes to the zinc coating [18]. This mechanism usually creates preferential pathways through areas of higher porosity and causes locally accelerated corrosion. Therefore, the rate of corrosion of HDG steel has been established to be essentially linear [16].

3. THE LAYERS PROTECTING THE BASE METALS

3.1. The Protective Paint Systems

As mentioned earlier, industrially important metals are often protected by a Cr-containing pretreatment, a Cr-containing primer and a topcoat, as shown in Fig. 3. It is, however, well known that the chromate in the pretreatment

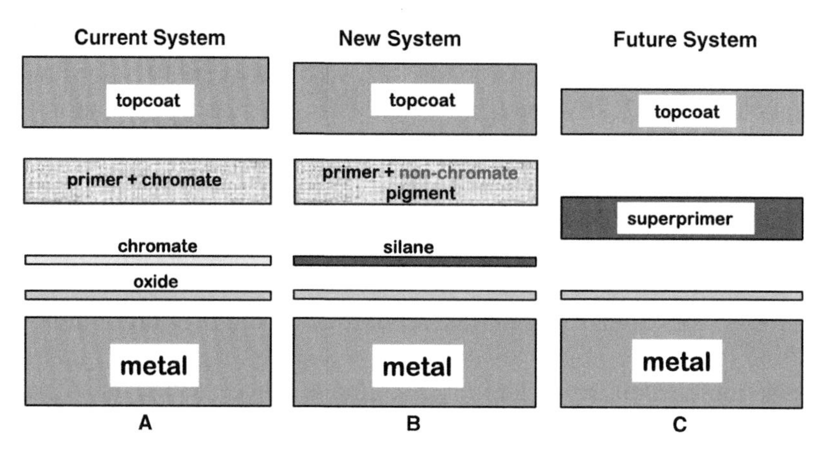

Fig. 3. Concept of the superprimer system [7].

layer as well as in the primer layer is highly toxic and environmentally undesirable [17, 18]. Therefore, possibilities to replace the chromate in both layers are being explored all over the world.

3.2. The Chromate Pretreatment Layer

The chromate pretreatment layer, which is also called the chromate conversion coating (CCC) varies in thickness depending on the chemistry of the process and the application method used. The CCC layer is, however, usually not thicker than a few microns, which in coating weight is somewhere between 5 and 25 mg/m^2, expressed as Cr [19]. This CCC layer improves the adhesion between the metal and the primer, it aids in the protection of scratches and defects and it also protects cut edges of the metal to some extent [20]. The hexavalent chromate in the CCC layer is known for its low solubility and the self-healing effect, which means that it only leaches out on demand when the base metal has been scratched [21].

3.3. The Primer Layer Containing the Corrosion Protective Inhibitors

The primer paint layer is, however, much thicker than the CCC layer. For instance in aerospace applications the primer coating is usually 25 μm thick, in coil coatings it is around 8 or 20 μm depending on the type of coating [22–24]. Still in many applications the primer coating is loaded with chromate, typically strontium chromate, because its solubility is optimal in primer coatings [6].

Naturally, when comparing the film thicknesses of the CCC and primer layer it is obvious that the replacement of the chromate in the paint layer has a more substantial effect on reducing the amount of chromate that may leach/dissolve out into the environment than replacing the pretreatment layer. The first step in improving the conventional three layer paint system shown in scheme A was to replace the CCC layer with a Cr-free pretreatment and the Cr-containing primer with a primer loaded with Cr-free inhibitors. This resulted in scheme B [25, 26]. In recent years attempts have been made to incorporate the silane into the primer, which results in 2-in-1 primers, where the pretreatment layer is built in the primer paint layer. This idea was first introduced by van Ooij et al., who have investigated these types of silane-containing Cr-free primers on aluminum alloys, HDG steel and CRS [7–11, 27–30].

The function of the duplex paint coatings on all systems shown in Fig. 3 is to protect the metal by providing a good barrier against, e.g., water, electrolytes and pollutants. It is well known that in typical duplex systems, the most important layer for active corrosion protection of the metal is the primer layer [20, 23]. When the primer is loaded with chromates (scheme A) then the chromate in the primer coating protects the metal in the same way as in the thinner CCC layer. Due to its low and selective solubility, it leaches out from the primer only on-demand, when the painted metal gets scratched or torn. The chromate CrO_4^{2-} ion is a powerful oxidizing agent. In order to protect the metal the hexavalent Cr in CrO_4^{2-} gets reduced to trivalent state and passive trivalent $Cr(OH)_3$ is formed on the metal where the corrosion started. Remarkably for chromates they work in the whole pH region providing both anodic as well as cathodic protection to metals [6, 31, 32].

In summary, it is the chromate in the primer that actively inhibits corrosion of the metal. Several Cr-free inhibitors have been investigated as replacements for chromates in organic coatings. These can be classified as inorganic, organic and hybrid inorganic–organic inhibitors. Inorganic inhibitors studied for paint applications include nitrites, phosphates, molybdates, metaborates, silicates and cyanamides, but actually very few of these possess inhibitor properties suitable for use in paints [6]. Inorganic inhibitors work by leaching and dissolving out from the coating and acting as anodic or cathodic inhibitors. Organic inhibitors such as azoles typically function by forming a film on the metal and providing protection by sealing the location of corrosion. Hybrid inhibitors possess inhibition action of both the types of inhibitors. These types of Cr-free inhibitors are incorporated into both solvent- and water-based primers, because simultaneously as legislation is pressing paint producers to discontinue the use of chromates, there is pressure to reduce the amount of volatile organic compounds (VOCs) in paints as well.

It is, however, not so easy to just replace the chromate in an existing paint with a Cr-free inhibitor, because none of the Cr-free inhibitors are active over the whole pH range as chromate is and they also do not possess similar optimal solubility properties as, e.g., strontium chromate does. Therefore, for the Cr-free primers to perform equally to Cr-containing primers they need to contain a mixture of Cr-free inhibitors to cover the entire pH-range and the paint layer matrix needs to be designed so that they are able to leach out on demand to protect the base metal [6].

The future system (scheme C) involves only two layers, the primer and the topcoat. The VOC problem is solved by dispersing the resin, the cross-linker, the silane and the Cr-free inhibitors into a water-based solution [7–11, 27–31]. The silane improves the adhesion of the primer to the metal, the Cr-free inhibitors incorporated in the primer protect the metal from corrosion and the functionality of the silane-containing polymer matrix ensures good adhesion to the subsequently applied topcoat [33].

3.4. The Topcoat

The topcoat which is usually ~50 μm thick in many paint systems, e.g., aerospace coatings as well as coil coatings, provides a decorative and durable exterior surface to the metal. In applications such as aerospace coatings the whole coating system including the topcoat needs to withstand large temperature variations (from −65 to +150°C) and also corrosive hydraulic fluids, which sets high demands on the topcoat as well as the entire coating system [34].

3.5. Silane Pretreatments

Numerous types of alternatives for the CCC have been proposed, some of these are based on silicates and/or silanes [35–48], acids of Ti and Zr [49–52], molybdates and permanganates [53–54] and rare earth metals such as Ce [55–58].

Silane-based pretreatments, based on trialkoxysilanes have been shown to be competitive replacements for CCCs [59]. They provide good corrosion protection to the metal substrate by improving adhesion between the primer and the substrate and imparting hydrophobicity to the metal surface. The silane on the metal is mainly in the form of siloxane Si-O-Si, which is an inorganic polymer network. The thickness of the siloxane layer is between 0.1 and 3 μm [38–41].

Trialkoxysilanes can be divided into two categories: mono-silanes and bis-silanes. Mono-silanes have the general formula: $R'\ (CH_2)_nSi(OR)_3$, where R' is an organo-functional group and R is an alkyl group and n is the number of CH_2 groups. R' represents an organo-functional group such as chlorine, primary or secondary amines, or vinyl. Typically the value of n is around 3, but individual moieties can vary. Both the type of R' the group and the value of n, have a strong influence on whether a particular monomer is water-soluble, but in general most of them are. As can be seen from the general formula, mono-silanes have only a single leg of alkoxy groups. Bis-silanes have the general formula: $(OR)_3Si(CH_2)_nR'(CH_2)_nSi(OR)_3$. They have, therefore, two legs of alkoxy groups which can hydrolyze in the presence of water to produce silanol groups $Si(OH)_3$ [38–41]. The silanol groups can further condense with each other to form siloxane, –Si-O-Si–. The first reaction is known as the hydrolyzation and the second as the condensation of silanes [38–41]. Because of the described difference in the structures of mono- and bis-silanes the latter result in more densely formed Si-O-Si networks than the former, which further result in good barrier properties of the coatings and therefore better corrosion resistance of the metal. Silanes used in pretreatments and superprimers are shown in Fig. 4 [7–11, 27–30].

On some metals, such as aluminum, the condensation of the silanol groups may occur with the metal substrate, which results in metallo-siloxane (–Me-O-Si–) on the interface. These metallo-siloxane bonds are, however, hydrolytically unstable, which means that the metallo-siloxane groups are in a reversible equilibrium with silanols and metal-hydroxide groups. Hence, to maintain the adhesion of the silane film to the metal and for adequate corrosion protection, the silane film should be made hydrophobic enough to prevent water permeation into the metallo-siloxane layer [59].

If the bis-silane has an organo-functional group, then it can bond chemically to the primer layer. If it does not contain a functional group then a mono-silane can be added to the solution to impart organic functionality. Thus, combinations of silanes (mixtures) usually perform better than films of silanes alone. Unfortunately, many of the bis-silanes are not water-soluble which limits their immediate industrial use to those silanes with water stabilizing functional groups such as amines. Regardless, certain optimal mixtures have been developed, clearly showing the potential of silane deposited films to be able to give equivalent performance to chromate underneath paint systems and even as stand alone passivation treatments [38–41].

For silane treatments under paint a silane solution concentration of 2% is typical and for bare corrosion protection purposes a solution concentration

bis-sulfur (bis-[3-(triethoxysilylpropyl)] tetrasulfide)

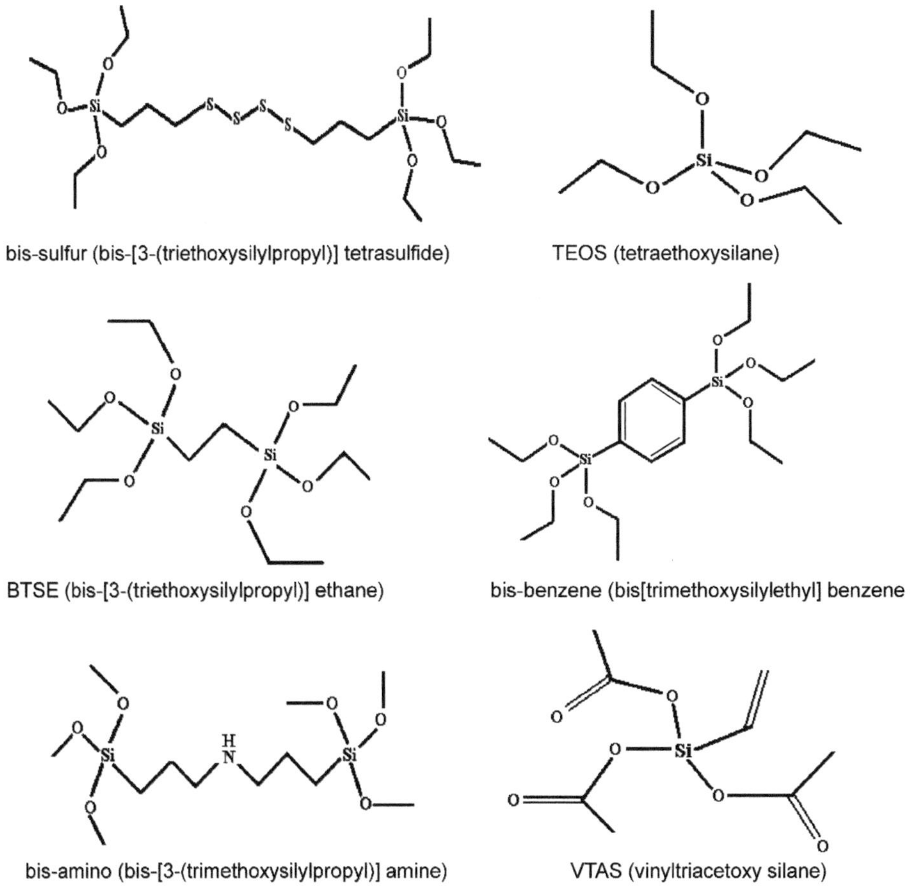

TEOS (tetraethoxysilane)

BTSE (bis-[3-(triethoxysilylpropyl)] ethane)

bis-benzene (bis[trimethoxysilylethyl] benzene

bis-amino (bis-[3-(trimethoxysilylpropyl)] amine)

VTAS (vinyltriacetoxy silane)

Fig. 4. Chemical structures of commonly used silanes in pretreatments and superprimers [7–11, 27–30].

of 5% is usually used [40–43]. For bare corrosion protection the hydrophobic silanes; bis-1,2-(triethoxysilyl) ethane (BTSE) and bis-[triethoxysilylpropyl] tetrasulfide (bis-sulfur) have already demonstrated their corrosion protectiveness for many metals [37–43]. Unfortunately, their hydrophobic nature requires a large amount of organic solvents such as ethanol or methanol in the preparation of the silane solutions.

A solvent-based silane system which has proven to give excellent corrosion resistance in un-painted state is the mixture of the bis-sulfur and bis-aminosilanes with the ratio 3:1. This system is able to protect many metals, for instance, aluminum, HDG steel and Mg alloys. Fig. 5 presents the scanned

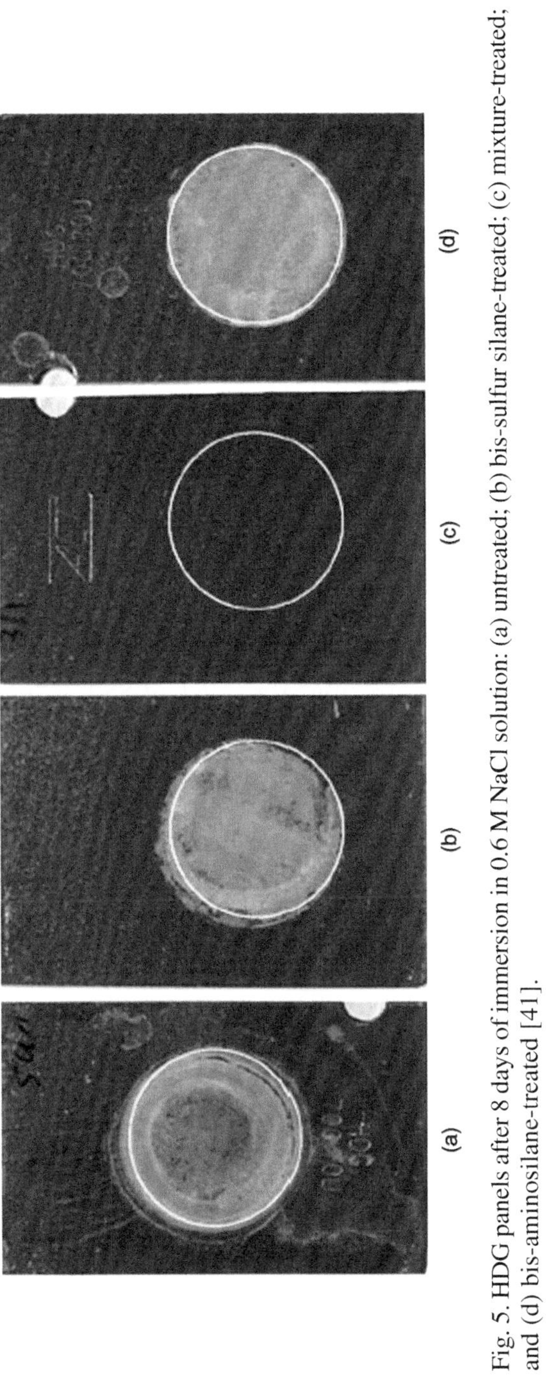

Fig. 5. HDG panels after 8 days of immersion in 0.6 M NaCl solution: (a) untreated; (b) bis-sulfur silane-treated; (c) mixture-treated; and (d) bis-aminosilane-treated [41].

images of silane-treated HDG panels after EIS measurements. It is clearly seen that no corrosion is shown on the mixture-treated HDG panel (Fig. 5c), indicating that the combination provides good corrosion protection for this metal.

One of the main objectives in recent years has been to develop water-soluble silane solution mixtures with corrosion protection properties as powerful as the systems based on ethanol–water solutions. In this respect a universal water-soluble silane system based on bis-amino and vinyltriacetoxy silanes was invented by Zhu and van Ooij [38]. They studied this mixture on several metals.

The bis-amino/VTAS mixture is, by itself, quite stable and hydrolyzes readily in the aqueous mixture. The mixture works well, because with the addition of a small amount of bis-aminosilane, the VTAS solution becomes less acidic and the condensation of SiOH can be effectively suppressed. A likely mechanism is that the secondary amine groups in the bis-aminosilane form a more stable hydrogen bond with silanol groups than the one between silanols themselves. As a result, condensation of silanols is prevented in the solution.

Fig. 6 displays the salt spray testing (SST) results of AA 6061-T6 treated with the bis-amino/VTAS mixture, as compared with untreated and chromated panels. It is seen that the bis-amino/VTAS mixture treated AA 6061-T6 surface does not show any sign of corrosion after 336 h of SST, while the chromated surface exhibits a certain degree of discoloration. The bare AA 6061-T6 surface on the other hand, has corroded heavily. Performance tests also showed

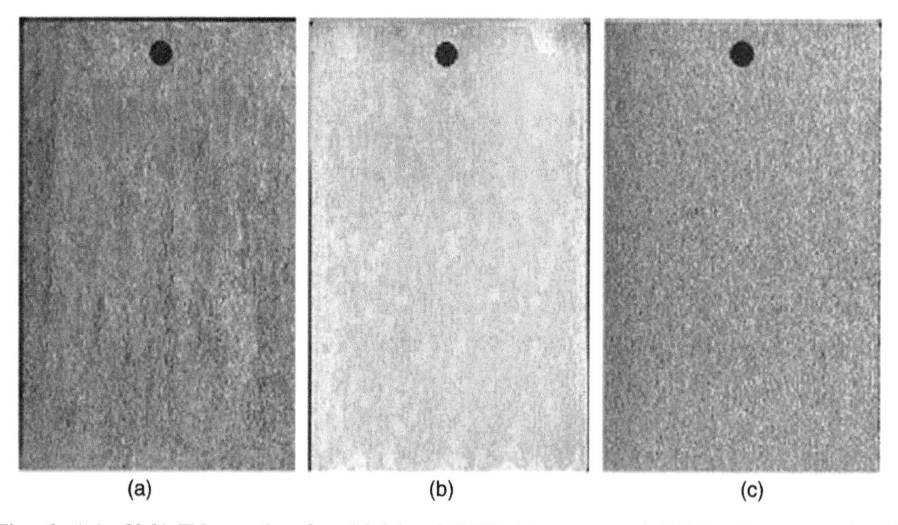

Fig. 6. AA 6061-T6 panels after 336 h of SST: (a) untreated (20 h of exposure); (b) chromated (Alodine-series); and (c) silane-treated (bis-aminosilane/VTAS = 1.5/1, 5%, pH 3.7) [38].

that the water-based bis-aminosilane/VTAS mixture provides comparable corrosion protection in the painted state, e.g., with polyurethane and polyester powder-paints on aluminum alloys, as compared with chromates [38].

4. SUPERPRIMERS ON METALS

4.1. The Concept of Superprimers

One major drawback to the widespread use of silane films as pretreatment or passivating treatments is that despite their hydrophobicity, eventually moisture reaches the metal–silane interface. The hydrolysis reaction that allowed the formation of Si-O-Si and Si-O-Me bonds is reversible, especially if the substrate's metal hydroxide is somewhat soluble [39–41]. This enables a mechanism by which the coating can be undermined. The use of bis-silanes result in higher concentration of both Si-O-Si as well as Si-O-Me bonds, which can mitigate this reversible effect to varying degrees [59]. The first step in making the silane films more robust was to incorporate nano-sized fillers and inhibitors into the films, which resulted in more protective hybrid silane films [60]. The next step included the idea of mixing silanes, resins and inhibitors. The advantage of incorporating resins into the silane films played the major improving role, because the resin makes the resulting film more hydrophobic and additionally the coating layer can be substantially increased compared with silane mixture films, which become brittle upon increasing the layer thickness above 3 μm. The resin-containing silane films can, however, be made to a thickness of 25 μm [7–11, 27–30].

The ingredients of a typical superprimer are:

- a binder based on a major resin or resins and minor binders;
- cross-linkers, e.g., amine adducts for epoxies;
- silane or silanes;
- non-chromate pigments and fillers;
- additives.

4.2. Major and Minor Binders

The superprimer formulation is usually based on an ionic or non-ionic aqueous resin dispersion. The solid content of the water-borne resin is often around 55 wt% and it may contain small amounts of organic solvents. Typical resins used in the primers are epoxies, acrylates and polyurethanes. The minor binder is incorporated in order to improve a particular coating property such as cold formability. Table 1 gives an overview of the family of superprimers based on different resin–cross-linker–silane–pigment combinations invented and investigated in our laboratory [7–11].

Table 1
The family of superprimers based on different resin–cross-linker–silane–pigment combinations [30]

Substrate	For AA 2024-T3 and AA 7075-T6	For AA 2024-T3	For AA 2024-T3	HDG steel
Resin system	Epoxy–acrylate	Novolac epoxy–polyurethane	Polyurethane	Two epoxies + polyurethane
Cross-linker	Isocyanate silane	Amine adduct	None	Amine adduct
Silane	Bis-sulfur	Bis-sulfur	BTSE	Bis-sulfur or bisbenzene
Typical pigment	Zinc phosphate	Zinc phosphate	Calcium zinc molybdate (CZM)	$NaVO_3$+ Corrostain (Ca, Zn, P, Si and O)

The epoxy used in the epoxy–acrylate system and in the epoxy-based system for HDG steel, is a typical DGEBA type epoxy formulated from epichlorohydrin and bisphenol A [11, 61]:

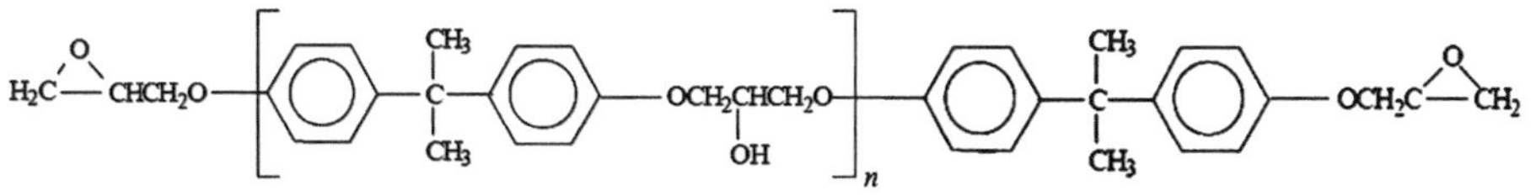

The acrylate in the epoxy–acrylate formulation is an anionic dispersion of a polyacrylate copolymer based on methylacrylate for which a small amount of the acrylate groups has been replaced by acrylic groups as follows [61]:

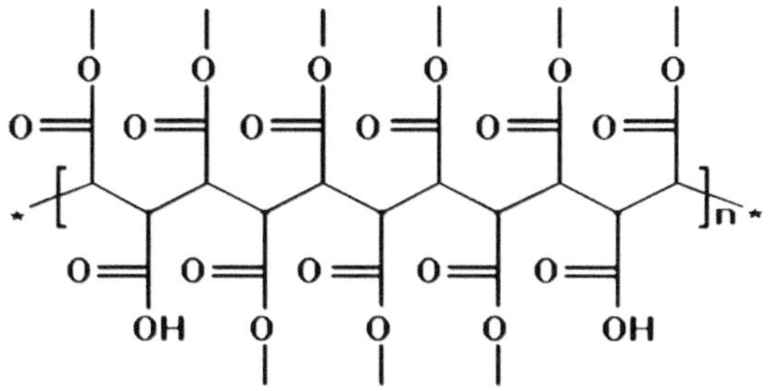

The bisphenol A novolac epoxy resin is a non-ionic aqueous dispersion of a polyfunctional aromatic epoxy resin. It contains reactive epoxide functionality and is intended for high performance applications which require maximum chemical and solvent resistance and/or elevated temperature service. This thixotropic dispersion contains no organic solvent and is completely water reducible. Upon evaporation of water, the novolac epoxy coalesces to form a clear, continuous, tacky film at ambient temperature, while in combination with a suitable cross-linking agent, it will form a clear, highly cross-linked, tough, chemical resistant film. The generic structure of a novolac epoxy is given below [62].

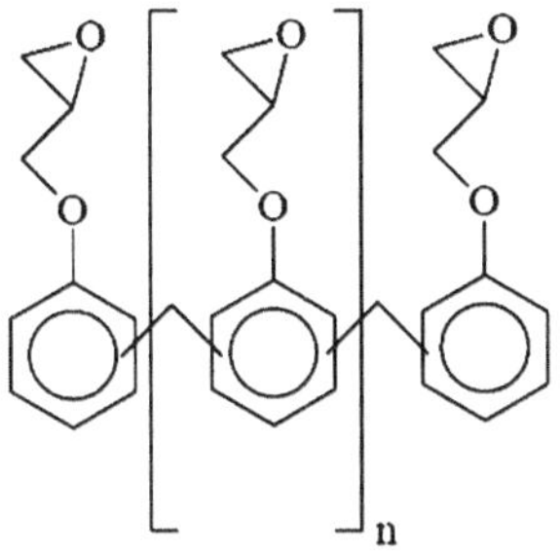

Polyurethane coatings are known for their flexibility, abrasion resistance, acid rain resistance, gloss retention and impact resistance. As can be seen from Table 1 polyurethanes are used as major and minor binders in superprimer formulations [11, 62, 63]. Polyurethanes are synthetized by reacting polyols with polyisocyanates. The structure of a simple linear polyurethane is shown below:

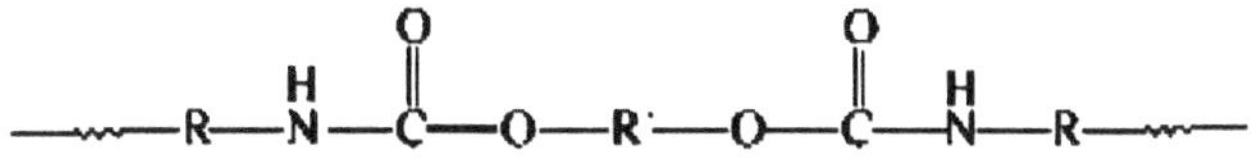

Polyurethane

4.3. Cross-Linkers

Some of the resins introduced may form a film of their own but it is preferable to react them with a hardener or curing/cross-linking agent in order to produce a cross-linked resin network that cures fairly fast in either room temperature (RT) or at elevated temperatures. For instance, the epoxy–acrylate system can be cross-linked with an isocyanate-based silane with a structure shown below [61]:

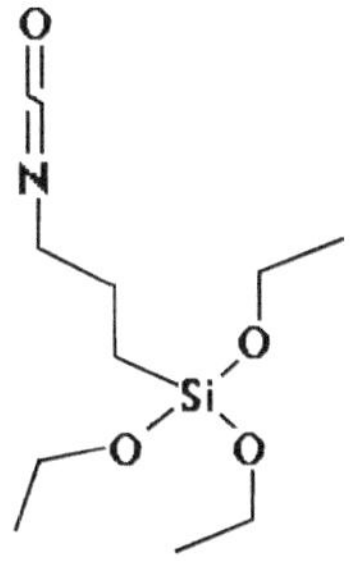

Polyamines, polyamides, organic acids, anhydrides, boron trifluoride and tertiary amine catalysts are among the more frequently used curing agents for epoxy resins. The curing agents open up the oxirane ring and act as bridges, binding the epoxy polymers into a dense three-dimensional network. Among the types of curing agents mentioned above, the first two types, i.e., polyamines and polyamides, are widely used in anticorrosive primers. The polyamines are suitable for superprimers [11].

Most of the amines, may irritate skin, possess a noxious odor or emit corrosive fumes. Hence they are mostly sold as modified variations with reduced vapor pressures and tendency to irritate. One approach to formulate amine cross-linkers with higher equivalent weight and lower toxic hazard is to make 'amine-adducts' by reacting standard liquid bisphenol A epoxy with excess of a multifunctional amine. A variety of amines can be used to provide adducts with a range of cure rates and pot lives. A simple epoxy–amine adduct is shown below. In two of the superprimer formulations water-borne epoxy–amine adducts are used as the curing agents [11, 62].

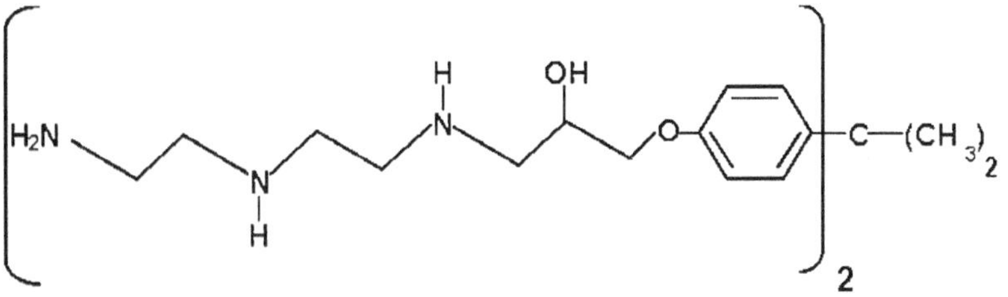

4.4. Silanes

Both mixtures of silanes as well as individual silanes have been investigated for use in superprimers. Usually, hydrophobic bis-silanes such as BTSE and bis-sulfur silane, shown in Fig. 4, work well in the primers.

4.5. Chromate-Free Inhibitors and Performance of Superprimer Coatings

Inorganic pigments have widely been investigated as inhibitors in organic coatings. The inorganic inhibitors are based on salts of $A_m^{n+} B_n^{m-}$, or basic salts of $A_m^{n+} B_{n-z}^{m-}$, where n, m = 2 or 3, and A^{n+} is Zn(II), Ca(II), Sr(II), Al(III), Ba(II), Mg(II), while B^{m-} is CrO_4^{2-}, PO_4^{3-}, MoO_4^{2-}, BO_2^-, $(SiO_3^{2-})_n$, $n>1$, HPO_3^{2-}, $P_3O_{10}^{5-}$, NCN^{2-}, CO_3^{2-} or various combinations of the same. Additionally, OH^- is a constituent of basic salts. It is the anion that is mainly responsible for the anodic or cathodic protection of the metal. The cations'

Table 2
Specific quality parameters for corrosion inhibitor pigments [6]

Quality parameters	Exclusive values
Solubility in water	<2 g/100 ml
pH of saturated solution	7–9.5
Specific gravity	1.5–5
Particle size distribution average	2–6 μm
Solubility in organic medium	Practically insoluble
Vapor pressure at 20°C	<1/10000 mmHg
Melting point	>100°C

contribution to inhibitive performances of pigments is secondary, however, they determine essential quality parameters such as solubility, hydrolysis pH and storage capacity for inhibitor species of pigments [6]. Specific quality parameters for corrosion inhibitor pigments are summarized in Table 2.

John Sinko also defined the following selected parameter values of pigment grade inhibitors: C_i, $B^{m-}/A_m^{n+}B_n^{m-}$ weight ratio; $g_{i_{sp}}$, specific gravity of pigment (g/cm^3); $c_{i_{sat}}$, solubility of pigment (mmol/l); $e_{i_{sp}}$, specific inhibitor capacity of pigment (mmol/cm^3) and I_i, inhibitor activity parameter.

The specific inhibitor capacity of a pigment is conveniently defined by

$$e_{i_{sp}} = C_i \frac{g_{i_{sp}}}{W_{B_i^-}} \tag{1}$$

where $W_{B_i^-}$ is the molar weight of B^{m-}. The inhibitor activity parameter is given by

$$I_i = n \frac{c_{i_{sat}}}{c_{i_{crt}}} \tag{2}$$

where $c_{i_{crt}}$ is the critical concentration of B^{m-}, i.e., the minimum concentration of a distinct inhibitor species necessary to maintain passivity on a metal exposed to aqueous environment. The inhibitor activity parameter should be $1 < I_i < 100$ in order for the pigment inhibitor to be suitable for use in organic coatings. Overall the parameters listed above should be close to the values of strontium chromate in order for the Cr-free pigment to be active as an inhibitor in paints. Sinko listed these parameters for some phosphates, orthophosphates, molybdates, metaborates, silicates, cyanamides and nitrites. Only two namely $ZnMoO_4$ and $Ba(BO_2)_2 \cdot H_2O$ came close to strontium chromate. For the others the inhibitor activity parameter was substantially lower, except for $NaNO_2$, for which it was extremely high (83000) [6]. Conclusively, none of the Cr-free pigments have the unique features that the chromate anion especially combined with the strontium cation possesses. It is thus understandable that in order to try to come even close to the protective action of $SrCrO_4$, mixtures of Cr-free inorganic pigments are used in anticorrosive primers.

However, since little can be done to modify Cr-free anions and mixtures thereof, another approach is to modify the chemistry, structure and properties of the organic coating. The first step is to incorporate the Cr-free pigments into a water-borne primer paint coating that could be expected to be more hydrophilic in nature compared with organic coatings deposited from solvent-based formulations. Another step is to further change the chemistry of the polymer coating by incorporating silane coupling agents into the resin matrix. This has been our approach, to modify the resin matrix with silanes and to deposit the networks from water-dispersed formulations. The siloxane in the resin matrix is able to play with the water penetrating into the coating and to adjust the water content in such a way that an optimal protective electrolytic environment is maintained in the coating and at the coating/ metal interface. In this environment Cr-free pigments may act as built in reservoirs of inhibitor species, even if they might not have suitable properties to act as active inhibitors in traditional organic coatings made from solvent-borne paints. Fig. 7 shows ASTM B-117 salt spray test results of superprimer coatings containing Cr-free pigments on aluminum alloys and HDG steel [30].

The systems shown in Fig. 7 are compared with Cr-containing control systems. Optimized coating systems all performed well over 2000 h in the test, which is usually a typical requirement in the industry. The polyurethane-based system on AA 6061 even survived 4000 h in ASTM B-117 (Fig. 7d). This gives ample proof of that the modification of the resin matrix might be a key factor in order to make the Cr-free pigments work on metals in a similar way as the strontium chromate works in traditional coatings.

Organic inhibitors, ion exchanged inhibitors and coated inhibitors such as plasma-polymerized inhibitors have also been investigated as pigments in silane pretreatments and/or superprimers. More on this subject can be found in Refs. [8, 61–63].

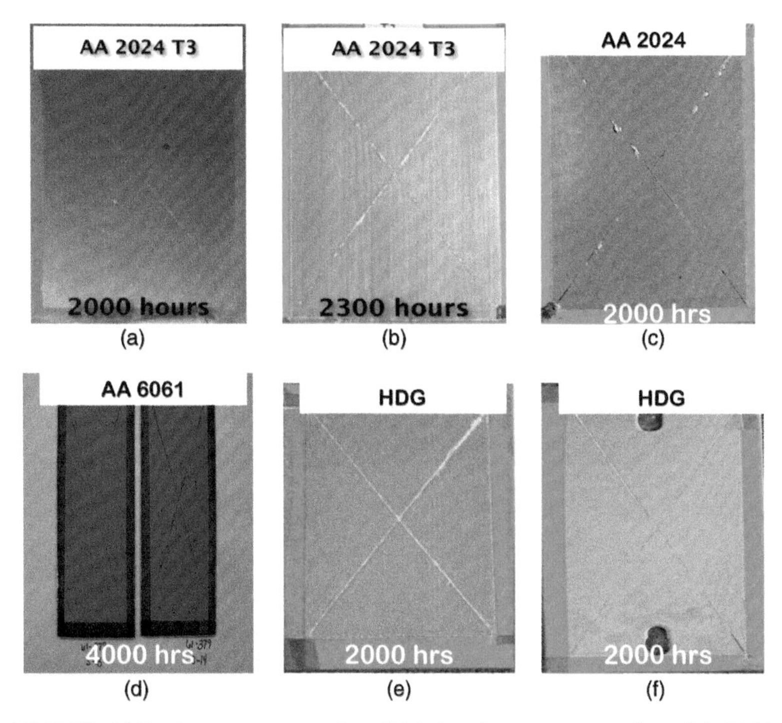

Fig. 7. ASTM B-117 salt spray test results of (a) the chromate control on AA 2024-T3; (b) the acrylate–epoxy based system on AA 2024-T3; (c) the novolac epoxy–polyurethane based coating on AA 2024-T3; (d) the polyurethane-based coating on AA 6061; (e) the epoxy-based system on HDG; and (f) the chromate control on HDG [30].

4.6. The Corrosion Protective Mechanism of Superprimers Including the Pigments

The superprimers developed in our laboratory have been characterized with various techniques such as Fourier transform infrared (FTIR) spectroscopy, nuclear magnetic resonance (NMR) spectroscopy, X-ray reflection methods, water and electrolyte uptake measurements, scanning electron microscope (SEM) combined with energy dispersive X-ray (EDX) analysis and time-of-flight secondary ion mass spectroscopy (TOF-SIMS). The characterization results have revealed how the coating ingredients react with each other and how the resulting coating on the metal protects the metal from corrosion [7–11, 27–30, 61–64].

The characterization results have shown that the coating components react with each other forming an interpenetrating network (IPN). The siloxane formed in the primer especially protects the metal–primer interface as it

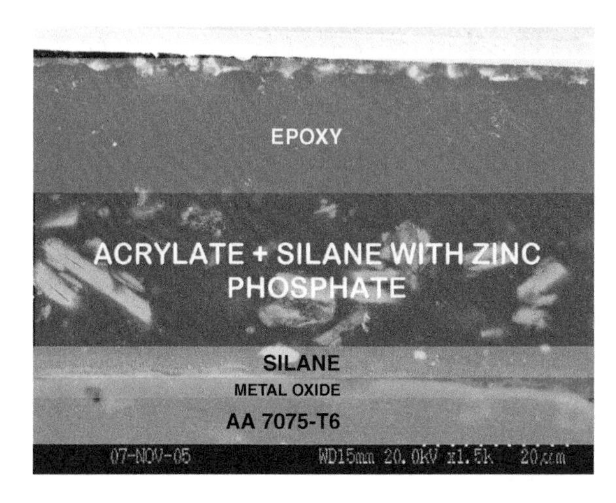

Fig. 8. The superprimer on AA 7075-T6 consists of three layers, which are self-assembled upon drying of the coating [30].

deposits as a cross-linked layer in this interface. The silane also assures good adhesion both to the substrate and the topcoat of the system [30].

The superprimer network is fairly hydrophilic. X-ray reflection results of silane–resin films have shown that these siloxane-containing primers actually contain small voids that attract water into the coating. As the water penetrates the voids, it does not make the coating swell [64]. It only allows the less soluble Cr-free pigments to move in the voids and leach out on-demand to protect the metal from corrosion. This on-demand protection behavior has been studied by exposing scribed panels to corrosive environments and subsequently the surfaces and the cross-sections of the samples have been examined by SEM/EDX for residues of pigments and corrosion products [7–11, 27–30, 61–63].

Some of the superprimers are more or less homogeneous, which means that the formulation ingredients have homogeneously reacted with each other and are evenly distributed in the coating [11, 62]. The epoxy–acrylate coating forms, however, a self-assembled three-layer coating, as shown in Fig. 8. The mechanism by which this coating system protects the aluminum alloy is illustrated in Fig. 9 [30].

The layered structure in Fig. 8 consists of a silane-rich layer at the metal/primer interface, a hydrophilic acrylate–silane–pigment layer in the center and a hydrophobic epoxy-rich layer on top. As illustrated in Fig. 9 the zinc

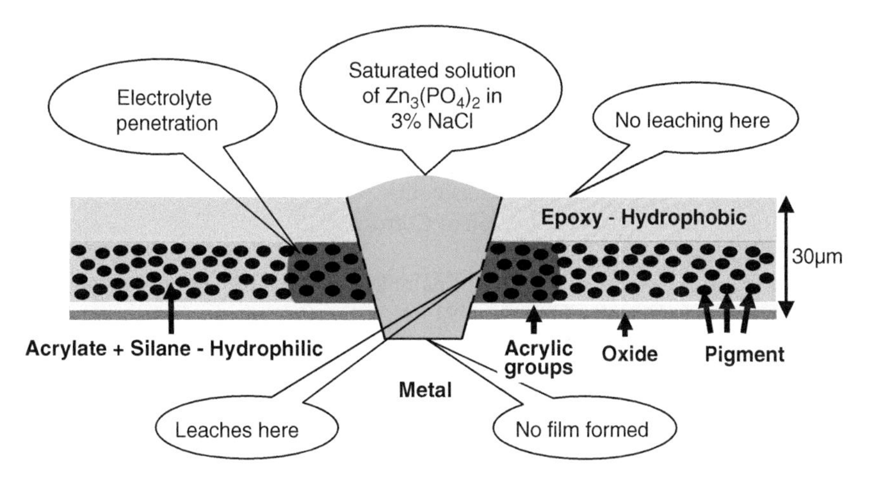

Fig. 9. Principle of the corrosion inhibiting mechanism of the epoxy–acrylate–silane super-primer containing the zinc phosphate pigments, which protect the metal on-demand [30].

phosphate can only leach out from the hydrophilic acrylate–silane–zinc phosphate layer, creating a reservoir of saturated zinc phosphate in the salt solution while the hydrophobic epoxy layer protects the acrylate-containing layer and thereby the remainder of the coated metal [30].

5. SUMMARY/CONCLUSIONS

It has been shown here that the hexavalent chromium can already be eliminated from the pretreatment step of several metals. In the interface between the metal and the primer the Cr^{6+} mainly improves adhesion and can, therefore, be replaced with silanes and other Cr-free replacement options.

The chromate pigment in the primer is more difficult to replace, but there is a new class of primers based on silane technology that have been tailored to boost the performance of less soluble Cr-free pigments like zinc phosphate. The chemical structure of these primers is unique, as they contain molecular voids, which enable lateral and vertical movement of pigments in the coating when the coating is exposed to corrosive conditions. This future technology also simplifies the coating process as the pretreatment step can be eliminated by incorporating the silane into the primer. The steps of the future process consist of cleaning, priming and topcoating without the need of environmentally hazardous pigments or VOCs.

REFERENCES

[1] Report by CC Technologies Laboratories, Inc. to Federal Highway Administration (FHWA), Office of Infrastructure Research and Development, Corrosion Costs and Preventive Strategies in the United States, Report FHWA-RD-01-156, September 2001. Available at http://corrosioncost.com/.
[2] D.A. Jones, Principles and Prevention of Corrosion, Prentice-Hall, Inc., Upper Saddle River, NJ, 1996.
[3] W.J. van Ooij and A. Sabata, Effect of paint adhesion on the underfilm corrosion of painted precoated steels, Corrosion'91 (NACE), Cincinnati, OH, M-15, March, Paper No. 517, pp. 1–17, Houston, TX, 1991.
[4] J.F.H. Eijnsbergen, Duplex Systems. Hot-Dip Galvanizing Plus Painting, Elsevier, Amsterdam, 1994.
[5] S. Maeda, Prog. Org. Coat., 28 (1996) 227.
[6] J. Sinko, Prog. Org. Coat., 42 (2001) 267.
[7] W.J. van Ooij, A. Seth, T. Mugada, G. Pan and D.W. Schaefer, A Novel Self-Priming Coating for Corrosion Protection, Proceedings of the 3rd International Surface Engineering Congress, Orlando, FL, August 2–4, 2004 (on CD).
[8] A. Seth and W.J. van Ooij, J. Mat. Eng. Perf., 13 (2004) 292.
[9] K. Suryanarayanan, M.S. Thesis, University of Cincinnati, Chemical and Materials Engineering, 2006.
[10] T. Mugada, M.S. Thesis, University of Cincinnati, Chemical and Materials Engineering, 2006.
[11] C. Shivane, M.S. Thesis, University of Cincinnati, Chemical and Materials Engineering, 2006.
[12] http://hyperphysics.phy-astr.gsu.edu/Hbase/chemical/corrosion.html
[13] F.C. Porter, in "Corrosion Resistance of Zinc and Zinc Alloys" (Philip A. Schweitzer, ed.), Marcel Dekker, Inc., New York, NY, 1994.
[14] J. West, Basic Corrosion and Oxidation, Ellis Horwood Limited, Chichester, 1986.
[15] J.R Davis, Corrosion of Aluminum and Aluminum Alloys, ASM International, Materials Park, OH, 2000.
[16] J.L. Hostetler, Corrosion Resistance of Unpainted Galfan Steel, Galfan Constant Improvement Program, GTRC, 1992.
[17] U.S. Public Health Service, Agency for Toxic Substances, Toxicological Profile for Chromium, Report No. ATSDR/TP-88/10, July, 1989.
[18] S.M. Cohen, Corrosion, 51 (1995) 71.
[19] P.J. Mitchell, Sheet Metal Industries, 2 (1987) 74.
[20] P. Puomi, H.M. Fagerholm and A. Sopanen, Anti-Corros. Methods Mater., 48 (2001) 160.
[21] M.W. Kendig, Corros. Sci., 34 (1993) 41.
[22] A.K. Chattopadhyay and M.R. Zentner, Aerospace and Aircraft Coatings, Federation Series on Coatings Technology, Federation of Societies for Paint Technology, Philadelphia, PA, 1990.
[23] P. Puomi, Ph.D. Dissertation, Department of Physical Chemistry, Abo Akademi University, Finland, 2000.

[24] B.M. Rosales, A.R. Di Sarli, O. de Rincon, A. Rincon, C.I. Elsner and B. Marchisio, Prog. Org. Coat., 49 (2004) 209.
[25] T.F. Child and W.J. van Ooij, Chemtech, 28 (1998) 26.
[26] D. Zhu, Ph.D. Dissertation, Department of Material Science and Engineering, University of Cincinnati, 2002.
[27] A. Seth and W.J. van Ooij, "Optimization of One-Step, Low-VOC, Chromate-Free Novel Primer Coatings Using Taguchi Method Approach" (P. Zarras, ed.), ACS Symposium Series Book, in press.
[28] A. Seth and W.J. van Ooij, in "A Novel, Low-VOC, Chromate-Free, One-Step Primer System for the Corrosion Protection of Metals and Alloys" (K.L. Mittal, ed.), presented at the Fifth International Symposium on Silanes and Other Coupling Agents, Toronto, Canada, June 22–24, 2005; and has been accepted for publication in Silanes and Other Coupling Agents.
[29] L. Yang, N. Simhadari, A. Seth and W.J. van Ooij, "Novel Corrosion Inhibitors for Silane Systems on Metals" (K.L. Mittal, ed.), presented at the Fifth International Symposium on Silanes and Other Coupling Agents, Toronto, Canada, June 22–24, 2005; and has been accepted for publication in Silanes and Other Coupling Agents..
[30] A. Seth, W.J. van Ooij, P. Puomi, Z. Yin, A. Ashirgade, S. Bafna and C. Shivane, Novel, One-Step, Chromate-Free Coatings Containing Anticorrosion Pigments for Metals – An Overview and Mechanistic Study, Prog. Org. Coat., 2006, available on-line from December 22, 2006.
[31] R.L. Howard, I.M. Zin, J.D. Scantlebury and S.B. Lyon, Prog. Org. Coat., 37 (1999) 83.
[32] T. Prosek and D. Thierry, Prog. Org. Coat., 49 (2004) 209.
[33] A. Ashirgade, P. Puomi, W.J. van Ooij, S. Bafna, A. Seth, C. Shivane and Z. Yin, 'Novel, One-Step, Chromate-Free Coatings Containing Anticorrosion Pigments for Metals that can be Used in a Variety of Industries', paper published on a conference CD, Eurocorr 2006 in Maastricht, Sept. 24–28, 2006.
[34] G.P. Bierwagen and D.E. Tallman, Prog. Org. Coat., 41 (2001) 201.
[35] V. Subramanian and W.J. van Ooij, Corrosion, 54 (1998) 204.
[36] V. Subramanian, Ph.D. Dissertation, Department of Material Science and Engineering, University of Cincinnati, 1999.
[37] W.J. van Ooij and V. Subramanian, to the University of Cincinnati, US Pat. 6,261,638, July 17, 2001, Method of Preventing Corrosion of Metals Using Silanes.
[38] W.J. van Ooij and D. Zhu, Prog. Org. Coat., 49 (2004) 42.
[39] D. Zhu and W. J. van Ooij, J. Adhesion Sci. Technol., 16 (2002) 1235.
[40] D. Zhu and W. J. van Ooij, Corros. Sci., 45 (2003) 2177.
[41] D. Zhu and W. J. van Ooij, Electrochim. Acta, 49 (2004) 1113.
[42] A. Franquet, Characterization of Silane Films on Aluminum, Ph.D. Thesis, Vrije Universiteit Brussel, Brussel, 2001–2002.
[43] A. Franquet, H. Terryn and J. Vereecken, Surf. Interf. Anal., 36 (2004) 681.
[44] A. Franquet, J. De Laet, T. Schram, H. Terryn, V. Subramanian, W.J. van Ooij and J. Vereecken, Thin Solid Films, 348 (2001) 37.
[45] A., Franquet, C. Le Pen, H. Terryn and J. Vereecken, Electrochim. Acta, 48 (2003) 1245.
[46] A. Franquet, H. Terryn and J. Vereecken, Appl. Surf. Sci., 211 (2003) 259.

[47] T. van Schaftighen, C. Le Pen, H. Terryn and F. Hörzenberger, Electrochim. Acta, 49 (2004) 2997.
[48] A.M. Cabral, R.G. Duarte, M.F. Montemor and M.G.S. Ferreira, Prog. Org. Coat., 54 (2005) 322.
[49] J. Karlsson, Alternatives to Chromates for Pretreatment of Metal Coated Steel, Licentiate Thesis, Centre for Industrial Engineering and Development (CITU), Dalarna University, 1996.
[50] D. Deck and D.W. Reichgott, Metal Finishing, 9 (1992) 29.
[51] P. Puomi, H.M. Fagerholm, J.B. Rosenholm and R. Sipilä, Surf. Coat. Technol., 115 (1999) 79.
[52] N. Tang, W.J. van Ooij and G. Gorecki, Prog. Org. Coat., 30 (1997) 255.
[53] F. Mansfeld and Y. Wang, Mater. Sci. Eng. A, 198 (1995) 51.
[54] L. Guangyu, N. Liyuan, L. Jianshe and L. Zhonghao, Surf. Coat. Technol., 176 (2004) 215.
[55] J.H. Osborne, Prog. Org. Coat., 41 (2001) 217.
[56] A.E. Hughes, J.D. Gorman and P.J.K. Paterson, Corros. Sci., 38 (1996) 1957.
[57] V. Poulain, J.-P. Petitjean, E. Dumont and B. Dugnoille, Electrochim. Acta, 41 (1996) 1223.
[58] A.L. Rudd, C.B. Breslin and F. Mansfeld, Corrosion Sci., 42 (2000) 275.
[59] W.J. van Ooij, D. Zhu, M. Stacy, A. Seth, T. Mugada, J. Gandhi and P. Puomi, Tsinghua Sci. Technol., 10 (2005) 639.
[60] W.J. van Ooij, V. Palanivel and D. Zhu, Prog. Org. Coat., 47 (2003) 384.
[61] A. Seth, Ph.D. Thesis, University of Cincinnati, Chemical and Materials Engineering, 2006.
[62] A. Ashirgade, M.S. Thesis, University of Cincinnati, Chemical and Materials Engineering, 2006.
[63] S. Bafna, M.S. Thesis, University of Cincinnati, Chemical and Materials Engineering, 2007.
[64] P. Wang and D.W. Schaefer, Characterization of Epoxy-Silane Films by Combined Scattering Techniques, presented at the American Crystallographic Association's Annual Meeting held in Honolulu, Hawaii, 2006.

Thermodynamics, Solubility and Environmental Issues
T.M. Letcher (editor)

Chapter 7

The Behavior of Iron and Aluminum in Acid Mine Drainage: Speciation, Mineralogy, and Environmental Significance

Javier Sánchez España

Mineral Resources and Geology Division, Geological Survey of Spain, Rios Rosas 23, 28003 Madrid, Spain

1. INTRODUCTION

The behavior of iron and aluminum in acid mine drainage (AMD) has been thoroughly studied during the last three decades [1–5] and the minerals controlling their solubility have been identified and investigated in detail [6–14]. Thus, ferric iron is known to be mostly dissolved under very acidic conditions ($pH < 2$), and above this pH, Fe(III) is usually hydrolyzed and precipitated as different ochreous minerals such as jarosite ($pH \sim 2$), schwertmannite (pH 2.5–4), or ferrihydrite ($pH > 5$). Fe(II) is much more soluble than Fe(III), and remains in solution below $pH \sim 8$. Then, hydrolysis and precipitation tend to occur and highly amorphous hydrous iron oxides (called "*green rust*" in the AMD literature) are formed upon neutralization. Aluminum is normally conservative below pH 4.5–5.0, and tends to precipitate above this pH in the form of several oxyhydroxysulfates like hydrobasaluminite, basaluminite, and/or hydroxides like gibbsite [1, 9, 11–14].

The precipitation of Fe(III) and Al compounds can retain, by adsorption and/or coprecipitation, many other toxic elements present in the mine effluents like As, Pb, Cr, Cu, Zn, Mn, Cd, Co, Ni, or U. Therefore, Fe(III) and Al act simultaneously as (i) strong buffering systems of the AMD solutions (at pH 2.5–3.5 and 4.5–5.0, respectively) and (ii) natural scavengers of toxic trace elements.

In this chapter, we describe the typical speciation of Fe(III) and Al in AMD, as well as the solid phases which act as solubility controls for these metals at different pH ranges. Finally, some examples of their environmental significance from the Iberian Pyrite Belt (IPB) are reported.

2. GEOCHEMISTRY AND MINERALOGY OF IRON AND ALUMINUM IN AMD

2.1. Formation of AMD and Dissolution of Metal Ions

In the presence of oxygen and water, pyrite is quickly oxidized by the overall reaction [15–18]:

$$FeS_2 + \frac{7}{2}O_2 + H_2O \Rightarrow Fe^{2+} + 2SO_4^{2-} + 2H^+ \quad (1)$$

Ferrous iron is then oxidized by one of the following overall reactions:

$$Fe^{2+} + \frac{1}{4}O_2(g) + H^+ \Rightarrow Fe^{3+} + \frac{1}{2}H_2O \quad (pH < 3) \quad (2)$$

$$Fe^{2+} + \frac{1}{4}O_2(g) + \frac{5}{2}H_2O \Rightarrow Fe(OH)_3(s) + 2H^+ \ (pH > 3) \quad (3)$$

Finally, Fe(III) enhances the further oxidation of pyrite by the reaction:

$$FeS_2 + 14Fe^{3+} + 8H_2O \Rightarrow 15Fe^{2+} + 2SO_4^{2-} + 16H^+ \quad (4)$$

Reactions (1) and (2) are usually catalyzed by iron-oxidizing bacteria (such as *Acidithiobacillus ferrooxidans*, *Leptospirilum ferrooxidans*, or *Ferrimicrobium* ssp.) which increase the oxidation rate by up to 10^5 over the abiotic rate [2–5]. Field and laboratory studies have shown that the ferrous iron dissolved in mine effluents is normally oxidized at rates of between 10^{-5} and 10^{-9} (average $\sim 10^{-7}$) mol L^{-1} s^{-1} [2, 5, 19, 20]. These rates depend on factors like (1) water temperature, (2) dissolved oxygen content, and (3) density of bacterial populations. The iron-oxidizing microbes are usually present at numbers of $\sim 10^3$–10^8 mL^{-1} [4, 19, 21].

Reactions (1) and (4) result in strongly acidified solutions which, in addition to dissolve large amounts of Fe(II) and SO_4^{2-}, also leach important quantities of many other major and trace metals from the mineralizations and the host rocks, including Al, Mg, and Ca (with minor Na, K, and Ba) from the accompanying aluminosilicates, carbonates, and sulfates, in addition to Cu, Zn, Mn, As, Pb, or Cd from other sulfides (sphalerite, chalcopyrite, galena, arsenopyrite) and sulfosalts (tetrahedrite-tennantite). Iron and aluminum are usually the most abundant metal cations of those present in AMD [5, 9, 12, 22].

2.2. Ionic Speciation

The pH-dependent distribution of dissolved ionic species of Fe and Al has been traditionally modeled for pure water (at 25°C, 1 bar), where the hydroxo complexes (Fe–OH and Al–OH species) are dominant [17, 18]. However, AMD normally shows high concentrations of the SO_4^{2-} ion, which strongly determines the complexation of Fe and Al, that is usually dominated by metal-sulfate and metal-bisulfate species of the type Fe–SO_4^{2-} and Al–SO_4^{2-}. Thus, at typical conditions of pH and sulfate concentration in AMD solutions (e.g., pH ~2.5–3.5 and 0.1 M SO_4^{2-}), Fe(III) and Al(III) are chiefly present as sulfate complexes ($FeSO_4^+$, $Fe(SO_4)_2^-$, $AlSO_4^+$, $Al(SO_4)_2^-$), with a minor presence of the free aqueous ions (Fe^{3+}, Al^{3+}), and with the hydroxyl-containing ionic complexes ($FeOH^{2+}$, $Fe(OH)_2^+$, $Fe_3(OH)_4^{5+}$, etc.) having very minor or negligible presence [12, 14]. However, this ionic speciation strongly varies with pH and the activity of the sulfate anion, as shown in Fig. 1, which plots the distribution of different ionic species of Fe(III) and Al in a solution with 0.1 M SO_4^{2-} as a function of pH. This figure shows that the free aqueous cations (Fe^{3+}, Al^{3+}) are only dominant at extremely acidic conditions (pH < 1.0), while the metal-sulfate complexes are dominant in the pH ranges 1–4.5 for Fe(III) and 1.5–6 for Al. As pH increases, the sulfated ionic complexes are progressively replaced by hydroxide forms (e.g., $Fe(OH)_2^+$, $Al(OH)_3^0$, $Al(OH)_4^-$), which become dominant at near-neutral conditions (pH > 5 for Fe(III) and pH > 6 for Al).

Such Fe(III) and Al speciation has important implications for the geochemical evolution of AMD. At typical conditions of pH (1.5–4) and sulfate concentrations (0.01–0.1 M SO_4^{2-}), the mineralogy of the precipitates will be dominated by sulfates and hydroxysulfates, instead of oxides or hydroxides, as described in the next section. Further, this metal speciation can imply important differences in their sorption behavior. For example, at sulfate activities of $\sim 10^{-4}$ to 10^{-2}, aluminum is present as free aqueous cation, being essentially conservative. However, at higher activities of the SO_4^{2-} anion (from 10^{-2} to $>10^{-1}$), Al forms bisulfate anionic species ($Al(SO_4)_2^-$) which can be sorbed onto positively charged mineral surfaces at low pH [14].

2.3. Mineral Phases Controlling the Solubility of Fe(III) and Al

Both Fe(III) and Al show amphoterism, with a solubility minimum at near-neutral conditions, and enhanced solubility at acidic and basic conditions [5, 17, 18]. Thus, at pH values well below 2, both iron and aluminum are chiefly dissolved, and these metals may only precipitate from highly concentrated brines in evaporative pools, where oversaturation of secondary sulfates is eventually reached. As pH increases (for example, by partial neutralization

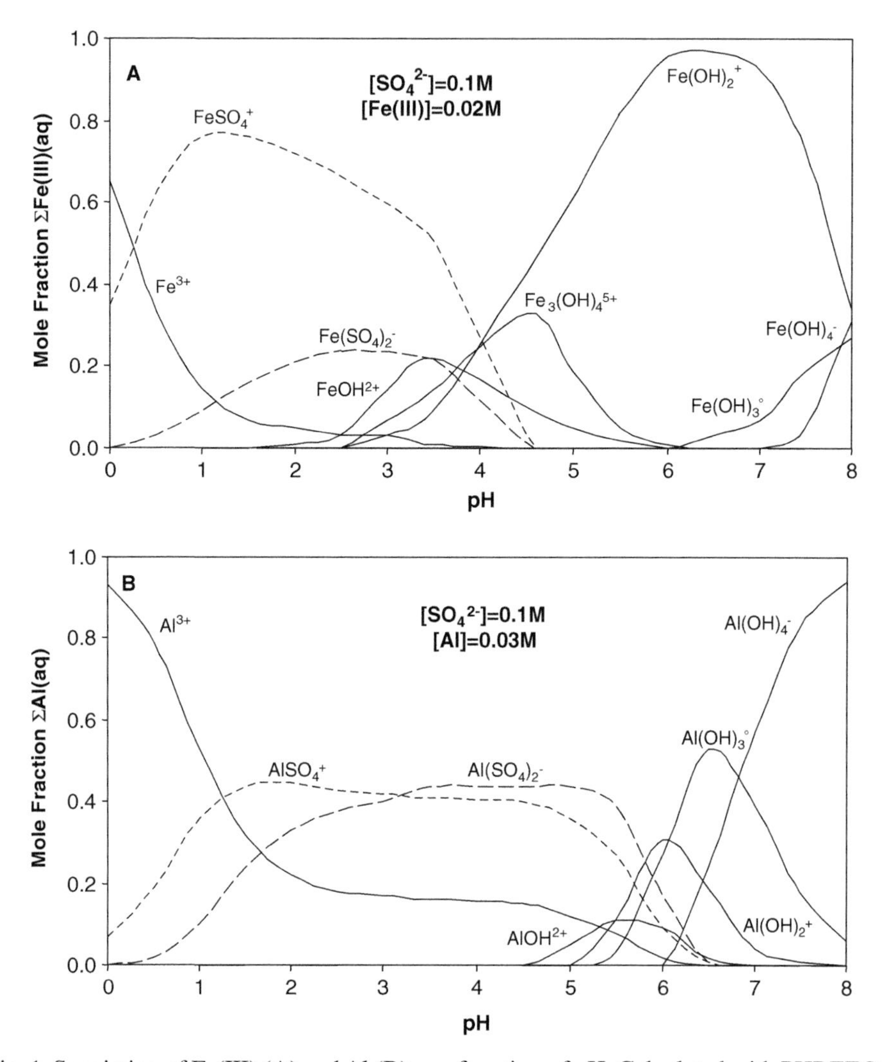

Fig. 1. Speciation of Fe(III) (A) and Al (B) as a function of pH. Calculated with PHREEQC 2.7 (25°C, 1 bar). The concentrations used as input to the program (0.1 M SO_4^{2-}, 0.02 M Fe(III), and 0.03 M Al) were selected from the average value of more than 70 acidic mine waters of the Iberian Pyrite Belt [12, 45].

during mixing with moderately alkaline waters), iron is hydrolyzed and precipitates at pH > 2 (first hydrolysis constant is $pK_1 = 2.2$), whereas Al tends to be hydrolyzed at pH ~ 4.5–5.0 [5, 9, 12].

The most common Fe(III) and Al minerals associated to, and precipitating from, AMD, along with their respective formulae and usual pH

Table 1
Common secondary minerals of iron and aluminum precipitating from AMD

Mineral	Formula	pH range	Reference
Soluble iron sulfate salts			
Melanterite	$Fe^{II}SO_4 \cdot 7H_2O$	<1	[12, 33–41]
Rozenite	$Fe^{II}SO_4 \cdot 4H_2O$	<1	[12, 33–41]
Szomolnokite	$Fe^{II}SO_4 \cdot H_2O$	<1	[12, 33–41]
Copiapite	$Fe^{II}Fe_4^{III}(SO_4)_6(OH)_2 \cdot 20H_2O$	2–3	[12, 33–41]
Coquimbite	$Fe_2^{III}(SO_4)_3 \cdot 9H_2O$	2–3	[12, 33–41]
Rhomboclase	$(H_3O)Fe^{III}(SO_4)_2 \cdot 3H_2O$	2–3	[12, 33–41]
Halotrichite	$Fe^{II}Al_2(SO_4)_4 \cdot 22H_2O$	2–3	[12, 33–41]
Iron hydroxides/hydroxysulfates			
Jarosite	$KFe_3^{III}(SO_4)_2(OH)_6$	<2	[5–14]
Natrojarosite	$NaFe_3^{III}(SO_4)_2(OH)_6$	<2	[5–14]
Hydronium jarosite	$(H_3O)Fe_3^{III}(SO_4)_2(OH)_6$	<2	[5–14, 37]
Schwertmannite	$Fe_8^{III}O_8(SO_4)(OH)_6$	2–4	[5–14]
Goethite	α-$Fe^{III}O(OH)$	2.5–7	[5–14]
Ferrihydrite	$Fe_5^{III}HO_8 \cdot 4H_2O$	5–8	[5–14]
Aluminum hydroxides/hydroxysulfates			
Gibbsite	γ-$Al(OH)_3$	>4.5–5.0	[1, 5, 9–14]
Alunite	$KAl_3(SO_4)_2(OH)_6$	3.5–5.5	[1, 5, 9–14]
Jurbanite	$Al(SO_4)(OH) \cdot 5H_2O$	<4	[1, 5, 9–14, 29–32]
Basaluminite	$Al_4(SO_4)(OH)_{10} \cdot 5H_2O$	>4.5–5.0	[1, 5, 9–14, 29–32]
Hydrobasaluminite	$Al_4(SO_4)(OH)_{10} \cdot 12\text{–}36H_2O$	>4.5–5.0	[9–14]

ranges of occurrence, are given in Table 1. This table includes from highly crystalline sulfate salts (e.g., melanterite, rozenite, copiapite, halotrichite), which are formed by evaporative processes from very acidic brines in small pools or in the margins of AMD-impacted streams, to nearly amorphous compounds of Fe(III) (schwertmannite, ferrihydrite) and Al (basaluminite and hydrobasaluminite, gibbsite). The typical XRD patterns of some of these solids are provided in Fig. 2.

2.3.1. Fe(III) Hydroxysulfates

The solid formed by precipitation of Fe(III) at pH 2.0–4.0 during mixing and neutralization of the AMD solutions is usually schwertmannite [6–9]. This mineral is a poorly crystallized iron oxyhydroxysulfate whose formation from acid-sulfate waters can be described as follows [8]:

$$8Fe^{3+} + SO_4^{2-} + 14H_2O \Leftrightarrow Fe_8O_8(SO_4)(OH)_6 + 22H^+ \quad (5)$$

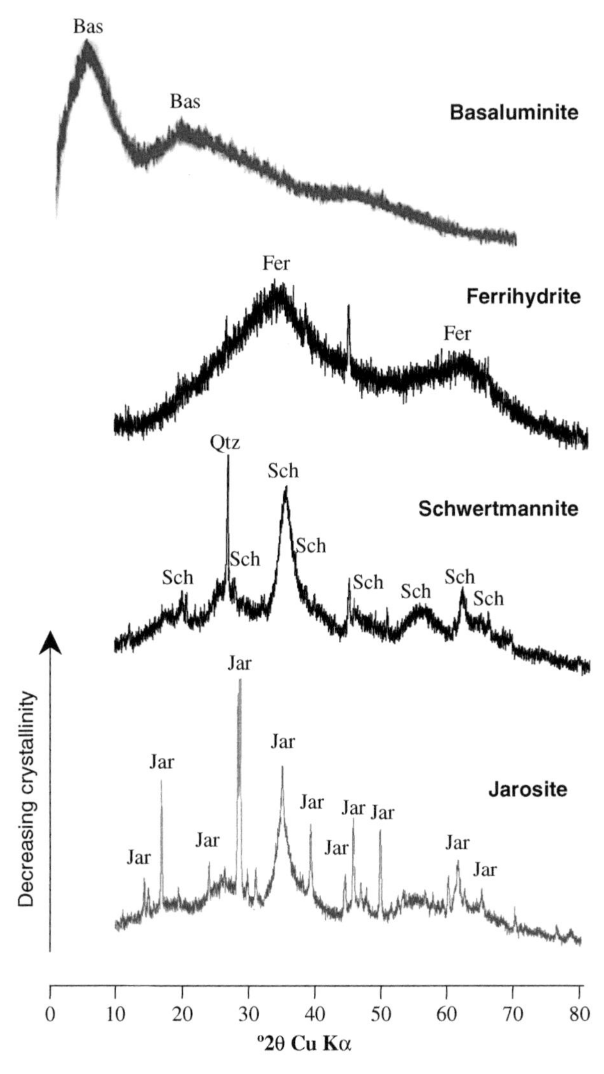

Fig. 2. XRD patterns of selected Fe(III) and Al precipitates found in AMD of the IPB (compiled from Refs. 13, 14 with kind permission of Springer Science and Business Media).

In addition to schwertmannite, some other minerals of Fe(III) are commonly recognized in the AMD systems, with a close relation between their occurrence and the water pH (Table 1). Jarosite is usually favored to precipitate from very acidic solutions, normally at $pH < 2$. Schwertmannite precipitates near the discharge points at pH 2.0–4.0, whereas ferrihydrite usually forms in fluvial environments (as in the confluences between AMD and unpolluted rivers) at

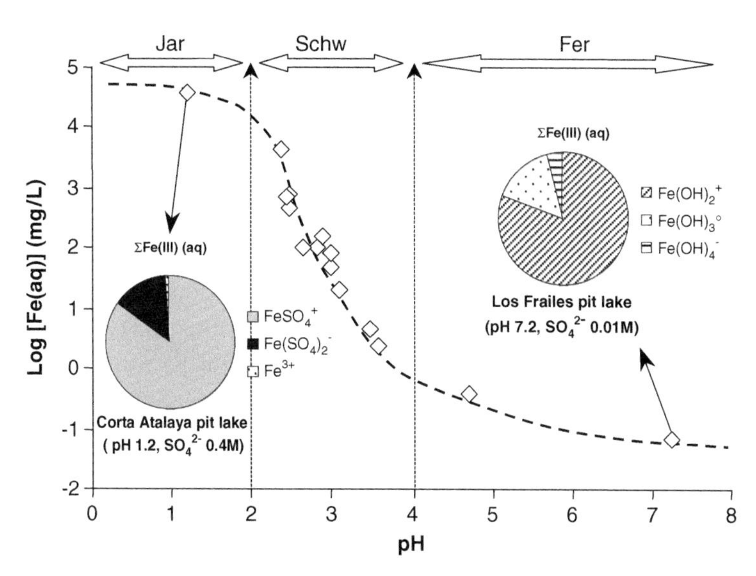

Fig. 3. Concentration of total dissolved iron as a function of pH in the pit lakes of the IPB. The mineral phases controlling the solubility of Fe(III) at different pH ranges are indicated (Jar, jarosite; Schw, schwertmannite; Fer, ferrihydrite). The ionic speciation of Fe(III) (calculated with PHREEQC 2.7) is also indicated for the two end-members of the compositional range (Corta Atalaya and Los Frailes). Data taken from Ref. 45.

pH > 5 [6–9, 12–14]. These three minerals are meta-stable with respect to goethite, which is the most stable form of Fe(III) at low temperature [1]. Goethite can also precipitate from AMD (especially at lower redox conditions and/or sulfate content [5–9]), although its presence in the particulate fraction of AMD waters is very minor in relation to schwertmannite or ferrihydrite.

The concentration and/or activity of dissolved Fe(III) is thus controlled by different mineral phases depending on the pH. This is illustrated in Fig. 3, which plots the pH-dependent variation in the concentration and speciation of dissolved iron in different mine pit lakes of the IPB. Among the 15 pit lakes studied, jarosite was only observed to precipitate in Corta Atalaya, so that this mineral controls the apparent equilibrium which seems to exist between dissolved and particulate Fe(III) in this lake. Conversely, ferrihydrite is the mineral form under which Fe(III) precipitates in Los Frailes pit lake. Most lakes, however, seem to be at or near equilibrium with respect to schwertmannite, which not only controls the solubility of Fe(III), but also buffers the systems at pH 2.5–3.5 (through reaction (5)), and sorbs toxic trace elements like As [5–14, 24–28].

2.3.2. Al Hydroxysulfates

If compared with the Fe(III) solids, aluminum minerals formed at pH ~ 4.5–5.0 are less well characterized. These solids are nearly amorphous to XRD, although two broad reflections (near 7° and 20° 2θ) are usually recognized by XRD (Fig. 2). This diffraction pattern, along with their chemical composition, suggests that the Al precipitates probably consist of poorly ordered hydroxysulfates with composition intermediate between basaluminite and hydrobasaluminite [1, 5, 9, 14, 29–32]. Calculations of saturation indices [14] suggest precipitation of alunite in some cases, although this mineral is rarely found as a direct precipitate. Finally, aluminite has been also suggested to precipitate from waters with high concentrations of Al and sulfate (0.1 and 0.4 M, respectively [14]).

Hydrobasaluminite is meta-stable with respect to basaluminite, which forms by dehydration of the former [9]. Basaluminite is a common Al hydroxysulfate in mine drainage environments, although it also tends to be transformed to alunite during maturation or heating [1, 9, 29–32]. The hydrolysis of Al^{3+} to form hydrobasaluminite is written as follows:

$$4Al^{3+} + SO_4^{2-} + 22 - 46H_2O \Leftrightarrow Al_4(SO_4)(OH)_{10} \cdot 12 - 36H_2O + 10H^{+} \quad (6)$$

As discussed below, the formation of this mineral during neutralization can also imply a significant removal of toxic trace elements (e.g., Cu, Zn, Cr, U) from the aqueous phase.

3. ENVIRONMENTAL SIGNIFICANCE

3.1. Evaporative Sulfate Salts as Temporal Sinks of Metals and Acidity

The formation of efflorescent, evaporative sulfates from AMD waters is especially abundant during the dry season. The mineralogy of these soluble sulfates is closely associated with their spatial distribution and the pH of the brines from which these salts are precipitated [12, 34–41]. Thus, the Fe(II)-sulfates like melanterite, rozenite, or szomolnokite are dominant in isolated and highly concentrated pools near the pyrite sources, under conditions typical of green, ferrous AMD with very low pH (normally below 1). On the other hand, mixed Fe(II)–Fe(III) and/or Fe(III)–Al sulfates like copiapite, coquimbite, or halotrichite are common in the margins of rivers impacted by AMD, where the Fe and Al concentrations of the evaporating waters are lower, and the pH is higher (typically between 2 and 3). These sulfates have been observed to follow a paragenetic sequence of dehydration and mineralogical maturation, with melanterite > rozenite > szomolnokite > copiapite > coquimbite > rhomboclase > halotrichite [5, 34–41].

Chemical analyses of mixtures of these sulfates have reported very high concentrations of trace metals like Cu and Zn (up to several percent units in some instances [12, 36]). Because these sulfates are highly soluble, the first rainstorm events taking place in the early autumn usually imply the re-dissolution of large amount of these salts accumulated during the spring-summer, and the subsequent incorporation of these toxic metals to the rivers. This seasonal pattern of background Cu and Zn values during the summer, followed by sharp increases in their concentrations during the first rainfalls in autumn, has been reported in many mine districts [42, 43]. The dissolution of these salts can also imply a rapid acidification and conductivity increase for the affected streams, as they release important amounts of sulfate and free acidity to the water [5, 12]. Therefore, seasonal cycles of precipitation/re-dissolution of salts may imply environmental consequences for water quality.

3.2. Natural Attenuation of Metal Concentrations

Downstream from the sources, the geochemical evolution of AMD is usually controlled by (1) oxidation of Fe(II) to Fe(III), (2) progressive pH increase and dilution of metal concentrations by mixing with pristine waters, (3) hydrolysis and precipitation of different metal cations as pH increases, and (4) sorption of different trace elements (As, Pb, Cr, Cu, Zn, Mn, Cd) onto the solid surfaces of precipitated metal hydroxides/hydroxysulfates [14]. The pH-dependent sequences of precipitation and adsorption are very similar and follow the order: Fe(III) > Pb > Al > Cu > Zn > Fe(II) > Cd [5, 23]. The overall result of these processes represents a mechanism of natural attenuation with environmental benefits for the water quality [12–14, 44].

As an example of this self-mitigating capacity, the graphs provided in Fig. 4 show the spatial evolution of metal loadings in an acidic effluent emerging from a waste-pile in Tharsis mine (SW Spain) [44]. To study the natural attenuation of metal contents provoked by precipitation and sorption, and not by dilution, metal loadings were calculated from measured metal concentrations and flow rate measurements in several sampling stations, so that dilution effects were not taken into account. This effluent emerged with a pH of 2.2, and subsequently converged with a number of small creeks of unpolluted water, thus provoking a slight but progressive pH increase up to values around 5. This pH increase allowed the precipitation of Fe(III) (mostly schwertmannite) during the first 11 km of the AMD course ($pH < 3$),

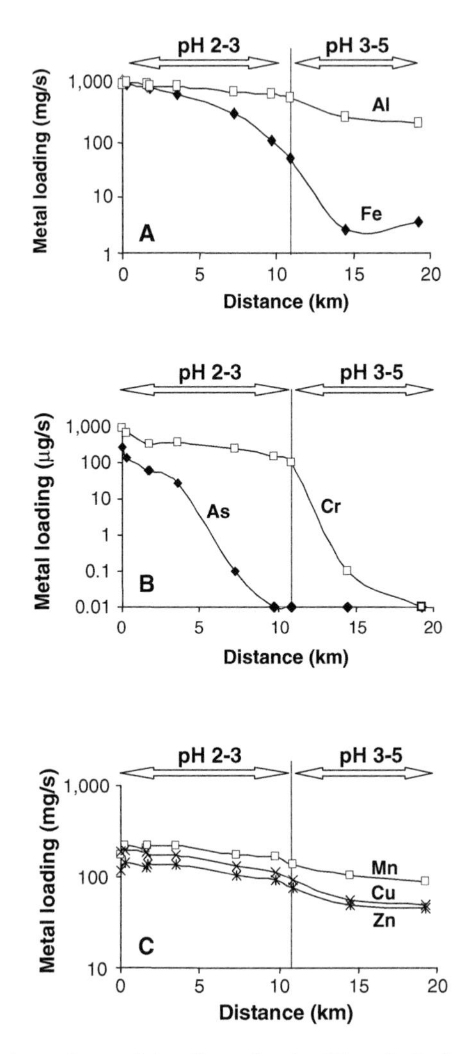

Fig. 4. Spatial evolution of metal loadings in the Tharsis-Dehesa Boyal acidic mine effluent (Tharsis mine, IPB): (A) Fe and Al; (B) As and Cr; (C) Cu, Zn, and Mn. Metal loadings calculated from analyzed aqueous metal concentrations and measured flow rates at distinct sampling stations downstream from the source point. Modified from Ref. 44 with kind permission of Springer Science and Business Media.

and of Fe(III) and Al compounds (Al as amorphous hydrobasaluminite) during the next 9 km of the stream course (pH 3–5). The evolution shows that the precipitation of Fe(III) resulted in a strong decrease of the total iron loading to around one-tenth of its initial content at only 10 km from the source.

The precipitation of Al compounds in the last reach also accounted for an important loss of the Al loading. Interestingly, the Fe and Al solids acted as efficient sorbents for a number of trace elements, though these elements showed different affinities for these mineral phases depending on their ionic charge and the water pH. Thus, elements like arsenic (present as $HAsO_4^{2-}$) showed a strong tendency to be adsorbed by the ferric iron colloids at low pH (<3), having been totally removed from the water at less than 10 km from the discharge point. Other trace metals such as Cr (present as $HCrO_4^-$) showed a clear affinity for sorption onto the Al compounds formed at pH ~4.5–5, which also implied a total scavenging of this toxic element at 20 km from the source. Finally, divalent metal cations like Cu^{2+}, Zn^{2+}, and Mn^{2+} behaved conservatively until pH ~5, above which they were only slightly removed by sorption onto the Al minerals during the final reach.

Such metal scavenging may imply important water quality improvements at a basin scale. For example, it was estimated that only 1% of the total Fe and As dissolved load and ~20–40% of the total Al, Mn, Cu, Zn, Cd, Pb, and SO_4 dissolved load initially released from the mine sites would have been transferred from the Odiel river basin (including more than 25 studied mines) to the Huelva estuary in 2003 [12]. Thus, most of the iron and arsenic load would have remained in solid form in the vicinity of the pyrite sources within the mines, mainly as ochreous Fe(III) precipitates and sulfate minerals covering the stream channel, whereas the rest of metal cations (Al, Cu, Zn, Mn) can migrate further downstream.

3.3. The Concept of Mineral Acidity

A remarkable consequence of the presence of Fe(III) and Al in AMD is related with the concept of *mineral* or *potential* acidity, which is the acidity released during the hydrolysis of different metal cations like Fe(III), Al, Cu, Zn, Mn, or Fe(II). The common pH measurements provide a good indicator of the proton (or *free*) acidity of AMD (i.e., the acidity consisting in free protons (H^+) present in the solution). However, it is useful to carry out titrations or neutralization experiments to see the acidity distribution over a given pH range.

An example of titration curve of a typical AMD containing Fe(III) and Al is given in Fig. 5. This curve shows two steep inflection points (followed by their respective *plateaus*) at pH ~3.0 and 4.5. The mineral acidity released during the hydrolysis of Fe(III) and Al, respectively, provokes a strong buffering of the waters. The existing *plateaus* at pH > 3.5 and >5.0 suggest that Fe(III) and Al have been almost totally hydrolyzed and precipitated at

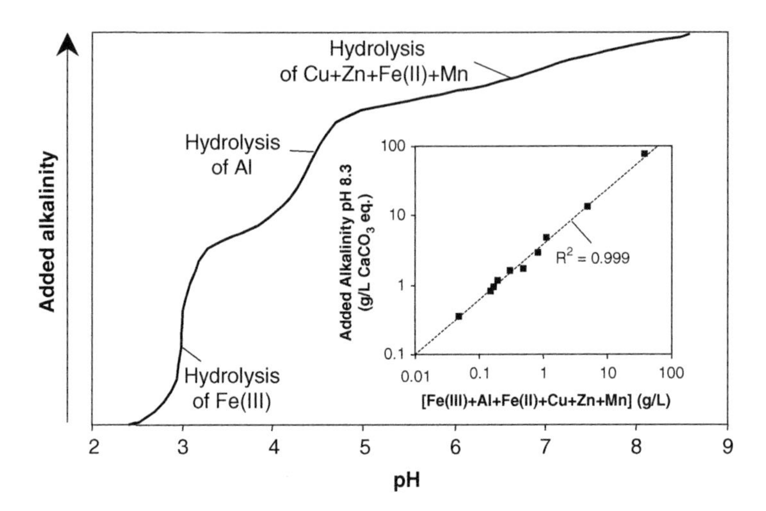

Fig. 5. Typical titration curve of an acidic mine water. The inset shows the correlation between the acidity released during titration of waters from acidic mine pit lakes of the IPB to a target pH of 8.3 (i.e., total acidity) and the respective concentrations of major cations (Fe(III) + Al + Fe(II) + Cu + Zn + Mn). Data taken from Refs. 14 and 45.

these pH values. The gentle slope existing in the pH range 5–9 indicates the progressive hydrolysis of other metal cations present in the waters, such as Cu, Zn, Mn, and/or Fe(II). An excellent correlation between the acidity measured at pH 8.3 (i.e., total acidity) in several mine waters and their respective concentrations of [Fe(III) + Al + Cu + Zn + Fe(II) + Mn] is also shown in Fig. 5.

An overall conclusion from the examination of these curves is that the mineral (total) acidity resulting only from the hydrolysis of Fe(III) and Al can be as high as 10 times the *free* acidity of these waters (i.e., that neutralized before the hydrolysis of metals) [12]. These calculations highlight the importance that the mineral acidity may have on any attempt of treatment or remediation initiative, especially if based on neutralization.

4. CONCLUSIONS

The water chemistry of acidic effluents draining abandoned metal and coal mines is usually characterized by low pH and high sulfate concentrations, and commonly includes high contents of iron and aluminum. This aqueous composition determines not only the complexation of both metals (dominated by sulfate and bisulfate ionic species), but also the mineralogy of their hydrolysis products, which mainly consist of oxyhydroxysulfates and highly soluble sulfate salts. The former are highly amorphous and have small particle size and high specific surface area, physical properties which make them efficient adsorbants of toxic trace metals, whereas the latter are highly crystalline and may incorporate large amounts of Cu and Zn in their crystalline lattices, so that their re-dissolution during rainstorm events is of environmental concern. Iron and aluminum represent, in addition, strong buffering systems of AMD solutions and hence additional sources of (mineral) acidity which must be taken into account in remediation or treatment attempts, either from a technical or an economic perspective.

REFERENCES

[1] D.K. Nordstrom, Geochim. Cosmochim. Acta, 46 (1982) 681.
[2] D.K. Nordstrom, U.S. Geol. Surv. Water-Supply Pap., 2270 (1985) 113.
[3] D.K. Nordstrom, Int. Geol. Rev., 42 (2000) 499.
[4] D.K. Nordstrom, Miner. Assoc. Can. Short Course, 31 (2003) 227.
[5] D.K. Nordstrom and C.N. Alpers, Rev. Econ. Geol., 6A (1999) 133.
[6] J.M. Bigham, L. Carlson and E. Murad, Miner. Mag., 58 (1994) 641.
[7] J.M. Bigham, U. Schwertmann, L. Carlson and E. Murad, Geochim. Cosmochim. Acta, 54 (1990) 2743.
[8] J.M. Bigham, U. Schwertmann, S.J. Traina, R.L. Winland and M. Wolf, Geochim. Cosmochim. Acta, 60 (1996) 2111.
[9] J.M. Bigham and D.K. Nordstrom, Rev. Miner. Geochem., 40 (2000) 351.
[10] J. Yu, B. Heo, I. Choi and H. Chang, Geochim. Cosmochim. Acta, 63 (1999) 3407.
[11] J.J. Kim and S.J. Kim, Environ. Sci. Technol., 37 (2003) 2120.
[12] J. Sánchez-España, E. López Pamo, E. Santofimia, O. Aduvire, J. Reyes and D. Barettino, Appl. Geochem., 20-7 (2005) 1320.
[13] J. Sánchez-España, E. López-Pamo, E. Santofimia, J. Reyes and J.A. Martín Rubí, Water, Air Soil Pollut., 173 (2006) 121.
[14] J. Sánchez-España, E. López-Pamo, E. Santofimia, J. Reyes and J.A. Martín Rubí, Aquat. Geochem., 12 (2006) 269.
[15] D.K. Nordstrom, Soil Sci. Soc. Am. Spec. Pub., 10 (1982) 37.
[16] P.C. Singer and W. Stumm, Science, 167 (1970) 1121.
[17] W. Stumm and J.J. Morgan, Aquatic Chemistry, 3rd edn., Wiley, NY, USA, 1996.
[18] D. Langmuir, Aqueous Environmental Geochemistry, Prentice-Hall, Inc., Upper Saddel River, NJ, 1997.
[19] N. Wakao, K. Hanada, Y. Sakurai and H. Shiota, Soil Sci. Plant Nutr., 24 (1978) 491.
[20] J. Sánchez-España, E. López-Pamo and E. Santofimia, J. Geochem. Explor., (2006), doi; 10.1016/j.gexplo.2006.08.010.
[21] D.B. Johnson, Water Air Soil Pollut., 3 (2003) 47.
[22] G.S. Plumlee, K.S. Smith, M.R. Montour, W.H. Ficklin and E.L. Mosier, Rev. Econ. Geol., 6B (1999) 373.

[23] D.A. Dzombak and F.M.M. Morel, Surface Complexation Modeling: Hydrous Ferric Oxide, 393 pp., Wiley, New York, 1990.
[24] S. Regenspurg and S. Peiffer, Appl. Geochem., 20 (2005) 1226.
[25] M. Leblanc, B. Achard, D. Ben Othman and J.M. Luck, Appl. Geochem., 11 (1996) 541.
[26] L. Carlson, J.M. Bigham, U. Schwertmann, A. Kyek and F. Wagner, Environ. Sci. Technol., 36 (2002) 1712.
[27] C. Casiot, G. Morin, F. Juillot, O. Bruneel, J.C. Personné, M. Leblanc, K. Duqesne, V. Bonnefoy and F. Elbaz-Poulichet, Water Res., 37 (2003) 2929.
[28] K. Fukushi, M. Sasaki, T. Sato, N. Yanese, H. Amano and H. Ikeda, Appl. Geochem., 18 (2003) 1267.
[29] F. Adams and Z. Rawajfih, Soil Sci. Soc. Am. J., 41 (1977) 686.
[30] F.A. Bannister and S.E. Hollingworth, Am. Miner., 33 (1948) 787.
[31] H. Basset and T.H. Goodwin, J. Chem. Soc., (1949) 2239.
[32] T. Clayton, Miner. Mag., 43 (1980) 931.
[33] D.K. Nordstrom, C.N. Alpers, Proc. Natl. Acad. Sci. U.S.A., 96 (1999) 3455.
[34] D.K. Nordstrom, C.N. Alpers, C.J. Ptacek and D.W. Blowes, Environ. Sci. Technol., 34 (2000) 254.
[35] J.L. Jambor, D.K. Nordstrom and C.N. Alpers, Rev. Miner. Geochem., 40 (2000) 303.
[36] T. Buckby, S. Black, M.L. Coleman and M.E. Hodson, Miner. Mag., 67-2 (2003) 263.
[37] F. Velasco, A. Alvaro, S. Suarez, M. Herrero and I. Yusta, J. Geochem. Explor., 87-2 (2005) 45.
[38] R.J. Bowell and J.V. Parshley, Chem. Geol., 215 (2005) 373.
[39] J.M. Hammarstrom, R.R. Seal II, A.L. Meier and J.M. Kornfeld, Chem. Geol., 215 (2005) 407.
[40] H.E. Jamieson, C. Robinson, C.N. Alpers, R.B. McCleskey, D.K. Nordstrom and R.C. Peterson, Chem. Geol., 215 (2005) 387.
[41] R.M. Joeckel, B.J. Ang Clement and L.R. VanFleet Bates, Chem. Geol., 215 (2005) 433.
[42] C.N. Alpers, D.K. Nordstrom and J.M. Thompson, Am. Chem. Soc. Symp. Ser., 550 (1994) 324.
[43] M. Olías, J.M. Nieto, A.M. Sarmiento, J.C. Cerón and C.R. Cánovas, Sci. Total Environ., 333 (2004) 267.
[44] J. Sánchez-España, E. López-Pamo, E. Santofimia, J. Reyés and J.A. Martín-Rubí, Environ. Geol., 49 (2005) 253.
[45] J. Sánchez-España, E. López-Pamo and E. Santofimia, IGME, 190 pp., Unpublished report, 2005.

PART III

RADIOACTIVE WASTES

Thermodynamics, Solubility and Environmental Issues
T.M. Letcher (editor)

Chapter 8

An Evaluation of Solubility Limits on Maximum Uranium Concentrations in Groundwater

Teruki Iwatsuki[a] and Randolf C. Arthur[b]

[a]Mizunami Underground Research Laboratory, 1-64 Yamanouchi, Akeyo-Cho, Mizunami-Shi, Gifu, 509-6132 JNC, Japan
[b]Monitor Scientific, LLC, 3900S, Wadsworth Blvd. 555, Denver, CO 80235, USA

1. INTRODUCTION

This paper presents a summary of on-going work by the Japan Atomic Energy Agency (JAEA) addressing solubility constraints on maximum uranium (U) concentrations in natural waters. The impetus for this work comes from the fact that similar constraints are assumed in most international performance assessments (PA) of deep geologic repositories for the permanent disposal of high-level radioactive wastes (HLW) [1]. Numerical models and computer codes used in PA predict how a repository's engineered and natural barriers to radionuclide migration[1] would respond to a variety of plausible scenarios governing the initial repository environment and its possible future evolution. Solubility constraints adopted in these scenarios allow for gradual increases to occur in the aqueous concentration of a given radioelement released from the waste until the solubility of a corresponding solid is reached. These solids are generally assumed to be idealized pure phases, such as simple oxides or hydrous oxides of the actinides and fission products in HLW [2]. The solution then equilibrates with the solid, and the aqueous concentration of the radioelement is thereafter fixed at this solubility

[1]Engineered barriers generally include a waste form (spent nuclear fuel or glass containing radioactive wastes) encapsulated within a metallic container (e.g., steel and copper), which may be surrounded by a low-permeability, clay-rich buffer and backfill. Natural barriers include the repository host rock and a volume of rock between the repository and biosphere. The engineered barriers and immediately adjacent host rock are referred to as the near field [1].

limit. In reality, sorption, co-precipitation and/or solid–solution reactions involving various minerals in the engineered barriers and host rock, or secondary minerals produced by the interaction of these materials with groundwater, are more likely to control radioelement concentrations in a HLW repository. Estimated solubilities are useful, however, because they represent credible and defensible upper bounds on these concentrations. This is conservative (i.e., pessimistic) for PA purposes because releases of radioelements to the biosphere tend to increase or decrease in proportion to their predicted concentrations in the near field.

JAEA have been evaluating such solubility limits specifically for U using field data characterizing the Tono Uranium Deposit in central Japan. The Tono Uranium Deposit provides evidence that U, which is an important component of HLW, is effectively immobilized within the deposit in a chemically reducing environment over long periods of time at depths within a few hundred meters of oxidizing conditions at the Earth's surface. This natural system is analogous in these respects to expected conditions in a HLW repository. A comprehensive and critically evaluated database characterizing the geology, hydrogeology, hydrochemistry and geomicrobiology of the deposit and its environs has been developed as a result of numerous field studies and drilling campaigns carried out in the Tono area over the past two decades. These data provide a rare opportunity to test whether maximum U concentrations in groundwaters circulating through the deposit are controlled by the solubility of various minerals that have been assumed to be solubility controlling in HLW repository.

The geology of the Tono deposit is briefly summarized below. This is followed by a discussion of bounding constraints on geochemical parameters that are believed to control the aqueous-speciation and solubility behaviour of U in groundwaters of the Tono area. This behaviour is evaluated in Section 4, and preliminary conclusions are summarized in Section 5.

2. GEOLOGIC SETTING OF THE TONO URANIUM DEPOSIT

Itoigawa [3], Yusa et al. [4], Shikazono and Utada [5] and Sasao et al. [6] described the regional geology of the Tono District, which includes the Tono Uranium Deposit. Basement granitic rocks overlain by essentially flat-lying sedimentary formations are distinguishing features of the region (Fig. 1).

The Toki Granite was emplaced ~72 Ma. The upper part of the granite, extending a few hundred meters below the palaeo-erosional surface, is a moderately fractured and relatively permeable zone. The granite has been

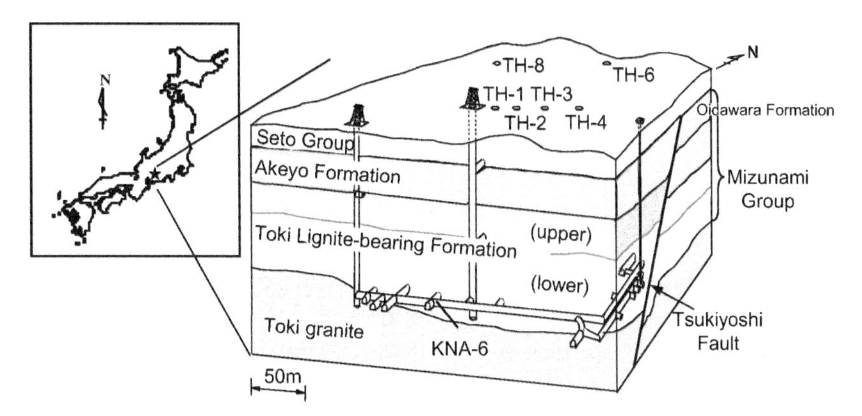

Fig. 1. Overview of the geology of the Tono Uranium Deposit showing locations of selected boreholes, shafts and drifts in the Tono Mine.

locally altered by early hydrothermal fluids, which circulated through the granite as it cooled, and later by dilute and saline groundwaters. These solutions are distinguished by carbon and oxygen isotope ratios in calcites lining fractures in the granite [7].

An unconformity separates the Toki Granite from overlying sedimentary formations of the Miocene Mizunami Group (27–15 Ma). The Mizunami Group includes, in ascending order, the Lower Toki Lignite-bearing Formation, Upper Toki Lignite-bearing Formation (also referred to as the Hongo Formation [6]), Akeyo Formation and Oidawara Formation. The first two formations consist of carbonaceous fluvial-lacustrine sediments. The Akeyo and Oidawara Formations consist of alternating shallow to deep marine siltstones and sandstones.

The Mizunami Group is unconformably overlain by Pliocene to Pleistocene rocks of the Seto Group (5–0.7 Ma). These rocks consist of poorly consolidated fluvial sediments containing clays, silts and conglomerates with rhyolite or chert clasts.

The Tsukiyoshi Fault cuts the Toki Granite and Mizunami Group in the vicinity of the Tono Mine (Fig. 1). This fault was initially activated ~17 Ma with 10–20 m normal displacement. It was reactivated ~15–12 Ma with 40–50 m reverse displacement [6].

Starting ~1.5 Ma and continuing to the present, the Tono region has risen by ~150–300 m as a result of 500 m of uplift and 200–350 m of erosion. Earlier periods of uplift and subsidence correlate with respective periods of sediment deposition from freshwater (Upper and Lower Toki Lignite-bearing

Formations and Seto Group) and from brackish and marine waters (Akeyo Formation and Oidawara Formation) [4].

The Tono Uranium Deposit is located within the Lower Toki Lignite-bearing Formation. It is a sandstone type deposit with mineralization localized along channel structures that follow the palaeo-erosional surface of the basement granite. The mineralization lies above this surface in a 2–5 m thick zone covering an area several hundred meters wide and 2–3 km long. Uranium enrichments may have occurred during periods of erosion resulting in the unconformity between the Upper Toki Lignite-bearing Formation and Akeyo Formation, or that between the Oidawara Formation and Seto Group [4]. A single fission-track age indicates that the deposit formed ~10 Ma [8].

The Toki Granite appears to be the source of U in the Tono Uranium Deposit [9, 10]. Relatively oxidizing groundwaters are believed to have leached U from the upper fractured and relatively permeable zone. These solutions then transported U into the overlying Lower Toki Lignite-bearing Formation, where the porewaters are chemically reducing. This transition from oxidizing to reducing conditions appears to be a key factor causing U to be enriched in the Tono Uranium Deposit.

Uranium deposition is associated with a variety of minerals and organic substances [11]. Some of the U is sorbed by carbonaceous materials (including lignite), dioctahedral micas and clay minerals, some is co-precipitated in calcite and possibly other authigenic minerals, and some is precipitated mainly as uraninite or coffinite ($USiO_4$) in amorphous, botryoidal (probably cryptocrystalline) or crystalline form [6, 11–14]. Isotopic ratios among natural decay-series nuclides (^{238}U, ^{234}U, ^{230}Th, ^{226}Ra, ^{210}Pb) suggest that U has been locally redistributed within the deposit during the past several hundred thousand years [15]. Such remobilization could result from local variations in groundwater chemistry, and may have been more or less continuous up to the present time [16].

3. GEOCHEMICAL CONSTRAINTS ON URANIUM SOLUBILITY

Uranium geochemistry is controlled by the ambient redox environment, pH and concentrations of inorganic and organic complexing ligands [17, 18]. Constraints on Eh, pH and carbonate concentrations are interpreted below based on hydrochemical and mineralogical data obtained in field studies of the Tono Uranium Deposit and nearby vicinity. Iwatsuki et al. [19] and

Arthur et al. [20] provide tabulations and assessments of the analytical quality and representativeness of these data.

3.1. Redox Environments

Redox potentials are plotted in Fig. 2 as a function of vertical distance between the sampling location and the unconformity between the Lower Toki Lignite-bearing Formation and Toki Granite (Fig. 1). This unconformity appears to be an important hydrogeological boundary separating aquifers in the overlying sedimentary rocks from those of the upper weathered zone and deeper regions of the Toki Granite [11]. As can be seen, redox conditions in groundwaters of the Tono region are distinctly layered. Reducing waters are present in sedimentary formations of the Mizunami Group and in the Toki Granite at depths greater than ~400 m. Relatively oxidizing waters lie between these two reducing horizons in the upper, weathered zone of the granite. The available in situ Eh measurements in

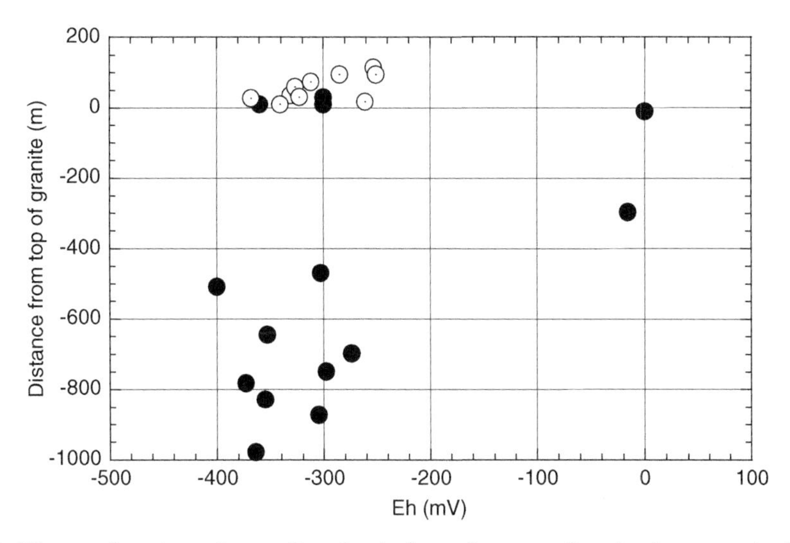

Fig. 2. Eh as a function of sampling depth from the unconformity between the Lower Toki Lignite-bearing Formation and Toki Granite (modified from Arthur et al. [20] with permission from Geological Society Publishing House). The plotted data include in situ Eh measurements (solid symbols), and calculated potentials assuming equilibrium for the SO_4^{2-}/HS^- half-cell reaction (open symbols). In situ Eh measurements in porewaters above the unconformity are from the Lower Toki Lignite-bearing Formation. Calculated Eh values are for porewater samples from this formation, and from overlying formations in the Mizunami Group.

porewaters of the Lower Toki Lignite-bearing Formation agree reasonably well with potentials in other sedimentary porewaters that were estimated assuming equilibrium for the SO_4^{2-}/HS^- couple.

Redox interpretations based on the chemistry of contemporaneous groundwaters, as discussed above, are compatible with mineralogical observations in the Tono area. Relatively oxidizing conditions in sedimentary formations at depths shallower than ~30 m below ground level are indicated qualitatively by the presence of ferric oxyhydroxides. Conversely, pyrite, which is stable only under reducing conditions, is present in these rocks at depths greater than ~60 m. Shikazono and Nakata [16] suggest that diagenesis of the Upper and Lower Toki Lignite-bearing Formations caused framboidal pyrite to precipitate as a result of microbially mediated sulphate reduction.

Fig. 3 illustrates a conceptual model of operative redox environments in the Tono region since the present geological structure formed (i.e., at least over the past few tens of thousands of years) [19, 21]. Microbial sulphate reduction, oxidation of organic matter and pyrite precipitation appear to be dominant reactions controlling the redox chemistry of sedimentary porewaters.

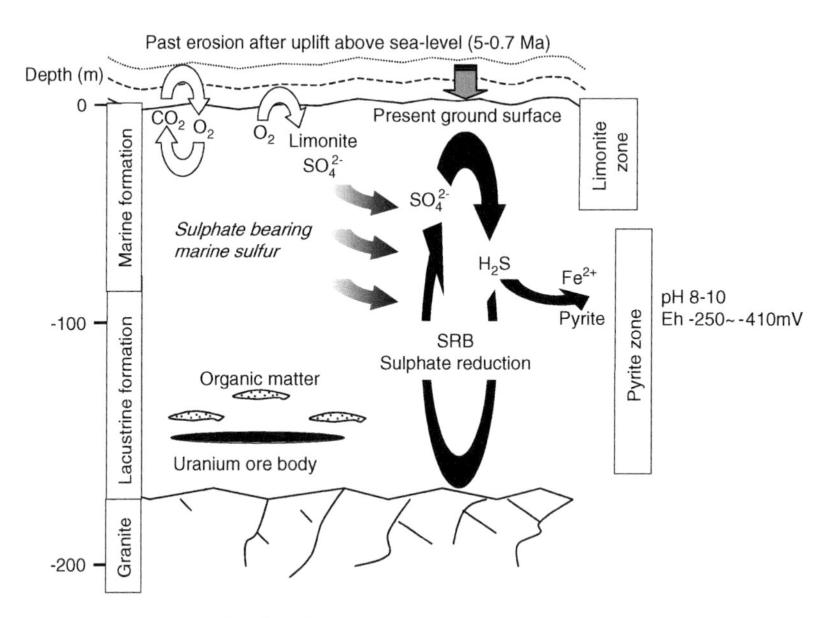

Fig. 3. Conceptual model of redox environments in and around the Tono Uranium Deposit [19].

3.2. Carbonate Equilibria and pH

The pH and carbonate content of groundwaters in the Tono area appear to be buffered by heterogeneous equilibria involving calcite [$CaCO_3$(s)], and possibly other carbonate minerals [11, 22]. Fig. 4 plots variations in pH and $p_{CO_2(g)}$, which has been calculated using analyses of pH and total carbonate concentrations in modern groundwaters of the Tono region [20]. The close correlation that is evident between these parameters includes groundwaters from a variety of rock types, which suggests that the variations in pH and $p_{CO_2(g)}$ may be controlled by a common reaction.

This possibility has been evaluated using a highly simplified model of calcite–water equilibria at 25°C [20]. It was assumed in this model that dissolved carbonate concentrations are controlled by calcite solubility and that Ca concentrations are fixed by charge balance in an open system with respect to $p_{CO_2(g)}$. Model results are represented by the curve in Fig. 4, which indicates that the predictions are in reasonable agreement with the empirical trend in pH-$p_{CO_2(g)}$ observed in deep groundwaters. However, the agreement is not as good for relatively acidic shallow groundwaters. This is consistent with the observation that these solutions are generally from Quaternary alluvium and sediments of the Seto Group, where calcite is notably absent [5]. With the exception of the shallow groundwaters, these model results and empirical observations suggest that calcite equilibrium is an important constraint on the pH and carbonate content of Tono groundwaters.

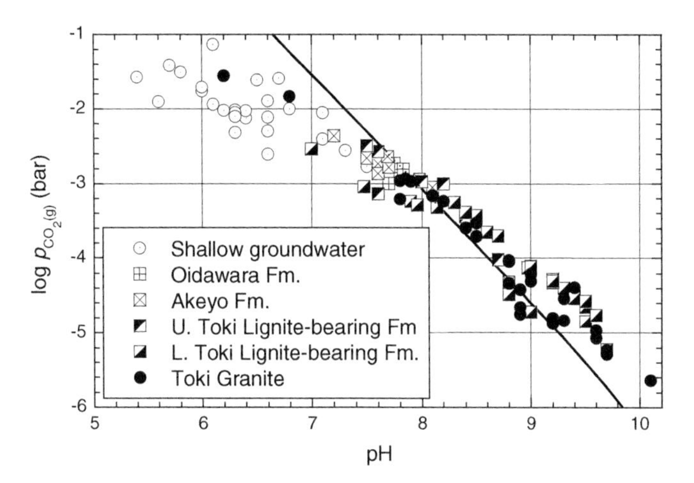

Fig. 4. Comparison of regional trends among pH and $p_{CO_2(g)}$ in Tono groundwaters with calculated variations in these parameters assuming simple calcite buffering in a closed system (line) (modified from Arthur et al. [20] with permission from Geological Society Publishing House).

This conclusion is also supported by other mineralogical evidence. The Akeyo Formation and Oidawara Formation contain abundant marine shell fossils composed dominantly of carbonate (most likely in the form of aragonite, which is a metastable polymorph of calcite). These fossils exhibit dissolution and re-crystallization textures at depths shallower than ~20–60 m below ground level, but these textures are not evident in deeper sections of these formations, nor do they appear in the underlying Upper and Lower Toki Lignite-bearing Formations. This suggests that the geochemical environment deep within the Mizunami Group has been compatible with the preservation of carbonate minerals since the Miocene. Carbonates tend to equilibrate rapidly with groundwater (e.g., Langmuir [23]) and it is therefore reasonable to assume that the buffering of pH and $p_{CO_2(g)}$ shown in Fig. 4 was as important in these Miocene groundwaters as it appears to be today.

4. EVALUATION OF URANIUM SOLUBILITY

The aqueous-speciation and solubility behaviour of U in groundwaters associated with the Tono Uranium Deposit is evaluated in this section using the geochemical constraints discussed above. Supporting calculations were carried out using the Geochemist's Workbench software package [24] and a thermodynamic database described by Arthur et al. [20, 25].

4.1. Aqueous Speciation of Uranium

Arthur et al. [20] defined two solution compositions that may be considered as being representative of relatively oxidizing and reducing groundwaters in the Tono region. The oxidizing solution represents groundwaters from the upper weathered and fractured zone of the Toki Granite (Fig. 2). For illustrative purposes it was assumed that $p_{O_2(g)}$ is fixed at 10^{-47} bar over the pH range 5–10 (see Fig. 4). Assuming equilibrium for the water-dissociation half cell [$4H^+ + 4e^- + O_2(g) = 2H_2O(l)$], this assumption constrains Eh to vary linearly with increasing pH from +0.25 V at pH 5 to −0.06 V at pH 10 [17].

The other groundwater is a strongly reducing solution representing groundwaters of the Lower Toki Lignite-bearing Formation (and other formations of the Mizunami Group), which, as discussed above, is the host formation of the Tono Uranium Deposit. For these solutions it was assumed that $p_{O_2(g)} = 10^{-8}$ bar, which gives a corresponding Eh range between −0.25 V at pH 5 and −0.54 V at pH 10.

Constraints on U and complexing-ligand concentrations in both representative groundwaters were bounded by the hydrochemical data compiled by Iwatsuki et al. [19]. A value of 0.5 ppb U was assumed to be representative of groundwaters in the Tono region (values range from 0.01 to 28 ppb). Upper concentration bounds on inorganic carbon (150 ppm as HCO_3^-.), F^- (20 ppm) and HPO_4^{2-} (0.1 ppm) were assumed in order to consider the maximum effects these ligands could have on U speciation. Scoping calculations indicated that the concentrations of other inorganic ligands (SO_4^{2-} and NO_3^-) and simple organic ligands (acetate, citrate, oxalate and lactate) are too low in Tono groundwaters to significantly affect U speciation. The organic-complexation behaviour of U in Tono groundwaters is the subject of on-going investigations by JAEA.

Fig. 5a and b summarize the results of speciation calculations for the two representative groundwaters. As can be seen in Fig. 5a, U speciation in the reducing groundwater is dominated by the neutral species $U(OH)_4(aq)$ at $pH > 6$. Other species having concentrations >1% of total U over the pH range 5–10 include $UF_4(aq)$, UF_3^+ and UF_5^-.

Uranium speciation is somewhat more complex in the reference oxidizing groundwater (Fig. 5b). Three uranyl carbonate complexes dominate when $pH > 6$. Scoping calculations indicate that phosphate complexes become important over the neutral to slightly acidic pH range as F^- concentrations decrease. Such calculations also indicate that the hydroxide complex, $UO_2(OH)_3^-$, is important at $pH > 8.4$ in groundwaters having carbonate concentrations near the minimum value observed in the Tono region (15 ppm), but it does not exceed 2% of total U in the high-carbonate reference water. Polynuclear, multi-ligand complexes, such as $(UO_2)_2(CO_3)_3(OH)_3^-$, are unimportant in Tono groundwaters because total U concentrations are too low.

In summary, the aqueous speciation of U in solutions that are believed to be representative of the relatively oxidizing and reducing groundwaters of the Tono region is dominated by $U(OH)_4(aq)$ under reducing conditions, and by the uranyl mono-, di- and tri-carbonato complexes, and (or) $UO_2(OH)_3^-$, under relatively oxidizing conditions. Uranous fluoride or phosphate complexes, or their uranyl counterparts, are important in these respective solutions only if pH is less than ~6 or 7.

4.2. Solubility Controlling Phases

Iwatsuki et al. [19] calculated saturation indices for various U minerals using available analyses of Tono groundwaters having both a measured U concentration and either an in situ Eh measurement or a calculated Eh value based on the SO_4^{2-}/HS^- redox couple [20]. The dimensionless saturation index, *SI*, is given by:

$$SI = \log(\mathrm{IAP}/K),$$

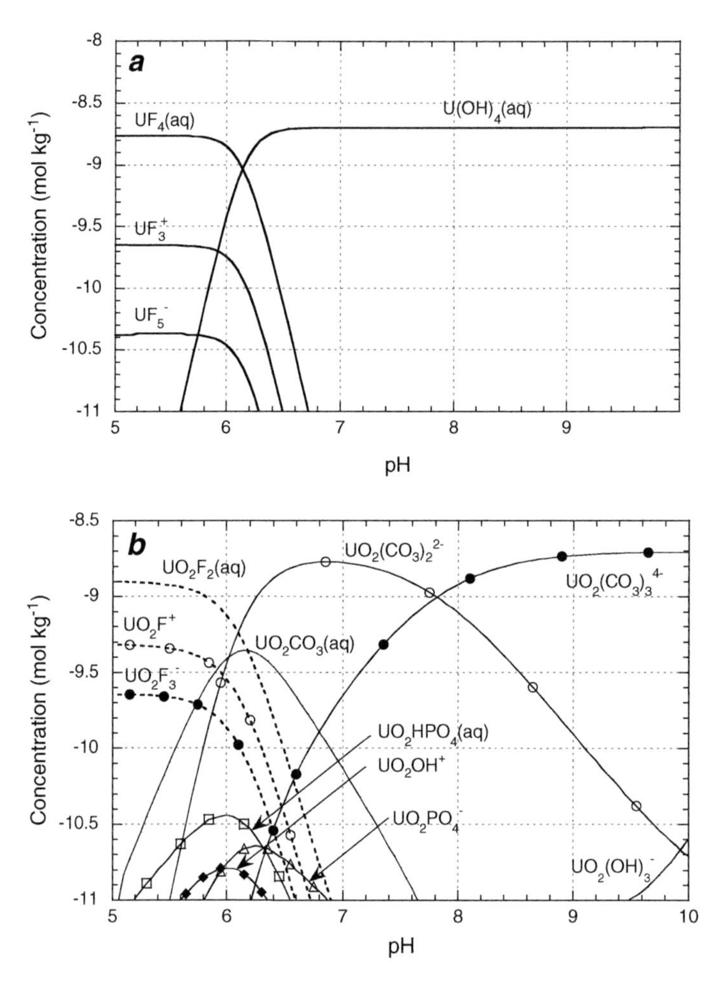

Fig. 5. Aqueous speciation of U in representative Tono groundwaters under reducing conditions (a) and relatively oxidizing conditions (b).

where IAP stands for the ion-activity product for a balanced dissolution reaction involving the mineral of interest and K denotes the corresponding equilibrium constant [23]. SI values less than 0 indicate a thermodynamic potential for dissolution and positive values indicate a corresponding potential for precipitation. Equilibrium is indicated when $SI = 0$. Given inherent uncertainties in thermodynamic data and those associated with groundwater sampling and analysis, it is generally assumed that such equilibrium is indicated when $SI = 0 \pm 0.5$. The U minerals considered include those that are

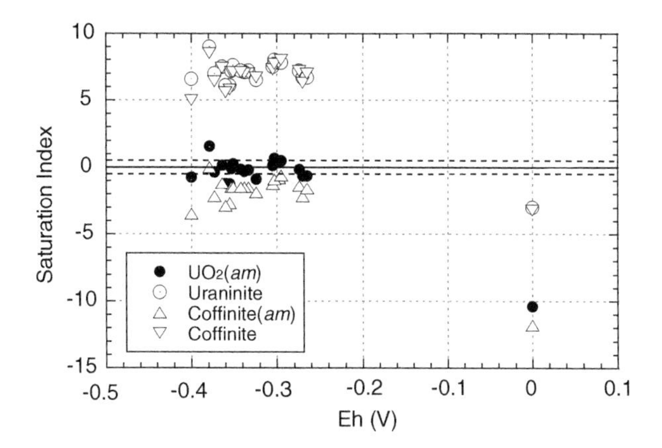

Fig. 6. Calculated saturation indices for selected U minerals [20].

common occurrences in U deposits. Associated thermodynamic data refer to pure crystalline or amorphous phases, and are believed to be reliable [19]. Redox potentials estimated using the SO_4^{2-}/HS^- couple were included in the evaluation because these estimates appear to be reasonably consistent with Eh measurements obtained using a Pt electrode (Fig. 2).

Results are shown in Fig. 6 for uraninite and coffinite, and for their amorphous counterparts UO_2(am) and coffinite(am). As can be seen, most of the reducing groundwaters have saturation indices for UO_2(am) equal to 0 ± 0.5, indicating that these solutions are effectively at equilibrium with this solid. The other reducing waters are either slightly supersaturated ($SI > 0.5$) or slightly undersaturated ($SI < 0.5$) with respect to UO_2(am). These small deviations from equilibrium could reflect additional uncertainties in the thermodynamic properties of UO_2(am) arising from variations in the crystallinity of this solid, which are observed over laboratory time scales [26, 27]. Most of the reducing groundwaters are undersaturated ($SI < 0.5$) with respect to coffinite(am), and strongly supersaturated with respect to uraninite and coffinite. The relatively oxidizing groundwater (Eh = 0 V) is strongly undersaturated with respect to uraninite, coffinite, UO_2(am) and coffinite(am). Other U minerals considered by Iwatsuki et al. [19], including the higher oxides (β-U_4O_9, β-U_3O_7,U_3O_8, schoepite), ningyoite, uranophane, rutherfordine, soddyite and autunite, are generally either highly supersaturated or highly undersaturated in reducing and oxidizing groundwaters of the Tono region. The *SI* results thus indicate that maximum U concentrations in groundwaters of the Tono region appear to be controlled by the solubility of UO_2(am). This includes porewaters in the Lower Toki Lignite-bearing Formation, which hosts the Tono Uranium Deposit, and in porewaters of the other formations in the Mizunami Group that are chemically similar.

It is worthwhile emphasizing that UO_2(am) rather than its crystalline counterpart, uraninite, appears to be solubility controlling in this natural system. This distinction is further illustrated by the Eh—pH diagrams shown in Fig. 7a and b. Both diagrams are drawn assuming that the activity of the relevant U aqueous species = $10^{-8.7}$, which corresponds to a representative total U concentration equal to 0.5 ppb (see Section 4.1). The figures cover a range of CO_2(g) partial pressures that are representative of groundwaters in the Tono region (Fig. 4). As can be seen, the stability field of UO_2(am) is much smaller than that of uraninite over this range. This suggests that if uraninite were solubility controlling, aqueous U concentrations would have to much lower at a given pH (i.e., by roughly eight orders of magnitude) than is actually observed. A comparison of the figures leads to the same (albeit not independent) conclusion drawn from the *SI* calculations: redox conditions in Tono groundwaters (indicated by the symbols) are compatible with solubility control by of UO_2(am).

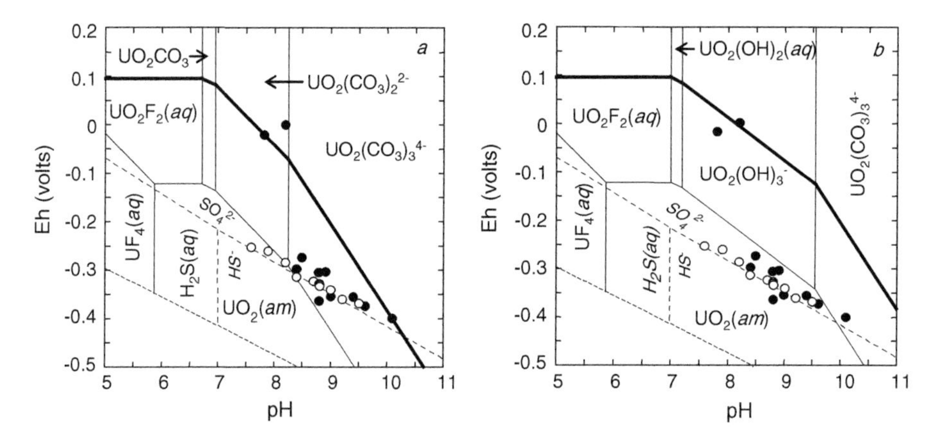

Fig. 7. Stability relations among U minerals and aqueous species at 25°C, 0.1 bar, with $p_{CO_2(g)} = 10^{-3}$ (a) and 10^{-5} bar (b) [19]. Activities of U, S and F species + $10^{-8.7}$, 10^{-5} and $10^{-3.3}$, respectively. Solid lines delineate stability boundaries for UO_2(am) and predominance fields for aqueous U species. Bold lines indicate uraninite stability boundaries. Dashed lines refer to predominance fields for aqueous species of S. Open symbols denote measured or estimated redox conditions in sedimentary formations in the Mizunami Group. Solid symbols refer to their counterparts in groundwaters of the Toki Granite.

This conclusion seems inconsistent with mineralogical observations indicating that uraninite is in fact present in the Tono Uranium Deposit (Section 2). A possible explanation for this apparent dichotomy is the experimental observation of Neck and Kim [28] that uraninite surfaces in contact with an aqueous phase at pH >3 may be coated with a thin layer of UO_2(am). Under such conditions, uraninite dissolution is effectively irreversible, and solubility is controlled by the amorphous surface layer. Additional experimental studies and observations of relevant natural systems are needed to test this hypothesis.

Berke [14] attempted to confirm the presence of UO_2(am) in rock samples from the Lower Toki Lignite-bearing Formation using a variety of mineralogical techniques. The selected samples were not from the Tono Uranium Deposit, but they did have U concentrations above background levels. Berke [14] found that U deposition is associated with fine-grained, poorly crystalline matrix minerals, either as poorly crystalline or amorphous U minerals or sorbed onto other non-U phases such as Ti-oxides. A definite conclusion regarding the crystalline or amorphous nature of any U minerals in these samples could not be made due to the poor crystallinity of the matrix.

4.3. Solubility Constraints on Maximum Uranium Concentrations

Solubility limits on U concentrations in groundwaters of the Tono Uranium Deposit can be estimated using the aqueous-speciation and solubility constraints discussed above. These constraints suggest that the solubility of UO_2(am) will mainly be controlled by variations in Eh, pH and dissolved carbonate concentrations. Fig. 8a–c plot calculated UO_2(am) solubilities as a function of Eh at pH 8, 9 and 10, respectively. Three curves are shown in each figure corresponding to the nominal, maximum and minimum $p_{CO_2(g)}$ values given by the best-fit curve through a selection of the pH-$p_{CO_2(g)}$ data shown in Fig. 4 for porewaters in the Lower Toki Lignite-bearing Formation [19]. Also shown in each figure are Eh values corresponding to the HS^-/SO_4^{2-} and pyrite/goethite equilibrium redox couples. These values are considered because the available Eh measurements in groundwaters of the Lower Toki Lignite-bearing formation, the presence of detectable sulphide and sulphate-reducing bacteria in these groundwaters [22], and the presence of authigenic pyrite in this formation [5], all point to reducing conditions that may be reasonably bounded by redox potentials corresponding to the HS^-/SO_4^{2-} and pyrite/goethite couples [19, 20]. The results shown in Fig. 8a–c suggest that UO_2(am) solubilities could vary by nearly three orders of magnitude ($10^{-6.1}$–$10^{-8.7}$ molal) over the selected range of $p_{CO_2(g)}$, Eh and pH values that are considered to be representative of conditions in the

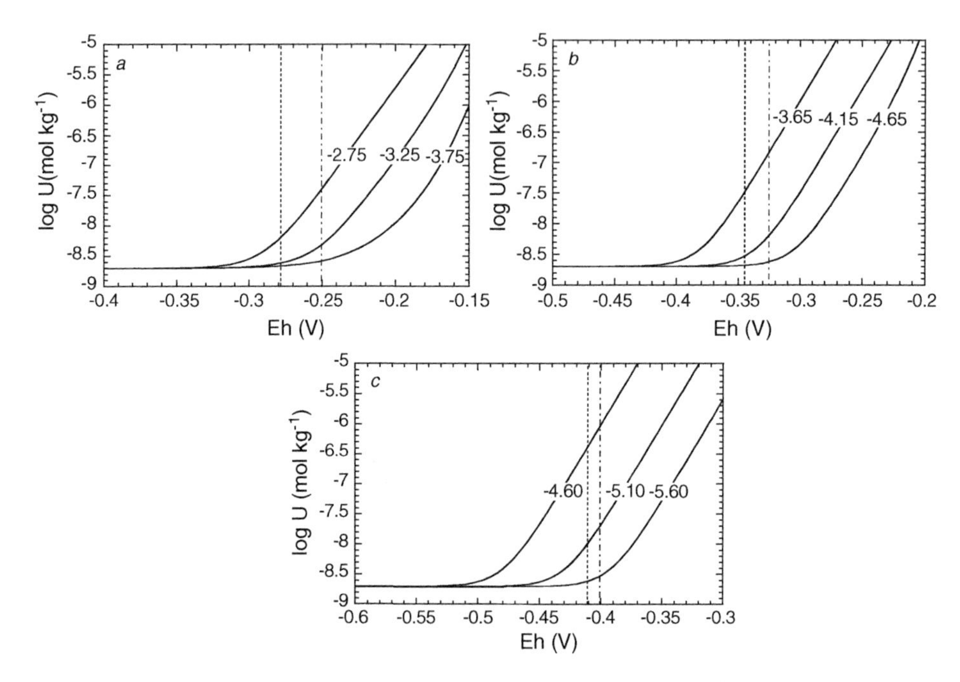

Fig. 8. Solubility of UO_2(am) as a function of Eh and $p_{CO_2(g)}$ at 25°C and 1 bar. Vertical lines indicate Eh values for the pyrite/goethite (dotted-dashed line; drawn assuming log $a_{SO_4^{2-}} = -4$) and SO_4^{2-}/HS^- (dashed line) redox couples at pH 8 (a), pH 9 (b) and pH 10 (c).

Tono Uranium Deposit. Local variations in these parameters in time and space could lead to a cycle of dissolution and re-precipitation of UO_2(am), which may be consistent with the isotopic evidence among natural decay-series nuclides noted in Section 2.1 suggesting that U has been locally remobilized in the deposit during the past several hundred thousand years. Such mobilization/re-precipitation of U would be most sensitive to local variations in $p_{CO_2(g)}$.

5. CONCLUSIONS

Natural systems that may be analogous in certain respects to a geologic repository for nuclear wastes provide a useful basis for testing assumptions, models and concepts used in repository performance assessments. The present study demonstrates that maximum U concentrations in groundwaters of the Tono Uranium Deposit appear to be limited by the solubility of the amorphous, hydrous oxide, UO_2(am). This conclusion, drawn from observations of a geologic system that has evolved over long periods of time, supports the assumption adopted in many international performance assessments that the solubility of UO_2(am), or an analogous UO_2 solid, would limit the aqueous concentrations of U released from a HLW repository over similar time scales [2]. Equilibrium calculations suggest that UO_2(am) solubilities could vary roughly between 10^{-6} and 10^{-9} mol kg^{-1} over the range of environmental conditions presently observed in the Tono Uranium Deposit and environs. Further study is needed to better characterize the organic geochemistry of Tono groundwaters and related impacts on the aqueous-speciation and solubility behaviour of U.

REFERENCES

[1] D. Savage (ed.), The Scientific and Regulatory Basis for the Geological Disposal of Radioactive Waste, Wiley, Chichester, UK, 1995.
[2] I.G. McKinley and D. Savage, J. Contam. Hydrol., 21 (1996) 335.
[3] J. Itoigawa, Monographs of the Mizunami Fossil Museum 1 (1980) 1.
[4] Y. Yusa, K. Ishimaru, K. Ota and K. Umeda, Proceedings of paleohydrogeological methods and their applications for radioactive waste disposal, OECD/NEA, p. 117, 1993.
[5] N. Shikazono and M. Utada, Mineral. Dep., 32 (1997) 596.
[6] E. Sasao, K. Ota, T. Iwatsuki, T. Niizato, R.C. Arthur, M.J. Stenhouse, W. Zhou, R. Metcalfe, H. Takase and A.B. Mackenzie, Geochem. Explor. Environ. Anal., 6 (2006) 5.
[7] T. Iwatsuki, H. Satake, R. Metcalfe, H. Yoshida and K. Hama, Applied Geochem., 17 (2002) 1241.
[8] Y. Ochiai, M. Yamakawa, S. Takeda and F. Harashima, CEC Nuclear Science and Technology Series, EUR, 11725 (1989) 126.
[9] K. Doi, S. Hirono and Y. Sakamaki, Econ. Geol., 70 (1975) 628.
[10] Y. Sakamaki, IAEA-TECDOC, 328 (1985) 135.
[11] JNC, JNC TN 1400 2000-002, Japan Nuclear Cycle Development Institute, Japan, 2000.
[12] K. Komuro, M. Yamamoto, S. Suzuki, T. Nohara and S. Takeda, Mining Geol., 40 (1990) 44.
[13] K. Komuro, Y. Otsuka and M. Yamamoto, Mining Geol., 41 (1991) 177.
[14] M.A. Berke, M.Sc. Thesis, University of California, Riverside, CA, 2003.
[15] T. Nohara, Y. Ochiai, T. Seo and H. Yoshida, Radiochim. Acta, 58/59 (1992) 409.
[16] N. Shikazono and M. Nakata, Res. Geol., Special Issue 20 (1999) 55.
[17] R.M. Garrels and C.I. Christ, Solutions, Minerals and Equilibria, Harper Row, New York, 1965.
[18] D.L. Langmuir, Geochim. Cosmochim. Acta, 42 (1978) 547.
[19] T. Iwatsuki, R.C. Arthur, K. Ota and R. Metcalfe, Radiochim. Acta, 92 (2004) 789.
[20] R.C. Arthur, T. Iwatsuki, E. Sasao, R. Metcalfe, K. Amano and K. Ota, Geochem. Explor. Environ. Anal., 6 (2006) 33.

[21] Y. Murakami, Y. Fujita, T. Naganuma and T. Iwatsuki, Microbes Environ., 17 (2002) 63.
[22] T. Iwatsuki, K. Sato, T. Seo and K. Hama, Proc. Mater. Res. Soc. Symp., 353 (1995) 1251.
[23] D.L. Langmuir, Aqueous Environmental Geochemistry, Prentice Hall, Upper Saddle River, NJ, 1997.
[24] C.M. Bethke, Geochemical Reaction Modeling, Oxford University Press, Oxford, UK, 1996.
[25] R.C. Arthur, H. Sasamoto, M. Shibata, M. Yui and A. Neyama, JNC TN8400 99-079, Japan Nuclear Cycle Development Institute, Japan; 1999.
[26] T. Yajima, Y. Kawamura and S. Ueta, Mater. Res. Soc. Proc., 353 (1995) 1137.
[27] D. Rai, A.R. Felmy, N.J. Hess and D.A. Moore, Radiochim. Acta, 82 (1998) 17.
[28] V. Neck and J.I. Kim, Radiochim. Acta, 89 (2001) 1.

Thermodynamics, Solubility and Environmental Issues
T.M. Letcher (editor)

Chapter 9

Leaching from Cementitious Materials Used in Radioactive Waste Disposal Sites

Kosuke Yokozeki

Civil Structure and Materials Group, Kajima Technical Research Institute, 19-1-2 Tobitakyu, Chofu-shi, Tokyo 182-0036, Japan

1. INTRODUCTION

Cementitious materials are being studied for use in waste disposal facilities for radioactive waste discharged from nuclear power stations and the like. To suppress leakage of radioactive nuclides, the facilities should have long-term durability, that is, durability over a period quite distinctly different from the performance period of ordinary structures, such as a period of several thousands or several tens of thousands of years, which needs to be explained logically as well as technically [1–5]. Furthermore, assessment methods for materials and structures, and development of high durability materials are anticipated considering aspects of design service life and importance of structures.

This report describes the degradation of concrete structures, that is, mainly structures over thousand years and about ten thousand years old with a history longer than that of Portland cement, due to soft water. In the underwater environment of radioactive waste repositories especially, as shown in Fig. 1, cement hydrates leach into the groundwater. An overview of the methods to predict degradation and methods to suppress degradation, with the focus on the phenomenon of degradation of cementitious material, is described here.

2. RADIOACTIVE WASTE DISPOSAL SITE AND CONCRETE

Radioactive waste disposal facilities are investigated for construction in deep strata 50–1000 m under the ground [1–5]. As shown in Fig. 2, the use of clayey materials such as bentonite, and cementitious materials, such as concrete and

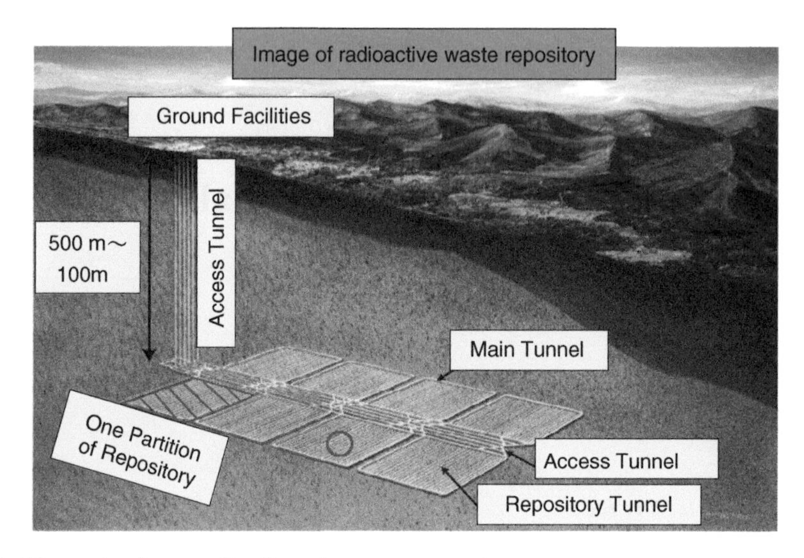

Fig. 1. Illustrative image of radioactive waste repository.

mortar either independently or in combination as back filling materials or engineered barrier materials, has been studied. Radioactive waste disposal facilities are generally constructed in a calm and stable environment where degradation does not occur easily. Therefore, the study of degradation over the long term, such as leaching of components into groundwater, is more important for engineered barrier materials than severe degradation over the short term [6]. On the other hand, when clayey materials such as bentonite and cementitious materials are combined, concerns of various problems due to their mutual interaction also arise. That is, the ion exchange reaction between the calcium that leaches from the cement hydrates and the sodium in the bentonite brings about cation exchange (of Na ion with Ca ion) in the bentonite, giving rise to the possibility of degradation in the swelling behavior of bentonite or degradation in its imperviousness. The degradation due to sulfate ions discharged from the bentonite is also a cause for concern in cementitious materials. Furthermore, the alkali that leaches from the cement hydrate changes the pH of the surrounding environment, degrading the protective ability of the metallic waste container, and also affecting the nuclide migration rate.

3. LEACHING FROM CEMENTITIOUS MATERIALS

The deterioration of concrete due to leaching has been reported in Sweden when a dam deteriorated because of soft water in the 1920s, and in Scotland

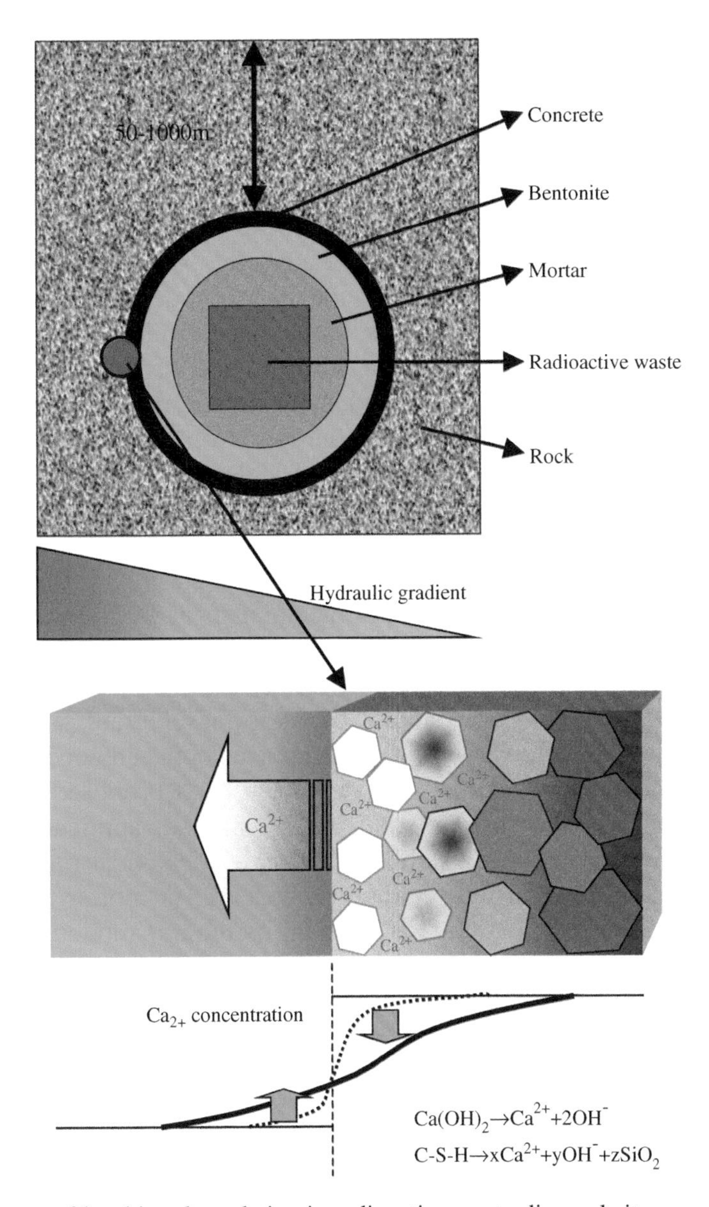

Fig. 2. Images of leaching degradation in radioactive waste disposal site.

when the strength of dams 10–60 years after construction deteriorated with the passage of time [7]. Results of organizational studies of a similar nature have also been reported from South Africa [8]. Additionally, deterioration of concrete has also been reported because of contact with soft water of low hardness such as the secondary lining of tunnels and water treatment facilities. Studies have also been carried out on the effects of safety especially the effects of leached components into drinking water in water supply facilities [9]. On the other hand, in the 1980s, the long-term soundness assessment of radioactive waste repositories became an important topic, and the leaching of components from concrete has become a popular research topic [10].

Both short-term accelerated tests in the laboratory and analytical methods have been proposed for the research on the long-term degradation of concrete used in radioactive waste repositories. Past experimental investigations have focused on the analysis of solutions with the aim of simulating the pore solution into which hydrates leach, and many batch-type experimental investigations have been carried out with the focus on solid–liquid solubility equilibrium [6, 11]. However, it was not possible to directly evaluate degradation with the passage of time from the results of such investigations. New conditions had to be tentatively set up, such as assumptions regarding inflows of water from external sources to allow for evaluation with respect to the passage of time. Typically, these experimental investigations involved determining the solid-phase components of the hydrates and the components of the pore solution. Moreover, investigations were carried out to evaluate changes in physical properties as the balance of solid-phase–liquid-phase components varied, and also to show up the differences between accelerated tests and actual phenomena. Although models have been proposed specifically for paste in analytical investigations [11, 12], methods applicable to actual structures have not yet been established. This is because of problems such as: (1) there are practically no models of concrete or mortar that take the aggregate into account; (2) there is no clear method of modeling diffusion coefficients; (3) there is no method for estimating physical properties; and (4) no verifications against actual structures have been carried out.

4. METHOD FOR PREDICTING DURABILITY OF CONCRETE [13–16]

Although various kinds of accelerated test methods [17] have been studied for evaluating and predicting degradation of concrete due to leaching, such methods require long periods of time that have never been experienced before; thus, finally, the dependence on analytical methods has become a necessity. Analytical prediction methods may be broadly divided into three types: (1) methods using empirical formula [18], (2) methods using diffusion equations [12], and (3) methods using geochemical mass transfer analysis [19].

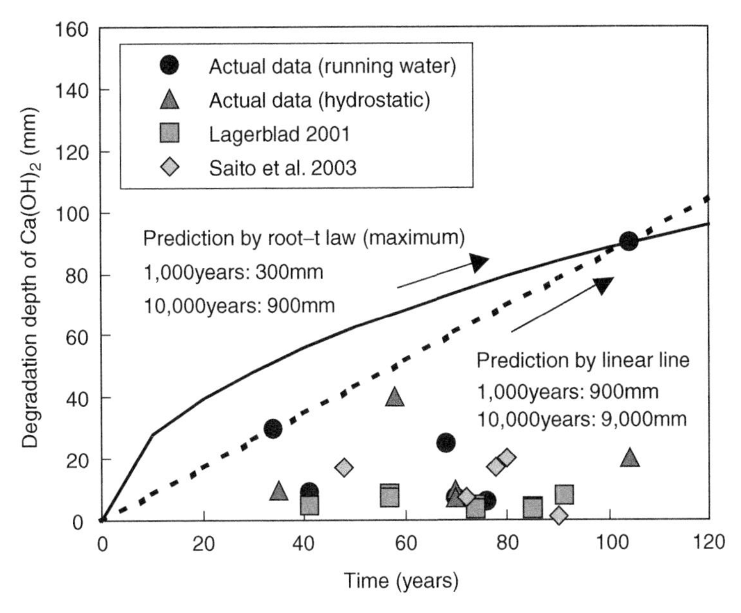

Fig. 3. Time-dependent degradation depths by Ca leaching.

Deterioration of existing structures due to leaching has been occurring without any doubt in structures acted upon by soft water such as water supply facilities, canals, dams, and vessel-type structures. Fig. 3 shows the results of studies on concrete structures of age in the range of 30–100 years (depth to which calcium hydroxide has decreased). Even after 100 years, the degradation thickness is a few tens of millimeters, and the degradation rate was extremely slow. Consequently, very few problems are anticipated in ordinary structures. The maximum value of these investigation results is extrapolated to estimate the degradation after 1000 years or 10000 years. It can be observed that when the data are subjected to linear regression and to regression on the root-t law, the degradation depth becomes three times at 1000 years, while it increases to 10 times at 10000 years.

Thus, an analysis method with a much better estimation accuracy is necessary. For a more detailed analysis, a method using the diffusion equation or the geochemical mass transfer analysis must be used. The biggest difference in these two methods is the chemical reaction model. The method is frequently used for predicting neutralization or salt attack of concrete structures in the civil engineering fields. Regarding mass transfer, the law of conservation of mass relating to the solid-phase element concentration C_{Pi} and the concentration of each ion in pore solution shown in Eq. (1) is used (details are omitted for both Eqs. (2) and (3) because of the lack of space). In Eq. (1), the left hand side expressing diffusion, convection, electric migration, and sink terms, the terms to be finally considered, should be decided from the grade of the study and the environment of the structure to be studied.

$$\frac{\partial(\theta \cdot C_i)}{\partial t} = \frac{\partial}{\partial x}\left(D_{\mathrm{eff},i} \cdot \theta \cdot \frac{\partial C_i}{\partial x}\right) - \frac{\partial(v_\mathrm{d} \cdot C_i)}{\partial x} - \frac{\partial(\theta \cdot u_i \cdot C_i)}{\partial x} - \frac{\partial C_{\mathrm{Pi}}}{\partial t} \tag{1}$$

However, in cementitious material, research related to the diffusion coefficient, the permeation coefficient, and quantification of the electric mobility has just made a start, and models corresponding to various materials and environmental conditions have not yet been constructed as of this stage. In the past, apparent diffusion coefficients were used including aggregates as shown on the left side of Fig. 4, but in practice, substances such as pore solution and ions did not move within the aggregates or the hydrates themselves, but they can move in capillary porosity and transition zones. Thus, an effective diffusion coefficient $D_{i,\mathrm{eff}}$ calculation equation is proposed as

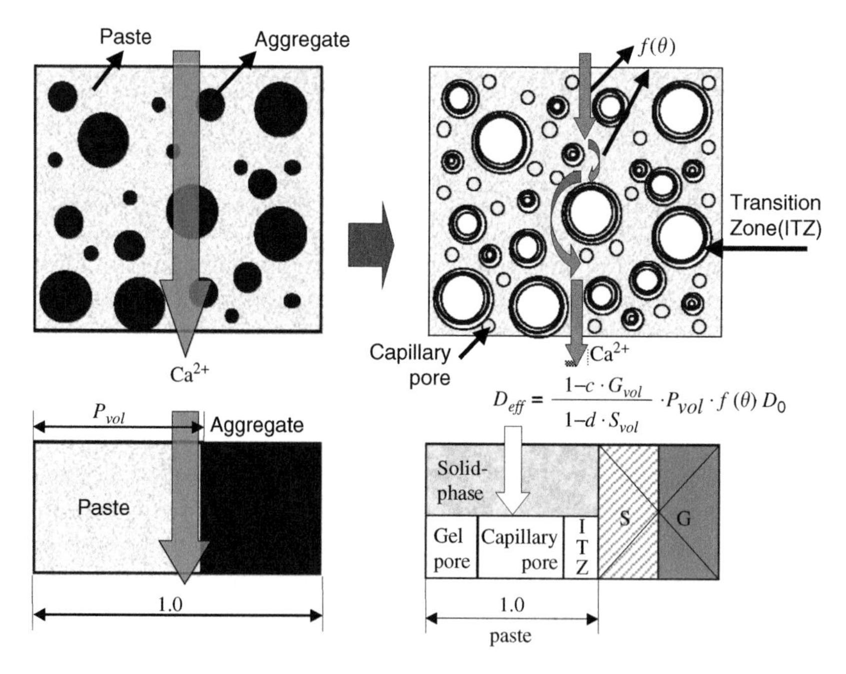

Fig. 4. Pore and micropore torsion modeling.

shown in Eq. (2) taking the electrochemical theory for ion transfer in aqueous solution, that is, by taking the Nernst–Planck equation as the basis, and considering the effects of concentration according to the Debye–Hückel theory, the effects of torsion of the porosity or aggregate and transition zone, and the increase in porosity with leaching. This enables a wide range of parameters such as type of cement, hydration, temperature, aggregate amount, and ion type to be considered. Fig. 5 shows the experimental results of porosity and diffusion coefficient, together with the results predicted based on Eq. (2). The diffusion coefficient decreases with the decrease in porosity; however, the relationship is not linear, but curve, suggesting that the assessment accuracy is satisfactory.

Moreover, the diffusion coefficient of each ion is different according to Eq. (2). However, in practice, the ion balance does not collapse because of diffusion; the mass exists maintaining electric neutrality at all times. Thus, the mutual interaction between the electric ions can be expressed by the diffusion coefficient by assuming that the pore solution always maintains its electrically neutral condition. The electric mobility u_i in this case is given by Eq. (3). This enables the mass transfer considering the mutual electric interaction of each ion to be treated within the law of conservation of mass of Eq. (1).

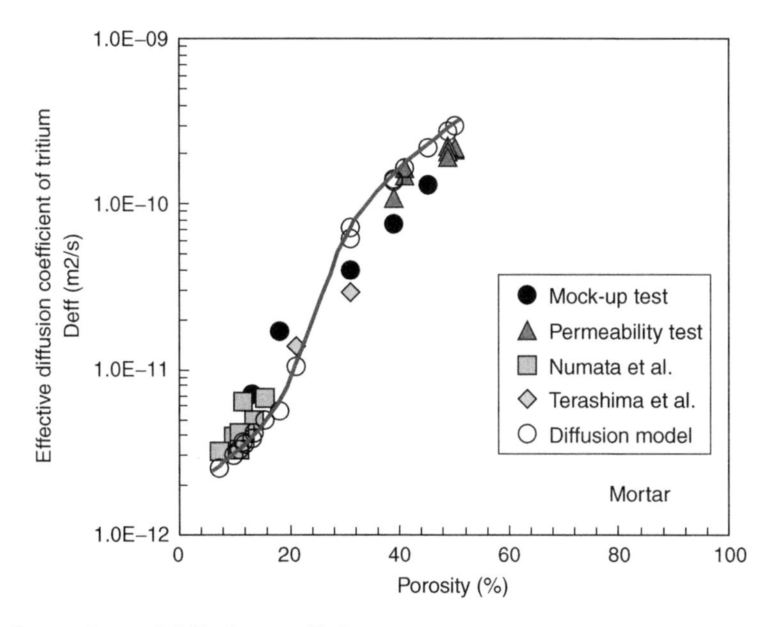

Fig. 5. Comparison of diffusion coefficients.

$$D_{i,\text{eff}} = \frac{1 - c \cdot G_{\text{vol}}}{1 - d \cdot S_{\text{vol}}} \cdot P_{\text{vol}} \cdot \{0.001 + 0.07 \cdot (\theta_{\text{total}} - \theta_{\text{gel}})^2$$
$$+ 1.8 \cdot (\theta_{\text{total}} - \theta_{\text{gel}} - 0.18)^2 \cdot H(\theta_{\text{total}} - \theta_{\text{gel}} - 0.18)\}$$
$$\cdot \frac{k \cdot T}{Z_i \cdot e} \cdot \frac{\lambda_i}{F} \cdot \left(1 - \ln 10 \cdot C_i \cdot \frac{0.51 \cdot Z_i^4}{4\sqrt{I}(1 + \sqrt{I})^2} \right) \tag{2}$$

$$u_i = Z_i \cdot D_{i,\text{eff}} \cdot \frac{\sum_i Z_i \cdot D_{i,\text{eff}} \cdot (\partial C_i / \partial x)}{\sum_i Z_i^2 \cdot D_{i,\text{eff}} \cdot C_i} \tag{3}$$

For modeling the dissolution reaction of cement hydrate, the method of using a thermodynamic database related to chemical reactions based on the experimental results for dissolution equilibrium of cement hydrates shown in Fig. 6 is also available. However, that is not applicable to concrete made of complex hydrates that are changeable as in the dissolution of cement hydrate. Moreover, the reaction also varies considerably depending on the type and

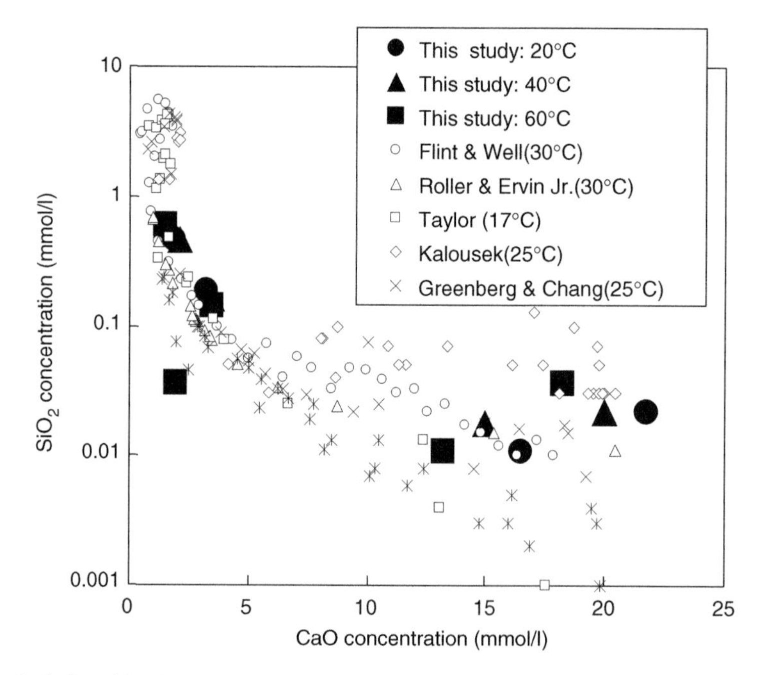

Fig. 6. Relationship between liquid-phase calcium ion and liquid-phase silica ion concentration.

mix proportion of the cement; therefore, it cannot be treated easily from the aspects of theoretical chemistry. Consequently, dissolution tests of several kinds of cement under different temperature conditions were carried out. Calcium, which is proportionately large, or calcium silica ratio (Ca/Si) was taken as the parameter of cement component, so that the dissolution behavior can be easily expressed. Fig. 7 shows the dissolution equilibrium model of calcium in the solid and the liquid phases. Substituting this relationship in the sink term, and calculating the mass transfer per unit time of the components in the liquid phase by the difference method, the change in mass in the solid phase can be calculated.

Fig. 8 shows an example of comparison of the analysis results and actually measured results of solid-phase calcium concentration after leaching. The actually measured results are for concrete of age 30–100 years in various kinds of environments such as distribution reservoir in a water supply facility, building foundation structure in contact with groundwater, dam, and so on, similar to Fig. 3. Variation exists in actually measured results also, and although calculated accuracy cannot be discussed unconditionally, satisfactory predictions can be made depending on various conditions.

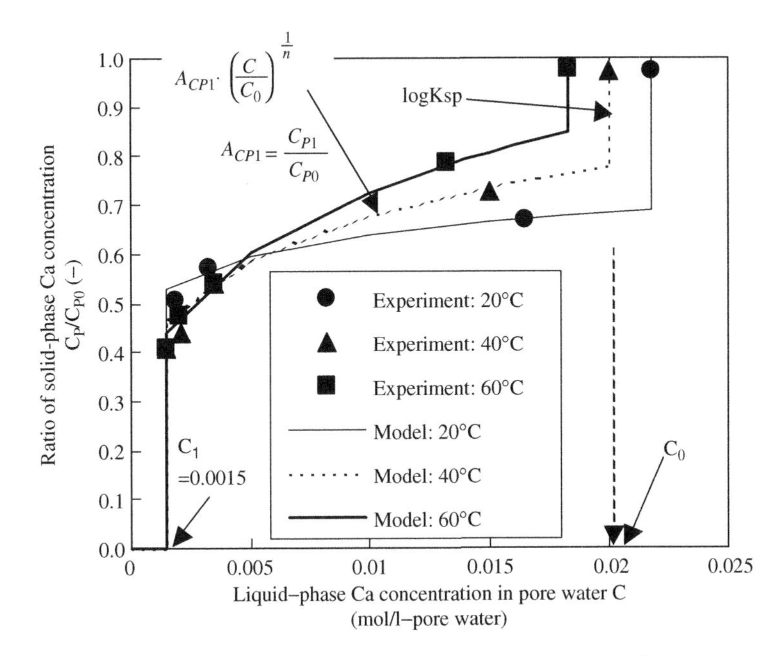

Fig. 7. Equilibrium model of solid-phase calcium concentration and liquid-phase calcium ion concentration.

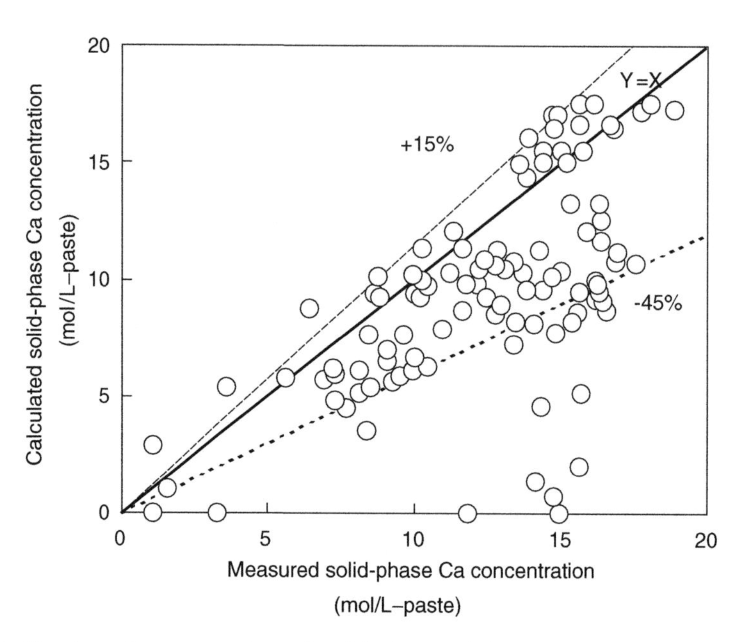

Fig. 8. Verification of concentration of solid-phase calcium.

Since the past, it has been said in relation to the changes in mechanical properties after leaching that if the calcium hydrate decreases by 1%, the compressive strength decreases by 1.5% [21]. However, the calcium hydrate in cement paste is ~30%, and the strength of the part in which the calcium hydrate was completely lost was not known. Fig. 9 shows a comparison of Vickers hardness and concentration of solid-phase calcium for the core from actual structures is suggesting good correlation. It is also known that the Vickers hardness has a good correlation with compressive strength. Accordingly, if the calcium concentration in the solid phase is known, the strength characteristics after leaching can be predicted. Fig. 10 shows the distribution of solid-phase calcium concentration in laboratory leaching tests [20]. It can be observed that the higher the value of *W/C*, the deeper is the degradation; degradation progresses due to diffusion at the corners since degradation has occurred above the two directions at the corners. The estimated Vickers hardness and the measured values can be compared by substituting the solid-phase calcium concentration measured in these tests in the relationship of Fig. 9. This comparison is shown in Fig. 11. The Vickers hardness can thus be estimated with good accuracy by this method.

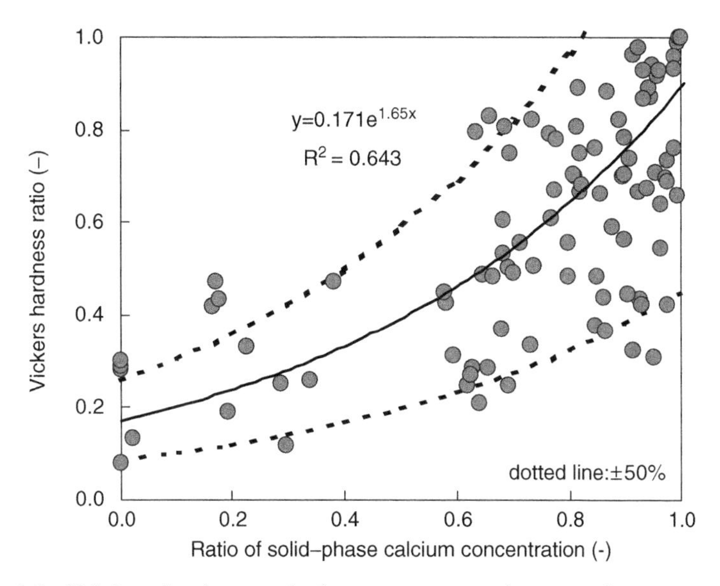

Fig. 9. Model of Vickers hardness ratio (paste, mortar, and concrete).

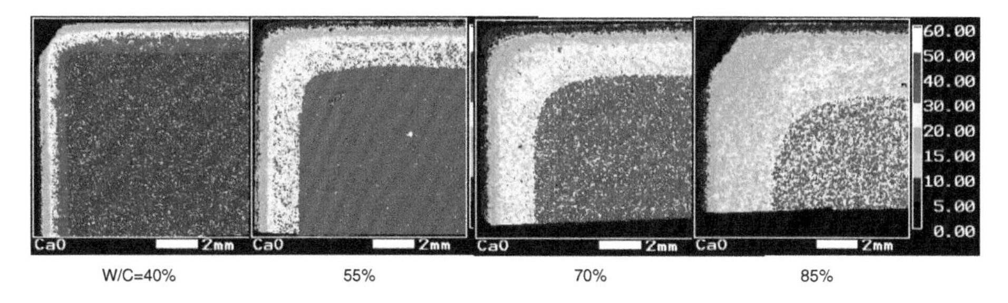

Fig. 10. Calcium concentration distributions after leaching.

5. MEASURES AGAINST LEACHING DEGRADATION

It is known from the past that hydrates leach from concrete. Taylor [21] has proposed various measures against leaching by using Pozzolanic materials, alumina cement, dense concrete, carbonated concrete, autoclave curing, and so on. Even ancient concrete used 5000 years ago in China [22] did not use steel; the cement close to today's low-heat Portland cement had carbonated, and had helped to maintain the long-term soundness of the material. Based on such information, the leaching resistance of carbonated concrete and Pozzolan material was evaluated. The dissolution equilibrium relationships

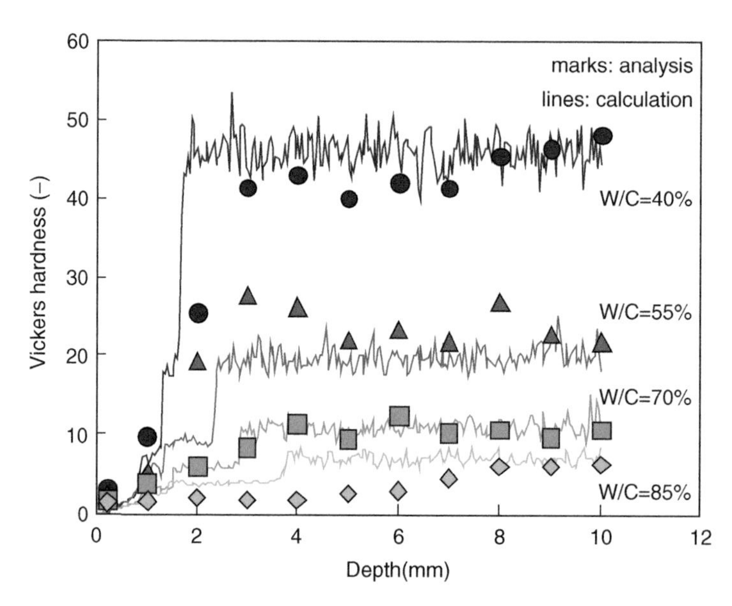

Fig. 11. Verification of predicted Vickers hardness.

of ordinary Portland OPC, concrete LPC+FA using Pozzolan material, and carbonated concrete HDC using γ-C_2S shown in Fig. 12 were used.

Fig. 13 shows the degradation prediction results when Pozzolan materials were used. The degradation depth of low-heat Portland cement was smaller compared to that of ordinary cement; the effect due to calcium hydroxide consumption by fly ash and the effect due to micropore torsion of fly ash concrete were large. In cements that included Pozzolan materials in which hydrates of lower solubility are formed, the degradation depth can be reduced to as much as one-fifth that of ordinary cement at an age of 5000 years. Fig. 14 shows the predicted results of the effects of carbonation curing on calcium leaching. Compared to water curing, the degradation depth at 10000 years can be reduced to about one-fifth, as suggested in the figure. This is considered to be due to the calcium hydroxide that has changed to calcium carbonate with low solubility.

Based on these predicted results, it was concluded that the leaching amount can be suppressed by using appropriate Pozzolan materials, carbonation curing, and using low water to binder ratio. If cement is used with low-heat Portland cement, γ-$CaO \cdot 2SiO_2$ (γ-C_2S), fly ash, and silica fume as admixtures, and carbonation curing is performed at the initial stage of hydration, the porosity can be reduced significantly, the calcium leaching

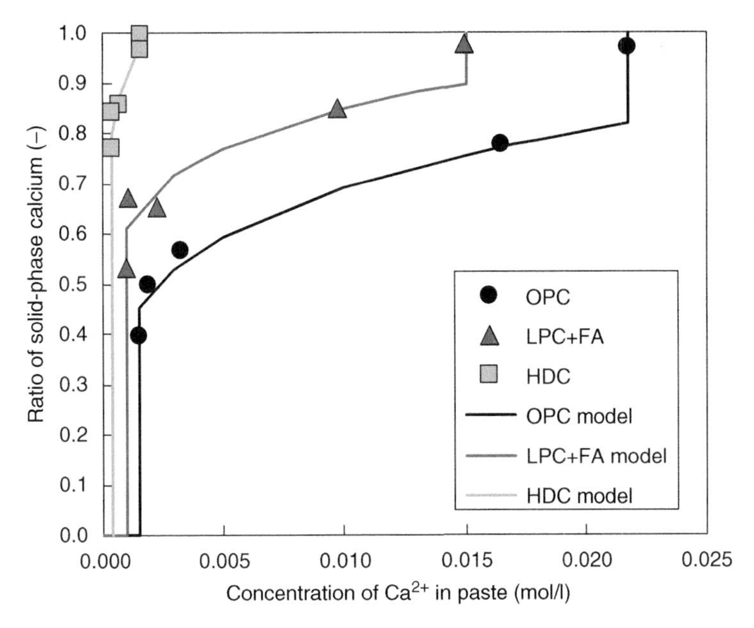

Fig.12. Equilibrium of calcium ion in cement paste.

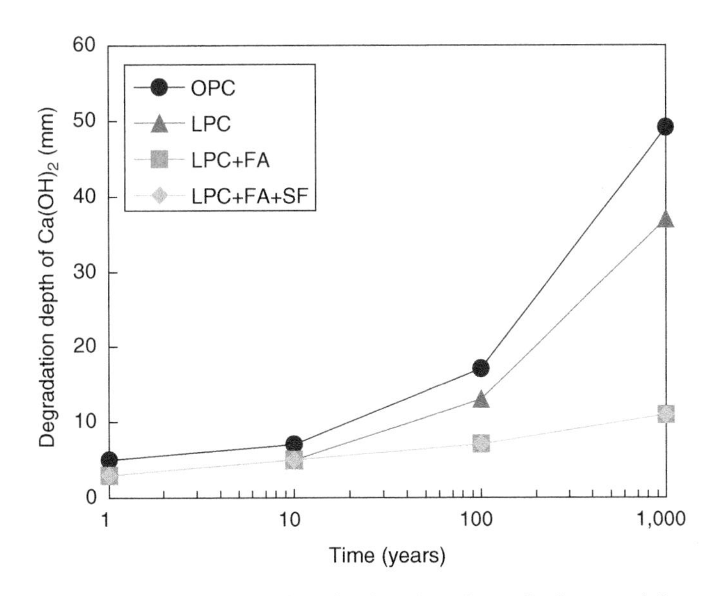

Fig. 13. Predicted results of degradation depth, when Pozzolanic materials are used.

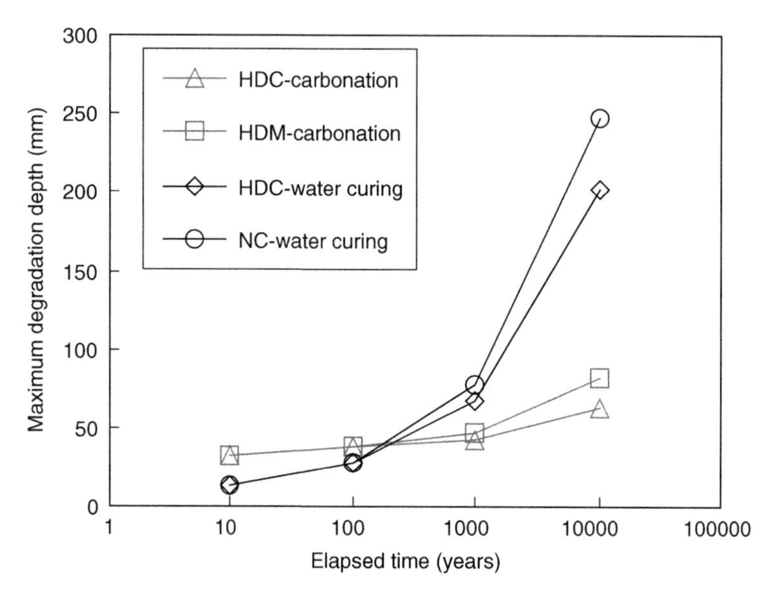

Fig. 14. Predicted results of degradation depth, when carbonation-cured materials are used.

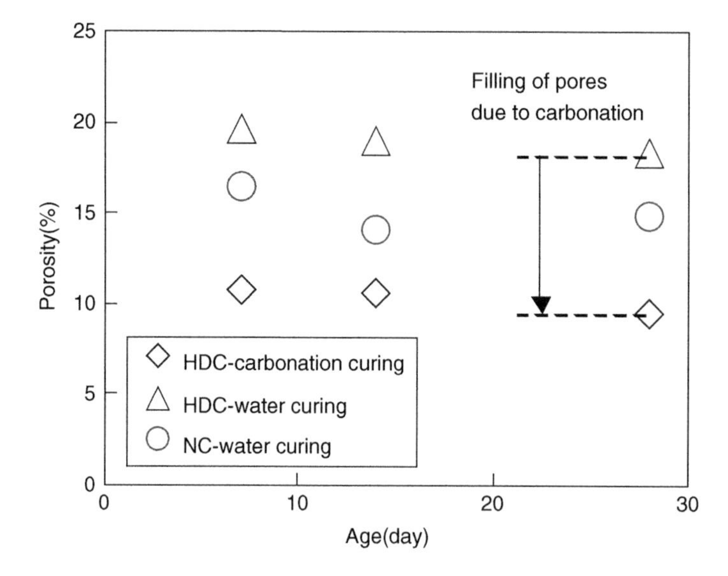

Fig. 15. Reduction of porosity by carbonation curing of HDC using γ-2CaO · SiO_2.

amount can be reduced considerably, and the pH value can be reduced (suppresses the degradation of bentonite and rock in the surroundings), as shown in Figs. 15–17 [23, 24]. Especially, γ-C_2S does not show any hydration reaction, and since it has the property of reacting with carbon dioxide gas,

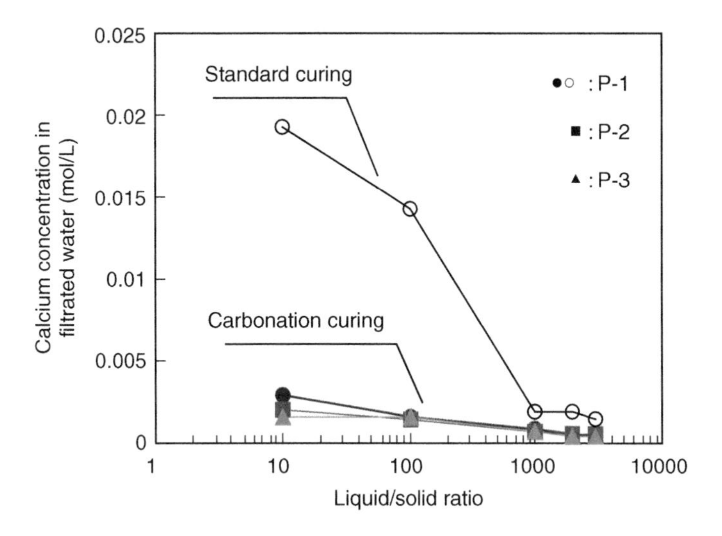

Fig. 16. Calcium concentration in immersed water on leaching test.

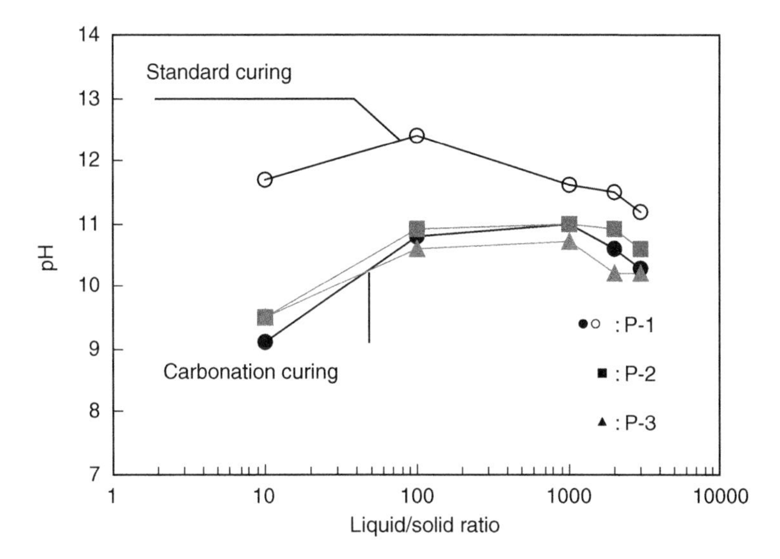

Fig. 17. pH value in immersed water on leaching test.

it had not been used conventionally with cementitious material; however, its use is anticipated in a big way in the future, considering low solubility and low porosity aspects. By using such leaching suppression techniques, the performance required in radioactive waste repositories, such as reduction due to: (1) load-bearing capacity, (2) water-sealing ability, (3) chemical stability, and (4) solubility on natural barriers, is likely to be achieved. Moreover, this material can also improve the mass transfer suppression performance by reducing the volume of porosity significantly and by ensuring long-term chemical stability. Thus, resistance to salt attack and suppression of neutrality is improved dramatically, and it is a technique that can be applied not only to radioactive waste disposal, but also to a wide range of fields in civil engineering.

6. CONCLUSIONS

Although degradation owing to solubility of cement hydrates has been known since the past, quantitative evaluation related to leaching until now has been inadequate because of lack of information such as cementitious material is a composite material that includes changeable hydrates and is a porous material, various kinds of reactions occur because of the environmental conditions in the surroundings, and so on. However, ever since studies on the construction of radioactive waste repositories in ground started, research related to leaching of cement components from concrete has accelerated. Henceforth, the understanding and modeling of solubility characteristics and mass transfer mobility of cement hydrates will play important roles, and the development of materials and construction methods that satisfy the required performance, such as confinement of radioactive nuclides, is anticipated.

In the near future, many countries in the world will have to start disposal of wastes with a higher concentration of radioactive waste than has been disposed until now. Moreover, aging of many concrete structures such as dams, water supply facilities, and underground foundation structures is likely to progress. Only 200 years have elapsed since the development of cement concrete, but with mankind's intelligence, structures with a durability in the order of tens of thousands of years are not impossible to build.

REFERENCES

[1] M. Hironaga, The view and technology of concrete for the disposal of radioactive in the future, Concrete J. (JCI), 37(3) (1999) 3 (in Japanese).
[2] M. Furuichi, Research on sealing systems of high level waste repository, Doctor degree thesis from Hokkaido University, 2000 (in Japanese).
[3] Japan Nuclear Cycle Development Institute, Technical reliability of high-level nuclear waste stratum disposal in Japan, JNC TN1400 99-020, 1999 (in Japanese).
[4] Japan Nuclear Cycle Development Institute, The Federation of Electric Power Companies of Japan Joint Team: Concept Research Report on TRU Waste Disposal, JNC TY1400 2000-001, TRU TR-2000-01, 2000 (in Japanese).

[5] A. Tamura and Y. Akiyama, The cement and concrete used in low level radioactive waste disposal facility, Cement & Concrete, December (620) (1998) 30 (in Japanese).
[6] Y. Furusawa, Research trends on degradation of concrete caused by calcium leaching and modeling, Concrete J. (JCI), 35(12) (1997) 29 (in Japanese).
[7] P.J. Mason, The effect of aggressive water on dam concrete, Construction Building Mater., 4(3) (1990) 115.
[8] Y. Ballim, Deterioration of concrete in soft waters: a review and limited laboratory study, SAICE J., Third Quarter (1993) 1.
[9] A. Neville, Effect of cement paste on drinking water, Mater. Struct., 34 (240) (2001) 367.
[10] A. Atkinson and J.A. Hearne, An Assessment of the Long-Term Durability of Concrete in Radioactive Waste Repositories, UK Atomic Energy Authority Report, AERE-R-11465, 1984.
[11] U. Berner, A Thermodynamic Description of the Evolution of Pore Water Chemistry and Uranium Speciation During the Degradation of Cement, PSI-Bericht No. 62, 1990.
[12] M. Buil, E. Revertegat and J. Oliver, in "A Model to the Attack of Pure Water or Undersaturated Lime Solutions on Cement", Stabilization and Solidification of Hazardous, Radioactive, and Mixed Wastes, Vol. 2, STP 1123 (T.M. Gilliam and C.C. Wiles, eds.), ASTM, Philadelphia, (1992) 227.
[13] K. Yokozeki, K. Watanabe, Y. Furusawa, M. Daimon, N. Otsuki and M. Hisada, Analysis of old structures and numerical model for degradation of concrete by calcium ion leaching, Concrete Library Int., December (40) (2002) 209.
[14] K. Yokozeki, K. Watanabe, D. Hayashi, N. Sakata and N. Otsuki, Modeling of ion diffusion coefficients in concrete considering with hydration and temperature effects, Concrete Library International, December (42) (2003) 105.
[15] K. Yokozeki, K. Watanabe, N. Sakata and N. Otsuki, Modeling of leaching from cementitious materials used in underground environment, clay in natural and engineered barriers for radioactive waste confinement, Appl. Clay Sci., 26 (2004) 293.
[16] K. Yokozeki, Long-term durability design of 1,000-year level on leaching of cement hydrates from concrete, Doctor degree thesis from Tokyo Institute of Technology, 2004 (in Japanese).
[17] Japan Society of Civil Engineer Concrete Committee, Leaching and reaction due to chemical attack, Concrete Eng. Ser., 53, JSCE, Tokyo, (2003) (in Japanese).
[18] S. Kamali, B. Gérard and M. Moranville, Modeling the leaching kinetics of cement-based materials – influence of materials and environment, Cement & Concrete Composites, 25 (2003) 451.
[19] W. Pfingsten, Experimental and modeling indications for self-sealing of a cementitious low- and intermediate-level waste repository by calcite precipitation. Radioactive waste management and disposal, Nucl. Technol., 140 (2002) 63.
[20] K. Yokozeki, K. Watanabe, N. Sakata and N. Otsuki, Prediction of changes in physical properties due to leaching of hydration products from concrete, J. Adv. Concrete Technol., 1(2) (2003) 161.
[21] W.H. Taylor, Concrete Technology and Practice, American Elsevier Publishing Company, Inc., New York, 1965.

[22] K. Asaga and Y. Furusawa, Reproduction of ancient cement used 5000 years ago in China, Cement & Concrete, November (633) (1999) 1 (in Japanese).
[23] K. Watanabe, K. Yokozeki, E. Sakai and M. Daimon, Improvement durability of cementitious materials by carbonation curing with γC_2S, Proc. Jpn. Concrete Inst., 26-1 (2004) 735 (in Japanese).
[24] K. Watanabe, K. Yokozeki, R. Ashizawa, N. Sakata, M. Morioka, E. Sakai and M. Daimon, High durability cementitious material with mineral admixtures and carbonation curing, Waste Manag., 26 (2006) 752.

PART IV

AIR, WATER, SOIL AND REMEDIATION

Thermodynamics, Solubility and Environmental Issues
T.M. Letcher (editor)

Chapter 10

Solubility of Carbon Dioxide in Natural Systems

Justin Salminen[a,b], Petri Kobylin[b] and Anne Ojala[c]

[a]*Department of Chemical Engineering, University of California, Berkeley, CA 94720, USA*
[b]*Laboratory of Physical Chemistry and Electrochemistry, Helsinki University of Technology, FIN-02015 TKK, Finland*
[c]*Department of Ecological and Environmental Sciences, University of Helsinki, FIN-15140 Lahti, Finland*

1. CARBON DIOXIDE: A NATURAL REAGENT

Carbon dioxide and its carbonate minerals play an important role in environmental chemistry and atmospheric physics. In natural waters, atmospheric CO_2 has a significant influence on pH, which varies from alkaline seawaters to acidic low mineral lakes, rivers, and soil water. In freshwaters and in oceans the equilibrium relationships between the carbon dioxide, the chemical and biological components, temperature, and the pH are complex functions. Chemical thermodynamics provide quantitative relationships between chemical energy, ionic reactions, solubilities, speciation, pH, and alkalinity. In natural systems these relationships are also complex functions of chemical and biological effects.

In this chapter we look at the basic relationships between CO_2 in natural waters that are important in the environment and also investigate CO_2 utilizing processes in the environment. As an example of natural systems we consider a boreal low mineral lake and discuss the effects of acidification of aquatic ecosystems in context of multiphase thermodynamics.

In low-mineral freshwaters with low buffer capacity, atmospheric CO_2 has a significant influence on pH. The effect of increased atmospheric CO_2 composition on the pH of seawater is small due to its large buffering capacity. Seawater pH is ca. 8.2, while low-mineral lakes have often low buffering capacities and can exhibit a pH as low as 3, partly due to anthropogenic acidification. Besides acid rain caused by sulphur- and nitric oxide emissions, this lowering of the pH can be due to CO_2.

Green plants and certain other organisms (e.g., cyanobacteria) synthesize carbohydrates from carbon dioxide and water using light as an energy source. In this photosynthesis process, oxygen is released as a by-product. Due to anthropogenic emissions and release of ancient photosynthates from fossil sediments, the concentration of CO_2 in the atmosphere is increasing. As shown by direct monitoring and various proxy data, the CO_2 composition in the atmosphere has increased 100 ppm to the present 380 ppm in just over 100 years [1]. The present rate of 0.5% annual increase is a very rapid change in a geological timescale. New energy and environmental technologies are focussing on minimizing CO_2 emissions [2–4].

Approximately 98% of the carbon in the ocean–atmosphere system is in the oceans in the form of dissolved carbon and calcite. Estimated annual ocean–atmosphere exchange of carbon is ~100 GT(C), which is many times the annual amount of 7 GT(C) of anthropogenic CO_2 emissions to the atmosphere [1].

Calcite, with its vast reservoirs on the ocean floor, plays a major role in the oceanic CO_2 balance, acting as an effective buffer against pH changes. Consequently the change in pH is small as CO_2 or other acids dissolve in the ocean or fresh waters with calcite buffer. The growth and erosion of coral material (calcium carbonate) is sensitive to pH fluctuations [5]. The lowering of pH in the sea causes dissolution of coral material. Inorganic chemical acidification, introduced in natural aqueous systems, causes dissolution of solid calcium carbonate [6, 7].

Besides being an elemental part of the biosphere CO_2 is utilized as a reactive substance in modern industrial processes where solubility plays an important role [8–10]. The weakly acid character of CO_2 is used in the neutralization and acidification of slightly alkaline aqueous mixtures. By controlling the pH with CO_2 gas one can control the precipitation and dissolution of solid carbonates. Biological activity is known to be highest, close to and around a neutral pH and at temperatures greater than 20°C. Many industrial aqueous processes operate at 20–50°C and close to a neutral pH, in conditions favourable to biochemical activity. CO_2 gas can be used in acidification of waters below the pH levels favouring undesirable biological activity. Aerobic decomposition of organic matter forms CO_2 gas while anaerobic decomposition – among other things – can produce ammonia, NH_3. While CO_2 lowers pH, ammonia increases pH. Because pH has a large influence on bioreactions, CO_2 can be used as controlling agent.

2. AQUEOUS SPECIATION OF CO_2

The physical equilibria for CO_2 dissolved in water is:

$$CO_2(g) \leftrightarrow CO_2(aq) \quad (1)$$

and the chemical equilibria are:

$$CO_2(aq) + H_2O \leftrightarrow H^+ + HCO_3^- \quad (2)$$

$$HCO_3^- \leftrightarrow H^+ + CO_3^{2-} \quad (3)$$

The dissociation reactions (2) and (3) define the acid–base chemistry of aqueous CO_2. These are related to the physical reaction in Eq. (1) as given by Henry's law.

In a closed system, the increase in the concentration of CO_2, x_{CO_2}, increases the partial pressure of CO_2. Scheme 1 shows the speciation of CO_2 in an aquatic solution as a function of pH.

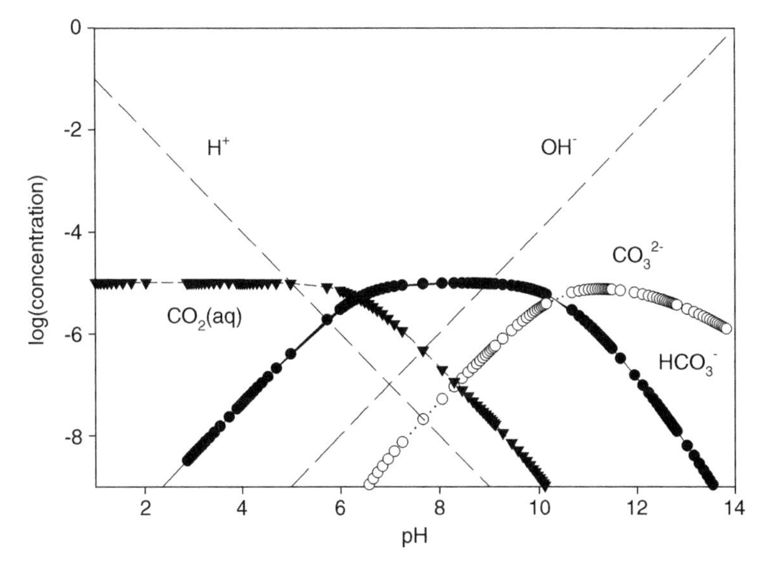

Scheme 1. Speciation of CO_2 as a function of pH.

In an aqueous metal–carbonate systems, the following main reactions occur where M^{2+} represent metal cations like Mg^{2+}, Ca^{2+}, Zn^{2+}, Cd^{2+}, Pb^{2+}. The overall reaction can be written as follows:

$$M^{2+}CO_3^{2-}(s) + 2H^+(aq) \leftrightarrow M^{2+}(aq) + H_2O + CO_2(g)\uparrow \tag{4}$$

The dissociation of a metal carbonate salt $M^{2+}CO_3^{2-}$ can be written as

$$M^{2+}CO_3^{2-}(s) \leftrightarrow M^{2+}(aq) + CO_3^{2-}(aq) \tag{5}$$

Eqs. (4) and (11) show that an increase of proton activity $a(H^+)$ in the solution raises the solubility of carbonate minerals and lowers the solubility of CO_2. A decrease in proton activity and subsequent increase in basicity, results in a precipitation of mineral carbonates. The decrease of the solubility of $M^{2+}CO_3$, e.g. can be caused by a common anion introduced by dissolved sodium or potassium carbonate or bicarbonate salts.

3. MULTIPHASE THERMODYNAMIC SYSTEM

Four equilibrium relations describe a multiphase aqueous system containing CO_2 and MCO_3. Equilibria for reactions in Eqs. (1–3) and (5) are written using total pressure P, gas-phase mole fraction y_i, liquid-phase molality m_i, activity a_i, fugacity coefficient ϕ_i, and activity coefficient γ_i, where i stands for a species that takes part in the chemical reaction. An activity coefficient model is needed to relate the liquid-phase activities of individual species to their molalities.

$$K_H = \frac{a_{CO_2(aq)}}{f_{CO_2(g)}} = \frac{m_{CO_2(aq)}\gamma_{CO_2(aq)}}{y_{CO_2(g)}\phi_{CO_2(g)}P} \tag{6}$$

$$K_1 = \frac{a_{H^+}a_{HCO_3^-}}{a_{CO_2(aq)}a_{H_2O}} = \frac{m_{H^+}\gamma_{H^+}m_{HCO_3^-}\gamma_{HCO_3^-}}{m_{CO_2(aq)}\gamma_{CO_2(aq)}m_{H_2O}\gamma_{H_2O}} \tag{7}$$

$$K_2 = \frac{a_{H^+}a_{CO_3^{2-}}}{a_{HCO_3^-}} = \frac{m_{H^+}\gamma_{H^+}m_{CO_3^{2-}}\gamma_{CO_3^{2-}}}{m_{HCO_3^-}\gamma_{HCO_3^-}} \tag{8}$$

$$K_{sp} = a_{M^{2+}} a_{CO_3^{2-}} = m_{M^{2+}} \gamma_{M^{2+}} m_{CO_3^{2-}} \gamma_{CO_3^{2-}} \tag{9}$$

Eq. (10) shows the relation between Henry's constant H_{CO_2}, equilibrium constant K_H, and the solubility of the neutral CO_2(aq):

$$K_H = \frac{x_{CO_2(aq)}}{p_{CO_2}} = \frac{1}{H_{CO_2}} \tag{10}$$

where, x is mole fraction and p the partial pressure. Henry's constant has units of pressure (MPa). Henry's law is strictly valid only at infinite dilution of CO_2 and at this concentration, the activity coefficient and fugacity coefficient have the numerical value of unity, assuming also, that the gas phase is ideal.

The solubility of a M^{2+} carbonate salt can be increased by adding an acid. Further, the M^{2+} carbonate salt can be precipitated by adding electrolytes with anions like OH^-, HCO_3, HS^-, CO_3^{2-}. The relation between solubility product, CO_2 partial pressure, proton activity, and M^{2+} ion activity is:

$$K_{sp} = a_{M^{2+}} p(CO_2) a_{H^+}^{-2} \tag{11}$$

Since CO_2 gas can be used as an acidifying agent in chemical or biochemical processes, it can also be used to dissolve or precipitate metal carbonate salts through these common-ion effects. The chemical state of a system is described by means of Gibbs energy G, chemical potential μ_i of species i, and mole amount n_i [11–14].

For the Gibbs energy change at constant T and P in a single phase is given by relation:

$$dG = \sum_i \mu_i dn_i \tag{12}$$

In chemically reactive systems, the change of chemical amounts, dn_i, can be expressed as:

$$dn_i = v_i d\xi \tag{13}$$

where v_i is the stoichiometric number of the species in the reaction and ξ the extent of the reaction in moles, and is given by:

$$\left(\frac{\partial G}{\partial \xi}\right)_{T,P} = \sum_i v_i \mu_i \tag{14}$$

The values of the stoichiometric coefficients are positive ($v_i > 0$) for products and negative ($v_i < 0$) for reactants.

At equilibrium, the Gibbs energy is at a minimum, which corresponds to a zero slope in the two-dimensional phase space, and the right-hand side of Eq. (15) is zero. Writing the chemical potential by means of standard state chemical potentials and activities:

$$\left(\frac{\partial G}{\partial \xi}\right)_{T,P} = \sum_i v_i \mu_i = \sum_i v_i \mu_i^\circ + RT \sum_i v_i \ln a_i = 0 \tag{15}$$

$$\sum_i v_i \mu_i^\circ = -RT \sum_i \ln a_i^{v_i} = -RT \ln \prod_{i=1} a_i^{v_i} \tag{16}$$

The thermodynamic equilibrium constant is defined by means of the product of activities:

$$K = \prod_i a_i^{v_i} \tag{17}$$

that is related to the standard molar Gibbs energy of a reaction:

$$\Delta_r G_m^\circ = -RT \ln K \tag{18}$$

The extensive quantity of Gibbs energy of the system is obtained as a sum of the chemical potentials multiplied by their chemical amounts over all the species and phases:

$$G = \sum_\alpha \sum_i n_i^\alpha \mu_i^\alpha \tag{19}$$

In the case where there exist one gas phase α, one liquid phase β, and invariant pure solid phases κ_i one can write:

$$\begin{aligned} G = {} & G^{\alpha}(T,P,n_i^{\alpha},\ldots,n_j^{\alpha}) + G^{\beta}(T,P,n_i^{\beta},\ldots,n_j^{\beta}) \\ & + G^{\kappa_1}(T,P,n_S^{\kappa_1}) + \ldots + G^{\kappa_n}(T,P,n_n^{\kappa_n}) \end{aligned} \tag{20}$$

The total Gibbs energy of a system is constructed by writing the chemical potential or the partial molar Gibbs energy μ_i for each species. The total Gibbs energy G^{α} of a gas phase is the sum of all the components in the gas phase:

$$G^{\alpha} = \sum_{n=1} n_n \left(\mu_n^{\circ} + RT \ln \left(\frac{y_n \phi_n P}{P^{\circ}} \right) \right) \tag{21}$$

The total Gibbs energy G^{β} of an aqueous liquid phase is the sum of the Gibbs energies of water, dissolved salts, and dissolved gases:

$$\begin{aligned} G^{\beta} = {} & n_w(\mu_w^{\circ} + RT \ln(x_w \gamma_w^x)) + \sum_{i=1} n_i \left(\mu_i^{\ominus} + RT \ln \left(\frac{m_i \gamma_i^m}{m^{\circ}} \right) \right) \\ & + \sum_{n=1} n_n \left(\mu_n^{\ominus} + RT \ln \left(\frac{m_n \gamma_n^m}{m^{\circ}} \right) \right) \end{aligned} \tag{22}$$

The total Gibbs energy G^{κ} of the solid phases is:

$$G^{\kappa} = \sum_{\kappa=1} n_{\kappa} \mu_{\kappa}^{\circ} \tag{23}$$

Chemical equilibrium in a closed system at constant temperature and pressure is achieved at the minimum of the total Gibbs energy, min(G) constrained by material-balance and electro-neutrality conditions. For aqueous electrolyte solutions, we require activity coefficients for all species in the mixture. Well-established models, e.g. Debye–Hückel, extended Debye–Hückel, Pitzer, and the Harvie–Weare modification of Pitzer's activity coefficient model, are used to take into account ionic interactions in natural systems [15–20].

4. MODELING NATURAL SYSTEMS

4.1. A Boreal Lake

The increasing interest in global carbon cycling and attempts to create extensive carbon budgets for terrestrial as well as aquatic systems, has put more emphasis on studies on lacustrine carbon cycling. Although lakes together with reservoirs cover only ~2% of the global land surface, they seem to have a significant role in carbon cycling. For instance, they have high carbon accumulation rates per surface area compared with oceans and due to the allochthonous carbon load from surrounding catchments they are usually supersaturated with CO_2 relative to the atmosphere. The carbon accumulation rates are especially high in small lakes in the boreal zone where lakes cover approximately 7% of the total land area [21]. Small lakes are numerous and for instance in Finland, the number of lakes with a surface area less than 0.01 km^2 is over 130000 [22]. Lakes can act as carbon sinks and also as sources of carbon and thus have an ambivalent character. The bulk of the carbon accumulating in the bottom sediments originates from autochthonous production, but part of the bound carbon – autochthonous as well as allochthonous – in the organic material is released as CO_2 at the end of its life cycle (Scheme 2).

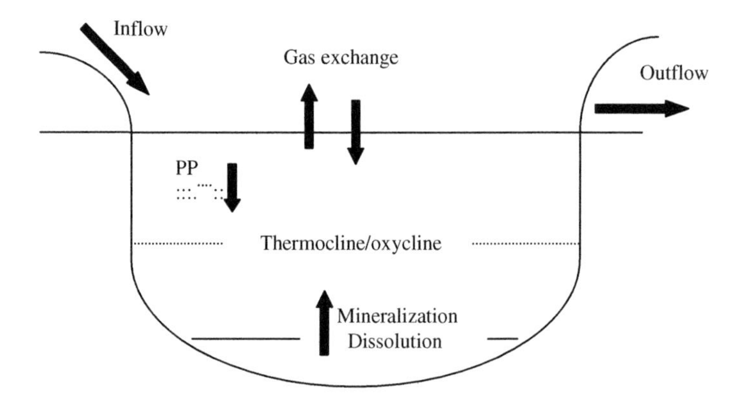

Scheme 2. A simple model for carbon cycling in a low mineral boreal lake. PP refers to primary production. Note that carbonate precipitation is not relevant in these systems located on the ancient Precambrian shield and is thus omitted from the figure. Note also that methanogenesis releasing CH_4 gas is an important mineralization process in anoxic hypolimnion and bottom sediment; at least part of methane is then oxidized to CO_2 in the water column through the activity of methanotrophic bacteria before escaping the system. During the stratification periods the biogenic gases accumulate in the hypolimnion and escape to the atmosphere mainly during the spring and fall mixing periods [25].

The release takes place mainly through biological mineralization but can also be released by non-biological photochemical processes. In boreal lakes the high concentration of dissolved organic carbon creates an effective shield against short wave length radiation and thus photodegradation is of minor importance [23]. The mineralized organic carbon shows increased alkalinity, defined as a sum of following ions:

$$C_{\text{Alk}} = C_{\text{HCO}_3^-} + 2C_{\text{CO}_3^{2-}} + C_{\text{OH}^-} - C_{\text{H}^+} \tag{24}$$

Alkalinity is expressed in millimole per cubic decimetre (mmol dm^{-3}) and mole per cubic decimetre (mol dm^{-3}) is equal to equivalents per litre (eq dm^{-3}), i.e. moles of protons neutralized by the alkalinity in litre of solution. Another commonly used unit is milligrams of $CaCO_3$ per litre (mg dm^{-3}) which is based on the neutralization reaction (4), where M^{2+} is Ca^{2+}. The relationship between alkalinity units is:

$$1\,\text{mol}\,\text{dm}^{-3} = 1\,\text{eq}\,\text{dm}^{-3}\ \text{and}\ 1\,\text{mmol}\,\text{dm}^{-3} = 50.045\,\text{mg}\,\text{CaCO}_3\,\text{dm}^{-3}$$

Decomposition of organic matter is a dynamic process. The oxidation of organic matter eventually produces CO_2 and in anaerobic environments microbiological process called methanogenesis produces methane (CH_4) gas [24].

In Scheme 3, the alkalinity concentration is plotted against pH from Lake Horkkajärvi, in southern Finland (61°13′N, 25°10′E). Lake Horkkajärvi is a truly meromictic lake where water column mixing periods, due to homothermal conditions, are absent [26]. In northern temperate zones, lakes usually circulate twice a year, i.e. in spring and fall, but in a meromictic lake permanent stratification separates the lake into two layers. The stabilising forces in Lake Horkkajärvi are high electrolyte concentration and the sheltered position of the landscape. The lake is 1.1 ha in area with a maximum and mean depth of 13 and 7 m, respectively. It is a humic lake greatly affected by the allochthonous carbon load; the water colour in units of mg Pt l^{-1} is >200.

The thermodynamic model shows the influence of the organic matter in the system. Using only dissolved atmospheric CO_2 yields higher pH values at a given alkalinity value. It is worth noticing that the biochemical reactions which result in shifting the curve to left are the same which lead to the meromixis through electrolyte accumulation.

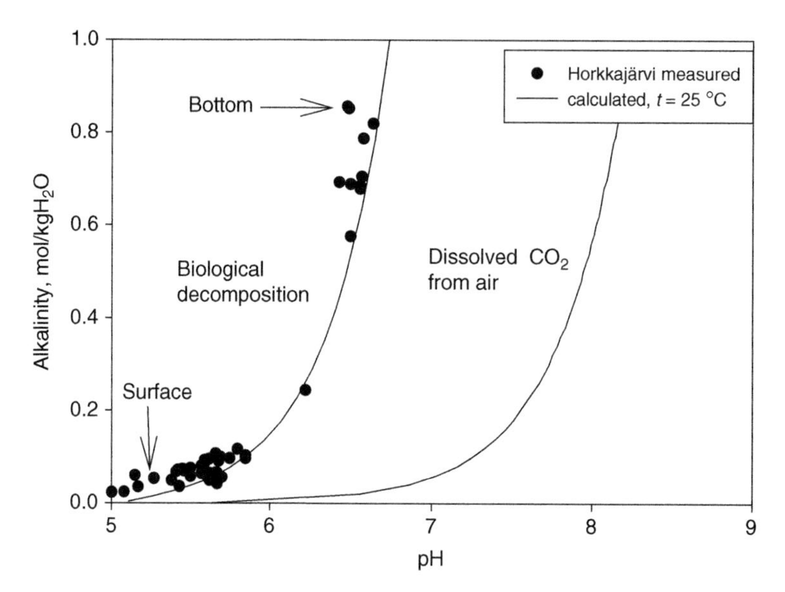

Scheme 3. Alkalinity as a function of pH in Lake Horkkajärvi. The experimental results from the lake are compared with thermodynamic calculations. The curve on the right-hand side shows the alkalinity–pH relation in a situation where CO_2 is dissolved from the air. The left-hand side curve shows experimental results and model calculation in a system where biological decomposition also takes place. The surface water (1–2 m) in the lake has been found to be well mixed but the bottom, at a mean depth of 7 m, is not well mixed. The figure shows that biological decomposition takes place at the bottom of the lake.

4.2. Anthropogenic Acidification of Surface Waters

Anthropogenic acidification, i.e. atmospheric deposition of weak solutions of sulphuric and nitric acids, is in general a serious threat to living organisms, particularly in sensitive low-mineral, nutrient-poor lakes in the Precambrian shield. These ecosystems have naturally low buffering capacities whereas in more fertile areas, calcium concentrations in the bedrock and soils are higher, helping to prevent acidification. The phenomena are displayed in Scheme 4, which shows the effect of sulphuric acid on the pH in non-buffered and in calcite-saturated solution.

In nature a slight increase in acidity of an aqueous system, can be noticed when CO_2 is dissolved. This raises a question on CO_2 dissolution into water, in the case of an aqueous system which has an increased H^+ concentration, i.e. a low pH value. Scheme 5 shows pH changes as a function of sulphuric acid addition in open and closed systems using artificial seawater with 3.5 wt.% salinity. The salt composition is also displayed.

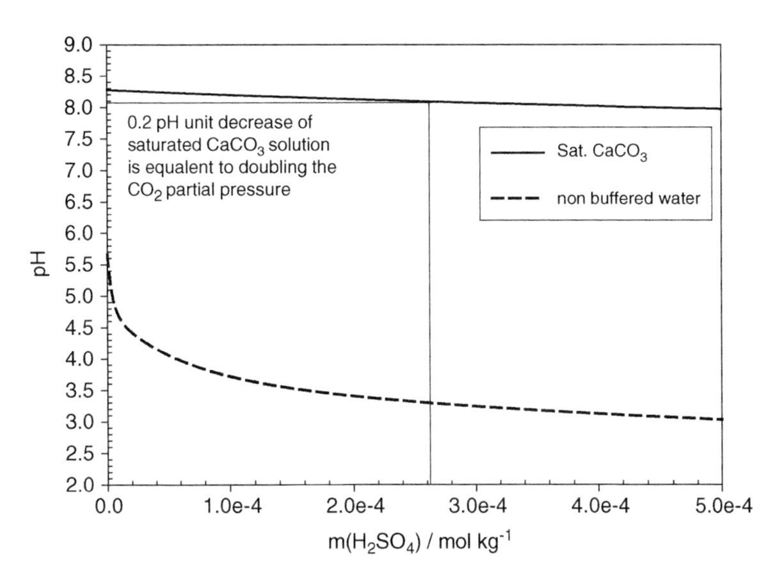

Scheme 4. pH decrease in saturated $CaCO_3$ solution and non-buffered CO_2–water system at 25°C with the addition of H_2SO_4. According to the model, in saturated calcite solution circa. 0.2 decrease of pH in the solution is reached by doubling the CO_2 partial pressure from a value of $p(CO_2) = 3.80 \times 10^{-4}$ bar.

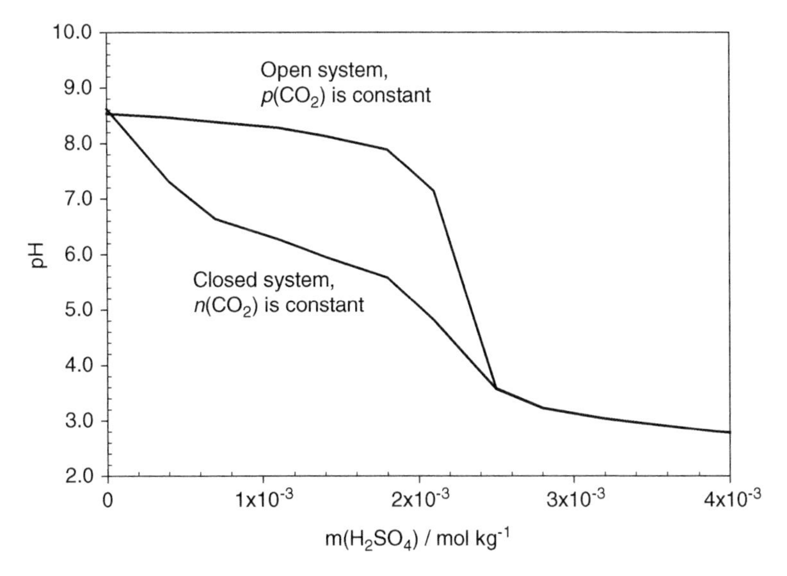

Scheme 5. Calculated pH change due to addition of sulphuric acid in open and closed systems. In open system $p(CO_2)$ is constant. In a closed system the chemical amount $n(CO_2)$ is constant. The composition of the artificial seawater was: H_2O (966 g), NaCl (26.5 g), $MgSO_4$ (3.3 g), $MgCl_2$ (2.4 g), $NaHCO_3$ (0.2 g), $CaCl_2$ (1.1 g), NaBr (0.08 g), and KCl (0.7 g).

In an open system the CO_2 partial pressure is kept constant and in a closed system the amount of CO_2 is constant. Thus, in open system CO_2 gas is allowed to leave the system and partial pressure of CO_2 equals atmospheric pressure (380 ppm). In closed system the total amount of CO_2 is constant and partial pressure of carbon dioxide $p(CO_2)$ will change. The calculated curves are comparable to the results obtained by Pilson [27]. In a closed system, even a small increase in sulphuric acid decreases the pH, which lowers the liquid's ability to form dissolved carbon and calcite. In an open system, the change of pH is smaller but important in natural systems. At higher acid concentrations, the pH of the open and closed systems becomes the same. The actual measured sulphuric acid concentration in natural waters is less than 2×10^{-4} mol kg$(H_2O)^{-1}$. In terms of aquatic ecosystems the CO_2 partial pressure is relatively constant despite the observed increase of atmospheric CO_2 concentration during the last 100 years and the ecosystems thus resemble the open system.

4.3. Solubility of $CaCO_3$ in Salt Solutions

The solubility of $CaCO_3$ as a function of added Na_2CO_3, $CaCl_2$, and $CaSO_4$ salts and a partial pressure of CO_2 initially at $p(CO_2) = 380$ ppm and 25°C yield different changes in $CaCO_3$ solubility and pH. Scheme 6 shows how the solubility of calcite changes as function of CO_2 partial pressure and

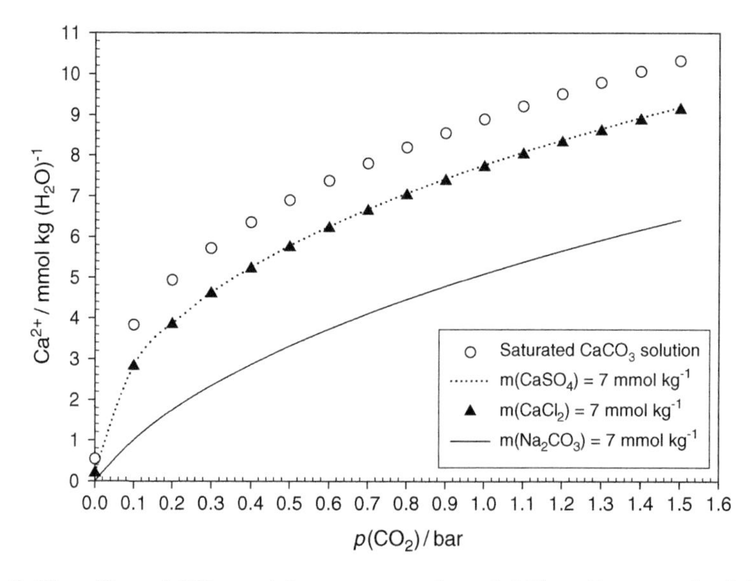

Scheme 6. The effect of CO_2 partial pressure on the solubility of saturated calcite solution with and without added common-ion salt. The multiphase thermodynamic model allows the determination of the solubilities of carbonates in different $p(CO_2)$s.

with added 7 mmol kg^{-1} common-ion salt, $CaSO_4$, $CaCl_2$, or Na_2CO_3. As expected Na_2CO_3 has the greatest effect on reducing the solubility of calcite in the system. The addition of $CaCl_2$ and $CaSO_4$ (0–7 mmol/kg) drops the pH from 8.25 to the values 7.78 and 7.77. The addition of Na_2CO_3 has the opposite effect increasing the pH from initial 8.25 to 9.24. Highly soluble Na_2CO_3 or $NaHCO_3$ salts can be used to precipitate M^{2+} carbonates with well-known common-ion technique [28].

Scheme 7 shows more comparisons between solubility of $CaCO_3$ as function of CO_2 partial pressure with different salt additions at 25°C. As can

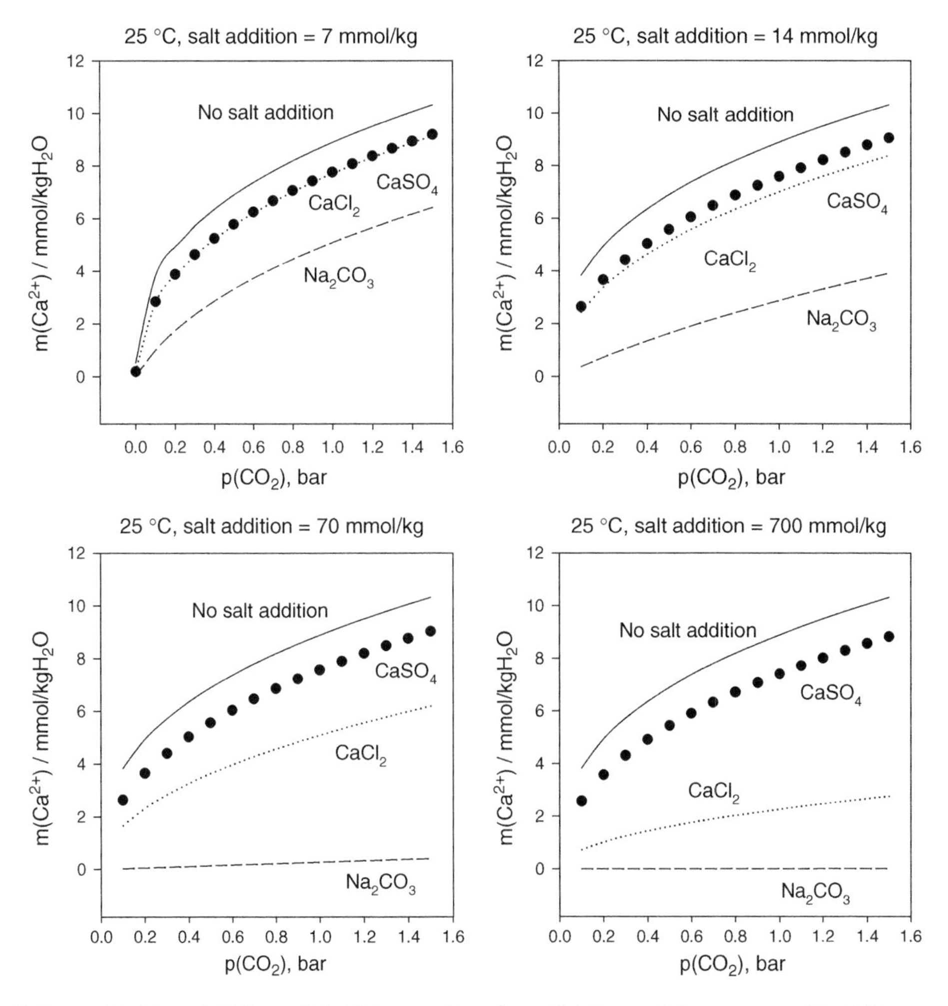

Scheme 7. The solubility of $CaCO_3$ as a function of CO_2 partial pressure with different added common-ion salts at 25°C.

be seen $CaSO_4$ additions higher than 7 mmol kg$(H_2O)^{-1}$ do not noticeably change the $CaCO_3$ solubility while $CaCl_2$ and Na_2CO_3 salt additions further decrease the solubility of $CaCO_3$. Increase in temperature results in decrease of $CaCO_3$ solubility.

5. CONCLUDING REMARKS

In low-mineral freshwaters with low buffering capacity, both atmospheric CO_2 and biological decomposition has a huge impact on the pH. The effect of atmospheric CO_2 on the pH of seawater is much smaller but multiphase equilibria follows from the interaction with dissolved and solid mineral carbonates.

The biological impact due to increased acidification of sea and dissolution of coral matter could be detrimental and irreversible. That is because the timescales of physical dissolution and biological recovery are different. Dissolution and precipitation processes of aqueous $CaCO_3$ solutions are controlled by acid–base chemistry; these processes are thermodynamically driven.

REFERENCES

[1] IPCC, in "The Scientific Basis" (J.T. Houghton, Y. Ding, D.J. Griggs, M. Noguer, P.J. van der Linden, X. Dai, K. Maskell and C.A. Johnson, eds.), Contribution of Working Group I to the Third Assessment Report of the Intergovernmental Panel on Climate Change, Cambridge University Press, Cambridge, UK, New York, NY, USA, 2001.
[2] Scientific American, Energy Solutions for Sustainable Future, September, 2006.
[3] M.I. Hoffert, K. Caldeira, G. Benford, D.R. Criswell, C. Green, H. Herzog, A.K. Jain, H.S. Kheshgi, K.S. Lackner, J.S. Lewis, H.D. Lightfoot, W. Manheimer, J.C. Mankins, M.E. Mauel, L.J. Perkins, M.E. Schlesinger, T. Volk and T.M.L. Wigley, Science, (1 Nov). 298(2002) 981. DOI: 10.1126/science.1072357 (in review).
[4] T. Takahashi, Science, 305 (2004) 352.
[5] J. Ruttimann, Nature, 442 (2006) 987.
[6] R.A. Freely, C.L. Sabine, K. Lee, W. Berelson, J. Kleypas, V.J. Farby and F.J. Millero, Science, 305 (2004) 362.
[7] C.L. Sabine, R.A. Feely, N. Gruber, R.M. Key, K. Lee, J.L. Bullister, R. Wanninkhof, C.S. Wong, D.W.R. Wallace, B. Tilbrook, F.J. Millero, T-H. Peng, A. Kozyr, T. Ono and F. Rios, Science (16 July) 305 (2004) 367.
[8] G.T. Hefter and R.P.T. Tomkins, The Experimental Determination of Solubilities, Wiley Series in Solution Chemistry, Vol. 6, Wiley, London, 2003.
[9] E.J. Beckman, Ind. Eng. Chem. Res., 42 (2003) 1598.
[10] S.A. Newman (ed.), Thermodynamics of Aqueous Systems with Industrial Applications, ASC symposium series 133, ACS, Washington, 1980.

[11] K. Denbigh, The Principles of Chemical Equilibrium, 3rd edn., Cambridge University Press, Cambridge, 1971.
[12] M.L. McGlashan, Chemical Thermodynamics, Academic Press, London, 1979.
[13] R.J. Silbey and R.A. Alberty, Physical Chemistry, 3rd edn., Wiley, New York, 2001.
[14] D. Kondepudi and I. Prigogine, Modern Thermodynamics – From Heat Engines to Dissipative Structures, Wiley, New York, 1998.
[15] P. Debye and E. Huckel, Phys. Z., 24 (1923) 185; H. Harned and B. Owen, The Physical Chemistry of Electrolyte Solutions, 2nd edn, Reinhold Publishing Corp., New York, 1950.
[16] K.S. Pitzer, Thermodynamics, 3rd edn, McGraw-Hill, New York, 1995.
[17] C. Harvie and J. Weare, Geochim. Cosmochim. Acta, 44 (1980) 997.
[18] C. Harvie, N. Möller and J. Weare, Geochim. Cosmochim. Acta, 48 (1984) 723.
[19] F.J. Millero, Geochim. Cosmochim. Acta, 45 (1983) 2121.
[20] F.J. Millero, Pure Appl. Chem., 57 (1985) 1015.
[21] P. Kortelainen, H. Pajunen, M. Rantakari and M. Saarnisto, Glob. Change Biol., 10 (2004) 1648.
[22] M. Raatikainen and E. Kuusisto, Terra, 102 (1990) 97, in Finnish.
[23] A.V. Vähätalo, M. Salkinoja-Salonen, P. Taalas and K. Salonen, Limnol. Oceanogr., 45 (2000) 664.
[24] K.M. Walter, S.A. Zimov, J.P. Chanton, D. Verbyla and F.S. Chapin III, Nature, 443 (2006) 71.
[25] P. Kankaala, J. Huotari, E. Peltomaa, T. Saloranta and A. Ojala, Limnol. Oceanogr., 51 (2006) 1195–1204.
[26] A. Hakala, Boreal Environ. Res., 9 (2004) 37.
[27] M.E.Q. Pilson, Introduction to the Chemistry of the Sea, Pearson Education POD, New Jersey, 1998.
[28] D. Langmuir, Aqueous Environmental Geochemistry, Prentice Hall, New Jersey, 1997.

Thermodynamics, Solubility and Environmental Issues
T.M. Letcher (editor)

Chapter 11

Estimation of the Volatilization of Organic Chemicals from Soil

Epaminondas Voutsas

Thermodynamics and Transport Phenomena Laboratory, School of Chemical Engineering, National Technical University of Athens, 9 Heroon Polytechniou Street, Zographou Campus, 157 80 Athens, Greece

1. INTRODUCTION

Soil contamination by organic chemicals, mainly pesticides, has been the subject of several studies, mainly due to its important role in the distribution of contaminants to other environmental compartments such as air, plants and the water table [1–6]. When a chemical substance is introduced into the soil, it may follow many different transport or loss pathways [7]. Some of the initially applied chemical is vapourized through the soil surface and some is transformed into intermediate products called metabolites. Furthermore some of the chemical's mass may be adsorbed onto the soil particulate surface and it may undergo further degradation. Chemicals can partition into the vapour phase and may then be transported by the mechanism of advection and/or gaseous diffusion within the gas filled portion of the soil voids. Solutes diffuse laterally into soil regions occupied by stagnant waters and may be absorbed into interior surfaces. The solute is also available in the solution phase for uptake by the plant roots. Finally, a part of the initially applied chemical may leach below the vadose zone to groundwater by the mechanism of advection and dispersion.

Volatilization and air transport are the principal means for widespread dispersion of pesticides and other organic chemicals in the atmosphere. As defined by Lyman et al. [8] volatilization is the process by which a compound evaporates into the atmosphere from another environmental compartment. In the case where the latter is soil, volatilization may be the most important mechanism for the loss of chemicals from it and their transfer to the air. This is more pronounced for chemicals that may persist in the environment for many years due to their resistance to transformation by biological or physical degradation processes.

Most studies on the volatilization from soils have been conducted for pesticides. Since the early sixties post-application losses of pesticides by volatilization have been increasingly recognized as a pathway for environmental contamination and also as a process limiting their effectiveness [9]. Volatile soil losses of pesticides have been measured for several pesticides in field situations, and were found to vary from as low as 1.1% of soil-surface-applied simazine in 24 days to 90% of soil-surface-applied trifluralin lost in 6 days [4]. Pesticides include several groups of compounds such as insecticides, herbicides, fungicides, rodenticides and fumigants, consisting of several hundred individual chemicals of different kinds with a wide range of properties. As pointed out by Lyman et al. [8], other than their use, there is apparently little which distinguishes pesticides from other organics, and therefore, it can be assumed that the observations based on pesticides are applicable to organic chemicals in general.

The rate at which a chemical volatilizes from soil is controlled by simultaneous interactions between soil properties, chemical's properties and environmental conditions. Soil properties that affect volatilization include soil water content, organic matter, porosity, sorption/diffusion characteristics of the soil, etc.; chemical's properties that affect volatilization include vapour pressure, solubility in water, Henry's law constant, soil adsorption coefficient, etc.; and finally, environmental conditions that affect volatilization include airflow over the surface, humidity, temperature, etc. Volatilization rate from a surface deposit depends only on the rate of movement of the chemical away from the evaporating surface and its vapour pressure. In contrast, volatilization of soil-incorporated organic chemicals is controlled by their rate of movement away from the surface, their effective vapour pressure at the surface or within the soil, and their rate of movement through the soil to the vapourizing surface.

A comprehensive model for the estimation of the volatilization of organic chemicals from soil surfaces should take, of course, explicitly into account all the above factors. Models developed for estimating volatilization rates are based on equations describing the rate of movement of the chemical to the surface by diffusion and/or by convection, and away from the surface through the air boundary layer by diffusion [10–16]. Additionally, the part of the chemical in soil that will be lost by volatilization depends on the resistance of the chemical to degradation. The application of most of these models requires, however, a number of input information, e.g. adsorption isotherm coefficients, diffusion coefficients in soil etc., which in most cases are not experimentally known and neither can they be accurately predicted.

In the following section some physicochemical properties of chemicals that affect the volatilization from soil are briefly introduced. Next, the factors influencing the volatilization process of a chemical from the soil and methods for measuring volatilization fluxes are discussed. Following, models that estimate the rate of volatilization of chemicals from soil are presented, and finally, some thermodynamic aspects of persistent organic chemicals and the concept of equilibrium partitioning are discussed.

2. PHYSICOCHEMICAL PROPERTIES OF CHEMICALS

2.1. Vapour Pressure

Vapour pressure of a pure compound is the pressure characteristic at any given temperature of a vapour in equilibrium with its liquid or solid form. Vapour pressure is a measure of the ability of a compound to bond with itself; compound molecules that bond well with each other will have a low vapour pressure (less tendency to escape to the vapour phase), while poorly bonding compounds will have a high vapour pressure.

From a thermodynamic standpoint, this can be viewed from the well-known Clausius–Clapeyron expression:

$$\frac{\mathrm{d}\ln P_{\mathrm{VP}}}{\mathrm{d}T} = \frac{\Delta H_{\mathrm{vap}}}{RT^2} \tag{1}$$

where ΔH_{vap} is the heat of vapourization, i.e. the energy needed to break intermolecular bonds in the pure condensed liquid. Since breaking bonds requires the input of energy, we would expect the vapour pressure of a compound to increase with increasing temperature. For solids, the direct conversion process of a solid to a vapour is called sublimation. The analogous property in Eq. (1) is the heat of sublimation, rather than the heat of vapourization. This is also a function of temperature, and a unit increase in temperature will produce a larger change in vapour pressure of solid–gas equilibrium than the comparable liquid–vapour equilibrium. This would be expected considering the stronger intermolecular bonding implied by the solid versus liquid state. Several models for vapour pressure prediction have been proposed in the literature [8, 17–19].

2.2. Henry's Law Constant and Air-to-Water Partition Coefficient

Henry's law states that when a liquid and a gas are in contact, the weight of the gas that is dissolved in a given quantity of liquid is proportional to the pressure of the gas above the liquid (partial pressure). The value of the constant of proportionality is called the Henry's law constant (H) and is dependant on the solute–solvent pair, temperature and pressure. Henry's law constant is a measure of a chemical's volatility from a solvent (in this chapter it is water): the larger a compound's H value, the more volatile it is, and it is also easier to be transferred from the aqueous phase into air.

A reliable method for the prediction of H values for chemicals with environmental interest is the bond-contribution model of Meylan and Howard [20]. This model calculates the water-to-air partition coefficient of a compound, which is the reciprocal of the Henry's law coefficient, as the summation of the contributions of the individual bonds that comprise the compound. Values of bond contributions are given in the original publication of Meylan and Howard [20].

In environmental applications the air–water partition coefficient (K_H) is also used, which is in fact the Henry's law constant. K_H is defined as:

$$K_H = \frac{P_\mathrm{i}}{C_\mathrm{aq}} \tag{2}$$

where P_i is the partial pressure of the chemical and C_aq the concentration of the chemical in the aqueous phase. Using standard units of atmospheres for partial pressure and moles per litre for concentration, the air–water partition coefficient has units of atm L mol^{-1}.

2.3. Octanol–Water Partition Coefficient

Octanol–water partition coefficient (K_ow) is a very valuable parameter with numerous environmental applications, where it is used as a primary characterizing parameter since it represents a measure of the tendency of a compound to move from the aqueous phase into lipids. K_ow, is defined as following:

$$K_\mathrm{ow} = \frac{C_\mathrm{op}}{C_\mathrm{w}} \tag{3}$$

where C_op and C_w are the concentrations, in g L^{-1}, of the species in the octanol-rich phase and in the water-rich phase respectively. It should be noted that in octanol/water partitioning at 25°C, the water-rich phase is essentially pure water (99.99 mol% water) while the octanol-rich phase is a mixture of octanol and water (79.3 mol% octanol). Numerous models have been proposed in the literature for the estimation of K_{ow} values. Sangster [21] and Howard and Meylan [22] have reviewed the predictive capabilities of various such models. The atom fragment contribution (AFC) model [23] seems to be better than the other methods. In the AFC model a compound is divided into appropriate atoms or fragments and values of each group are summed together (sometimes with structural correction factors) to yield the log K_{ow}. It must be noted that the limitation of all models that belong to the atom/fragment contribution approach is that they give the K_{ow} value only at 25°C.

2.4. Soil Organic Carbon–Water Partition Coefficient

The soil organic carbon–water partition coefficient (K_{oc}) is a parameter that is used to express the extent to which an organic chemical partitions itself between the soil and solution phases (i.e., dissolved in the soil water). K_{oc} is defined as the ratio of the chemical's concentration in a state of sorption (i.e., adhered to soil particles) and the solution phase. Thus, for a given amount of a chemical, the smaller the K_{oc} value, the greater the concentration of the chemical in solution. Chemicals with a small K_{oc} value are more likely to leach into groundwater than those with a large K_{oc} value. Sorption for a given chemical is greater in soils with a higher organic matter content, and thus their leaching is thought to be slower in those soils than in soils lower in organic matter. Because experimental K_{oc} data are not available for all chemicals in use, numerous correlations with some other property such as octanol–water partition coefficient, solubility in water, bioconcentration factor have been proposed [8, 24].

A reliable model for predicting K_{oc} values is the one proposed by Meylan et al. [25], which is based on the first order molecular connectivity index of the compound. Meylan et al. [25] report that their model accounts for 96% of the variation in measured log K_{oc} values for the training set (189 chemicals) and 86% of the variation for the validation set (205 chemicals).

2.5. Solubility in Water

Water solubility is a determining factor in the fate and transport of a chemical in the environment as well as the potential toxicity of a chemical. Of the various parameters that affect soil/air partitioning, water solubility is one of the most important. The solubility of a chemical in water may be defined as the maximum amount of the chemical that will be dissolved in pure water at a specified temperature. Above this concentration, two phases will exist if the organic chemical is a solid or a liquid at the system temperature: a saturated aqueous solution and a solid or liquid organic phase. A great deal of literature is available concerning the estimation of water solubility, especially using regression-derived equations and fragment constant approaches. For a review of such estimation methods see Refs. [8, 22].

3. FACTORS INFLUENCING VOLATILIZATION

The rate at which a chemical volatilizes from soil is influenced by many factors, including soil properties, chemical properties and environmental conditions.

Soil properties include clay content, organic matter content, soil drainage, soil pH, water content, structure, porosity and density [8]. Soil clay mineralogy influences chemical adsorption and co-precipitation and affects contaminant solubility and mobility. Soil organic matter is an important adsorbent for many inorganic and organic soil contaminants. Soil drainage affects the direction and rates of water and chemical movement within a soil. Soil pH is probably the most important transient property affecting inorganic contaminant solubility on a wide range of soils. Subsoil acidity may either increase or decrease the subsoil's capacity to retard downward contaminant movement, depending on pH-dependant binding of the contaminant. Soil water content affects volatilization since it is competing with the chemical for absorption sites on the soil. Soil structure and porosity refer to the spatial distribution and total organization of the soil subsystem. Although models assume the soil to be homogenous in structure, real soil varies greatly in composition down its profile and can also contain many cracks and fissures, which could greatly influence transport of chemicals to the soil surface and thus their volatilization.

Vapour pressure, solubility in water, Henry's law constant, diffusion constants are some of the chemical's properties that influence the volatilization rate. Ryan et al. [26] suggested that chemicals with dimensionless Henry law constant greater than 10^{-4} should be readily volatilized from soil. Wang and Jones [27] suggested that volatilization of organic chemicals from soil is related to the ratio of the Henry's law constant to the octanol–water partition coefficient. Later Harner and Mackay [28] suggested the use of the octanol–air partition coefficient to describe the partitioning of organic chemicals between soil and air, considering that most organic chemicals partition into organic phases for which octanol is a recognized surrogate. Woordow and Seiber [29] have shown good correlation of the logarithm of volatilization flux with the logarithm of the quantity $P_{VP}/K_{oc}S$, where P_{VP} is the vapour pressure, K_{oc} the organic carbon–water partition coefficient and S the solubility of the chemical in water.

Atmospheric conditions such as wind speed, temperature and relative air humidity are some environmental conditions that influence the volatilization rate. Increased wind speed tends to reduce the resistance to gas phase transport. This has the effect of increasing the dry gaseous deposition flux to the soil, but can also clearly increase the revolatilization of the chemical from the soil [5, 30]. Temperature has a profound effect on the vapour/soil partitioning. Volatilization rates are influenced by soil and ambient air temperature mainly through its effect on vapour pressure, i.e. increase of temperature increases vapour pressure. If the relative humidity of the air is not 100%, increases in airspeed will hasten the drying of soil. This indirect effect alters the soil water content, which as it was previously mentioned has an effect on volatilization [30, 31].

As discussed by Lyman et al. [8] all the above can be categorized in factors affecting (a) the movement away from the evaporating surface into the atmosphere, (b) the vapour density of the chemical and (c) the rate of movement to the evaporating surface.

3.1. Chemical Movement from the Soil Surface into the Atmosphere

The rate of movement away from the evaporating surface is a diffusion-controlled process [32]. The vapourizing substance is transported by molecular diffusion through an air layer that is close to the evaporating surface and is relatively still. The thickness of this air layer depends on the airflow rate and the temperature [33]. Diffusion away from the surface is related to the vapour density and the molecular weight of the chemical substance. Changes in temperature can influence the vapour density of the substance and subsequently the diffusion process. The molecules of the chemical substance volatilize independently with respect to water molecules.

3.2. Vapour Density

Vapour density is the concentration of a chemical in the air. When the concentration reaches a maximum, the vapour is saturated. The vapour density of a compound in the soil air determines the volatilization rate. Some chemicals require only a low total soil concentration to have a saturated vapour in moist soil. Thus, weakly absorbed compounds may volatilize rapidly, especially if applied only to soil surface. If the chemicals are incorporated into the soil, the concentration at the evaporating surface of the soil particles is reduced and the total volatilization rate decreases [10].

The nature of the chemical is an important factor that affects vapour density. Chemical's properties such as vapour pressure, solubility in water, etc. are important properties of a chemical that influence its vapour density. The chemical's absorption capacity also affects vapour density and the volatilization rate. Absorption of the chemical by the organic and inorganic soil components may be the result of chemical absorption (Coulombic forces), physical absorption (van der Waals forces) and hydrogen bonding [30]. Absorption reduces the chemical activity below that of the pure compound and affects the vapour density and subsequently the volatilization rate [10]. This is because the concentration of the compound present in a desorbed state in solution in the soil water controls the vapour density of the compound in soil air.

Soil water content affects volatilization losses by competing with the chemical for absorption sites on the soil [8, 30, 34, 35]. When the soil surfaces are saturated with just a molecular layer of water, the vapour density of a weakly polar compound in the soil air is greatly increased and any additional soil water will not influence the tendency of the compound to leave its sorbed site.

In general an increase of temperature on a soil-chemical system should cause an increase in the volatilization rate, through an increase of the equilibrium vapour density [30]. However, complicating factors may alter the expected result and an increase in temperature may not lead to an increase in volatilization rate. In addition, low temperature may not eliminate entirely the volatilization process since diffusion may continue even in frozen soil [35].

3.3. Chemical Movement Towards the Evaporating Surface

Volatilization of a soil incorporated chemical is mainly dependant on desorption of the pesticide from the soil and its upward movement to the soil surface. There are two general mechanisms whereby chemicals move to the evaporating surface: diffusion and mass flow. The total rate of movement is the sum of diffusion and mass flow [36].

Diffusion occurs whenever a concentration gradient is present. An increase in the difference of the concentration gradient will increase the diffusion rate. The depleted chemical on the soil surface due to volatilization will be replaced by additional chemical, which will be diffused upwards. Diffusion may occur along the vapour phase, the air–water interface, the water–water pathway and the water–solid interface. Even though only small quantities of the solution are in the vapour phase in respect to the amount absorbed on the soil, the amount of total mass transported by diffusion through the vapour phase is approximately equal to that of non-vapour phases. This is because the coefficients of diffusion in air are thousands times greater than those for water or surface diffusion [8]. The diffusion rate of a chemical is dependant on the temperature, soil bulk density, chemical substance concentration, soil water, organic matter and clay contents.

Mass flow occurs as a result of external forces. The chemical that is incorporated into soil is dissolved or suspended in water, present in the vapour phase or absorbed on solid mineral or organic components of the soil. Thus, mass flow of the chemical is the result of the mass flow of water and soil particles that the molecule is associated with. Mass flow due to air movement in soil is considered negligible [36]. When water evaporates from the soil surface the suction gradient produces results in an upward water movement. The upward movement of the chemical in the soil solution by mass flow with the evaporating water is called the wick effect. The magnitude of the wick effect is related to water evaporation rate, vapour pressure of the chemical and the chemical's concentration in the soil solution. The wick effect enhances much more the volatilization in wet soils because the degree of enhancement is related to the water evaporation rate. In general, when the water's evaporation rate is higher than the volatilization rate of the chemical, i.e. when the soil surface is dry, the chemical will accumulate on the soil surface. On the other hand, if the soil is rewetted again, then the chemical will volatilize.

3.4. Techniques for Measuring Volatilization Fluxes from Soil

Experimental measurements of concentrations and volatilization fluxes have been carried out for a wide range of organic chemicals, mainly pesticides, in the laboratory or in the field [37–41]. For laboratory measurements, the experimental apparatus usually comprises a chamber, e.g. a jar, in which the soil is placed. In the chamber air at specified velocity is drawn across or blown over, the soil surface or in some cases through the soil. The exhaust air is trapped, e.g. in air filters, polyurethane foam plugs, etc., and analysed mainly by gas chromatography. The laboratory experiments can be carefully controlled to provide useful information about the range of factors affecting volatilization fluxes in real field conditions. However, there are problems with designing laboratory volatilization experiments which accurately simulate the environment. For example, most laboratory studies use treated soils, e.g. sieved or artificially contaminated. A sieved soil is easier to handle, but the volatilization fluxes from a sieved soil are likely to be very different from an intact soil structure. Furthermore, in laboratory experiments it is usual practice to pre-clean the incoming air to the volatilization chamber, because otherwise it would be impossible to determine if the presence of a compound in the exhaust was due to volatilization from the soil, or whether it was already in the inlet air. In the field, however, the overlying air will contain some of the volatilizing compound and a concentration gradient will exist between the soil and air.

Field measurements of organic chemical concentrations and volatilization fluxes have been carried out for pesticides applied to soil. There are several different available techniques and these have been reviewed by Majewski et al. [42]. The basis of these techniques is to measure the vertical pesticide vapour gradient by measuring air concentrations at several different heights above the soil surface and to relate this gradient to the volatilization flux using theoretical techniques and field measurements such as wind speed, wind profiles, air temperature, etc. Air sampling and chemical analyses are the same as in the case of laboratory studies.

4. ESTIMATION OF VOLATILIZATION OF CHEMICALS FROM SOIL

As mentioned by Mackay [43], the best environmental model is the least bad set of simplifying assumptions that yields a model which is not too complex, but at the same time sufficiently detailed to be useful. A comprehensive model for the estimation of the volatilization of organic chemicals from soil surfaces should take, of course, explicitly into account all the factors mentioned in the previous section. The complexity and difficulty of incorporation of the above factors into a single model clearly indicate the expected shortcomings and uncertainties of the theoretical studies of soil volatilization. This is because the approximations that are used in the existing developed models do not sufficiently explain or cover the great number of complexities during the volatilization process.

Many different models have been developed in order to estimate the quantity of a chemical that is volatilized from soil on specific time intervals after its application. The models appear to be valid when compared with results of laboratory experiments and conditions. When, however, they are compared with field conditions they are often subjected to errors because the boundaries and/or environmental and atmospheric conditions are not well defined and predictable as they are under the controlled and easily manipulated laboratory conditions.

The selection of the most appropriate model will be dictated by the nature of the chemical and the particular set of the conditions. The models presented here are distinguished in two categories: (a) those applied to chemicals that are incorporated and distributed into soil and (b) those applied to chemicals volatilized from the soil surface, i.e. chemicals not incorporated into soil.

4.1. Models Applicable to Chemicals Incorporated into Soil

Several models have been proposed for estimating volatilization of chemicals incorporated into soil, but not all of them are applicable to a given situation. Moreover, no simple model for the estimating the volatilization rate for chemicals distributed and incorporated into soil, is available. The application of these complex models requires a number of input data, e.g. adsorption isotherm coefficients, diffusion coefficients, environmental properties, which in most cases are not experimentally known nor can be accurately predicted.

4.1.1. Wet Soil

For wet soil, where water is flowing up through the soil column to the soil surface and is evaporating there, the relatively simple model of Hamaker [35] and the more complex one of Jury et al. [12] can be used for estimating the volatilization rate.

4.1.1.1. The model of Hamaker In Hamaker's model for the case of non-zero water flux, the total loss of chemical per unit area over some time, Q_t, is calculated from the following expression:

$$Q_t = \frac{P_{\mathrm{VP}}}{P_{\mathrm{H_2O}}} \frac{D_{\mathrm{V}}}{D_{\mathrm{H_2O}}} f_{\mathrm{w,V}} + c_{t_0} f_{\mathrm{w,L}} \tag{4}$$

where P_{VP} refers to the vapour pressure of the chemical, $P_{\mathrm{H_2O}}$ the vapour pressure of water, D_{V} the diffusion coefficient of the chemical, $D_{\mathrm{H_2O}}$ the diffusion coefficient of water, $f_{\mathrm{w,V}}$ the vapour flux of water, $f_{\mathrm{w,L}}$ the liquid flux of water and c_{t_0} the initial concentration of the chemical in the soil solution.

4.1.1.2. The model of Jury et al. For the case of non-zero water flux, the concentration of the chemical in the gas phase as a function of time and depth, $c(z,t)$, is given by:

$$c(z,t)=\left(\frac{c_{t_0}-\gamma}{\varepsilon}\right)\left[1-0.5\left(1-\operatorname{erf}\left(\frac{z+(V_E t/\varepsilon)}{2\sqrt{D_E t/\varepsilon}}\right)\right)\right.$$
$$\left.-0.5\exp\left(-\frac{V_E z}{D_E}\right)\left(1-\operatorname{erf}\left(\frac{z-(V_E t/\varepsilon)}{2\sqrt{D_E t/\varepsilon}}\right)\right)\right] \quad (5)$$

where c_{t_0} is the initial total concentration of the chemical in soil; $\gamma = \beta \cdot \rho_b$, where β is an adsorption isotherm parameter and ρ_b the soil bulk density; $\varepsilon = \rho_b \cdot H \cdot \alpha + \theta \cdot H + \eta$, where H is the Henry's law constant, α an adsorption isotherm parameter and η the soil air content; V_E the effective convention velocity; D_E the effective diffusion coefficient where $D_E = D_g + H \cdot D_l$, and D_g and D_l are the diffusion coefficients in the gas and liquid phase respectively and erf(x) the error function of value x.

The flux at the surface for any time is given by the equation:

$$f=-(c_{t_0}-\gamma)\sqrt{\frac{D_E}{\pi \varepsilon t}}\exp(-w^2)-V_E\left(\frac{c_{t_0}-\gamma}{2\varepsilon}\right)[1+\operatorname{erf}(w)] \quad (6)$$

where $w=\dfrac{V_E{}^2 t}{4D_E}$

Finally, the total chemical loss, Q_t, can be found by integrating Eq. (6).

4.1.2. Dry Soil

In the case where the soil is dry, or if no water is flowing in the soil column because of reduced evaporation at the surface, the volatilization is controlled by slow diffusion. Two methods can be used in order to calculate the volatilization in the case of no water flux: the relatively simple model of Hartley [32] and the more complex one of Jury et al. [12].

4.1.2.1. The model of Hartley Hartley proposed a method that is based on an analysis of the heat balance between the evaporating chemical and air. The flux is expressed as:

$$f=\frac{\rho_{max}(1-h)/\delta}{1/D_v+(\Delta H_{vap}{}^2\rho_{max}M_w/kRT^2)} \quad (7)$$

where ρ_{max} refers to the chemical's saturated vapour concentration, h the humidity of the air, δ the thickness of the stagnant layer through which the chemical must pass, D_v the vapour diffusion coefficient, ΔH_{vap} the chemical's latent heat of vapourization, M_w the chemical's molecular weight, k the thermal conductivity of air, R the gas constant and T the temperature.

For less volatile compounds Eq. (7) is simplified to the following form:

$$f = \frac{D_v \rho_{max}(1-h)}{\delta} \tag{8}$$

The total chemical loss in both cases, Q_t, over a given time period, can be calculated as the product of the flux and the time.

4.1.2.2. The model of Jury et al. According to this model, for the case of no water flux, the concentration of the chemical in the gas phase as a function of time and depth, $c(z,t)$, is given by the following equation:

$$c(z,t) = \left(\frac{c_{t_0} - \gamma}{\varepsilon}\right) \operatorname{erf}\left(\frac{z\sqrt{D_E t/\varepsilon}}{2}\right) \tag{9}$$

The chemical's flux at the surface for any time is given by the equation:

$$f = -(c_{t_0} - \gamma)\sqrt{\frac{D_E}{\pi \varepsilon t}} \tag{10}$$

Finally, the total chemical loss, Q_t, can by found by integrating Eq. (10). Notations are the same as in Eqs. (5) and (6).

4.1.3. Comments on the Performance of the Models

Apart from the models mentioned here, many others have been proposed in the literature, e.g. [6, 11, 13–16]. As mentioned before, no simple model is available, and for many of the complex models, input information is sometimes not available. Furthermore, special care should be paid in the application of the models, especially when applied to chemicals other than those for which the model was developed. For example, Lyman et al. [8] applied the model of Jury et al. for dry soil (described in Section 4.1.2.2) for the case of trichloroethylene, which requires ten input parameters. The estimated total amount of trichloroethylene lost per day, for an initial total concentration of $0.05\,g\,cm^{-3}$, was found to be $1.6\,g\,cm^{-2}$. This value is considered very high, since in a 1-cm^2 soil column the model indicates that trichloroethylene would have to be removed entirely to a depth of 32 cm.

4.2. Models Applicable to Volatilization of Chemicals from the Soil Surface

For estimating the rate of volatilization from soil surface, i.e. for chemical not incorporated and distributed into soil, two simple methods can be used: the so-called Dow method [8] developed by scientists of the Dow Chemical company and the one by Voutsas et al. [44]. Application of both methods does not require special chemical properties such as diffusion coefficients, nor environmental properties such as soil moisture, soil type, temperature, etc.

4.2.1. The Dow Method

This method was developed in Dow Chemical Company using a limited dataset of nine chemicals. Dow method is a very simple and fast method for the estimation of the half-life for depletion of an organic chemical from the soil surface, $t_{1/2}$, which is given by the following equation:

$$t_{1/2} = 1.58 \times 10^{-8} \left(\frac{K_{oc} \times S}{P_{VP}} \right) \tag{11}$$

where $t_{1/2}$ is the half-life for depletion of the chemical from the soil surface, K_{oc} the chemical's soil adsorption coefficient, S the solubility of the chemical in water and P_{VP} the chemical's vapour pressure.

Assuming that the volatilization process follows first order kinetics, the concentration of the chemical, at any time, can be calculated by the following expression:

$$c(t) = c_{t_0} e^{-k_V t} \tag{12}$$

where c_{t_0} is the concentration at $t = 0$, t the time since the chemical's application and k_v the volatilization rate constant $k_v = \ln 2 / t_{1/2}$.

As discussed by Voutsas et al. [44], despite the empirical nature of the Dow model, it can be considered as a two step equilibrium partitioning model: compound from a sorption site on the soil particles → compound in the soil water → compound in the atmospheric air.

4.2.2. *The Methods of Voutsas et al.*

Voutsas et al. [44] have examined several simple correlations for the estimation of the half-life for the depletion of an organic chemical from a soil surface to air. They proposed different correlations that have been presented for wet and dry soils.

4.2.2.1. Wet soil For wet soils Voutsas et al. proposed two models. The first, Eq. (13), requires the knowledge of the soil/organic carbon partition coefficient and the Henry's law constant and the second, Eq. (14), the vapour pressure of the chemical involved.

$$t_{1/2} = 0.033965\left(\frac{K_{oc}}{H}\right)^{0.53787} \tag{13}$$

where K_{oc} is the chemical's soil adsorption coefficient, H the Henry's law constant of the chemical in water, and

$$t_{1/2} = 0.132\ P_{VP}^{-0.3817} \tag{14}$$

where P_{VP} is the chemical's vapour pressure.

If reliable experimental P_{VP} values are not experimentally available, the use of Eq. (13) is suggested. In cases when Henry's law constant values are not available, predicted ones by the bond-contribution method of Meylan and Howard [20] can be used. Also, due to uncertainties in the experimental K_{oc} values, e.g. for lindane the experimental K_{oc} values reported from different sources are in the range of 686–12400, while for trifluralin in the range of 1200–13700, Voutsas et al. proposed that predicted values by the group-contribution model of Meylan et al. [25] should be used.

4.2.2.2. Dry soil For dry soils Voutsas et al. presented similar expressions such as those for wet soils:

$$t_{1/2} = 1.3216\left(\frac{K_{oc}}{H}\right)^{0.3108} \tag{15}$$

and

$$t_{1/2} = 1.7543 P_{\mathrm{VP}}^{-0.2688} \tag{16}$$

where notations are the same as in Eqs. (13) and (14). The presence of the Henry constant in Eq. (15) in the absence of water indicates the empirical character of the correlation.

4.2.3. Comments on the Performance of the Models

Table 1 presents experimental and calculated $t_{1/2}$ values from the Dow model and the models of Voutsas et al. for wet soils. The models of Voutsas et al. give better results than Dow method with Eq. (13) and are considered to be in general superior to Eq. (14). Using logarithmic values, Eq. (13) accounts for 50% variation in the measured $t_{1/2}$ data yielding a mean absolute error lower than 0.5, and represents a substantial improvement over the Dow method that accounts for just the 23% variation, yielding a mean absolute error of 0.8. Table 2 presents a comparison between experimental and predicted chemicals' per cent volatilization after 1 and 5 days application, which is defined as:

$$\%\,\text{volatilized} = 100 - \frac{c_x}{c_{t_0}} \times 100 \tag{17}$$

where c_x is the chemical's concentration at $t = x$ days, calculated from Eq. (12) and c_{t_0} the chemical's concentration at $t = 0$ days.

Table 3 presents experimental and calculated $t_{1/2}$ values from the Dow model and the models of Voutsas et al. for dry soils. The results for dry soils are poorer than those obtained for moist soils. Order-of-magnitude agreement between experimental and predicted $t_{1/2}$ values should be expected in this case. Also, Table 4 presents chemicals' per cent volatilization after 1 and 5 days application for the same chemicals.

5. THERMODYNAMICS OF PERSISTENT ORGANIC CHEMICALS: THE EQUILIBRIUM PARTITIONING APPROACH

Persistent (slowly bio-degrading) organic chemicals, such as pesticides, PCBs, etc. occur throughout the environment, and in a matter of years, by many different transport mechanisms, they appear in all environmental compartments: air, water, soil, sediment, aerosols, water-suspended particulates and vegetative and animal biota. Mackay [43] developed the so-called

Table 1
Chemicals, chemical properties, experimental and predicted half-life values for moist soils

Chemical	P_{VP} (mmHg)	K_{oc}	S (mg L^{-1})	$t_{1/2}$, experimental (days)	$t_{1/2}$, calculated (days)		
					Dow method	Eq. (13)	Eq. (14)
trans-Chlordane	5.03E-05	8.67E + 04	0.056	10.7	1.5	9.0	5.8
cis-Chlordane	3.60E-05	8.67E + 04	0.056	11.3	2.1	9.0	6.6
Pendimethalin	3.00E-05	2.62E + 03	0.3	107	0.4	12.1	7.0
Methyl parathion	3.50E-06	5.23E + 02	37.7	13.5	89.0	16.1	16.0
Carbofuran	4.85E-06	7.09E + 01	320	24	73.9	35.7	14.1
DDT	1.6E-07	2.20E + 05	0.01	55.4	119.7	38.6	51.8
Dieldrin	5.89E-06	1.06E + 04	0.195	17.4	5.5	6.8	13.1
Endrin	3.00E-06	1.06E + 04	0.25	19.1	14.0	8.7	16.9
Heptachlor	4E-04	5.24E + 04	0.18	8.4	0.4	2.6	2.6
Heptachlor epoxide	1.95E-05	5.26E + 03	0.20	13.8	0.9	3.1	8.3
Lindane	4.2E-05	3.38E + 03	7.3	4.8	9.3	5.3	6.2
Chlorpyrifos	2.03E-05	6.83E + 03	1.12	3	6.0	10.4	8.2
Diuron	6.90E-08	1.36E + 02	42	12	1308.0	134.5	71.5
Dinoseb	7.50E-05	3.54E + 03	52	26	38.8	19.9	5.0
Endosulfan	1.73E-07	2.20E + 04	0.325	5.4	653.0	3.7	50.3
p,p'-DDD	1.35E-06	1.52E + 05	0.09	12	160.1	35.8	23.0
Fenpropimorph	2.63E-05	2.69E + 04	4.3	11.2	69.5	24.1	7.4
Metolachlor	3.14E-005	2.92E + 02	530	88	77.9	43.0	6.9
Atrazine	2.89E-07	2.30E + 02	34.7	162	437.1	77.9	41.4
Trifluralin	4.58E-05	9.68E + 03	0.184	5	0.6	1.9	6.0
Dacthal (DCPA)	2.50E-06	2.83E + 02	0.5	49.7	0.9	2.2	18.2
EPTC	2.4E-02	2.58E + 02	375	1.19	0.06	0.72	0.55
Alachlor	2.05E-05	1.85E + 02	240	161	34.2	35.1	8.1
Fonofos	3.38E-04	8.36E + 02	15.7	11.4	0.6	2.12	2.8

Note: Sources of experimental data can be found in ref. [44].

Table 2
Experimental and predicted per cent volatilization from moist soils

Chemical	% Volatilized after 1 day, experimental	% Volatilized after 5 days, experimental	% Volatilized after 1 day, calculated			% Volatilized after 5 days, calculated		
			Dow method	Eq. (13)	Eq. (14)	Dow method	Eq. (13)	Eq. (14)
trans-Chlordane	6.3	27.7	37.0	7.4	11.3	90.1	32.0	45.0
cis-Chlordane	6.0	26.4	28.1	7.4	10.0	80.8	32.0	40.9
Pendimethalin	0.7	3.2	82.3	5.6	9.4	100.0	24.9	39.1
Methyl parathion	5.0	22.6	0.8	4.2	4.2	3.8	19.4	19.5
Carbofuran	2.9	13.5	0.9	1.9	4.8	4.6	9.3	21.8
DDT	1.2	6.1	0.6	1.8	1.3	2.9	8.6	6.5
Dieldrin	3.9	18.1	11.8	9.7	5.2	46.8	39.9	23.3
Endrin	3.6	16.6	4.8	7.7	4.0	21.9	32.9	18.5
Heptachlor	7.9	33.8	82.3	23.4	23.4	100.0	73.6	73.6
Heptachlor epoxide	4.9	22.2	53.7	20.0	8.0	97.9	67.3	34.1
Lindane	13.5	51.4	7.2	12.3	10.6	31.1	48.0	42.8
Chlorpyrifos	20.6	68.5	10.9	6.5	8.1	43.9	28.3	34.5
Diuron	5.6	25.1	0.1	0.5	1.0	0.3	2.5	4.7
Dinoseb	2.6	12.5	1.8	3.4	12.9	8.5	16.0	50.0
Endosulfan	12.1	47.4	0.1	17.1	1.4	0.5	60.8	6.7
p,p'-DDD	5.6	25.1	0.4	1.9	3.0	2.1	9.2	14.0
Fenpropimorph	6.0	26.6	1.0	2.8	8.9	4.9	13.4	37.4
Metolachlor	0.8	3.9	0.9	1.6	9.6	4.4	7.7	39.5
Atrazine	0.4	2.1	0.2	0.9	1.7	0.8	4.4	8.0
Trifluralin	12.9	50.0	68.5	30.6	10.9	99.7	83.9	43.9
Dacthal (DCPA)	1.4	6.7	53.7	27.0	3.7	97.9	79.3	17.3
EPTC	44.2	94.6	100.0	61.8	71.6	100.0	99.2	99.8
Alachlor	0.4	2.1	2.0	2.0	8.2	9.6	9.4	34.8
Fonofos	5.9	26.2	68.5	27.9	21.9	99.7	80.5	71.0

Table 3
Chemicals, chemical properties, experimental and predicted half-life values for dry soils

Chemical	P_{VP} (mmHg)	K_{oc}	S (mg L^{-1})	$t_{1/2}$, experimental (days)	$t_{1/2}$, calculated (days)		
					Dow method	Eq. (15)	Eq. (16)
Atrazine	2.89E-07	2.30E + 02	34.7	599	437.1	115.5	100.4
Simazine	2.21E-08	1.49E + 02	6.20	1112.00	660.5	134.2	200.4
Trifluralin	4.58E-05	9.68E + 03	0.184	9.5	0.6	13.3	25.7
Dieldrin	5.89E-06	1.06E + 04	0.195	1433	5.5	28.3	44.7
Chlordane	9.75E-06	8.67E + 04	0.056	70.3	7.9	33.3	39.0
Heptachlor	4E-04	5.24E + 04	0.18	4.40	0.4	16.3	14.4
Lindane	4.2E-05	3.38E + 03	7.3	7.48	9.3	24.4	26.3
Toxaphene	6.69E-06	9.93E + 04	0.55	39	129.0	66.5	43.2
EPTC	2.4E-02	2.58E + 02	375	3.3	0.1	7.7	4.8
Alachlor	2.05E-05	1.85E + 02	240	69	34.2	72.9	31.9

Note: Sources of experimental data can be found in ref. [44].

Table 4
Experimental and predicted per cent volatilization from dry soils

Chemical	% Volatilized after 1 day, experimental	% Volatilized after 5 days, experimental	% Volatilized after 1 day, calculated			% Volatilized after 5 days, calculated		
			Dow method	Eq. (15)	Eq. (16)	Dow method	Eq. (15)	Eq. (16)
Atrazine	0.12	0.58	0.16	0.60	0.69	0.79	2.96	3.39
Simazine	0.06	0.31	0.10	0.52	0.35	0.52	2.55	1.71
Trifluralin	7.04	30.57	68.50	5.08	2.66	99.69	22.94	12.62
Dieldrin	0.05	0.24	11.84	2.42	1.54	46.75	11.53	7.46
Chlordane	0.98	4.81	8.40	2.06	1.76	35.51	9.88	8.50
Heptachlor	14.58	54.51	82.32	4.16	4.70	99.98	19.15	21.39
Lindane	8.85	37.08	7.18	2.80	2.60	31.11	13.24	12.35
Toxaphene	1.76	8.50	0.54	1.04	1.59	2.65	5.08	7.71
EPTC	18.95	65.01	99.90	8.61	13.45	100.00	36.24	51.42
Alachlor	1.00	4.90	2.01	0.95	2.15	9.64	4.64	10.30

level I fugacity model, which is based on the assumption that a persistent chemical will be in thermodynamic phase equilibrium in all environmental compartments over the time period which it persists.

For the case of soil/air partitioning and speaking the language of thermodynamics, for having equilibrium of a chemical between soil and air, the following standard thermodynamic equation has to be satisfied:

$$f_i^{soil}(x_i^{soil},T,P) = f_i^{air}(x_i^{air},T,P) \tag{18}$$

where f and x are the fugacity and mole fraction of species i, in the soil and air phases. Fugacity is a thermodynamic property that describes a chemical's tendency to escape from one phase to another and has units of pressure. For example if a chemical has a fugacity in the soil phase greater than that in air then there will be a tendency of the chemical to move from the soil to the air phase in order to establish equilibrium conditions.

Mackay [43] proposed the calculation of fugacities in the various environmental compartments through the concentrations and fugacity capacities of the compounds in the compartments. According to this approach:

$$f_i^{compartment}(c_i^{compartment},T,P) = \frac{c_i^{compartment}}{MW \times z_i^{compartment}} \tag{19}$$

where: $c_i^{compartment}$ is the concentration of the chemical in the compartment (g m^{-3}), MW the molecular weight of the chemical (g mol^{-1}), $z_i^{compartment}$ the fugacity capacity of the chemical in the compartment (mol m^{-3} Pa^{-1}) and R the gas constant (8.314 J mol^{-1} K^{-1}).
The fugacity capacity for air is calculated using:

$$z_i^{air} = \frac{1}{RT} \tag{20}$$

The fugacity capacity of soils is estimated using:

$$z_i^{soil} = \frac{f_{oc}\rho_s K_{i,oc}}{H_i} \tag{21}$$

where f_{oc} is the faction of organic carbon in soil (typical 7%), ρ_s the soil density (typical 1.5 g cm^{-3}), K_{oc} the soil organic carbon–water partition coefficient and H_i the Henry's law constant.

Cousins and Jones [45] have used soil to air fugacity ratios to investigate the equilibrium in soil and air for some semi-volatile organic compounds. Fugacity ratios near indicate an equilibrium between soil and air, fugacity ratios greater than unity indicate a tendency of the chemical to volatilize from soil, while fugacity ratios less than unity, indicates a tendency of the chemical to remain in the soil and a capacity of the air to supply the soil with more chemical. Using this approach Cousins and Jones [45] have found that for the case of spiked soils, the fugacity ratios were all larger than one, which consisted of the observed losses during the field experiments. On the other hand, for the unspiked soils most of the chemicals and especially PCBs were close to the equilibrium, which explained the small uptake of PCBs in the unspiked soils.

Of course, the equilibrium partition approach has some limitations: (a) it considers only a phase equilibrium situation and does not take into account the rate of transport between air and soil, (b) the fine distribution of compounds within surface soils may influence the calculation of fugacities and (c) it provides only a snap-shot for a given set of environmental conditions.

REFERENCES

[1] W.F. Spencer and M.M. Gliath, Environ. Sci. Technol., 3 (1969) 670.
[2] A.W. Taylor, J. Air Pollut. Contam. Assoc., 28 (1978) 922.
[3] G.H. Willis, L.L. McDowell, S. Smith, L.M. Southwick and E.R. Lemon, Agron. J., 72 (1980) 627.
[4] A.W. Taylor and D.E. Glotfelty, in "Evaporation From Soils and Crops" (R. Grover, ed.), Environmental Chemistry of Herbicides, CRC Press, Boca Raton, FL, 1988.
[5] I.T. Cousins, A.J. Beck and K.C. Jones, Sci. Total Environ., 228 (1999) 5.
[6] M.M. Hantush, M.M. Marino and M.R. Islam, J. Hydrol., 227 (2000) 66.
[7] M.M. Hantush, R.S. Govindaraju, M.M. Marino and Z. Zhang, J. Hydrol., 260 (2002) 58.
[8] W.J. Lyman, W.F. Reehl and D.H. Rosenblatt, Handbook of Chemical Property Estimation Methods, American Chemical Society, Washington, DC, 1990.
[9] G.H. Willis, L.L. McDowell, L.A. Harper, L.M. Southwick and S. Smith, J. Environ. Qual., 12 (1983) 80.
[10] W.F. Spencer, W.J. Farmer and M.M. Gliath, Residue Rev., 49 (1973) 1.
[11] R. Mayer, L. Letey and W. Farmer, Soil Sci. Soc. Am. Proc., 38 (1974) 563.
[12] W. Jury, R. Grover, W. Spencer and W. Farmer, Soil Sci. Soc. Am. J., 44 (1980) 445.
[13] W. Jury, W. Spencer and W. Farmer, J. Environ. Qual., 12 (1983) 558.
[14] W. Jury, W. Spencer and W. Farmer, J. Environ. Qual., 13 (1984) 567.
[15] W. Jury, W. Spencer and W. Farmer, J. Environ. Qual., 13 (1984) 573.
[16] W. Jury, W. Spencer and W. Farmer, J. Environ. Qual., 13 (1984) 580.

[17] B.E. Poling, J.M. Prausnitz and J.P. O'Connell, The Properties of Gases and Liquids, 5th edn., McGraw-Hill, New York, 2001.
[18] E. Voutsas, M. Lampadariou, K. Magoulas and D. Tassios, Fluid Phase Equilib., 198 (2002) 81.
[19] Ph. Coutsikos, E. Voutsas, K. Magoulas and D. Tassios, Fluid Phase Equilib., 207 (2003) 263.
[20] W. Meylan and P. Howard, Environ. Toxicol. Chem., 10 (1991) 1283.
[21] H. Sangster, Octanol–Water Partition Coefficients: Fundamentals and Physical Chemistry, Wiley, New York, 1997.
[22] P.H. Howard and W.M. Meylan, Prediction of Physical Properties, Transport, and Degradation for Environmental Fate and Exposure Assessments, Syracuse Research Corporation, Syracuse, NY, 1998.
[23] W.M. Meylan and P.H. Howard, J. Pharm. Sci., 84 (1995) 83.
[24] J.R. Baker, J.R. Mihelcic and E. Shea, Chemosphere, 41 (2000) 813.
[25] W. Meylan, P. Howard and R. Boethling, Environ. Sci. Technol., 26 (1992) 1560.
[26] J. Ryan, R. Bell, J. Davidson and G.A. Connor, Chemosphere, 17 (1988) 2299.
[27] M.-J. Wang and K.C. Jones, Environ. Sci. Technol., 28 (1994) 1843.
[28] T. Harner, D. Mackay and K.C. Jones, Environ. Sci. Technol., 29 (1995) 1200.
[29] J.E. Woordow and J.N. Seiber, Environ. Sci. Technol., 31 (1997) 523.
[30] W. Guenzi and W. Beard, in "Volatilization of Pesticides" (W. Guenzi, ed.), Pesticides in Soil and Water, Soil Science Society of America, Madison, WI, 1974.
[31] B. Grass, B.W. Wenclawiak and H. Rudel, Chemosphere, 28 (1994) 491.
[32] G.S. Hartley, in "Evaporation of Pesticides", Pesticidal Formulations, Research, Physical and Colloidal Chemical Aspects, Advances in Chemistry Series, 86, American Chemical Society, Washington, DC, 1969.
[33] J.V. Parochetti and G.F. Warren, Weeds, 14 (1966) 281.
[34] L.W. Petersen, D.E. Rolston and P. Moldrup, J. Environ. Qual., 23 (1994) 799.
[35] J. Hamaker, in "Diffusion and Volatilization" (C. Goring and J. Hamaker, eds.), Organic Chemicals in the Soil Envinonment, Marcel Dekker, New York, 1972.
[36] L. Letey and W. Farmer, in "Movement of Pesticides in Soil" (W.D. Guenzi, ed.), Pesticides in Soil and Water, Soil Science Society of America, Madison, WI, 1974.
[37] W.F. Spencer and M.M. Gliath, J. Environ. Qual., 2 (1973) 284.
[38] B. Lindhardt and T.H. Christensen, Water Air Soil Pollut., 92 (1996) 375.
[39] R.G. Nash, J. Agric. Food Chem., 31 (1983) 210.
[40] H. Rudel, Chemosphere, 35 (1997) 143.
[41] A.I., Garcia-Valcarcel and J.L. Tadeo, J. Agric. Food Chem., 51 (2003) 999.
[42] M.S. Majewski, D.E. Glotfelty, U.K.T. Paw and J.N. Seiber, Environ. Sci. Technol., 24 (1990) 1490.
[43] D. Mackay, Multimedia Environmental Models. The Fugacity Approach, Lewis Publishers Inc., Chelsea, MI, 1991.
[44] E. Voutsas, C. Vavva, K. Magoulas and D. Tassios, Chemosphere, 58 (2005) 751.
[45] I.T. Cousins and K.C. Jones, Environ. Pollut., 102 (1998) 105.

Thermodynamics, Solubility and Environmental Issues
T.M Letcher (editor)

Chapter 12

Solubility and the Phytoextraction of Arsenic from Soils by Two Different Fern Species

Valquiria Campos

University of São Paulo, Rua Marie Nader Calfat, 351 apto 71 Evoluti, Morumbi, São Paulo 05713–520, SP, Brazil

1. INTRODUCTION

As soils are increasingly used in the society for purposes other than agriculture, the frequency and extent of soil contamination by toxic trace elements will increase. Phytoextraction on metals or metalloids is an uncommon phenomenon in terrestrial higher plants. In 2001, Ma et al. [1] reported that the *Pteris vittata* was extremely efficient in extracting arsenic from soils. The study by Ma was the first to report a fern *P. vittata* grown on a chromium–copper–arsenium-contaminated site as an arsenic hyper-accumulator. Other plant species have been reported by researchers as being able to remove trace elements [2–9]. Because of solubility, mobility, bioavailability, and toxicity of arsenic, studies on arsenic speciation and removal are essential in attempting to reduce the negative impact on the environment of arsenic.

Arsenic is a toxic element that can be found in anthropogenic wastes and some geochemical environments. Arsenic is a semi-metallic element and it also may be present as organometallic forms, such as methylarsinic acid and dimethylarsinic acid, which are active ingredients in many pesticides [10]. Levels of arsenic in soils (1–40 mg As kg^{-1}) may be elevated due to mineralization and contamination from industrial activity and agrochemicals [11]. Arsenic can be added to soils through the use of synthetic fertilizers and As-based pesticides. The continued application of these products can result in an accumulation of toxic residues that, even in relatively low concentrations, could compromise or limit the use of water. The complexation of arsenic by dissolved organic matter in natural environments prevents sorption and co-precipitation with solid-phase organic and inorganic compounds, thus increasing the mobility of arsenic in aquatic systems and soil [12]. Inorganic forms of arsenic can be present as either arsenate As (V) or arsenite As (III). Arsenate is tetrahedral oxyanions that can compete for adsorption sites on soil mineral surfaces. Arsenate is the predominant inorganic species of arsenic under oxidizing soil conditions and is retained in soils by adsorption reactions.

The main exposure to inorganic arsenic experienced by the general public is through ingestion. Long-term exposure to low concentrations of arsenic in drinking water can lead to skin, bladder, lung, and prostate cancer. Non-cancer effects of ingesting arsenic at low levels include cardiovascular disease, diabetes, and anemia, as well as reproductive and developmental, immunological, and neurological effects [13]. Substantial evidence led the International Agency for Research on Cancer to conclude that ingestion of inorganic arsenic can cause skin cancer [14]. Arsenic exposure is a potential health risk to local populations around gold mining areas in southeastern Brazil, where 20% of the total sample population showed elevated As concentrations (2.2–106 $\mu g\ L^{-1}$) [15]. In the Iron Quadrangle region the public supply of water comes mainly from local watersheds. However, in some places the population uses spring water or groundwater from areas close to closed mines, for human consumption. The main natural sources of arsenic in the Iron Quadrangle come from rocks that contain lode gold deposits. The anthropic sources of arsenic are contaminated refuse piles, soil, and sediments. It was found that total As concentration is 2980 $\mu g\ L^{-1}$ and As^{3+} is 86 $\mu g\ L^{-1}$ in water samples collected from underground gold mines, artesian wells, and springs in Ouro Preto and Mariana counties [16]. Speciation is important since mobility and toxicity of As (III) are supposed to be much higher than those of As (V). The toxicity of such compounds decreases in the order: arsine > arsenite > arsenate > alkyl arsenic acids > arsenium compounds and metallic arsenic [17].

Arsenic is distributed widely among plant species. In plant tissues its normal concentration is 0.01–5 mg kg^{-1} on a dry-weight basis [18]. The soil–plant system is an open system subject to inputs, such as contaminants and agrochemicals, and to losses, such as the removal of trace elements in harvested plant material, leaching, erosion, and volatilization. The factors affecting the amounts of trace elements absorbed by a plant are those controlling (a) the concentration and speciation of the element in the soil solution; (b) the movement of the element from the bulk soil to the root surface; (c) the transport of element from the root surface into the root; (d) its translocation from the root to the shoot. Plant uptake of mobile ions present in the soil solution is largely determined by the total quantity of the ions in the soil. In the case of strongly adsorbed ions, absorption is more dependent on the amount of root produced [19].

Certain species of plants have been found to accumulate very high concentrations of certain trace elements and these are referred as "hyper-accumulator" species. Plants can also intercept significant amounts of some elements through foliar absorption. Foliar absorption of solutes depends on the plant species, its nutritional status, the thickness of its cuticle, the age of the leaf, the presence of stomata guard cells, the humidity at the leaf surface, and the nature of the solutes.

P. vittata L. is characterized by having 1-pinnate-imparipinnate fronds, pinnae linear, not articulate to the rachis and with truncate base, free venation with simple or furcate veins, and trilete spores. The pinnae display a wide variation in size, angle of divergence from the rachis (acute or right angle), and shapes (straight to shortly falcate). Plants terrestrial or rupestral are generally found in sunny, open places, such as banks along highways and on city walls (like weeds), at ~2000 m. The typical plant's habit and morphological characteristics of *P. vittata*, respectively, can be observed in Figs 1 and 2; Linnaeus [20] classified ferns into genera based on the shape of the spores. It is native to the Old World (China) and was introduced in California, Mexico, Cuba, Bahamas, Dominica, Martinique, Barbados, Trinidad, Guiana, Peru, Brazil, and Argentina. It can be found in Brazil from the northern to the southern region [21].

A widespread and highly variable tropical/subtropical epiphytic fern, which adapts easily to cultivation and is popularly grown as a common house fern, is the golden polypody *Phlebodium aureum* (L.) J. Sm. It is an epiphytic fern, native to Florida, the Caribbean, Mexico, and Central and South America. Golden polypody is a member of the Polypodiaceae, the many-footed fern family. Polypodies grow by a creeping horizontal rhizome. The leaves arise alternately along the rhizome. Linnaeus originally named the fern *P. aureum* in 1753. The current generic name first came from Robert Brown (Brownian motion fame) in 1838, but was not applied as a generic name until made official by John Smith in 1841. *Phlebodium* is derived from the Greek *phlebodes*, "full of veins," and refers to the internal structure of the rhizome; *aureum* refers to the golden brown scales that cover the outside of the plant. The rhizomes creep along the ground or on the surface to which the climbing plant is attached. The fronds vary greatly in size and color, from gray-green, through silvery-green to rich, powdery, blue-green. In their native areas, they are commonly found growing on palm trunks or tree limbs, rarely terrestrial in habit [22]. The morphological characteristics of *P. aureum* can be observed in Fig. 3. The objective of this study was to evaluate the relationships between soil type and trace element removal status of the vascular plants *P. vittata* Linnaeus and *P. aureum* (L.)

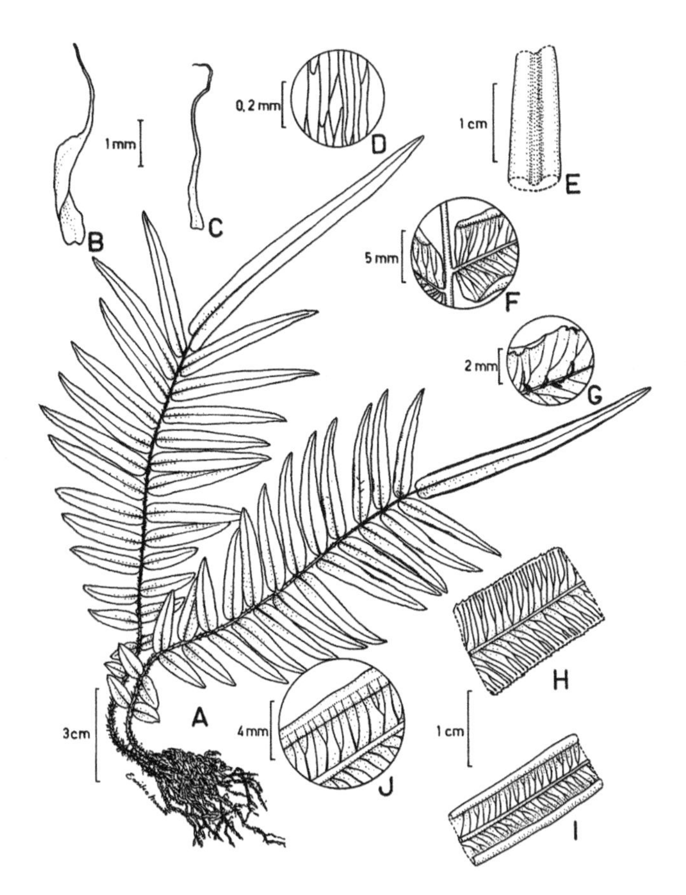

Fig. 1. *Pteris vittata* L. general characteristics [21]: (A) habit; (B and –C) rhizome scales; (D) detail of the cells of a rhizome scale; (E) stipe; (F) detail of non-articulated pinnae; (G) costa, abaxial view showing scales; (H) venation of the sterile lamina; (I) venation and pseudoindusium of the fertile lamina; (J) detail of the venation and pseudoindusium.

J. Sm. using the phytoremediation method. The interaction of soil-plant system is fundamental for controlling the migration of arsenic species to groundwater and its impact on drinking water.

2. MATERIALS AND METHODS

Interpretations are based on results of green house experiments with soils contaminated with arsenic. Two soils were used as adsorbents: Red and Yellow Latosols. Both *P. vittata* and *P. aureum* have relatively large biomass and are species of common ferns. The two species were planted individually into pots,

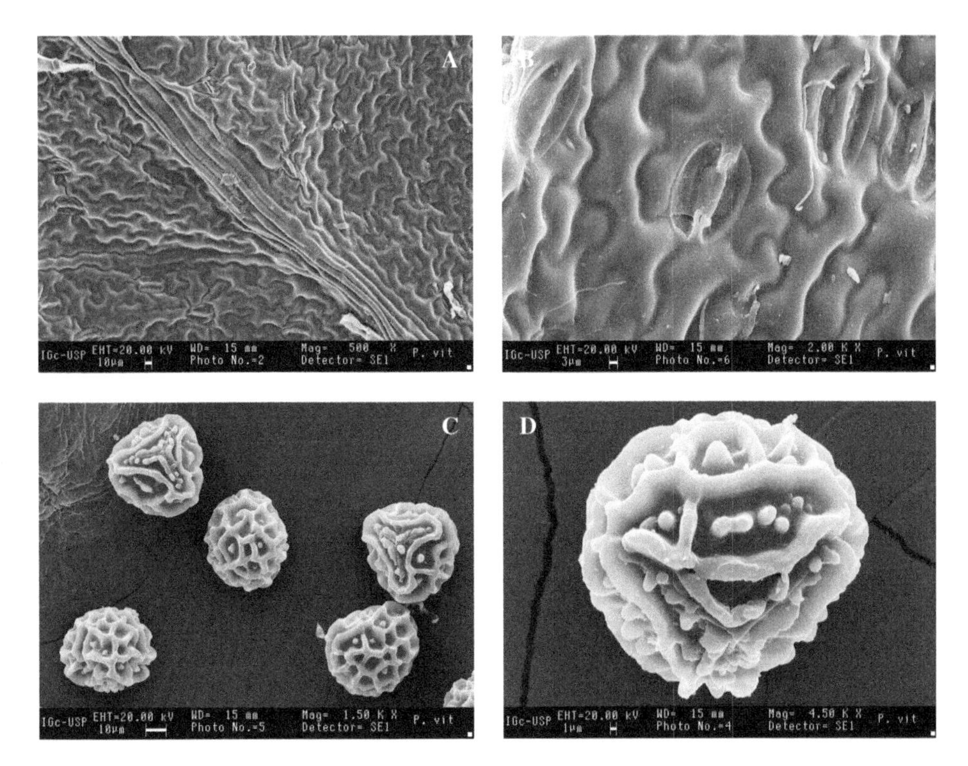

Fig. 2. Scanning electron microscope of the *Pteris vittata* L: (A) abaxial surface of a pinna showing hairs, stomata, costa, secondary veins, and free venation pattern (500×); (B) detail of abaxial surface of a pinna showing stomata and epidermal cells (2000×); (C) general view of the trilete spores (1500×); (D) detail of a trilete spore and its surface (4500×) having large areoles on the distal face and a well-developed equatorial flange.

each containing 250 g general purpose compost with Red and Yellow Latosols. The soil classification of Piracicaba, São Paulo, was based on the Brazilian Soil Taxonomy and Classification System [23], which assumes soil genesis in a higher order and considers morphological characteristics at another level for discrimination of soils regarding their potential for agricultural use.

Scanning electron microscope (SEM) studies are directed toward the delineation of systematic relationships of the genera. The morphological properties of the *P. vittata* and *P. aureum* may be recognized by microscopic observation, especially by SEM that reveals important aspects of tissue, size, and morphology.

Samples were collected from the top soil layer (0.0–0.30 m) from the Red and Yellow Latosols. Soil samples were air dried, crushed, sieved through

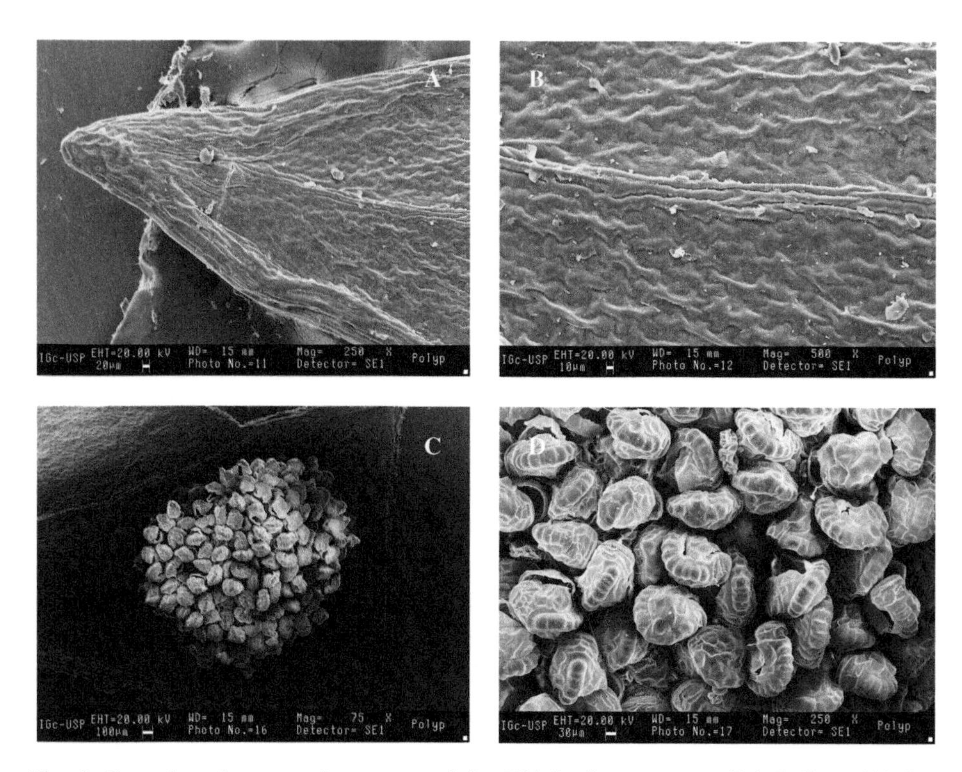

Fig. 3. Scanning electron microscope of the *Phlebodium aureum* (L.) J. Sm.: (A) free venation pattern (250×); (B) detail of abaxial surface of apinna and epidermal cells (500×); (C) fertile pinnule with sorus (75×); (D) sporangia with monolete spores. Monolete spores predominate in large genera such as Polypodiaceae (250×).

2-mm sieve, and reserved for subsequent analysis. Exchangeable cations were extracted from the soil by 1 M NH_4OAc [24], and then extractions were analyzed by an atomic absorption spectroscopy [25]. Cation exchange capacity (CEC) was determined by the ammonium saturation method at pH 7.0. Selected properties of the soils are given in Table 1.

Arsenic sorption is a function of several chemical factors, including solution pH, competing anion concentration, and Fe oxide concentration in the soil. Impact of soil oxalate extractable Fe contents on As sorption in two Piracicaba soils was determined previously (Table 2). The solubility of amorphous forms is greater than it is for highly crystalline forms. The amount of Fe in the soil solution was determined by the colorimetric method using *o*-phenanthroline [26].

Very low Fe_o content indicates small quantities of non-crystalline form. Free iron, extracted with dithionite (Fe_d), corresponds to 58.1 and 17.9 $g\,kg^{-1}$

Table 1
Chemical and physical analyses: the total organic matter (OM), pH, sulfur, phosphorus, exchangeable cations, cation exchange capacity (CEC), texture, and volume for soil classification

Soil	OM (g dm^{-3})	pH ($CaCl_2$)	S (mg dm^{-3})	P (mg dm^{-3})	K (mmol dm^{-3})	Ca (mmol dm^{-3})	Mg (mmol dm^{-3})	H + Al (mmol dm^{-3})	CEC (mmol dm^{-3})	Texture (%)	V (%)
1	28	4.3	78	2	1.2	11	2.5	94	110.1	Clay	13
2	19	4.8	26	2	0.9	13	5.7	27	46.9	Clay	37

Note: (1) Red Latosol; (2) Yellow Latosol.

Table 2
The amount of Fe (amorphous and crystalline forms)

Soil	Fe_o		Fe_d	
	%	g kg^{-1}	%	g kg^{-1}
1	1.80	18.42	5.81	58.1
2	0.6	6.12	1.79	17.9

Note: (1) Red Latosol; (2) Yellow Latosol.

of the total Red and Yellow Latosols, respectively. The relationship Fe_o/Fe_d goes from medium to very low (from 5.81–1.79 to 1.8–0.6 g kg^{-1}) indicating the dominance of crystalline Fe oxides in both soils in this study. The relationship clay/Fe_d decreased in Yellow Latosol and Fe oxides' solubility varies in hardened layers, as a consequence of pedological processes. In both soils, Fe extraction with oxalate (amorphous Fe oxides) decreases, while Fe extraction with dithionite increases with clay texture (Red Latosol 73% and Yellow Latosol 14%). The Red Latosol contains more iron and crystalline Fe oxides. Fe in the form of goethite and hematite (crystalline iron oxides) was estimated by X-ray diffraction analysis. To make the diagram more useful, the redox couples Fe $(OH)_3$/Fe (II) were also included in Fig. 4.

The experiments were conducted using trace elements solution containing different concentrations. Each treatment was replicated three times. The rhizome and frond were analyzed for inorganic arsenic. Total arsenic contents in soil and plant were determined by instrumental neutron activation analyses (INAA), Nuclear Reactor IEA-R1m. The sample irradiation was carried out at the nuclear reactor for 8 h and under thermal neutron flux of 2.7×10^{12} ncm^{-2} s^{-1}; subsequently, a decay time of 3 days and 3 h counting time were employed. The radionuclide activities were measured using a gamma-ray spectrometer comprising a hyper-pure GE detector and associated electronic system. The spectra of the samples were measured under the same experimental conditions of the standard photopeak of 657 keV of ^{76}As. The amount of anion absorbed was calculated as the difference between the anion concentration before and after reaction with the soil–plant systems.

3. RESULTS AND DISCUSSION

Redox conditions are an important regulator of aqueous biogeochemistry because the oxidation state of elements affects their solubility, adsorption

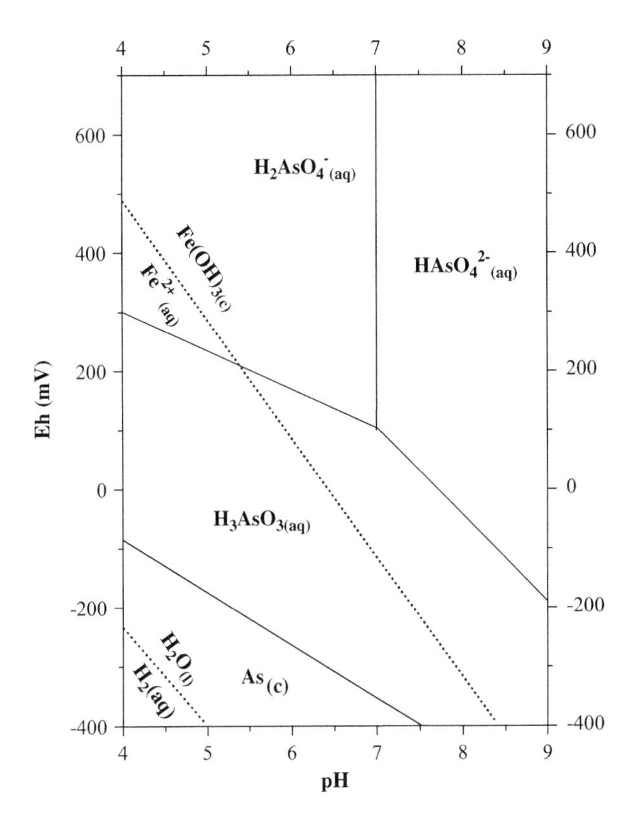

Fig. 4. Eh–pH diagram for the system As–H_2O; system activities of arsenic and iron were all taken to be 10^{-4}.

behavior, toxicity, and distribution of soil systems. The influence of redox potential on arsenic speciation and solubility was studied in two different soils: Red and Yellow Latosols. In the first experiment, some arsenic was reduced and released into solution before the solubilization of the ferric layer. Solubilization of arsenic was occurring simultaneously with the reduction of As (V) to As (III) after 30-day equilibration period under controlled redox and pH 7.0 conditions (Fig. 5).

Redox status, pH, and Fe presence affected the speciation and solubility of arsenic. The Yellow Latosol had a higher concentration of the As (V) than did the Red Latosol. At redox potentials of 200 and 400 mV, As (V) was the major dissolved arsenic species. The experimental data indicated that As solubility was mainly controlled by an iron phase. Under moderately reduced Red Latosol conditions (0 and −200 mV) in the Red Latosol, arsenite

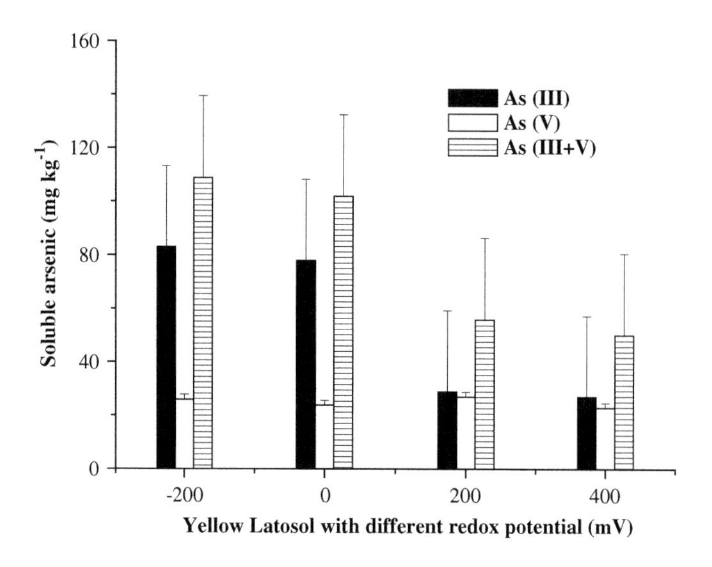

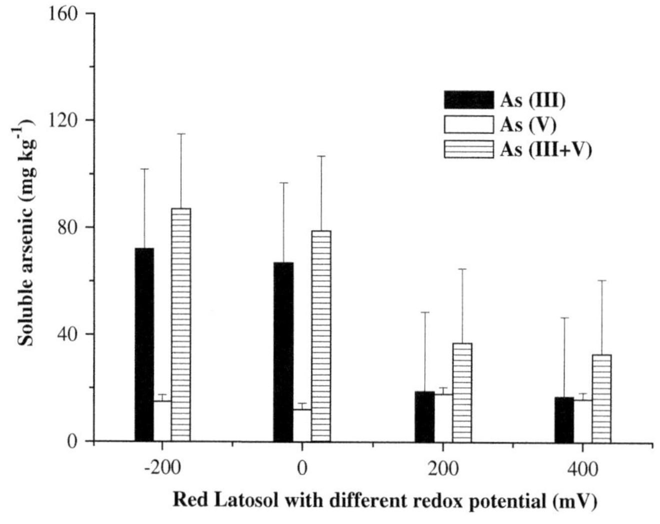

Fig. 5. Distribution of soluble arsenic species at two soils after 30 days of contact (pH 7.0 for −200, 0, 200, and 400 mV).

became the main dissolved arsenic species and As solubility increased. Arsenic was co-precipitated as arsenate with iron hydroxides according to the following equation:

$$H_2AsO^-_{4(aq)} + Fe(OH)_{3(aq)} + H^+_{(aq)} \rightleftharpoons FeAsO_4 \cdot 2H_2O_{(c)} + H_2O_{(l)}$$

The results indicate that Fe oxides and potential redox are the main soil properties controlling arsenic sorption rate. Yellow Latosol adsorbed less arsenic presumably because of its kaolinite content. Kaolin particles which had clean surfaces and low ferric oxide contents showed no uptake of As (V). The soluble arsenate was controlled by adsorption/co-precipitation in soils. Both parameters (redox potential and Fe presence) are important in assessing the fate of arsenic-containing compounds in soil; therefore, at higher redox levels arsenate were the predominant As species and their solubility was low. Due to the slow kinetics of the As (V) – As (III) transformation, a considerable amount of the thermodynamically unstable arsenate species was observed under reducing conditions.

In the second experiment, sodium arsenate and Na arsenite were added to the soils to the level of 120 mg As kg^{-1} soil. All the experiments were run in triplicate. Two vascular ferns were tested in this experiment (*P. vittata* Linnaeus and *P. aureum* (L.) J. Sm.). After 12 weeks of growth, the vascular plants were extracted and the soil suspensions were shaken for 1 h on a mechanical shaker, centrifuged, and filtered (Figs 6 and 7).

While the first experiment gave an idea of arsenic transformations under natural conditions (pH 7.0), the second experiment indicated similar transformations rates. The results in Fig. 6 indicate that in the second experiment, arsenic removal was influenced by the effect of the redox-pH chemistry of arsenic and the high concentrations of Fe present. In both the soils on reduction to −200 mV, the soluble arsenic content increased as compared to 400 mV. Similar mechanisms controlling As solubility were observed in both experiments.

4. CONCLUSIONS

The effects of arsenic uptake were shown to vary with in situ soil conditions. Arsenic removal and decontamination of the soils was more pronounced under reduced conditions. There was a great difference between the two fern species. The *P. vittata* showed a good absorption capacity for soluble arsenic. The results revealed that the majority of As was present as inorganic arsenite. In addition to *P. vittata*, the results show that *P. aureum* also accumulate arsenic under similar redox conditions but most of the arsenic was present as inorganic As (III).

The absorption status of *P. vittata* L. plants is a function the soil and pedological characteristics. Correlation analysis showed that most of the soluble arsenic was related to the type of soil. When the soil type and redox-pH 7.0

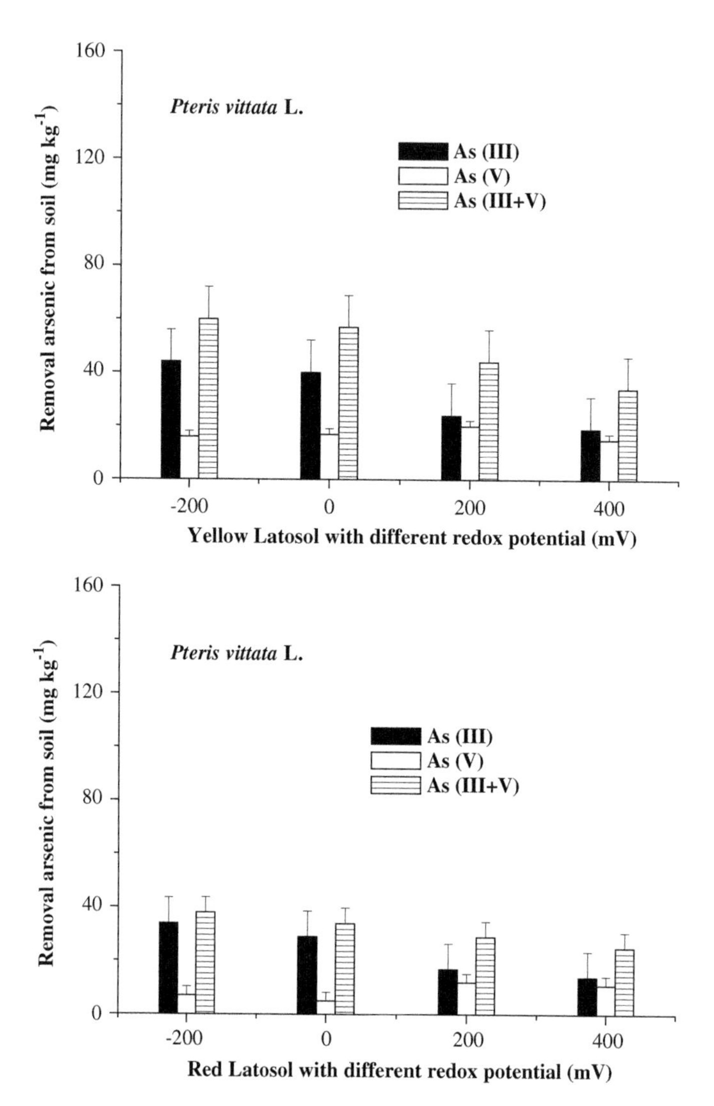

Fig. 6. Removal of soluble arsenic species by *Pteris vittata* L. after a 90-day equilibration period under controlled redox and pH conditions (pH 7.0 for −200, 0, 200, and 400 mV).

conditions were considered the Yellow Latosol showed a higher capacity for absorption than the Red Latosol. The absorption of arsenic species by the two ferns was significantly reduced by the presence of iron oxyhydroxides in the Red Latosol. This was attributed to competitive interactions. The presence of fern species in a system that contains iron oxides might have several important

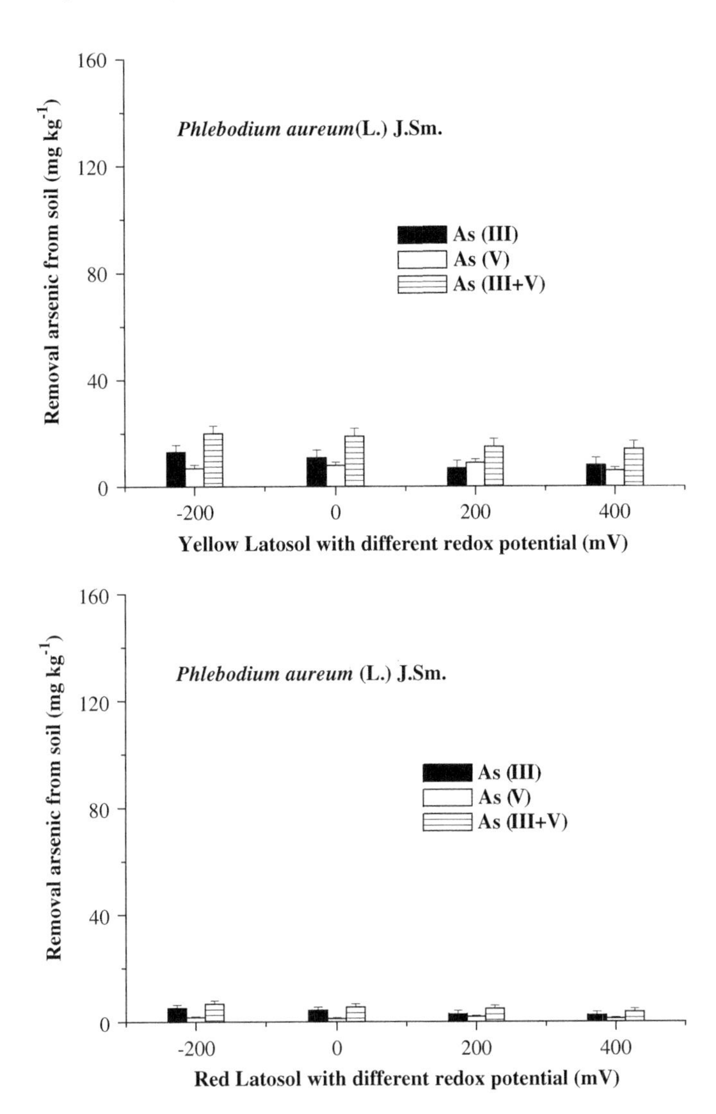

Fig. 7. Removal of soluble arsenic species by *Phlebodium aureum* (L.) J. Sm. after a 90-day equilibration period under controlled redox and pH conditions (pH 7.0 for −200, 0, 200, and 400 mV).

effects. The presence of iron tended to reduce arsenic absorption in all fern–soil systems. Moreover, the absorption of arsenic by *P. vittata* and *P. aureum* increases under reducing conditions. An alkaline pH, or the reduction of As (V) to As (III), released substantial proportions of arsenic into solution. Under moderately reduced soil conditions (0–100 mV) arsenic solubility was controlled by the dissolution of iron oxyhydroxides. Although it is very difficult to distinguish between adsorption and co-precipitation reactions without direct examination of the solid surfaces involved, consideration of the molar Fe/As ratios released on reduction suggests that arsenic was co-precipitated with the iron oxyhydroxides.

The rate of absorption within plants depends on the species, redox potential, and Fe presence. The *P. vittata* has a higher potential for in situ cleaning-up of arsenic-contaminated soils and suggests that As phytoextraction is a constitutive property in *P. vittata*.

ACKNOWLEDGMENTS

Financial support from Conselho Nacional de Desenvolvimento Científico e Tecnológico – CNPq (grant 300713/01-0) has made this work possible. The author is grateful to Dr. Isaac Jamil Sayeg (Laboratório de Microscopia Eletrônica de Varredura, Instituto de Geociências, Universidade de São Paulo). Thanks are also extended to Dr. Jefferson Prado (Instituto de Botânica do Estado de São Paulo), Mr. Luiz Alberto Campos, and Ms. Lourdes Lossurdo Campos for their assistance in the project.

REFERENCES

[1] L.Q. Ma, K.M. Komar, C. Tu, W. Zhang, Y. Cai and E.D. Kennelley, Nature, 409 (2001) 579.
[2] R.R. Brooks, Plants that Hyperaccumulate Heavy Metals, p. 380, CAB International, New York, 1998.
[3] A.J.M. Baker, S.P. McGrath, R.D. Reeves and J.A.C. Smith, in "Phytoremediation of Contaminated Soil and Water" (N. Terry and G. Bañuelos, eds.), p. 8, Lewis Publishers, Boca Raton, 2000.
[4] K. Francesconi, P. Visoottiviseth, W. Sridokchan and W. Goessler, Sci. Total. Environ., 284 (2002) 27.
[5] P. Visoottiviseth, K. Francesconi and W. Sridokchan, Environ. Pollut., 118 (2002) 453.
[6] S.P. McGrath, F.J. Zhao and E. Lombi, Adv. Agron., 75 (2002) 1.
[7] F.J. Zhao, S.J. Dunham and S.P. McGrath, New Phytol., 156 (2002) 27.
[8] M.X. Zheng, J.M. Xu, L. Smith and R. Naidu, J. Phys., 107 (2003) 1409.
[9] V. Campos and M.A.F. Pires, Commun. Soil Sci. Plant Anal., 35(15, 16) (2004) 2137.
[10] L.A. Smith, J.L. Means, A. Chen, B. Alieman, C.C. Chapman, J.S. Tixier, Jr., S.E. Brauning, A.R. Gavaskar and M.D. Royer, Remedial Options for Metals-Contaminated Sites, p. 221, Lewis Publishers, Boca Raton, 1995.
[11] M. Sadiq, Water Air Soil Pollut., 93 (1997) 117.

[12] V. Campos, Environ. Geol., 42 (2002) 83.
[13] A. Renzoni, N. Mattei, L. Lari and M.C. Fossi, Contaminants in the Environment, p. 286, Lewis Publishers, Boca Raton, 1994.
[14] International Agency for Research on Cancer, IARC Summary and Evaluation, IARC, Vol. 84, Lyon, 2004.
[15] J. Matschullat, R.P. Borba, E. Deschamps, B.R. Figueiredo, T. Gabrio and M. Schwenk, Appl. Geochem., 15 (2000) 181.
[16] R.P. Borba, B.R. Figueiredo and J.A. Cavalcanti, Rev. Escola Minas, 57 (2004) 45.
[17] World Health Organization, Environmental Health Criteria, Arsenic, p. 1774, Vol. 18, WHO, Geneva, 1981.
[18] A. Wild, Russells's Soil Conditions and Plant Growth, 11th edn., p. 364, Longman, London, 1988.
[19] N.W. Lepp, Effect of Heavy Metal Pollution on Plants, p. 275, Applied Science Publishers, London, 1981.
[20] C. Linnaeus, Species Plantarum, p. 516, 1st edn., Stockholm, 1753.
[21] J. Prado and P.G. Windisch, Boletim Inst. Botânico, 13 (2000) 103.
[22] D.L. Jones, Encyclopedia of Ferns, Timber Press, Portland, 1987.
[23] Empresa Brasileira de Pesquisa Agropecuária, Sistema Brasileiro de Classificação de Solos, Brasília, p. 412, SPI/Embrapa-CNPS, Rio de Janeiro, 1999.
[24] G.W. Thomas, in "Methods of Soil Analysis" (A.L. Page, R.H. Miller and D.R. Keeney, eds.), 2nd edn., Vol. 9, American Society of Agronomy, Madison, 1982 (Part II).
[25] M.L. Jackson, Soil Chemical Analysis, Vol. 2, UW, Madison, 1969.
[26] R.L. Loeppert and W.P. Inskeep, Soil Sci. Soc. Am., 53 (1996) 639.

Thermodynamics, Solubility and Environmental Issues
T.M. Letcher (editor)

Chapter 13

Environmental Issues of Gasoline Additives – Aqueous Solubility and Spills

John Bergendahl

Department of Civil & Environmental Engineering, Worcester Polytechnic Institute, Worcester, MA, USA

1. INTRODUCTION

Gasoline is a complex mixture of many organic compounds, with various additives that provide for a suitable automobile fuel. It is produced through refining operations from crude oil, which is a natural blend of many different hydrocarbons found in underground reservoirs in certain parts of the world. The chemical composition of gasoline is variable as the crude oil feedstock varies in its properties from place-to-place and refineries are configured and operated differently.

The refining process consists of many steps occurring in series and parallel, which produce different useful fractions from crude oil feedstock. The fractions produced may include gasoline, kerosene, diesel, heavy gas oil, fuel oil, and others. The initial refining step is distillation, which separates the crude oil feedstock into fractions of increasing boiling points. Cracking, including catalytic cracking, coking, hydrocracking, steam cracking, thermal cracking, and visbreaking, can then break heavier fractions to lighter ones. Catalytic reforming is a step where the anti-knock characteristics of the naphtha fraction are increased, primarily by the formation of aromatic compounds. In catalytic reforming, precious metal catalysts on an alumina support are utilized for the reforming step. Catalytic reforming also results in isomerization, dehydrocyclization, hydrocracking, and dealkylation reactions. Further alkylation, polymerization, and isomerization processes to increase octane are part of the refining operation. Finishing steps such as caustic washing, oxidation of mercaptans, and desulfurization must be done to produce an acceptable gasoline fuel [1].

The final gasoline product produced by the refining operation is a mixture of aromatic and aliphatic organic compounds with the number of carbons ranging from C_4 to C_{12} [2]. The average molecular weight of the final gasoline blend varies from ~92 to 95 [2]. Blending operations ensure that the gasoline is within specifications.

Subsequent to the refining operation, constituents are added to gasoline to provide for better performance characteristics of gasoline as an automobile fuel. These additives include oxidation and corrosion inhibitors, demulsifiers, deposit controls, anti-icers, markers and dyes, and oxygenates and octane boosters [2]. The oxygenates and octane boosters are present in gasoline in much higher concentrations than the other additives and are therefore of the greatest environmental concern and will be discussed further.

2. COMMON OXYGENATES AND OCTANE BOOSTERS

In the pursuit of additives to gasoline to reduce detonation, tetraethyl lead and other lead constituents were found to be successful octane boosters. Lead compounds had been used in automotive gasoline from the 1920s to the 1970s [1]. Lead anti-knock compounds have mostly been phased out in gasoline due to the recognition of the health hazards of lead and the potential for catalyst fouling in automobile catalytic converters that were mandated in the U.S. to alleviate air pollution. In lieu of leaded octane boosters, various ethers and alcohols have been used in gasolines to provide anti-knock benefits.

In addition to providing octane boost, ethers and alcohols are also used as oxygenates to provide for "cleaner," more complete combustion of automobile fuels. In the U.S., the use of oxygenated gasoline and reformulated gasoline (RFG) is required in areas with air pollution problems. The required oxygen content of gasoline is specified, and ethers and alcohols can meet this oxygen specification as well as boost octane. Oxygenates themselves do not provide free, "excess" oxygen to the combustion process; the oxygenates require stoichiometrically less oxygen for complete combustion than the other major constituents in gasoline, thereby resulting in an overall excess of oxygen when used as an additive to improve the combustion process.

The ethers and alcohols that are added to gasoline as oxygenates and octane boosters include: methyl tertiary-butyl ether (MTBE), ethyl tertiary-butyl ether (ETBE), tertiary-amyl methyl ether (TAME), tertiary-amyl ethyl ether (TAEE), diisopropyl ether (DIPE), ethanol, methanol, isopropanol, and tertiary-butyl alcohol. Relevant properties of these compounds are listed in Table 1 and molecular structures are shown in Table 2.

Table 1
Properties of common gasoline oxygenates

Property	MTBE	ETBE	TAME	TAEE	DIPE	Methanol	Ethanol	Tertiary-butyl alcohol	Isopropanol
Molecular formula	$C_5H_{12}O$	$C_6H_{14}O$	$C_6H_{14}O$	$C_7H_{16}O$	$C_6H_{14}O$	CH_4O	C_2H_6O	$C_4H_{10}O$	C_3H_8O
Molecular weight	88.15	102.17	102.17	116.2	102.17	32.04	46.07	74.12	60.10
Pure liquid density [g/mL] @20°C	0.7404 [3]	0.7404 [3]	0.7703 [3]	0.70 [3]	0.75 [3]	0.7913 [4]	0.7894 [4]	0.7866 [4]	0.7855 [4]
Pure liquid viscosity	0.31 cSt @37.8°C [3]	0.528 cSt @40°C [3]	n/a	n/a	n/a	0.918 cP @20°C [5]	1.1 cP @40°C [3]	4.2 cP @25.5°C [3]	2.38 cP @20°C [1]
Vapor pressure [mmHg] @25°C [6]	250	124	75.2	n/a	149	127	59.3	40.7	45.4
Boiling point [°C] [5]	55.05	72.82	86.25	102.0	68.25	64.65	78.35	82.35	82.35
Henry's law constant [atm m^3/mol] @25°C [6]	5.87×10^{-4}	1.39×10^{-3}	2.68×10^{-3}	n/a	2.28×10^{-3}	4.55×10^{-6}	5.00×10^{-6}	9.05×10^{-6}	8.10×10^{-6}
log K_{ow}	1.24 [7], 0.94 [6]	1.920 [6]	1.920 [6]	n/a	1.52 [6]	−0.77 [6]	−0.31 [6]	0.35 [6]	0.05 [6]
log K_{om} [log L (kg)][a]	0.911	1.714	1.714	n/a	1.39	−0.49	−0.11	0.43	0.18

Table 1 (Continued)
Properties of common gasoline oxygenates

Property	MTBE	ETBE	TAME	TAEE	DIPE	Methanol	Ethanol	Tertiary-butyl alcohol	Isopropanol
Pure phase aqueous solubility [mg/L]	51000 @25°C [6]	12000 @20°C [6]	2640 @25°C [6]	4000 @20°C [3]	8800 @20°C [6]	1×10^6 @25°C [6]	1×10^6 @25°C [6]	1×10^6 @25°C [6]	1×10^6 @25°C [6]

[a]Calculated from $\log K_{om} = 0.82 \log K_{ow} + 0.14$ [8].

Table 2
Molecular structures of common gasoline oxygenates

MTBE	ETBE	TAME	TAEE	DIPE

Ethanol	Methanol	Isopropanol	Tertiary butyl alcohol

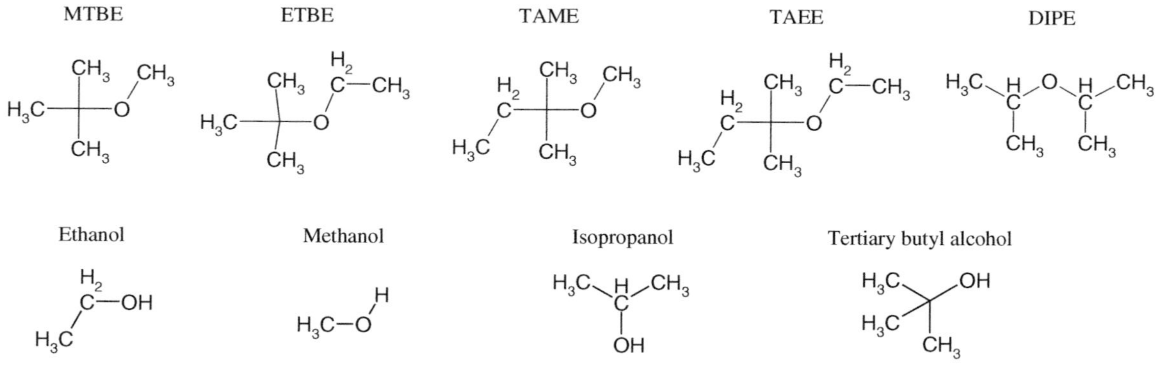

2.1. Ethers

Ethers are a class of organic compounds that consist of an oxygen bonded to two alkyl groups (R-O-R). MTBE is a colorless organic liquid with a high aqueous solubility, and it mixes well with gasoline. It has been used at up to 15% (v/v) in gasoline in the U.S. to successfully alleviate air pollution and provide octane, yet has resulted in contamination of many drinking water aquifers due to gasoline spills [9]. In 2004 alone, 47.6 million barrels of MTBE were added to gasoline by refiners [10]. It is manufactured by reacting isobutene with methanol in the presence of a catalyst, and boosts octane as it has a high octane number: 117 Research Octane Number (RON) and 101 Motor Octane Number (MON). It possesses 18.2% oxygen content and has a distinctive odor and taste which usually defines the regulatory level in drinking water. The odor threshold is reported to be 53 μg/L [11], and the taste threshold between 20 and 40 μg/L [11]. The scientific reports on the health effects of MTBE exposure are limited, controversial, and not conclusive [12]. Some studies indicated that very high doses of MTBE through inhalation and ingestion may cause cancerous and non-cancerous health problems, but the validity and applicability of these studies is still being debated [12]. The U.S. EPA has advised limiting MTBE levels to below 20–40 μg/L MTBE in potable water based on taste and odor concerns, and considers MTBE at high doses to be a potential human carcinogen. It is on EPA's Contaminant Candidate List for the possibility of regulation due to health effects [13]. Some degradation products of MTBE have recognized health concerns (e.g., formaldehyde).

ETBE is a colorless ether similar to MTBE, which is produced from isobutene and ethanol. It has a slightly greater octane rating than MTBE, with 118 RON and 101 MON. It has a lower vapor pressure providing for more flexibility in producing gasoline formulations as the vapor pressure of gasoline is limited by standards. The oxygen content of ETBE is 15.7% and the threshold for taste and odor is reported to be 47 and 13 μg/L, respectively [11].

TAME is manufactured by reacting methanol with tertiary-amylene. The octane ratings of TAME are lower than MTBE with RON of 112 and MON of 98. The oxygen content of TAME is 15.7%. The threshold for taste and odor is reported to be 128 and 27 μg/L, respectively [11].

TAEE is produced by reaction of ethanol with tertiary-amylene. It has a lower octane than MTBE and ETBE with an RON of 105 and an MON of 95.

DIPE is the result of reaction of water and propylene which produces isopropanol as intermediate, and DIPE as the final product [3]. DIPE has a lower octane rating than MTBE and ETBE with an RON of 105 and an MON of 95. It has an oxygen content of 15.7%.

2.2. Alcohols

The alcohols are a group of organic compounds containing hydroxyl groups (R-OH). Ethanol is a colorless organic liquid with infinite aqueous solubility. It is commonly added to gasoline at up to 10% (v/v) for octane enhancement and to produce an oxygenated fuel, but it is increasingly being used as an alternative fuel at high percentages (i.e., E85). It is useful as an octane booster as it has an RON of 123 and an MON of 96 [3]. Ethanol is derived through fermentation of sugars in agricultural stocks such as sugar cane and sugar beets, and from starches in crops like corn which are easily converted to sugars. Research is underway to cost-effectively obtain ethanol from cellulose-containing biomass, which is ubiquitous around the world. U.S. refineries used 74.1 million barrels of ethanol in 2004 [10]. Ethanol has great affinity for water and will absorb water into gasoline/ethanol mixtures. At high water concentrations, the ethanol can separate from gasoline into a separate alcohol/water phase, so ethanol is typically added to gasoline immediately preceding distribution.

In the event of an environmental release of gasoline containing ethanol, the ethanol may end up in surface waters and aquifers where it is readily biodegradable. While the human health effects from long-term exposure to ethanol in potable water are not completely known, ethanol itself is not expected to be a health concern at low levels. Humans have ingested ethanol-containing beverages for many generations, and health effects such as liver damage and fetus damage have been connected to consumption of beverages containing ethanol at much higher concentrations than expected in aquifers impacted by ethanol from gasoline. However, due to its co-solvency effect, high concentrations of ethanol may mobilize deleterious organics into water (discussed below).

Methanol is an organic compound with infinite solubility in water. It is used in gasoline as an oxygenate and octane booster, but has a greater phase separation tendency in comparison to ethanol (and the ethers). Co-solvents such as tertiary-butyl alcohol are used in gasoline/methanol mixtures to assist in keeping the mixture as a single phase. It has an RON of 123 and an MON of 91 [3]. Methanol is usually manufactured from natural gas via steam reforming. It is readily metabolized in the human body, and biodegradable in the environment.

Isopropanol is a clear colorless organic liquid with an RON of 121 and an MON of 98 [14], which is produced by the oxidation of propylene. In addition to its use as an oxygenate and octane booster, it is frequently used as a gasoline additive to assist in keeping water from forming a separate phase in gasoline.

Tertiary-butyl alcohol is commonly used as a co-solvent in gasoline/methanol blends and as an oxygenate and octane booster by itself. It is predominantly produced from isobutene, but may also be a product of fermentation reactions [15]. It is biodegradable in the environment, and is infinitely soluble in water.

3. RELEASES TO THE ENVIRONMENT

In the event of substantial spills of gasoline to the environment, the gasoline phase will move due to gravity until a barrier is encountered, as shown in Fig. 1. Gasoline typically has a relative density between 0.72 and 0.78 [1], significantly lower than water, so the barrier to downward movement is frequently the water table. Gasoline can persist as a light non-aqueous phase liquid (LNAPL) at the water table, with soluble constituents continuously dissolving into the surrounding groundwater (see Fig. 1). The gasoline LNAPL constituents may have a tendency to transfer to the gas phase in the unsaturated zone above the water table and LNAPL, but dissolution of the constituents from the LNAPL into flowing groundwater is typically the environmental concern as the groundwater plume created by dissolution can travel significant distances. Gasoline may also be trapped as residual in soil pores providing a continuous source for dissolution. Gasoline releases to the environment

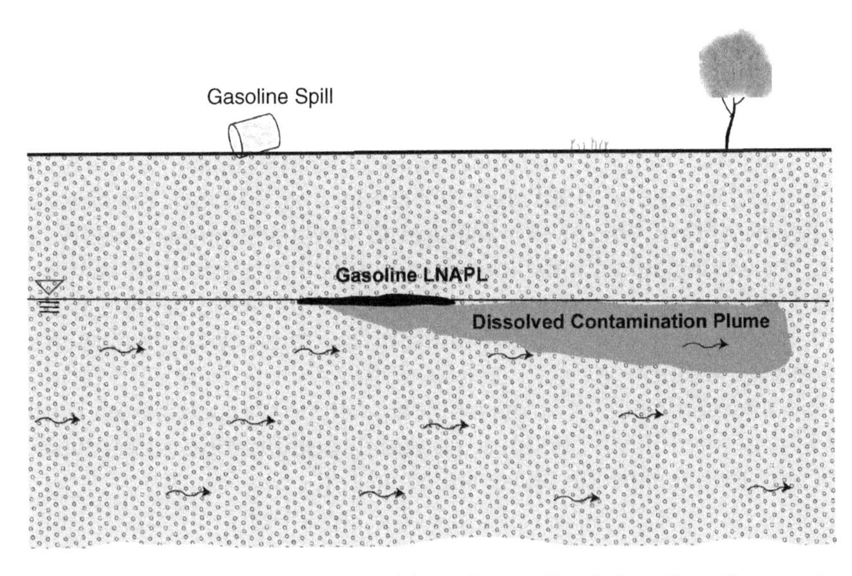

Fig. 1. Spill of gasoline forming LNAPL with continuous dissolution of gasoline constituents.

may also occur due to runoff from roadways and from small releases to surface waters.

3.1. Molecular Interactions in the Environment

Molecular interactions dictate the behavior of gasoline constituents and additives, and specifically drive partitioning of the gasoline components between the gasoline, air, water, and soil phases. Intermolecular forces that govern the behavior of these organics in the phases include: *van der Waals* attraction due to instantaneous induced polarization of electrons associated with the molecule, *polar interactions* due to the variable electronegativity of atoms comprising the molecule (i.e., permanent dipole), and *hydrogen bonding* from the difference in electronegativity between hydrogen and oxygen or nitrogen [8]. The overall (sum) of the intermolecular forces acting on an organic molecule in each of the phases present determines the tendency of the molecules of a substance to partition between the phases.

For an organic molecule to translate from an organic phase to the aqueous phase, the energetics of dissolution consist of enthalpic and entropic considerations. The enthalpy of dissolution is due to organic : organic, organic : water, and water : water molecular interactions during dissolution. Dissolution also brings about enthalpic changes from increased randomness from the mixing, cavity formation in the water phase, and structure formation of the water molecules around the organic molecule in the water phase [8].

For a gasoline LNAPL in contact with water, the fractions with the greatest aqueous solubility have a greater tendency to dissolve into the surrounding groundwater than the constituents with lower aqueous solubilities. Due to the molecular structures of the oxygenates, the oxygenates are much more soluble in water than the other gasoline constituents. For example, the aqueous solubility of benzene is 1780 mg/L [16], which is 29 times lower than the solubility of MTBE, the most widely used oxygenate until recently. Even TAME, the oxygenate with the lowest solubility, has a solubility that is 50% greater than benzene. Note that the aqueous solubility of the alcohols is infinite at room temperature, due to ability of the hydroxyl groups to hydrogen bond with water molecules.

The thermodynamics of dissolution of the oxygenates from gasoline LNAPL is defined using chemical potential. Equilibrium occurs when the chemical potentials in each phase are equal [8]:

$$\mu_{\text{gasoline}} = \mu_{\text{water}} \qquad (1)$$

for each constituent in the gasoline and water phases.

From the definition for chemical potential [8]:

$$\mu_0 + \mathrm{RT}\ln(\gamma_{\mathrm{gasoline}} x_{\mathrm{gasoline}}) = \mu_0 + \mathrm{RT}\ln(\gamma_{\mathrm{water}} x_{\mathrm{water}}) \quad (2)$$

where μ_0 is the standard state chemical potential, R the gas constant, T the temperature, x the mole fraction in the respective phase, and γ the activity coefficient in each phase.

As the standard state chemical potentials in each phase are equal (μ_0 in the gasoline phase $= \mu_0$ in the water phase), Eq. (2) can be simplified to:

$$x_{\mathrm{water}} = \frac{\gamma_{\mathrm{gasoline}} x_{\mathrm{gasoline}}}{\gamma_{\mathrm{water}}} \quad (3)$$

So the aqueous solubility of a constituent in contact with gasoline LNAPL containing the constituent is dependent on the amount of the constituent in the gasoline phase (x_{gasoline}), and the activity coefficients in the gasoline and water phases ($\gamma_{\mathrm{gasoline}}$, γ_{water}) which resolve the degree of non-ideality of the constituent in each phase. The behavior of a molecule in a phase (degree of non-ideality) results from net molecular interactions. And the greater the difference in molecular interactions of a constituent molecule from the molecules making up the phase, the greater the deviation from ideality (where $\gamma = 1$ for ideality). The current oxygenates possess either hydroxyl groups for the alcohols (R-OH) or ether groups (R-O-R). These groups result in a greater degree of non-ideality (due to the net molecular interactions) in the gasoline phase than many of the other organic compounds comprising the phase ($\gamma_{\mathrm{gasoline}} > 1$). And the oxygenates are more similar to the water molecules than many of the other constituents in the gasoline phase (manifested in lower γ_{water}), thereby having a greater affinity for the aqueous phase (lower γ_{water}).

Once gasoline constituents have dissolved into groundwater moving by the LNAPL, the constituents will partition between the water and any soil phase present. While the oxygenates have moieties that enhance aqueous dissolution, they are organic compounds and have some affinity for natural organic matter present in groundwater aquifers. Natural organic matter is the result of degradation of biological sources into recalcitrant residues called humic substances. While natural organic matter is not completely characterized or understood, it is composed of 40–50% carbon (m/m) [8] and has various surface functional groups such as hydroxyl, carboxy, and phenoxy groups.

Organic substances moving with water through the aquifer partition into the stationary natural organic matter phase, thereby retarding the rate of transport of the organic contaminant plume. The degree of retardation is a function of the amount of organic matter present, as well as the affinity of the contaminant for the stationary organic matter phase.

The partitioning of organic contaminants between water and *n*-octanol has become a recognized analytical method useful for determining the environmental fate and transport of organic contaminants. A contaminant with a greater K_{ow}, the octanol–water partition constant, will have a greater affinity for organic phases in the environment. Log K_{ow} values for the typical oxygenates are listed in Table 1. The ethers, specifically ETBE and TAME, have the greater K_{ow} values, and the alcohols have the lower values. Generally, the addition of groups composed of carbon and hydrogen to an organic molecule increases the value of K_{ow}, providing for an increased tendency of the organic molecule to partition into the organic phase in lieu of the aqueous phase. For the ethers and alcohols that contain R-O-R and R-OH groups, respectively, the presence of the oxygen atom decreases K_{ow} [8]. Therefore, the ether and alcohol oxygenates have an increased tendency to partition from the gasoline LNAPL phase to the aqueous phase due to the presence of these groups.

A strong correlation (a linear free energy relationship) exists between K_{ow} and the partitioning constant between organic matter in soil and water, K_{om} [8]. Values for log K_{om} are listed in Table 1. Partitioning of a contaminant between water and the natural organic matter in the soil is manifested in retardation of the contaminant transport in the aquifer in relation to the water advection, and is quantified with a retardation factor, R.

The governing equation describing contaminant transport with advection and sorption through an aquifer is:

$$\left[1+\frac{\rho_{\text{bulk}}}{\varepsilon}K_{\text{om}}f_{\text{om}}\right]\frac{\partial C}{\partial t}=-v_{\text{water}}\frac{\partial C}{\partial x} \tag{4}$$

where ρ_{bulk} is the bulk density of soil, ε the porosity, f_{om} the fraction of organic matter in the soil, C the concentration, t the time, v_{water} the velocity of water through the aquifer, and x the distance. If the quantity $[1 + (\rho_{bulk}/\varepsilon)K_{om}f_{om})]$ is equal to 1.0, the contaminant moves through the aquifer at the same velocity as the water. As the quantity increases above 1.0, the transport of the contaminant is retarded in the aquifer. The *retardation factor*, R, is defined as: $[1 + (\rho_{bulk}/\varepsilon)K_{om}f_{om})]$. Retardation factors calculated from this definition are listed in Table 3 for the common oxygenates as well as benzene and *m*-xylene, common gasoline constituents, for given soil conditions (ρ_{bulk} = 1.8 kg/L, ε = 0.30, and f_{om} = 0.1%). It can be seen that the alcohols (methanol, ethanol, tertiary-butyl alcohol, and isopropanol) and MTBE travel within ~5% of the velocity of the water, retarded by only a small amount. The other ethers have a slightly higher retardation factor with R = 1.311 for ETBE and TAME, and 1.147 for DIPE. The retardation factors for these oxygenates are lower than the retardation factors for benzene (R = 1.425) and *m*-xylene (R = 3.389), both constituents of interest in gasoline. Therefore, the gasoline oxygenates are typically much more mobile in groundwater than the other gasoline components.

Organic compounds with high aqueous solubilities (e.g., ethanol, methanol) may be present in water at high concentrations and may function as co-solvents with the ability to solubilize other organic gasoline constituents that are more harmful to human health and the environment. At high co-solvent concentrations, the surface of the solubilized organic molecules in the water is surrounded by the organic co-solvent as well as water, thereby affecting the net molecular interactions. The aqueous activity coefficient is influenced by the co-solvent in the aqueous phase, thereby solubilizing greater amounts of gasoline constituents than pure water alone. However, it is thought that the co-solvency effect only occurs to any significant extent above ~10% (v/v) ratio [8], in that below this volume ratio the co-solvents may be fully hydrated by water molecules. The enhancement of solvency in the presence of a co-solvent is an exponential function of the co-solvent concentration, so the ability of a co-solvent to enhance solubility drops off rapidly with concentration [8]. It is also a function of the hydrophobicity of the constituent (hydrophobic surface area) [8], so the more hydrophobic contaminates would be expected to be affected to a greater extent.

Investigation with saturated column experiments mimicking fuel spills indicated that the co-solvency characteristics of ethanol from a gasoline spill (with 10% ethanol) are minimal, with 9% increase in peak benzene concentration, 12% increase in toluene concentration, 19% increase in *m*-xylene concentration, 24% increase in 1,2,4-trimethylbenzene concentration, and 54% increase in octane concentration [17]. However, the results pointed to significant solubility enhancements when neat ethanol was used in the experiments – a solubility enhancement of up to 75× for benzene and 23000× for octane. And in other research, high aqueous concentrations of methanol (>8%) were found to increase the aqueous concentrations of benzene, toluene, ethylbenzene, xylenes (BTEX) in batch tests, while MTBE did not act as a co-solvent under the conditions tested [18]. The co-solvency effect of the oxygenates in the event of a gasoline spill is expected to be negligible

Table 3
Retardation factors for oxygenates transport for $\rho_{bulk} = 1.8$ kg/L, $\epsilon = 0.30$, and $f_{om} = 0.1\%$

Contaminant	MTBE	ETBE	TAME	DIPE	Methanol	Ethanol	Tertiary-butyl alcohol	Isopropanol	Benzene	*m*-Xylene
log K_{om} log (L/kg)	0.911	1.714	1.714	1.39	−0.49	−0.11	0.43	0.18	1.85[a]	2.6[a]
Retardation factor, R	1.049	1.311	1.311	1.147	1.002	1.005	1.016	1.009	1.425	3.389

[a]Average of values listed in ref. 16.

under the conditions of low (<15%) amendment of gasoline with these oxygenates, and sufficient groundwater flow in the impacted aquifer that will keep the co-solvent concentration fairly low (<10%) due to water replenishment. However, there may be conditions where a spill of pure or near pure alcohol such as E85 or M85 (or another organic that may act as a co-solvent) near an existing gasoline LNAPL may facilitate mobilization of other gasoline constituents in the aquifer.

4. CONCLUSIONS

Ethers and alcohols are added to gasoline to increase octane and provide an oxygenate in current gasoline formulations. These additives have different properties (molecular weights, vapor pressures, octanol–water partition coefficients), yet are very soluble in water and may be quite mobile in the environment when gasoline spills occur.

REFERENCES

[1] K. Owen and T. Coley, Automobile Fuels Reference Book, 2nd edn., 1995, Society of Automotive Engineers, Inc., Warrendale, PA.

[2] Chevron USA, Motor Gasolines Technical Review, Chevron USA, San Ramon, CA, 2006, http://www.chevron.com/products/prodserv/fuels/bulletin/motorgas/.

[3] M.A. Ali and H. Hamid, in "Properties of MTBE and Other Oxygenates" (H. Hamid and M.A. Ali, eds.), Handbook of MTBE and Other Gasoline Oxygenates, Marcel Dekker, Inc., New York, 2004.

[4] SAE, Surface Vehicle Recommended Practice, J312, Society of Automobile Engineers, Warrendale, PA, 2001.

[5] E.W. Lemmon, M.O. McLinden and D.G. Friend, in "Thermophysical Properties of Fluid Systems" (P.J. Linstrom and W.G. Mallard, eds.), NIST Chemistry WebBook, NIST Standard Reference Database Number 69, National Institute of Standards and Technology, Gaithersburg, MD, 2005, http://webbook.nist.gov.

[6] NIH, US National Library of Medicine, National Institutes of Health, Bethesda, MD, 2006, http://chem.sis.nlm.nih.gov/chemidplus.

[7] ATSDR, Toxicological Profile for Methyl *t*-Butyl Ether (MTBE), Agency for Toxic Substances and Disease Registry (ATSDR), US Department of Health and Human Services, Public Health Service, Atlanta, GA, 1996.

[8] R.P. Schwarzenbach, P.M. Gschwend and D.M. Imboden, Environmental Organic Chemistry, Wiley, New York, NY, 1993.

[9] R. Johnson, J. Pankow, D. Bender, C. Price and J. Zogorski, Environ. Sci. Technol., 34 (2000) 210A.

[10] S.C. Davis and S.W. Diegel, Transportation Energy Data Book, 25th edn., ORNL-6974, Oak Ridge National Laboratory, Oak Ridge, TN, 2006.

[11] US EPA, Achieving Clean Air and Clean Water: The Report of the Blue Ribbon Panel on Oxygenates in Gasoline, US Environmental Protection Agency, Washington, DC, EPA420-R-99-021, 1999.
[12] J.M. Davis, in "Health Risk Issues for Methyl *tert*-Butyl Ether" (A.F. Diaz and D.L. Drogos, eds.), Oxygenates in Gasoline, Environmental Aspects, ACS Symposium Series 799, American Chemical Society, Washington, DC, 2002.
[13] US EPA, Methyl tertiary-Butyl Ether, US Environmental Protection Agency, Washington, DC, 2006, http://www.epa.gov/mtbe/water.htm.
[14] D.L. Drogos and A.F. Diaz, in "Appendix A: Physical Properties of Fuel Oxygenates" (A.F. Diaz and D.L. Drogos, eds.), Oxygenates in Gasoline, Environmental Aspects, American Chemical Society, Washington, DC, 2002.
[15] Wikipedia, http://en.wikipedia.org/wiki/Butanol.
[16] API, Strategies for Characterizing Subsurface Releases of Gasoline Containing MTBE, API Publication No. 4699, American Petroleum Institute, Washington, DC, 2000.
[17] W.G. Rixey, X. He and B.P. Stafford, The Impact of Gasohol and Fuel-Grade Ethanol on BTX and Other Hydrocarbons in Ground Water: Effect on Concentrations Near a Source, API Soil and Groundwater Research Bulletin No. 23, American Petroleum Institute, Washington, DC, 2005.
[18] J.F. Barker, R.W. Gillham, L. Lemon, C.I. Mayfield, M. Poulsen and E.A. Sudicky, Chemical Fate and Impact of Oxygenates in Groundwater: Solubility of BTEX from Gasoline-Oxygenate Compounds, API Publication No. 4531, American Petroleum Institute, Washington, DC, 1991.

Thermodynamics, Solubility and Environmental Issues
T.M. Letcher (editor)

Chapter 14

Ecotoxicity of Ionic Liquids in an Aquatic Environment

Daniela Pieraccini[a], Cinzia Chiappe[a], Luigi Intorre[b] and Carlo Pretti[b]

[a]Dipartimento di Chimica Bioorganica e Biofarmacia, Università di Pisa, Via Bonanno 33, Pisa, Italy
[b]Dipartimento di Patologia Animale, Profilassi ed Igiene degli Alimenti, Università di Pisa, Viale delle Piagge 2, Pisa, Italy

1. INTRODUCTION

Ionic liquids (ILs) are considered highly promising, neoteric solvents. The large number of possible combinations of anions and cations, and thus the fine-tuning of their chemical properties, has given the chemical industry a new target-oriented reaction media for synthesis, extraction and catalysis [1, 2]. The non-volatility of ILs was once considered to be one of their most promising features. It is however not enough for defining them as green solvents. Other environment-related properties such as ecotoxicity, lipophilicity and biodegradability need to be assessed in order to fully understand the consequences of introducing ILs into the environment. Indeed, a moderately toxic but highly persistent compound may exert the same effect as a highly toxic but readily biodegradable one.

The effect of ILs on the environment may be approached from two different but related levels:

- Performing specific tests in order to get direct information on IL properties linked to the environment.
- Estimating (eco)toxicity, biodegradability and lipophilicity through structure–activity relationship, or structure–property correlations of ILs. Experimental data are used to corroborate predictions.

The delay in using ILs as solvents is due to the lack of information of their impact on the environment. However, this is changing and information on the environmental effects (both experimental and theoretical) of ILs have appeared in the literature over the past few years, showing an increasing interest in the subject [3]. With respect to this chapter, attention has been given to data published on the ecotoxicity of ILs towards aquatic environments, as well as on their biodegradability and lipophilicity. The aim of the work is to provide the reader with an understanding of the impact of ILs on the environment, pointing out, where possible, which structures will cause the least damage to the aquatic environment.

2. LIPOPHILICITY

Experimentally determined 1-octanol/water partitioning coefficients of several imidazolium salts have been reported by Brennecke and co-workers [4] and reproduced in Table 1 in logarithmic form. Despite differences in the absolute value of K_{ow}, which may be attributed to different concentration ranges investigated, the pattern is consistent with the expectations concerning the importance of the alkyl chain length and anion hydrophobicity. For anionic as well as cationic surfactants a relationship between acute fish lethality, expressed as log LC_{50}, and the hydrophobicity of a chemical compounds, expressed as log K_{ow}, has been reported [4]. In the case of ILs, only a few data points on partitioning coefficients as well as on toxicity profiles for fish have been reported, so a correlation is not possible at the moment.

Several methods, other than the partition coefficient, have been developed to assess the lipophilicity of ILs. In a recent paper, Stepnowski and Storoniak [5] used reversed phase chromatography to obtain information on cation partitioning (regardless of the identity of the anion), in order to simulate what happens when ILs enter an aquatic environment. All the ILs investigated showed a preferential partitioning into the aqueous phase. It is worth noting that, even if this behaviour may account for the fate of well solvated dissociated cations, it does not take into account possible interactions with biological membranes.

3. BIODEGRADABILITY

According to one of the principles of green chemistry, "*a chemical compound should be designed so that at the end of its function, it does not persist in the environment, but breaks down into innocuous degradation products*" [6].

Table 1
Hydrophobicities of imidazolium-based ionic liquids, expressed as log K_{ow}

Structure		log K_{ow}	Reference
Cation	Anion		
C_2H_5 N (+) N CH_3	[Tf_2N]	−1.05/−0.96	[3]
C_2H_5 N (+) N CH_3 CH_3	[Tf_2N]	−1.15/−0.92	[3]
C_4H_9 N (+) N CH_3	[Cl] [BF_4] [PF_6] [Tf_2N]	−2.4 −2.5 −1.7 −0.96/−0.2	[3]
C_6H_{13} N (+) N CH_3	[Tf_2N]	0.15/0.22	[3]
C_7H_{15} N (+) N CH_3	[Tf_2N]	0.13/0.25	[3]
C_8H_{17} N (+) N CH_3	[Tf_2N]	0.8/1.05	[3]

At present, two different kinds of studies have been reported to assess the biodegradability of ILs and thus their persistence in the environment:

- The evaluation of methods used to oxidise ILs. These methods are generally based on the assumption, which in the case of ILs has still to be verified, that any remaining oxidation products are more hydrophilic and, above all, are less toxic than the parent compound. According to Stepnowski and Zaleska [7] and Morawski et al. [8], the best results, in terms of oxidation, are achieved with a combination of UV light and a catalytic amount of an oxidant such as hydrogen peroxide or titanium dioxide. The efficiency of the oxidation process is related to the alkyl chain length of the cation and the presence of particular functional groups appended on the side chain of the anion moiety. As an example, 360 min is necessary to remove 85% of $[C_{14}mim]^+$.

- The determination of those structural modifications which result in readily biodegradable ILs. According to the detailed studies performed by Gathergood and Scammells [9−12], conventional ILs with non-functionalised alkyl chain are generally less biodegradable than cationic surfactants. A higher level of biodegradability may be obtained by introducing an ester in the alkyl chain of imidazolium salts (see Table 2). With respect to the anion, while $[bmim][C_8SO_4]$ underwent a modest amount of biodegradation after 28 days (25%), less than 5% biodegradation was achieved with other anions. As expected, the presence of ester functionality together with the octylsulphate anion had a synergic effect, and the biodegradability reached a maximum of 56% under the same conditions.

4. AQUATIC TOXICITY

Hydric bodies are the ultimate recipients of most toxic substances generated by industrial, agricultural and domestic activities and released into the environment. The management of chemical compounds entering natural waters is difficult because contaminants often enter an aquatic system from multiple or diffuse sources. Although aquatic ecosystems are adaptable, with a variety of physical, chemical and biological mechanisms by which chemical compounds may be assimilated without serious implications for endemic biota, when contaminants reach levels in excess of the assimilative capacity of the receiving waters, they may affect survival, development, growth, reproduction or behaviour of organisms.

4.1. Microorganism

Several groups have focussed their attention on the use of *Vibrio fischeri* as test organism to assess the effect of several ILs towards the aquatic environment [13–15]. *V. fischeri* is a rod-shaped bacterium, which possesses

Table 2
Biodegradability of ionic liquids by means of OECD 301D test

Structure			Closed bottle, OECD 301D (28 days) (%)	References
Cation	Functional groups	Anion		
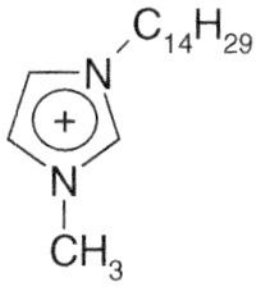 $C_{14}H_{29}$, N, +, N, CH_3		[Br], $[BF_4]$, $[PF_6]$	0–3	[9–12]
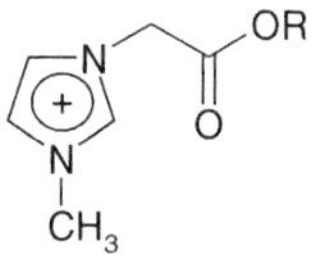 OR, N, +, O, N, CH_3	$R=C_1-C_8$	[Br]	16–32	[9–12]
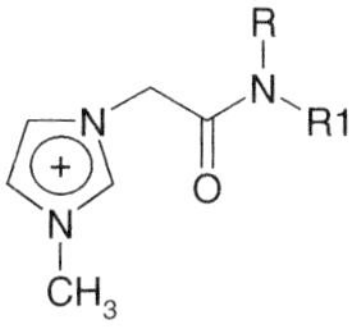 R, N, R1, N, +, O, N, CH_3	R=H; $R_1={}^n$Bu R=Me; $R_1={}^n$Bu $R=R_1$=Et	[Br]	<<5	[9–12]
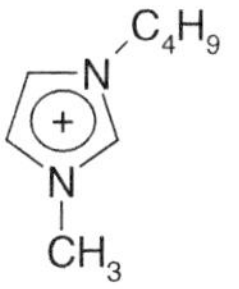 C_4H_9, N, +, N, CH_3		[Br], [Cl], $[BF_4]$, $[PF_6]$, $[Tf_2N]$, $[N(CN)_2]$	<<5	[9–12]
		$[C_8SO_4]$	25	[9–12]
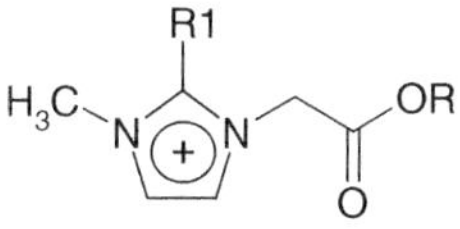 R1, H_3C, N, +, N, OR, O	$R=C_3$; R_1=H $R=C_3$; $R_1=C_1$	[Br]	23–24	[9–12]
	$R=C_5$; R_1=H $R=C_5$; $R_1=C_1$	[Br]	32–33	[9–12]
	$R=C_3$; R_1=H $R=C_3$; $R_1=C_1$	$[C_8SO_4]$	49–55	[9–12]
	$R=C_5$; R_1=H $R=C_5$; $R_1=C_1$	$[C_8SO_4]$	54–56	[9–12]

bioluminescent properties, and it is found predominantly in symbiosis with various marine animals, such as the bobtail squid. The tests that exploit this bacterium rely on the measurements of the decrease in luminescence from the bacterium itself (bioluminescence inhibition assay): a decrease in the light output is indeed linked to a decrease in respiration and is used as an indirect measure of the toxicity of the test compound. Because this method is very rapid and cost-effective, it has been widely used to determine the toxicity of pure substances as well as to monitor the quality of industrial effluents [16, 17]. Moreover, as bacteria have short generation times, they serve as ideal starting points for structure–activity relationship investigations.

Pioneering work in this field was done by Ranke et al. [13] who used the bioluminescence inhibition assay to determine acute toxicity of several imidazolium salts. The samples were carefully chosen in order to systematically study the influence of different structural modifications on the biological activity of the ILs (T-SAR approach). Subsequently, Docherty and Kulpa [14] extended the bioluminescence inhibition assay to pyridinium derivatives. Data published by these authors constituted the rationale for a quantitative structure–property relationship (QSPR) modelling recently performed by Maginn and co-workers [15], where structures other than imidazolium and pyridinium have been tested. Despite very promising results in predicting $\log_{10}$ EC_{50} for aromatic ILs, the correlation developed by the authors was less useful in modelling the behaviour of ammonium and phosphonium salts.

Although it is well known that anion modification leads to changes in the physico-chemical properties of ILs, all the works published so far showed no systematic effect of the anion, indicating that ILs' toxicity is largely driven by the alkyl chain branching and hydrophobicity of the cation (Fig. 1).

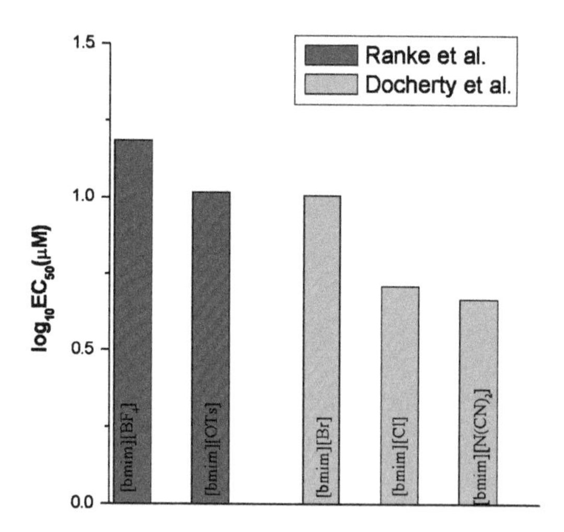

Fig. 1. Acute toxicity values of $[bmim]^+$ ionic liquids for *V. fischeri*, expressed as average log EC_{50} in micromolar.

The influence of the *n*-alkyl chain length in imidazolium-based tetrafluoroborate is clearly seen in Fig. 2, where $\log_{10} EC_{50}$ is reported as a function of the number of carbon atoms of the substituent in the 1 position: toxicity towards *V. fischeri* increased almost linearly with increasing alkyl chain length, R_1. Moreover, substituting an ethyl moiety for the methyl one attached to the second nitrogen atom leads to a further increase in toxicity. These results suggest that the use of R_1-methylimidazolium derivatives should be preferred to R_1-ethylimidazolium compounds. Generally, short chain derivatives seem to be less toxic. Docherty and Kulpa [14] studying 1-alkyl-3-methylpyridinium bromides also confirmed the trend of the increasing toxicity towards *V. fischeri* on increasing the alkyl chain length.

A trend similar to that reported for 1-alkyl-3-methylimidazolium salts was found by studying acute toxicity in *Photobacterium phosphoreum* [11]. Indeed, even in this case, the longer the alkyl chain resulted in a lower EC_{50} value. The estimated concentration, resulting in a 50% reduction of the light produced by the bacteria, ranged from 10 μM in the case of the [omim][Br] to 2200 μM for [bmim][Cl] (that is, from 0.95 to 3.34 in the logarithmic form), indicating a lower sensitiveness of this bacterium towards ILs when compared with *V. fischeri* (Table 3).

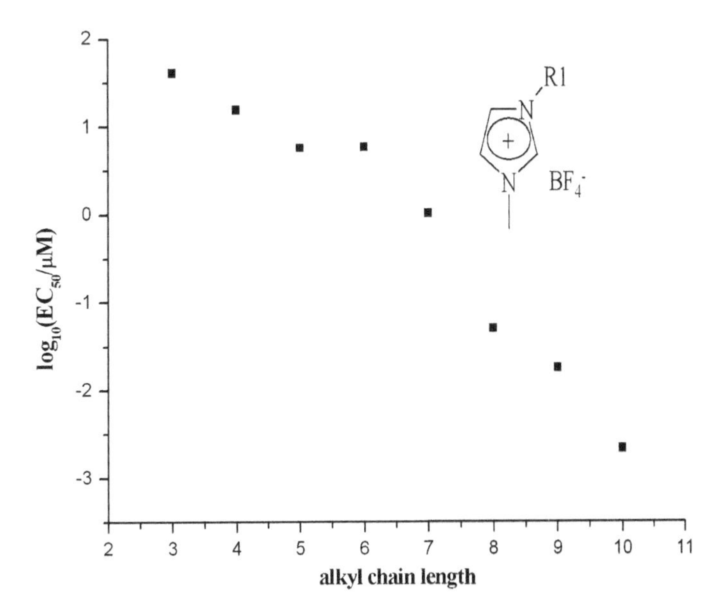

Fig. 2. Effect of alkyl chain length on acute toxicity of $[C_nMIM][BF_4]$ salts towards *V. fischeri*, expressed as average $\log_{10} EC_{50}$ in micromolar.

Table 3
Acute toxicity values for several classes of ionic liquids towards *V. fischeri*

Anion	Cation	Side chain	Endpoints	Range	References
[Cl], [Br], $[BF_4]$, $[PF_6]$, [OTs], $[N(CN)_2]$, $[Tf_2N]$	R2, N, +, N, R1	$R_1=C_3-C_{10}$ $R_2=C_1-C_2$	$\log_{10} EC_{50}$	μM: −2.67/1.61 ppm: (0.66/8710)	[13–15]
[Cl], [Br], $[N(CN)_2]$, $[Tf_2N]$	R3, R2, +, N, R1	$R_1=C_4-C_8$ $R_2=H; C_1$ $R_3=H; C_1$	$\log_{10} EC_{50}$	μM: −2.21/0.31 ppm: (1.17/538.40)	[14, 15]
[Br]	R1, R2, N^+, R1, R1	$R_1=C_1-C_4$ $R_2=C_1-C_6$	$\log_{10} EC_{50}$	μM: −0.54/>2 ppm: (72/>19500)	[15]
[Br], [diethylphosphate]	R1, R2, P^+, R1, R1	$R_1=C_4-C_6$ $R_2=C_2-C_{30}$	$\log_{10} EC_{50}$	μM: −0.29/0.41 ppm: (175/2030)	[15]
[Cl], $[Tf_2N]$	N^+, OH		$\log_{10} EC_{50}$	μM: 1.15/>2 ppm: (5424/13900)	[15]

4.2. Microalgae

Algae, which are associated with the plankton and represent the lowest step of most food chains, produce oxygen and are important in the cycling of nutrients. Algae also play a role in controlling the physical and chemical qualities of natural water. Because of their ecological importance, freshwater and saltwater algae may be used as biomonitors of environmental change: indeed, as phytoplanktons have such short life cycles, they respond quickly to environmental change and their diversity and density can indicate the quality of their habitat. It is therefore not surprising that aquatic ecologists have used freshwater algae to monitor and determine the impact of pollutants on ponds, lakes and rivers.

The first work reporting the effect of ILs on marine algae was performed by Latala et al. [18], who selected *Oocystis submarina* and *Cyclotella meneghiniana* as testing organisms. Following the T-SAR approach, among all the ILs available, 1,3-dialkylimidazolium tetrafluoroborate with different alkyl chains ($C_2 > R_1 > C_6$) or an aromatic substituent at the N_2 position have been selected. In this way it would have been possible, at least in principle, to correlate the activity of imidazolium salts to the length or type of side chain (Fig. 3).

Although no precise EC_{50} data have been published by the authors, several interesting features have emerged from this work. First, evaluation of the temporal relationships between algal population density and ILs' concentrations in batch cultures of the algae revealed that ILs effectively inhibit algal growth. Second, significant differences in the responses of the two species towards ILs have emerged. Whereas *Oocystis* cultures appeared to acclimatize to imidazolium ILs in concentration ranging from 5 to 50 μM (after 10 days they recovered their growth ability), the growth of *C. meneghiniana* was effectively inhibited through the period test, regardless of the ILs' concentration used. This difference in behaviour has been attributed by the authors to structural differences in the cell walls, which in the case of green algae is

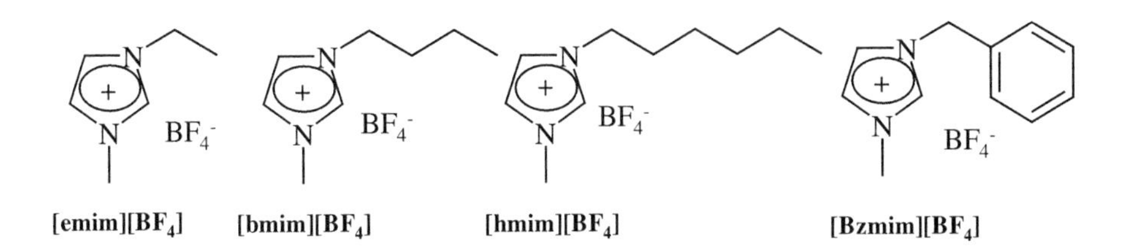

Fig. 3. Ionic liquids employed.

mainly cellulose, whereas in diatoms the cells are surrounded by siliceous frustule. It is however necessary to point out that the effect exerted by ILs on the algal growth is also related to the salinity of the medium, as the toxicity of ILs is reduced in saline water. As an example, the cell density of *Oocystis* was reduced to 50% after 3 days exposure to a 500 μM solution of [hmim][BF_4] but the reduction fell to 30% at a salinity of 8 PSC and was further reduced to 10% in higher saline waters. With respect to the structure of the ILs tested, it appears that the shortest alkyl-substituted derivative, [emim][BF_4], was the most effective in inhibiting the growth of the algae, even if a higher sensitiveness could be envisaged for the *O. submarina*: indeed, while a 500 μM solution of [emim][BF_4] caused an 80% inhibition for former, only 40% inhibition was recorded in the case of *C. meneghiniana*. Moreover, the structure–activity relationship obtained for marine algae (emim > bmim > hmim) is opposite to that obtained using other organisms, where the shorter alkyl chain length resulted in a reduction of the toxic effect. This is a clear example to demonstrate that in order to obtain reliable data on the environmental effect of ILs towards aquatic organisms, it is necessary to take into consideration not only each of the trophic levels, but also several species among each of the levels selected.

A detailed study on the effect of ILs on *Selenastrum capricornutum* has recently been published by Wells and Coombe [19] and is summarised in Table 4. *S. capricornutum* is probably the most widely applied organism used to perform freshwater ecotoxicological tests, thanks to its ease of culture, sensitiveness and quick response to the exposure of chemicals. The test, based on this microorganism, is a growth inhibition one, aimed at evaluating changes in number and biomass of the algal population.

Contrary to what was reported for *Oocystis* and *C. meneghiniana*, there seemed to be a similar pattern of toxicity between *S. capricornutum* and *Daphnia magna*: in both cases, the most and the least toxic salts were separated by more than four orders of magnitude in concentration. Moreover, even for these particular algae a strong correlation between the length of the alkyl side chain and toxicity was found: indeed, while the C_4 species may be classified as moderately toxic, derivatives bearing C_{12}, C_{16} and C_{18} alkyl substituents belonged to the class of highly toxic chemical compounds. These authors showed that with respect to the cation structure, a direct comparison is possible with imidazolium and pyridinium salts, with the latter being less toxic. Differences in the length and substitution of the alkyl chain in phosphonium and ammonium salts make it impossible to obtain a clear relationship between these ILs.

Table 4
Ecotoxicity of several ionic liquids to *S. capricornutum*

Anion	Cation	Side chain	Endpoints (48 h)	Range	References
[Cl], $[PF_6]$	R2, N, +, N, R1	$R_1=C_4; C_{12}; C_{16}; C_{18}$ $R_2=C_1$	EC_{50}	μM: 0.0038/158 ppm: 0.0011/45	[19]
[Cl]	+, N, R1	$R_1=C_4$	EC_{50}	μM: 368 ppm: 63	[19]
$[Tf_2N]$, $[MeSO_4]$	R1, R2, N^+, R1, R1	$R_1=[C_2O\ (C_2O)_4C_1]$ $C_{14}; R_2=C_1$ $R_1=C_8; R_2=C_1$	EC_{50}	μM: 0.089/0.11 ppm: 0.058/0.088	[19]
[Cl], $[(EtO)_2PO_2]$	R1, R2, P^+, R1, R1	$R_1=C_4; R_2=C_2$ $R_1=C_6; R_2=C_{14}$	EC_{50}	μM: 16.1/0.081 ppm: 6.2/0.042	[19]

4.3. Invertebrates

Most of the data published with respect to the toxic effect of ILs for invertebrates are focussed on the use of *D. magna* as test organisms [11, 15, 18, 20]. *D. magna* are freshwater crustaceans, which, thanks to their ease of growth and sensitivity to a variety of pollutants, have been exploited as model organisms in standard comparative toxicity bioassays. In the natural ecosystem there is an important link between the primary producers (microalgae) and the higher trophic levels such as fish and hence microalgae have been the subject of intensive ecological studies. Acute toxicity data, derived from experiments exploiting these organisms, are expressed as IC_{50} or log IC_{50}, that is, on the estimated concentration necessary to immobilize 50% of the crustacean population. Pioneering works have been performed by Lamberti and co-workers [20] who first demonstrated that even in the case of *D. magna*, the toxic effect exerted by imidazolium-based ILs is mainly due to the cationic moiety. The estimated IC_{50} ranged from 0.006 to 0.085 μM (-1.07 to -2.22 in the logarithmic form): these data therefore indicate that the *D. magna* are much more sensitive to the toxic effect of 1-butyl-3-methylimidazolium salts than are bacteria. Similar results have been reported by Gathergood and co-workers [11] who extended the study, evaluating the effect of elongated alkyl chains on the mobility of the crustacean population (Fig. 4). The quantitative structure–activity relationships obtained between toxicity and alkyl chain length provide evidence that biological activity is closely related to structural parameters that determine the hydrophobicity of the molecule and hence its K_{ow}. Once again, results published so far clearly suggest that the nature of the anion has only a small effect. ILs, not necessary belonging to the class of dialkylimidazolium salts, have also been tested by Maginn and co-workers [15], who used experimental data to determine which part of the molecule is responsible for the observed toxic effect through a QSPR analysis. In this way, a correlative and predictive equation was generated: in agreement with experimental data, the model predicted that acute toxicity of ILs for *D. magna* is highly dependent on the length of the alkyl chain attached to the aromatic nitrogen atoms. Moreover, toxicity was predicted to increase slightly with increasing number of aromatic nitrogen atoms on the cation: this indicates that ammonium salts are less toxic than pyridinium salts, which in turn are less toxic than imidazolium salts. Moreover, methylation of the aromatic cation generally reduces the toxic effect and hence picolinium salts are less toxic than pyridinium ones. A further improvement on the determination of the toxic effect of ILs for *D. magna* came from a study recently reported by Wells and Coombe [19], where salts belonging to the four most common classes of ILs

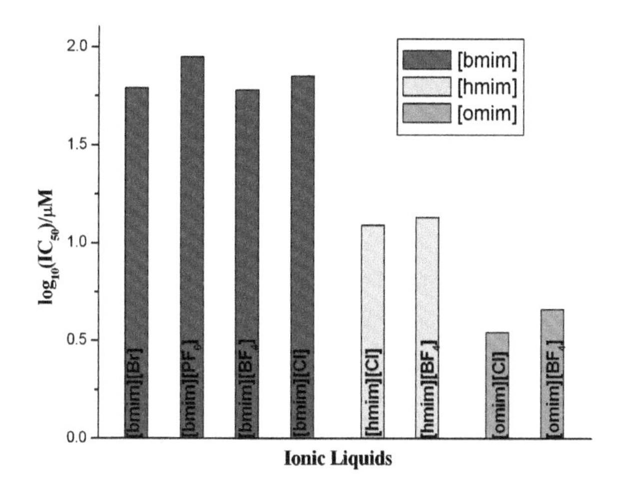

Fig. 4. Acute toxicity values of imidazolium-based ionic liquids for *D. magna*, expressed as average log EC_{50} in micromolar.

have been considered. All the data available are collected in Table 5, and confirm the strong correlation between toxicity and the alkyl side chain for each of the classes involved. IL toxicity to *D. magna* ranged from the highly toxic [C_{18}mim][Cl], with an $IC_{50} = 0.004\ \mu M/L$, to the less toxic 1-butyl-3,5-dimethylpyridinium bromide ($IC_{50} = 97\ \mu M/L$). With respect to other solvents, imidazolium salts generally proved to be much more toxic than methanol, acetonitrile and dichloromethane but less toxic than cationic surfactants [19].

Lamberti and co-workers [20] also investigated the chronic toxicity of imidazolium-based ILs towards *D. magna*. According to their results, all ILs, independent of the anion, negatively affected the reproductive output of the crustaceans, with the lowest number of neonates being produced at the higher concentrations of the IL tested.

Physa acuta is a common pulmonate snail that inhabits streams, ponds and lakes throughout North America and has been used as an organism to obtain more information on the effects of ILs on aquatic environments. Four different classes of compounds have been considered, namely imidazolium, picolinium, ammonium and phosphonium salts. According to data published by Lamberti and co-workers [21] and summarised in Table 6 the most toxic IL to *P. acuta* was 1-octyl-3-methylpyridinium bromide ($LC_{50} = 1.0$ ppm) and the least toxic was tetrabutylammonium bromide ($LC_{50} = 580.2$ ppm).

Table 5
Acute toxicity of different classes of ionic liquids for *D. magna*

Anion	Cation	Side chain	Endpoints (48 h)	Range	References
[Cl], [Br], [BF_4], [PF_6]	R3, R2, R1 (imidazolium)	$R_1=C_4-C_{18}$ $R_2=C_1$ $R_3=C_1$	$\log_{10} IC_{50}$	μM: −5.33/−1.07 ppm: 0.00017/19.91	[11, 15, 19, 20]
[Br], [Cl]	R3, R2, R1 (pyridinium)	$R_1=C_4-C_8$ $R_2=H; C_1$ $R_3=H; C_1$	$\log_{10} IC_{50}$	μM: −2.60/−1.01 ppm: 0.715/23.84	[15, 19]
[Br]	R2, R1 (4-dimethylaminopyridinium)	$R_1=C_6$ $R_2=C_1$	$\log_{10} IC_{50}$	μM: −3.28/−2.79 ppm: 0.15/0.41	[15]
[Br], [Tf_2N], [$MeSO_4$]	R1, R2, R1, R1 (N^+)	$R_1=R_2=C_4$ $R_1=[C_2O\ (C_2O)_4C_1]C_{14}$; $R_2=C_1$ $R_1=C_{8;}\ R_2=C_1$	$\log_{10} IC_{50}$	μM: −3.51/−1.53 ppm: 0.2/9.5	[15, 19]
[Br], [Cl], [$(EtO)_2PO_2$]	R1, R2, R1, R1 (P^+)	$R_1=R_2 = C_4$ $R_1=C_4;\ R_2=C_2$ $R_1=C_6;\ R_2=C_{14}$	$\log_{10} IC_{50}$	μM: −1.42/−3.85 ppm: 0.072/11	[19, 15]

Table 6
Median lethal concentration (LC_{50}) of different ionic liquids to *P. acuta* in 96 h acute toxicity bioassay

Anion	Cation	Side chain	Endpoints (96 h)	Range	References
[Br], [PF_6]	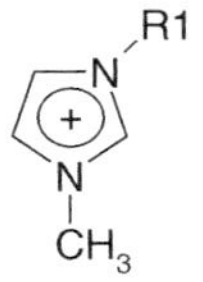 R1–N(+)N–CH_3	$R_1=C_4-C_8$	EC_{50}	μM: 0.03/1.04 ppm: 8.2/229	[21]
[Br]	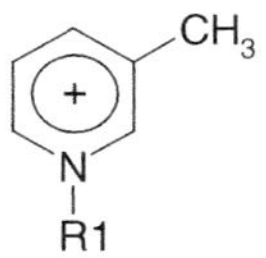 (+)N–R1, CH_3	$R_1=C_4-C_8$	EC_{50}	μM: 0.003/1.41 ppm: 1.0/325.2	[21]
[Br]	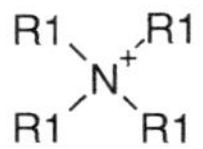 R1, R1, N^+, R1, R1	$R_1=C_4$	EC_{50}	μM: 1.80 ppm: 580.2	[21]
[Br]	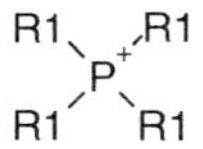 R1, R1, P^+, R1, R1	$R_1=C_4$	EC_{50}	μM: 0.6 ppm: 208	[21]

As in the case of other aquatic organisms (*D. magna*, *V. fischeri*, *P. phosphoreum*) a relationship was found relating alkyl chains to toxicity, regardless of the structure of the cation: the higher the number of carbon atoms appended to the aromatic nitrogen, the greater was the toxic effect. On the other hand, no consistent trend of toxicity was obtained by comparing imidazolium and picolinium salts bearing the same substituent: indeed, whereas [hmim][Br] was more toxic than the corresponding picolinium salts, the opposite was found for the octyl series. No appreciable differences could be interpreted for the butyl-derivatives.

The similarities between ILs and many cationic surfactants, together with the effect of increasing alkyl chain on the cation, seem to point to the fact that the toxic effect is exerted through a disruption of the phospholipid bilayer. Exposure of *Physa snail* to sub-lethal doses of ILs causes a reduction in movements as well as grazing rates of the organisms, with a typical non-linear dose−response relationship. The attempt of the snail to escape the toxicant, searching for clean water, may well account for the U-shaped relationships found by the authors [21].

4.4. Freshwater Fish

The importance of freshwater fish in ecotoxicology is a logical consequence of ecological and economic factors. At maturity, fish often occupy the niche of top predator in their ecosystem and at the same time, they are exploited as a food resource. Although they have often been adopted as the sentinel organisms for the health of the freshwater environment, at the time of writing only one paper has focussed on their use for the assessment of ILs' effect on aquatic environment. Acute toxicity of ILs to the zebrafish (*Danio rerio*) has recently been investigated by Pretti et al. [22] who measured the lethal effect of 15 widely used ILs, bearing different anions and cations, after 96 h of exposure. The results clearly revealed that ILs might cause a completely different effect on fish according to their chemical structure. Indeed 13 out of 15 of the ILs tested, that is, imidazolium, pyridinium and pyrrolidinium ILs, had a 96 h LC_{50} of greater than 100 mg/L and therefore can be regarded as non-lethal towards fish. On the other hand, mortality occurred when fish were treated with AMMOENG 100™ and AMMOENG 130™ (see Table 7), which showed similar LC_{50}, respectively, 5.9 mg/L and 5.2 mg/L. It is interesting to note that these values are remarkably lower than those reported for organic solvents and tertiary amines. Moreover, these data seemed to show that fish are less sensitive to the effect of ILs than species belonging to lower trophic levels. According to observations by the authors, treated fish

Table 7
Acute toxicity for *D. rerio*

Solvent		Limit test, LC_{50} (96 h) (mg/L)	Full test, LC_{50} (96 h)	Reported LC_{50} (mg/L)
Cation	Anion			
C_2H_5, N, (+), N, CH_3	$[EtSO_4]$, [OTs]	>100		
C_4H_9, N, (+), N, CH_3	$[PF_6]$, $[BF_4]$, $[Tf_2N]$, $[N(CN)_2]$, [OTf], $[NO_3]$	>100		
C_4H_9, N, (+), N, CH_3, CH_3	$[PF_6]$	>100		
(+), N, C_4H_9	$[Tf_2N]$	>100		
+, N, C_4H_9	$[Tf_2N]$	>100		
C_2H_5, C_2H_5, N^+, $CH_2CH_2(OCH_2CHCH_3)nOH$, CH_3, Cl^- AMMOENG 110™		>100		
C_2H_5, C_2H_5, N^+, $CH_2CH_2(OCH_2CHCH_3)nOH$, CH_2CH_2OH, $H_2PO_4^-$ AMMOENG 112™		>100		
C_2H_5, C_2H_5, N^+, $CH_2CH_2(OCH_2CH_2)nOH$, $CH_2CH_2(OCH_2CH_2)nOH$, MeSO4⁻ AMMOENG 100™		<100	5.9 (4.3; 8.0)	
C_2H_5, H_3C, N^+, Stearyl, Stearyl, Cl^- AMMOENG 130™		<100	5.2 (3.8; 7.0)	

Table 7 (Continued)

Solvent		Limit test, LC_{50} (96 h) (mg/L)	Full test, LC_{50} (96 h)	Reported LC_{50} (mg/L)
Cation	Anion			
MeOH				12700–29400
DCM				>100
AcN				>100
Aniline				Up to 100
Et_3N				44

Note: 96 h LC_{50} (mg/L; confidence limits 95%).

exposed to both AMMOENG 100™ and AMMOENG 130™ at concentrations of 10 mg/L showed reduction in general activity, loss of equilibrium, erratic swimming and staying motionless at mid-water for prolonged periods. Through a hystopathological examination of fish after exposure to ILs, it has been possible to hypothesize that skin alteration was the origin of the lethal effect, as is usual when fish are exposed to cationic surfactants. Although in principle it is not possible to identify which part of the molecule is responsible for the detrimental effect on *D. rerio*, as both the anion and the cation have been changed in these two latter compounds, data presented by the authors [22], together with the well known effect exerted by cationic surfactants, seem to suggest that modifications of the cationic moiety are responsible. These data are therefore in agreement with those published by other authors with respect to different trophic aquatic levels.

5. CONCLUSIONS

As shown in this chapter, preliminary information on the environmental fate of ILs is now available. It is known that imidazolium ILs, which normally biodegrade slowly, may be effectively decomposed by advanced oxidation techniques, and their biodegradation can be improved by the incorporation of one or more oxygen atoms in the alkyl chain. Ecotoxicity studies pointed out that at least for some of the most commonly employed IL salts, there will be a significant environmental risk associated with their use. ILs show a varying degree of toxicity according to the organism tested, but in each case, it has been possible to demonstrate that the cation plays an important role in determining the effect of the salts on aquatic environments. With few exceptions, structure–activity relationships showed that simple ammonium and cholinium salts are less toxic towards aquatic environment than pyridinium salts, which in turn are less toxic than imidazolium. Moreover, acute toxicity of ILs to several microorganisms and invertebrates is highly dependent on the length of the alkyl chain appended on the aromatic nitrogen, and is independent of the cation under investigation. The data available may therefore serve as the base for a rationale design of a new class of ILs, which have to be readily biodegradable and with an acceptable environmental profile. Obviously more detailed studies on the chronic effect associated with the exposure of ILs to the aquatic environment are necessary, together with an accurate risk profile of each of the metabolites derived from the degradation of these salts.

REFERENCES

[1] P. Wasserscheid and T. Welton, Ionic Liquids in Synthesis, Wiley-VCH, Verlag Gmbh & Co., 2003.
[2] Ionic Liquids Workshop Background, State-of-the-Art, and Academic/Industrial Applications, March 23–24, The University of Alabama, Tuscaloosa, AL, 2006.
[3] Ionic Liquids: Not Just Solvents Anymore or Ionic Liquids: Parallel Futures, 231st ACS National Meeting, March 26–30, Division of Environmental Chemistry, Atlanta, GA, 2006.
[4] L. Ropel, L.S. Belveze, S.N.V.K. Aki, M.A. Stadtherr and J.F. Brennecke, Green Chem., 7 (2005) 83.
[5] P. Stepnowski and P. Storoniak, Environ. Sci. Pollut. Res., 12 (2005) 199.
[6] P.A. Anastas and J.C. Warner, Green Chemistry: Theory and Practice, Oxford University Press, Oxford, 1998.
[7] P. Stepnowski and A. Zaleska, J. Photochem. Photobiol. A Chem., 170 (2005) 45.
[8] A.W. Morawski, M. Janus, I. Goc-Maciejewska, A. Syguda and J. Pernak, Pol. J. Chem., 79 (2006) 1929.
[9] N. Gathergood and P.J. Scammells, Aust. J. Chem., 55 (2002) 557.
[10] N. Gathergood, M.T. Garcia and P.J. Scammells, Green Chem., 6 (2004) 166.
[11] M.T. Garcia, N. Gathergood and P.J. Scammells, Green Chem., 7 (2005) 9.
[12] N. Gathergood, P.J. Scammells and M.T. Garcia, Green Chem., 8 (2006) 156.
[13] J. Ranke, K. Molter, F. Stock, U. Bottin-Weber, J. Poczobutt, J. Hoffmann, B. Ondruschka, J. Filser and B. Jastorff, Ecotoxicol. Environ. Saf., 58 (2004) 396.
[14] K.M. Docherty and F.C. Kulpa, Jr., Green Chem., 7 (2005) 185.
[15] D.J. Couling, R.J. Bernot, K.M. Docherty, J.K. Dixon and E.J. Maginn, Green Chem., 8 (2006) 82.
[16] Standard Test Method for Assessing the Microbial Detoxification of Chemically Contaminated Water and Soil Using Toxicity Test with a Luminescent Marine Bacterium, ASTM Designation: D 5660-96, 1996.

[17] K.L.E. Kaiser and V.S. Palabrica, Water Pollut. Res. J. Can., 26 (1991) 361.
[18] A. Latala, P. Stepnowski, M. Nedzi and W. Mrozik, Aquat. Toxicol., 73 (2005) 91.
[19] A.S. Wells and V.T. Coombe, Org. Proc. Res. Dev., 10 (2006) 794.
[20] R.J. Bernot, M.A. Brueseke, M.A. Evans-White and G.A. Lamberti, Environ. Toxicol. Chem., 24 (2005) 87.
[21] R.J. Bernot, E.E. Kennedy and G.A. Lamberti, Environ. Toxicol. Chem., 24 (2005) 1759.
[22] C. Pretti, C. Chiappe, D. Pieraccini, M. Gregori, F. Abramo, G. Monni and L. Intorre, Green Chem., 8 (2006) 238.

Thermodynamics, Solubility and Environmental Issues
T.M. Letcher (editor)

Chapter 15

Rhamnolipid Biosurfactants: Solubility and Environmental Issues

Catherine N. Mulligan

Department of Building, Civil and Environmental Engineering, Concordia University, 1455 de Maisonneuve Boulevard West, Montreal, Quebec H3G 1M8, Canada

1. INTRODUCTION

Surface active agents (surfactants) are amphiphilic compounds with two opposing portions, one part is hydrophilic and the other is hydrophobic [1]. They reduce the free energy of the system by replacing the bulk molecules of higher energy at an interface. By reducing the interfacial tension and forming micelles, surfactants or biosurfactants have shown many environmental applications including enhanced oil recovery, removal of heavy metals from contaminated soil and remediation of hydrophobic organic compounds from soil [2–4]. They are used in environmental applications to enhance solubility of organic or inorganic components for soil washing or flushing. Typical desirable properties include solubility enhancement, surface tension reduction and low critical micelle concentrations (CMC).

Yeast and bacteria can produce biosurfactants, biological surfactants from various substrates including sugars, oils, alkanes and wastes [5]. Some types of biosurfactants are glycolipids, lipopeptides, phospholipids, fatty acids, neutral lipids, polymeric and particulate compounds [6]. Most are either anionic or neutral, while only a few with amine groups are cationic. The hydrophobic part of the molecule is based on long-chain fatty acids, hydroxy fatty acids or α-alkyl-β-hydroxy fatty acids. The hydrophilic portion can be a carbohydrate, amino acid, cyclic peptide, phosphate, carboxylic acid or alcohol.

A wide variety of microorganisms can produce these compounds [7]. The CMCs of the biosurfactants generally range from 1 to 200 mg/L and their molecular masses are usually from 500 to 1500 Da [8]. They can be potentially effective with some distinct advantages over the highly used synthetic surfactants including high specificity, biodegradability and biocompatibility [9]. Biosurfactants are surfactants produced by bacterial and yeasts that are potentially more biodegradable than synthetic ones.

Biosurfactants have been tested in enhanced oil recovery and the transportation of crude oils [10]. They were demonstrated to be effective in the reduction of the interfacial tension of oil and water in situ, the viscosity of the oil, the removal of water from the emulsions prior to processing and in the release of bitumen from tar sands. The high molecular weight Emulsan has been commercialized for this purpose [11]. It contains a polysaccharide with fatty acids and proteins attached. Other high-molecular weight biosurfactants are reviewed by Ron and Rosenberg [12].

In this chapter, very effective, low molecular weight biosurfactants, called rhamnolipids are discussed. In each case, environmental applications will be examined including enhancing solubilization and biodegradation, soil and waste treatment (in situ and ex situ) and water treatment. Areas of potential research will be identified.

2. BACKGROUND ON RHAMNOLIPIDS

A group of biosurfactants that has been studied extensively is the rhamnolipids from *Pseudomonas aeruginosa* [13–15] and are commercially available (Jeneil Biosurfactant). Up to seven homologues have now been identified [16]. These biosurfactants can lower surface tensions of water to 29 mN/m. Critical micelle concentrations can vary from 10 to 230 mg/L, depending on the structure. Two types of rhamnolipids contain either two rhamnoses attached to β-hydroxydecanoic acid or one rhamnose connected to the identical fatty acid (Scheme 1). Type I, (R1) is L-rhamnosyl-β-hydroxydecanoyl-β-hydroxydecanoate, molecular mass = 504 Da; Type II, (R2) is L-rhamnosyl-β-L-rhamnosyl-β-hydroxydecanoyl-β-hydroxydecanoyl-β-hydroxydecanoate, molecular mass = 660 Da; Type III (R3) is one rhamnose attached to β-hydroxydecanoic acid and Type IV (R4) is two rhamnoses attached to β-hydroxydecanoic acid. *P. aeruginosa* can produce rhamnolipids from a wide variety substrates including C_{11} and C_{12} alkanes, succinate, pyruvate, citrate, fructose, glycerol, olive oil, glucose and mannitol [17]. Composition and yields depend on the fermentor design, pH, nutrient composition, substrate and temperature [18].

R1

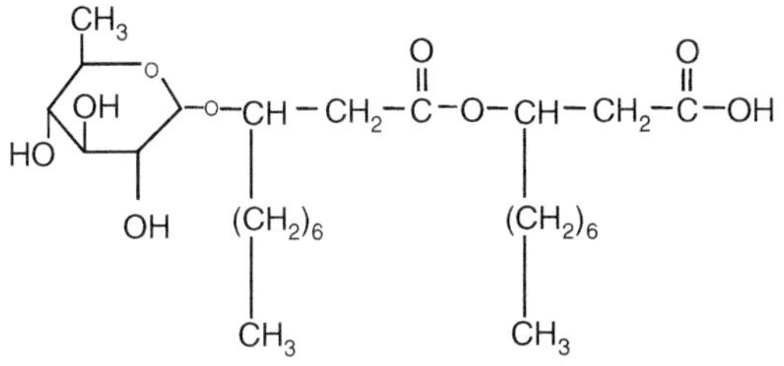

R2

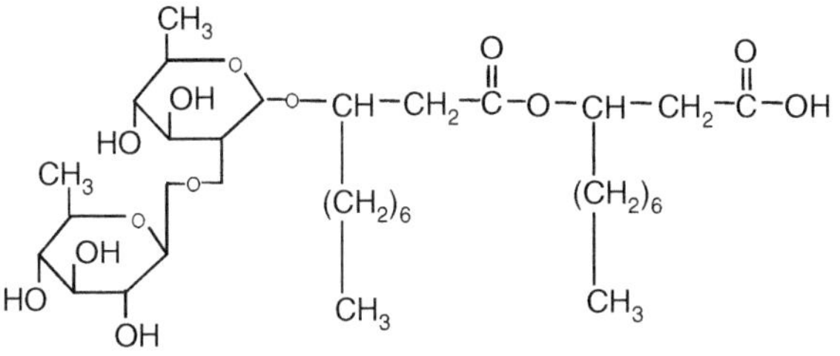

R3

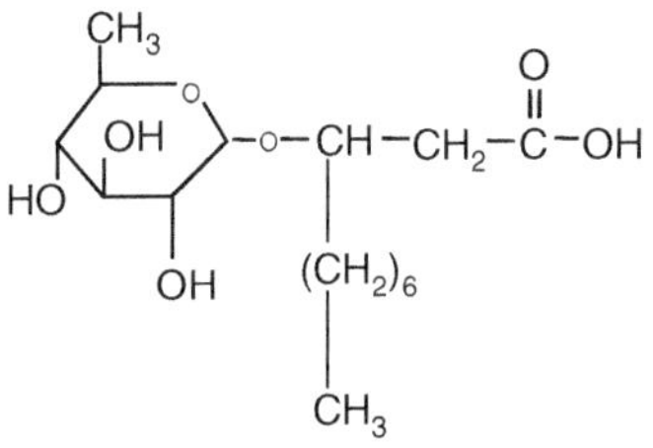

R4

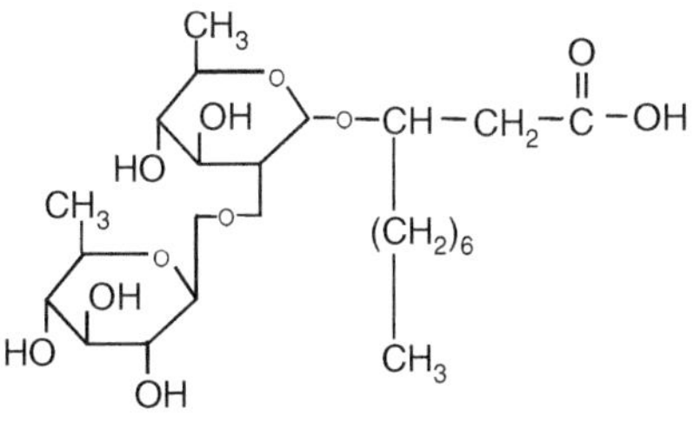

Scheme 1. Structure of four different rhamnolipids (R1–R4) produced by *P. aeruginosa*.

3. ENHANCED BIODEGRADATION OF RECALCITRANT COMPOUNDS

3.1. Aqueous Systems

Several studies have examined the effects of rhamnolipid on the biodegradation of organic contaminants; especially focussing on various hydrocarbons of low solubility. The low solubility and high hydrophobicity of organic compounds limits their ability to be transported into microbial cells and thus be biodegraded. Rhamnolipid addition can enhance biodegradation of various pure compounds such as hexadecane, octadecane, *n*-paraffin and phenanthrene in liquid systems [19] in addition to hexadecane, tetradecane, pristine, creosote and hydrocarbon mixtures in soils. Two mechanisms for enhanced biodegradation are possible. They include enhanced solubility of the substrate for the microbial cells, and interaction with the cell surface, which increases the hydrophobicity of the cell surface allowing hydrophobic substrates to associate more easily [20, 21]. Zhang and Miller [21] demonstrated that a concentration of 300 mg/L of rhamnolipids increased the mineralization of octadecane to 20% compared with 5% for the controls. Beal and Betts [22] showed that the cell surface hydrophobicity increased by the biosurfactant strain (50.5%) more than a non-biosurfactant producing strain (33.7%) during growth on hexadecane. The rhamnolipids also increased the solubility of the hexadecane from 1.8 to 22 μg/L. Therefore, both the changes in the cell surface and increased solubilization occurred simultaneously.

The spilling of oil into the sea is a major problem that can destroy coastlines. The Amoco Cadiz spill off the Britanny coast in 1978 and the Exxon Valdez near Prince William Sound in 1989 are examples of significant coastline contamination. Biosurfactants can be useful for oil spills since they would be less toxic and persistent than synthetic surfactants but fewer studies have been performed on the enhanced biodegradation of complex mixtures. Chakrabarty [23] showed that an emulsifier produced by *P. aeruginosa* SB30 could disperse oil into fine droplets which could enhance biodegradation. Shafeeq et al. [24] showed that biosurfactants were produced during biodegradation of a hydrocarbon mixture by *P. aeruginosa* S8. Chhatre et al. [25] found that four types of bacteria isolated from crude oil were able to degrade 70% of the Gulf and Bombay High Crude Oil. One of the isolates produced a rhamnolipid biosurfactant that enhanced biodegradation by emulsification of the crude oil.

Abalos et al. [26] studied the effect of the rhamnolipid produced by *P. aeruginosa* AT10 on Casablanca crude oil degradation. They determined that the biodegradation level could be increased from 32 to 61% at 10 days. However, the effect was particularly significant for isoprenoids and alkylated polycyclic aromatic hydrocarbons (PAHs) that increased in biodegradation from 16 to 70% and 9 to 44%, respectively. Micelles increased the apparent solubility of the oil and formed oil in water and oil emulsions.

Rahman et al. [27] showed that addition of rhamnolipid produced by *Pseudomonas* sp. DS10-129 along with poultry litter and coir pith enhanced

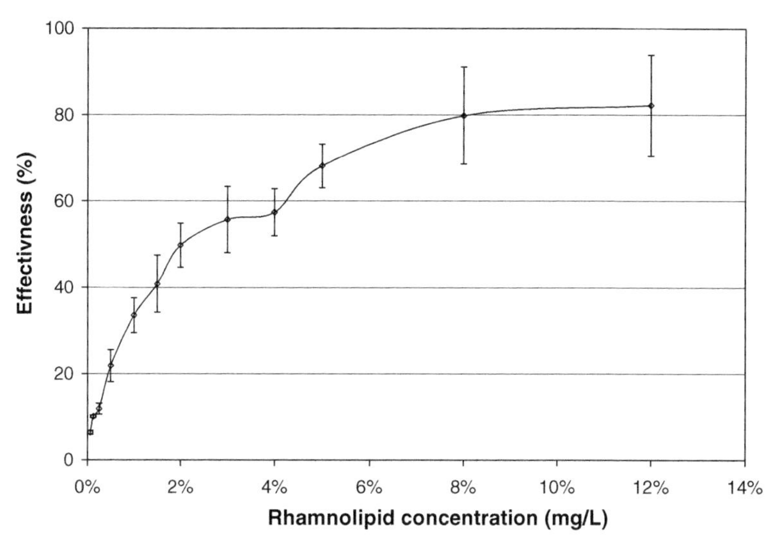

Fig. 1. Effect of rhamnolipid concentration (mg/L) on its effectiveness [31].

ex situ bioremediation of a gasoline-contaminated soil. Research with another strain, *P. marginalis*, also indicated that the produced biosurfactants solubilized PAHs such as phenanthrene and enhanced biodegradation [28]. Other research by Garcia-Junco et al. [29] indicated that addition of rhamnolipids led to attachment to the phenanthrene that enhanced bioavailability and hence degradation of the contaminant by *P. aeruginosa.*

The feasibility of biosurfactants for dispersing oil slicks was also examined by Holakoo and Mulligan [30]. At 25°C and a salinity of 35%, a solution of 2% rhamnolipids diluted in saline water and applied at a dispersant to oil ratio (DOR) of 1:2, immediately dispersed 65% of crude oil. Co-addition of 60% ethanol and 32% octanol with 8% rhamnolipids applied at a DOR of 1:8 improved dispersion to 82%. Dispersion efficiency decreased in fresh water and at lower temperatures.

Subsequent work by Dagnew [31] was performed to determine the ability of the rhamnolipid to act as a biodispersant for oil. The effect of rhamnolipid concentration was evaluated. Fig. 1 shows the dispersion of fresh crude oil at different rhamnolipid concentrations at a pH 7.5-buffered solvent. The dispersion of crude oil increased dramatically with an increase in rhamnolipid concentration. At 22 ± 2°C and 35% salinity, the rhamnolipid biosurfactant dispersed from 10 to 82% of oil into the water column when the rhamnolipid solution was applied at concentrations between 0.1 and 12%, respectively. In comparison to natural dispersion, the rhamnolipid

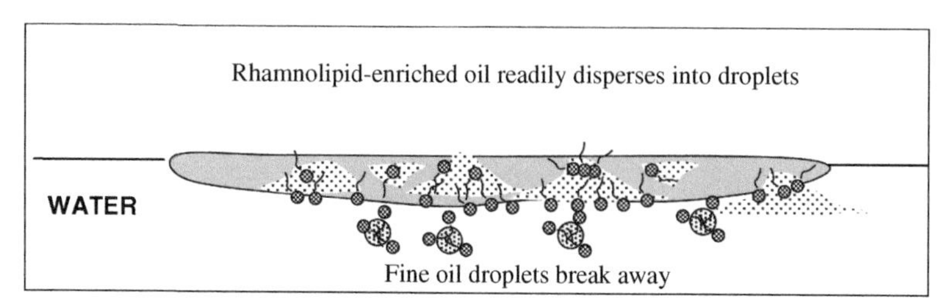

Fig. 2. Detailed mechanism of chemical dispersion (adapted from Fiocco and Lewis [34]).

biosurfactant enhanced dispersion of crude oil by more than 50 orders of magnitude, from 1.6 to 82%. The rhamnolipid was applied at a concentration greater than its CMC value (0.0035%), showing that the enhancement of crude oil dispersion is dominated by the physical biosurfactant interaction with hydrocarbon compounds which resulted in dispersion of the crude oil into the water column by encapsulation of hydrocarbon compounds into micellar aggregates (Fig. 2). Overall in terms of effectiveness, delivering the same rhamnolipid concentration either using concentrates at a lower DOR or less concentrated at a higher DOR showed no significant difference.

Mata-Sandoval et al. [32] compared the ability of the rhamnolipid mixture to solubilize the pesticides, trifluralin, coumaphos and atrazine, with the synthetic surfactant Triton X-100. The synthetic surfactant was able to solubilize approximately twice as much of all pesticides as the rhamnolipid. The biosurfactant seems to bind trifluralin tightly in the micelle and releases the pesticide slowly to the aqueous phase, which could have implications for microbial uptake. This approach utilizing micellar solubilization capacities and aqueous–micelle solubilization rate coefficients and micellar–aqueous transfer rate coefficients could be useful for future studies on microbial uptake. Addition of rhamnolipid in the presence of cadmium enabled biodegradation of the hydrocarbon naphthalene to occur as if no cadmium was present [33].

3.2. Soil Systems

A review by Maier and Soberon-Chavez [19] showed that hexadecane, tetradecane, pristine, creosote and hydrocarbon mixture in soils. Other researchers [35] compared the solubilization of naphthalene by a rhamnolipid, sodium dodecyl sulphate (SDS), an anionic surfactant and Triton X-100, a non-ionic surfactant. The biosurfactant increased the solubility of naphthalene by 30 times. However, biodegradation of naphthalene (30 mg/L) took 40 days in the presence of biosurfactant (10 g/L) compared with 100 h for an equal concentration of Triton X-100 (10 g/L). It appeared that the biosurfactant was used as a carbon source instead of the naphthalene which did not occur in the case of Triton X-100. In the presence of SDS naphthalene biodegradation did not occur.

Rhamnolipids from the UG2 strain were able to enhance the solubilization of four-ring PAHs in a bioslurry more significantly than three-ring PAHs and that the biosurfactants were five times more effective than SDS [36]. However, high molecular weight PAHs were not biodegraded despite surfactant addition. Providenti et al. [37] also studied the effect of UG2 biosurfactants on phenanthrene mineralization in soil slurries and showed that lag times decreased and degradation increased.

Noordman et al. [38] studied the effect of the biosurfactant from *P. aeruginosa* on hexadecane degradation. They determined that the biosurfactant could enhance biodegradation if the process is rate-limited as in the case of small soil pore sizes (6 nm), where the hexadecane is entrapped and is of limited availability. The rhamnolipid stimulates the release of entrapped substrates (if mixing conditions are low such as in a column) and enhances uptake by cells (if the substrate is available). This could then become important in the stimulation of bacterial degradation under in situ conditions.

Further experiments were carried out by Dean et al. [39] who investigated the bioavailability of phenanthrene in soils to two strains of microorganisms known to biodegrade this substrate. Either rhamnolipid was added or a biosurfactant-producing strain of *P. aeruginosa* ATCC 9027. Results were mixed and difficult to interpret. One strain (strain R) showed enhanced biodegradation when the surfactant was added but the other (strain P5-2) did not. Co-addition of the strain–strain enhanced mineralization of phenanthrene only by strain R. There seemed to be some interaction between the two strains, which will need further investigation. Addition of the rhamnolipid enhanced release of the phenanthrene but did not necessarily enhance biodegradation. A summary of the cases where rhamnolipid addition has enhanced biodegradation is shown in Table 1.

4. EX SITU WASHING

4.1. Hydrocarbon Removal

Besides studies on biodegradation, rhamnolipid surfactants have been tested to enhance the release of low solubility compounds from soil and other solids. They have been found to release three times as much oil as water alone from the beaches in Alaska after the Exxon Valdez tanker spill [40]. Removal efficiency varied according to contact time and biosurfactant concentration. Scheibenbogen et al. [41] found that the rhamnolipids from *P. aeruginosa* UG2 were able to effectively remove a hydrocarbon mixture from a sandy

Table 1
Summary of biodegradation studies with a positive effect of the rhamnolipid on biodegradation

Medium	Microorganism	Contaminant	Reference
Soil slurry	*P. aeruginosa* UG2	Hexachlorobiphenyl	[45]
Soil slurry	*P. aeruginosa* UG2	Phenanthrene	[37]
Soil	*P. aeruginosa* UG2	Aliphatic and aromatic hydrocarbons	[41]
Soil	*P. aeruginosa* UG2	Phenanthrene and hexadecane	[38, 46, 47]
Soil	*P. aeruginosa* #64	Phenanthrene, fluoranthrene, pyrene, pentachlorophenol benzo[a]pyrene	[48]
Soil	*P. aeruginosa* ATCC 9027	Napthalene and phenanthrene	[49]
Soil	*P. aeruginosa* ATCC 9027	Octadecane	[21]
Soil	*P. aeruginosa* ATCC 9027	Phenanthrene and cadmium	[33]
Soil	*P. aeruginosa* ATCC 9027 mono-rhamnolipid	Naphthalene and cadmium	[50]
Liquid	*Pseudomonas* di-rhamnolipid	Toluene, ethyl benzene, butyl benzene	[51]
Liquid	Mixed culture with rhamnolipid addition	Crude oil	[31]
Liquid	Mixed culture with rhamnolipid addition	Styrene	[56]

Adapted from Makkar and Rockne [44].

loam soil and that the degree of removal (from 23 to 59%) was dependent on the type of hydrocarbon removed and the concentration of the surfactant used. Van Dyke et al. [42] had previously found that the same strain could remove at a concentration of 5 g/L, approximately 10% more hydrocarbons from a sandy loam soil than a silt loam soil and that SDS was less effective than the biosurfactants in removing hydrocarbons. Lafrance and Lapointe [43] also showed that injection of low concentrations of UG2 rhamnolipid (0.25%) enhanced transport of pyrene more than SDS with less impact on the soil.

Bai et al. [52] showed that after only two pore volumes, a 500 mg/L concentration of rhamnolipid could remove 60% of the hexadecane at pH 6 with 320 mM sodium. They also showed that pH and ionic strength can influence solubilization. The addition of 500 mM of Na could increase hexadecane solubilization by 7.55-fold and 25-fold by 1 mM of magnesium by reducing interfacial tension. Decreasing the pH from 7 to 6 also reduced the interfacial tension.

Various biological surfactants were compared by Urum et al. [53] for their ability to wash a crude oil-contaminated soil. They included rhamnolipid, aescin, lecithin, saponin, tannin and SDS. Temperature (5, 20, 35 and 50°C), surfactant concentration (0.004, 0.02, 0.1 and 0.5%), surfactant volume (5, 10, 15 and 20 ml), shaker speed (80, 120, 160 and 200 strokes/min) and wash time (5, 10, 15 and 20 min) were evaluated. The conditions of 50°C and 10 min were optimal for most of the surfactants such as the rhamnolipid, SDS and saponin were able to remove more than 79% of the oil.

Southam et al. [54] studied the effect of biosurfactants on the biodegradation of waste hydrocarbons. To degrade hydrocarbons, bacteria must adsorb onto the surfactant–oil interface of 25–50 nm in thickness. Approximately 1% of all the biosurfactant was needed to emulsify the oil. This type of studying with a transmission electron microscope showed that the microorganisms were able to uptake nanometre-sized oil droplets during growth. More of this type of research is required to determine the mechanism of hydrocarbon metabolism and biosurfactant applications.

A completely different approach for oil cleanup was performed by Shulga et al. [55]. They examined the use of the biosurfactants in cleaning oil from coastal sand, and from the feathers and furs of marine birds and animals. The strain *Pseudomonas* PS-17 produced a biosurfactant/biopolymer that reduced the surface tension of water to 29.0 mN/m and the interfacial tension against heptane to between 0.01 and 0.07 mN/m. The biopolymer had a molecular weight of $3–4 \times 10^5$. The results indicated that this biosurfactant/biopolymer was able to remove oil from marine birds and animals.

The ability of the rhamnolipid to remove styrene from contaminated soil was evaluated by Guo and Mulligan [56]. It was shown that it was feasible to use rhamnolipid as a washing agent to remove styrene. The results show that more than 70% of removal could be achieved for 32,750 mg/kg styrene after 1 day and 88.7% removal after 5 days while a 90% removal was obtained for 16340 mg/kg of styrene after 1 day. In the preliminary tests, time was an important factor. Longer contact times (5 days) could achieve higher removal efficiencies (90%). The influence of biosurfactant concentration on styrene solubility is shown in Fig. 3 (unpublished data). It can be seen by the correlation factor that the correlation is quite linear. From the slope, a weight solubilization ratio of 0.29 g styrene is solubilized per gram of rhamnolipid added. After removal from the soil by rhamnolipid, more than 70% of the styrene could be biodegraded by an anaerobic biomass [56], in a combined soil flushing and leachate treatment.

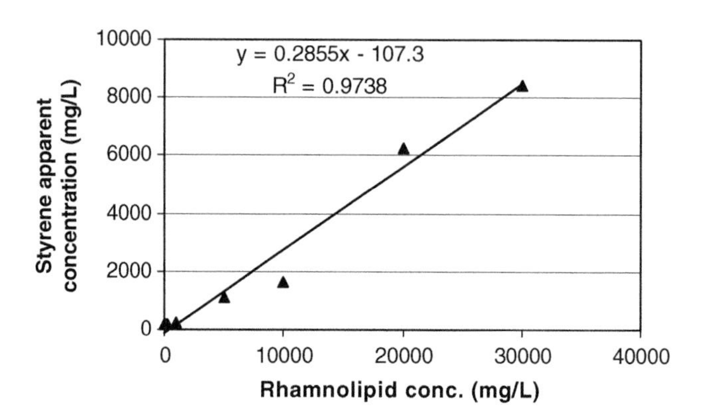

Fig. 3. Solubilization of styrene by rhamnolipids.

4.2. Heavy Metal Removal

Due to the anionic nature of rhamnolipids, they are able to remove metals from soil and ions such as cadmium, copper, lanthanum, lead and zinc due to their complexation ability [57–59]. More information is required to establish the nature of the biosurfactant–metal complexes. Stability constants were established by an ion exchange resin technique [60]. Cations of lowest to highest affinity for rhamnolipid were $K^+ < Mg^{2+} < Mn^{2+} < Ni^{2+} < Co^{2+} < Ca^{2+} < Hg^{2+} < Fe^{3+} < Zn^{2+} < Cd^{2+} < Pb^{2+} < Cu^{2+} < Al^{3+}$. These affinities were approximately the same or higher than those with the organic acids, acetic, citric, fulvic and oxalic acids. This indicated the potential of the rhamnolipid for metal remediation. Molar ratios of the rhamnolipid to metal for selected metals were 2.31 for copper, 2.37 for lead, 1.91 for cadmium, 1.58 for zinc and 0.93 for nickel. Common soil cations, magnesium and potassium, had low molar ratios, 0.84 and 0.57, respectively.

Mulligan [61] determined that the rhamnolipid ratio to metal ratio was ~3:1 in soil washing experiments with 0.01% but this ratio increased substantially as the concentration of surfactant increased. Solubilization in the micelle was also examined using the following equation:

$$K_m = \frac{X_{mic}}{X_{aq}}$$

where K_m = micelle–water partitioning coefficient which includes X_{mic} = MSR/(1+MSR), the mole fraction of total metal in the micelle, and where molar solubilization ratio (MSR) is the mole of metal in the micelle/mole of

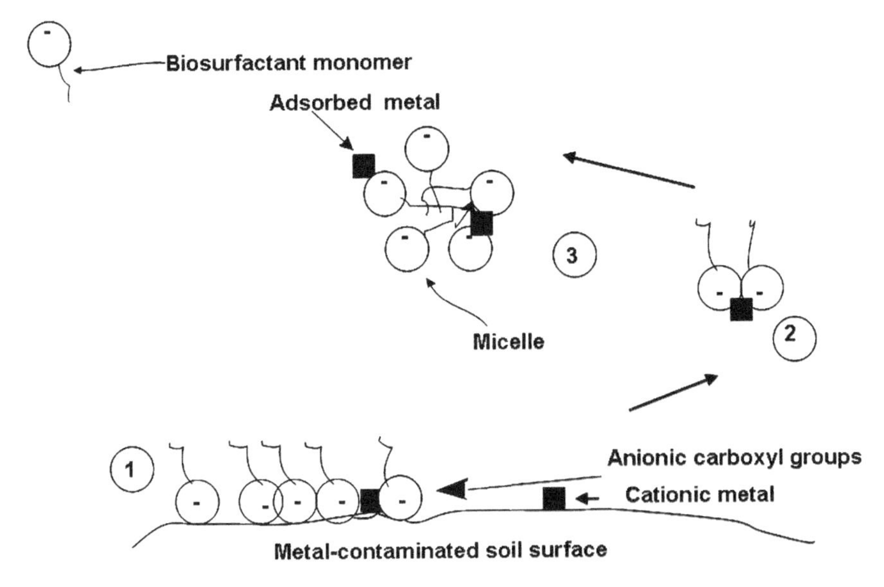

Fig. 4. Potential mechanism for metal removal by the rhamnolipid biosurfactant. (1) Adsorption of the biosurfactant on the soil surface and interaction via electrostatic attraction and solubilization of soil fractions containing the metal. (2) Removal of the metal from the soil surface. To maintain electrostatic neutrality, two carboxylic groups are needed per divalent metal. (3) Incorporation of the metal–biosurfactant complex into micelles (adapted from Mulligan et al. [68]).

Table 2
Relationship between K_m and K_{ow} for the rhamnolipid [61]

Rhamnolipid concentration (%)	K_{ow} for the total metal concentration	K_m
0.01	9	33.9
0.1	30	85.4

surfactant. X_{aq} is the mole fraction in the aqueous phase [62]. The overall mechanism for this is shown in Fig. 4.

From Table 2, it can be seen that for the rhamnolipid at lower biosurfactant concentrations, the ratio of metal in the micelle to that in the aqueous phase is lower. Concentrations of 1% rhamnolipid unfortunately made it difficult to determine the K_{ow} due to emulsification. Metals appeared to have higher affinities at low surfactant concentrations which agrees with solubilization trends. This also explains why multiple washes at low concentrations are successful.

Rhamnolipids were applied to a soil in the presence of oil contamination [15] and from sediments to remove heavy metals [63]. Although 80–100% of cadmium and lead can be removed from artificially contaminated soil, in field samples the results were more in the range of 20–80% [64]. Clay and iron oxide contents affected the efficiency of the biosurfactants but this has not been researched. Biosurfactant could be added as a soil washing process for excavated soil. Due to the foaming property of the biosurfactant, metal–biosurfactant complexes can be removed by addition of air to cause foaming and then the biosurfactant can be recycled through precipitation by reducing the pH to 2.

Neilson et al. [65] studied lead removal by rhamnolipids. A 10 mM solution of rhamnolipid removed about 15% of the lead after 10 washes. High levels of Zn and Cu did not impact lead removal. Similar to the studies on zinc by Mulligan et al. [15, 63], lead could be removed from the iron oxide, exchangeable and carbonate fractions. These removal levels are very low and the process could be improved if the biosurfactants could be recycled [65].

Rhamnolipids have also been added to another metal contaminated media, mining ores, to enhance metal extraction [66]. Batch tests were performed at room temperature. Using a 2% rhamnolipid concentration, 28% of the copper was extracted. Although concentrations higher than 2% extracted more copper, the rhamnolipid solution became very viscous and difficult to work with. Addition of 1% NaOH with the rhamnolipid enhanced the removal up to 42% at a concentration of 2% rhamnolipid but decreased at higher surfactant concentrations. Sequential extraction studies were also being performed to characterize the mining ore and to determine the types of metals being extracted by the biosurfactants. Approximately 70% of the copper was associated with the oxide fraction, 10% with the carbonate, 5% with the organic matter and 10% with the residual fraction. After washing with 2% biosurfactant, pH 6, for 6 days, it was determined that 50% of the carbonate fraction and 40% of the oxide fraction were removed.

A study by Massara et al. [67] was conducted on the removal of Cr(III) to eliminate the hazard imposed by its presence in kaolinite. The effect of addition of negatively charged biosurfactants (rhamnolipids) on chromium-contaminated soil was studied. Results showed that the rhamnolipids have the capability of extracting a portion of the stable form of chromium, Cr(III), from the soil. The removal of hexavalent chromium was also enhanced using a solution of rhamnolipids. Results from the sequential extraction procedure showed that rhamnolipids remove Cr(III) mainly from the carbonate, and oxide/hydroxide portions of the soil. The rhamnolipids had also the capability of reducing close to 100% of the extracted Cr(VI) to Cr(III) over a period of 24 days.

5. IN SITU FLUSHING APPLICATIONS

5.1. Hydrocarbon Removal

To simulate in situ flushing conditions, soil column experiments were performed by Noordman et al. [46]. They determined that 500 mg/L of rhamnolipids removed two to five times more phenanthrene than without the rhamnolipids. For optimal use of rhamnolipids in in situ soil remediation, they must not adsorb strongly to the soil. Noordman et al. [47], then, investigated adsorption of the biosurfactant to the soil. The unitless retardation factor, R, used in models for transport estimation was determined. $R = 1 + (\rho/\varepsilon)k_d$, where ρ is the soil density, ε the soil porosity and k_d the soil water partition coefficient. Without the rhamnolipid, R was found to be between 2 for naphthalene with silica to 700 for phenanthrene with octadecyl-derivatized silica. The addition of 500 mg/L of rhamnolipid decreased R by a factor of eight. To avoid adsorption, biosurfactant concentrations higher than the CMC should be used to limit interfacial hydrophobic adsorption of surfactant aggregates such as hemimicelles to the soil. Adsorption of the biosurfactant was not related to the organic content of the soil which indicates that adsorption is due to interfacial phenomena not partitioning. The biosurfactant most effectively enabled the transport of the more hydrophobic compounds while the less hydrophobic ones were retarded due to the sorption of surfactant admicelles (micelles adsorbed to the soil surface). The interactions could include ion exchange reactions with the anionic carboxylate portion of the rhamnolipid, complexation at the surface and hydrogen bonding of the rhamnose headgroups [69]. Further studies by Herman et al. [70] on the biosurfactant interaction mechanisms determined that low concentrations of rhamnolipids below the CMC enhance mineralization of entrapped hydrocarbons while concentrations above the CMC enhance hydrocarbon mobility.

Attempts are being made to determine non-aqueous phase liquid (NAPL) mole fraction and micelle–aqueous partition coefficients with rhamnolipids to understand equilibrium solubilization behaviour in surfactant-enhanced soil remediation situations [51]. A modification of Raoult's law was used for surfactant-enhanced solubilization but deviation from this ideal behaviour depended on the hydrophobicity of the compounds and the NAPL-phase mole fraction. Micelle–water partition coefficients were non-linear in relation to the NAPL-phase mole fraction. Also enhancements by the surfactant were greater than predicted for hydrophobic and less than predicted for the more hydrophilic compounds, possibly due to the preference of the hydrophobic micelle interior for hydrophobic compounds. Empirical relationships were developed between multicomponent NAPLs and the biosurfactant. Correlations would be incorporated into transport models after validation by soil column experiments.

The effect of rhamnolipids on the partitioning of the PAHs, naphthalene, fluorene, phenanthrene and pyrene from NAPLs was examined [71]. Enhanced partitioning of the PAHs, occurred even with humic acid–smectite clay complexes, with the exception of naphthalene. The rhamnolipids sorbed onto the solids increasing the amount of solid phase PAHs. These solids are typical of those found in the subsurface and thus indicate the potential for enhancement of the in situ partitioning of PAHs. The equation $dC/dt = k(C_{eq} - C)$ was used where C_{eq} is the equilibrium aqueous phase concentration and C the PAH in the aqueous and solid phases. At biosurfactant concentrations above the CMC, k values were lower since there was competition for the PAH between the sorbed biosurfactants and micelles.

A rhamnolipid produced by *P. putida* was evaluated for the desorption of phenanthrene from clay-loam soil. The biosurfactant (250 mg/L) improved desorption of the PAH substantially following a linear model and therefore a linear k (k_l) was determined [72]. Desorption with water (reference) gave a $k_{l,ref} = 139$ ml/g and with the biosurfactant, it was $k_l = 268$ ml/g. Therefore, there was close to a twofold improvement of the availability enhancement factor (AEF) $= k_l/k_{l,ref} = 1.93$ by the biosurfactant.

Studies by Cheng et al. [73] showed that solubilization of the PAHs (phenanthrene and pyrene) increased linearly as the concentration of the biosurfactant produced by the *P. aeruginosa* P-CG3 strain increased (Fig. 5) and from the strain ATCC 9027. The weight of phenanthrene and pyrene solubilized per gram of rhamnolipid was approximately 0.0025 and 0.0068, respectively, which is much less that shown earlier for styrene (Fig. 3). The size of the micelles increases as the amount of solubilized hydrocarbon increases [71]. Concentrations of the biosurfactant above the CMC in the aqueous phase increased desorption from the soil phase.

The capability of a rhamnolipid was evaluated for its ability to remove another organic contaminant, pentachlorophenol (PCP), from soil by foam production [2]. The stability of the rhamnolipid foam was excellent and the quality was 99%. When the foam was injected into contaminated soil with a 1000 mg/kg level of PCP, 60 and 61% of the PCP could be removed from a fine sand and sandy-silt soil, respectively. The advantage of the foam is

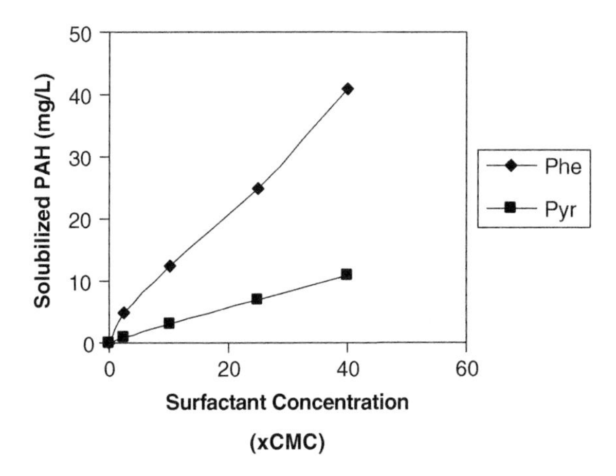

Fig. 5. Solubilization of phenanthrene(phe) and pyrene(pyr) at 55°C by a biosurfactant produced by the *P. aeruginosa* P-CG3 strain (adapted from Cheng et al. [73]).

that the fluid can be injected into the soil at low pressures which will avoid problems like soil heaving. The high-quality foams, such as those generated in this study, contain large amounts of air and thus large bubbles with thin liquid films. The foam then collapses easily and is thus less resistant to the soil upon introduction, resulting in lower soil pressures.

5.2. Heavy Metal Removal

Metal removal from a sandy soil contaminated with 1710 mg/kg of Cd and 2010 mg/kg of Ni was later evaluated [3]. Maximum removal was obtained by foam produced by 0.5% rhamnolipid solution, after 20 pore volumes. Removal efficiency for the biosurfactant foam was 73.2% of Cd and 68.1% of Ni. For the biosurfactant liquid solution, 61.7% Cd and 51.0% Ni were removed. This was superior to Triton X-100 foam which removed 64.7% Cd and 57.3% Ni and liquid Triton X-100 which removed 52.8% Cd and 45.2% Ni. Distilled water removed only 18% of both Cd and Ni. Concentrations of 0.5, 1.0 and 1.5% rhamnolipid at pH values of 6.8, 8 and 10 were also evaluated but did not show significant effects. For a 90% foam quality, the average hydraulic conductivity was 4.1×10^{-4} cm/s, for 95%, it was 1.5×10^{-4} cm/s and for 99%, it was 2.9×10^{-3} cm/s. Increasing foam quality decreases substantially the hydraulic conductivity. All these values are lower than the conductivity of water at 0.02 cm/s. This higher viscosity will allow better control of the surfactant mobility during in situ use. Therefore, rhamnolipid foam may be an effective and non-toxic method of remediating heavy metal, hydrocarbon or mixed contaminated soils. Further efforts will be required to enable its use at field scale.

Solutions to heavy metal-contaminated sediment require the understanding of the availability of heavy metals, interaction of the contaminants with soil and sediment particles, and metal retention mechanisms, which are complicated phenomena. The objective of the investigation by Dahrazma and Mulligan [74] was to study the removal of heavy metals (copper, zinc and nickel) from sediments by employing a biosurfactant rhamnolipid in continuous flow tests in a column. This type of setup since the configuration simulates the process of soil flushing. In the case of heavy metals, the flowing washing agent appeared to reduce the possibility of readsorption of the metals on to the soil and sediment. The present study shows that in the short term, the column tests can improve the removal of copper up to three times more than batch tests. In addition, since the distribution of heavy meals between soil and solute is the key to evaluating the environmental impact of the metals, in determining the mobility of metals and recommending reliable removal techniques, selective sequential extraction tests were used to determine the fraction of sediment from which the metal was removed. Exchangeable, carbonate, reducible oxide and organic fractions responded to washing techniques while residually bound contaminants are not economical or feasible to remove. This information is vital in proposing the most appropriate conditions for sediment washing.

6. MICELLAR ENHANCED ULTRAFILTRATION OF CONTAMINATED WATER

Using a technique called micellar enhanced ultrafiltration, Mulligan [61] studied the removal of various concentrations of metals from water by various concentrations of rhamnolipid by a 10000 Da molecular weight cutoff ultrafiltration membrane. Cadmium and zinc rejection ratios were optimal for 0.1% rhamnolipid at pH 6.7 (90% for copper, 85% for cadmium and 100% for zinc) for initial metal concentrations of 10, 20 and 10 mg/L respectively. For concentrations of 1% rhamnolipid, almost 100% rejection ratios were obtained. The addition of 0.4% oil as a co-contaminant decreased slightly the retention of the metals by the membrane. The ultrafiltration membranes also indicated that metals became associated with the micelles as the metals remained in the retentate and did not pass through the permeate. Assuming an average molecular weight of the rhamnolipids of approximately 500 Da, the micelles are composed of at least 20 monomers. Previous work had shown that the 10000 molecular weight cutoff membrane was more effective than the 30000 molecular cutoff membrane [75]. In summary, rhamnolipids were effective for hydrocarbon and heavy metal removal. Studies have not been performed at large scale, however.

7. CONCLUSIONS

Rhamnolipids have shown their potential for remediation of contaminated soil and water. Both organic and inorganic contaminants can be treated through desorption or biodegradation processes. The biosurfactants seem to enhance biodegradation by influencing the bioavailability of the contaminant through solubilization and affecting the surface properties of the bacterial cells. Due to their biodegradability and low toxicity biosurfactants such as rhamnolipids are very promising for use in remediation technologies. In addition, there is the potential for in situ production, a distinct advantage over synthetic surfactants. However, further research regarding prediction of their behaviour in the fate and transport of contaminants will be required. More investigation into the solubilization mechanism of both hydrocarbons and heavy metals by rhamnolipids is needed to enable model predictions for transport and remediation.

REFERENCES

[1] K. Tsujii, Surface Activity, Principles, Phenomena, and Applications, Academic Press, San Diego, CA, 1998.
[2] C.N. Mulligan and F. Eftekhari, Eng. Geol., 70 (2003) 269.
[3] C.N. Mulligan and S. Wang, Eng. Geol., 85 (2006) 75.
[4] K. Urum, T. Pekdemir and M. Gopur, Trans. Inst. Chem. Eng., 81B (2003) 203.
[5] S.C. Lin, J. Chem. Technol. Biotechnol., 63 (1996) 109.
[6] M. Biermann, F. Lange, R. Piorr, U. Ploog, H. Rutzen, J. Schindler and R. Schmidt, in "Surfactants in Consumer Products, Theory, Technology and Application" (J. Falbe, ed.), Springer-Verlag, Heidelberg, 1987.
[7] C.N. Mulligan, Environ Pollut., 133 (2005) 183.
[8] S. Lang and F. Wagner, in "Biosurfactants and Biotechnology" (N. Kosaric, W.L. Cairns and N.C.C. Gray, eds.), p. 21, Marcel Dekker, New York, 1987.
[9] D.G. Cooper, Microbiol. Sci., 3 (1986) 145.
[10] M.E. Hayes, E. Nestau and K.R. Hrebenar, Chemtech., 16 (1986) 239.
[11] Anonymous, Chemical Week, 135 (1984) January, 58.
[12] E.Z. Ron and E. Rosenberg, Curr. Opin. Biotechnol., 13 (2002) 249.
[13] K. Hitsatsuka, T. Nakahara, N. Sano and K. Yamada, Agric. Biol. Chem., 35 (1971) 686.

[14] L.H. Guerra-Santos, O. Käppeli and A. Feichter, Appl. Environ. Microbiol., 48 (1984) 301.
[15] C.N. Mulligan, R.N. Yong and B.F. Gibbs, Environ. Prog., 18 (1999) 50.
[16] A. Abalos, A. Pinaso, M.R. Infante, M. Casals, F. Garcia and M.A. Manresa, Langmuir, 17 (2001) 1367.
[17] M. Robert, M.E. Mercadé, M.P. Bosch, J.L. Parra, M.J. Espiny, M.A. Manresa and J. Guinea, Biotechnol. Lett., 11 (1989) 87.
[18] C.N. Mulligan and B.F. Gibbs, in "Biosurfactants, Production, Properties, Applications" (N. Kosaric, ed.), pp. 329–371, Marcel Dekker, New York, 1993.
[19] R.M. Maier and G. Soberon-Chavez, Appl. Microbiol. Biotechnol., 54 (2000) 625.
[20] G.S. Shreve, S. Inguva and S. Gunnan, Mol. Mar. Biol. Biotechnol., 4 (1995) 331.
[21] Y. Zhang and R.M. Miller, Appl. Environ. Microbiol., 58 (1992) 3276.
[22] R. Beal and W.B. Betts, J. Appl. Microbiol., 89 (2000) 158.
[23] A.M. Chakrabarty, Trends Biotechnol., 3 (1996) 32.
[24] M. Shafeeq, D. Kokub, Z.M. Khalid, A.M. Khan and K.A. Malik, MIRCEN J. Appl. Microbiol. Biotechnol., 5 (1989) 505.
[25] S. Chhatre, H. Purohit, R. Shanker and P. Khanna, Water Sci. Technol., 34 (1996) 187.
[26] A. Abalos, M. Vinas, J. Sabate, M.A. Manresa and A.M. Solanas, Biodegradation, 15 (2004) 249.
[27] K.S.M. Rahman, I.M. Banat, T.J. Rahman, T. Thayumanavan and P. Lakshmanaperumalsamy, Bioresource Technol., 81 (2002) 25.
[28] G. Burd and O.P. Ward, Biotechnol. Tech., 10 (1996) 371.
[29] M. Garcia-Junco, E. De Olmedo and J.-J. Ortega-Calvo, Environ. Microbiol., 3 (2001) 561.
[30] L. Holakoo and C.N. Mulligan, Proceedings of the Annual Conference of Canadian Society for Civil Engineering, June 5–8, Montreal, Canada, 2002.
[31] M. Dagnew, Rhamnolipid assisted biodegradation of crude oil spilled on water, Concordia University, M.A.Sc. Thesis, Montreal, Canada, 2004.
[32] J.C. Mata-Sandoval, J. Karns and A. Torrents, Environ. Sci. Technol., 34 (2000) 4923.
[33] P. Maslin and R.M. Maier, Biorem. J., 4 (2000) 295.
[34] R.J. Fiocco and A. Lewis, Pure Appl. Chem., 71 (1999) 27.
[35] C. Vipulanandan and X. Ren, J. Environ. Eng., 126 (2000) 629.
[36] L. Deschênes, P. Lafrance, J.-P. Villeneuve and R. Samson, Fourth Annual Symposium on Groundwater and Soil Remediation, September 21–23, Calgary, Alberta, 1994.
[37] M.A. Providenti, C.A. Fleming, H. Lee and J.T. Trevors, FEMS Microbiol. Ecol., 17 (1995) 15.
[38] W.H. Noordman, J.J.J. Wachter, G.J. de Boer and D.B. Janssen, J. Biotechnol., 94 (2002) 195.
[39] S.M. Dean, Y. Jin, D.K. Cha, S.V. Wilson and M. Radosevich, J Environ. Qual., 30 (2001) 1126.
[40] S. Harvey, I. Elashi, J.J. Valdes, D. Kamely and A.M. Chakrabarty, Biotechnology, 8 (1990) 228.
[41] K. Scheibenbogen, R.G. Zytner, H. Lee and J.T. Trevors, J. Chem. Technol. Biotechnol., 59 (1994) 53.

[42] M.I. Van Dyke, P. Couture, M. Brauer, H. Lee and J.T. Trevors, Can. J. Microbiol., 39 (1993) 1071.
[43] P. Lafrance and M. Lapointe, Ground Water Monit. Remed., 18 (1998) 139.
[44] R.S. Makkar and K.J. Rockne, Environ. Toxicol. Chem., 22 (2003) 2280.
[45] G. Berg, A. Seech, H. Lee and J. Trevors, J. Environ. Sci. Health, 7 (1990) 753.
[46] W. Noordman, N. Brusseau and D. Janssen, Environ. Sci. Technol., 31 (1998) 2211.
[47] W.H. Noordman, M.L. Brusseau and D.B. Janssen, Environ. Sci. Technol., 34 (2000) 832.
[48] W.L. Straube, C.C. Nestler, L.D. Hansen, D. Ringleberg, P.J. Pritchard and J.H. Jones-Meehan, Acta Biotechnol., 2–3 (2003) 179.
[49] Y. Zhang, W. Maier and R. Miller, Environ. Sci. Technol., 31 (1997) 3276.
[50] T.R. Sandrin, A.M. Chech and R.M. Maier, Appl. Environ. Microbiol., 66 (2000) 4585.
[51] J.E. McCray, G. Bai, R.M. Maier and M.L Brusseau, J. Contam. Hydrol., 48 (2001) 45.
[52] G. Bai, M.L. Brusseau and R.M. Miller, J. Contam. Hydrol., 30 (1998) 265.
[53] K. Urum, T. Pekdemir and M. Gopur, Trans. Inst. Chem. Eng., 81B (2003) 203.
[54] G. Southam, M. Whitney and C. Knickerboker, Int. Biodeterior. Biodegrad., 47 (2001) 197.
[55] A. Shulga, E. Karpenko, R. Vildanova-Martishin, A. Turovsky and M. Soltys, Adsorp. Sci. Technol., 18 (2000) 171.
[56] Y. Guo and C.N. Mulligan, in Hazardous Materials in the Soil and Atmosphere: Treatment, Removal and Analysis (R.C. Hudson, ed.), pp. 1–37, Nova Publishers, Hauppage, NY, 2006.
[57] D.C. Herman, J.F. Artiola and R.M. Miller, Environ. Sci. Technol., 29 (1995) 2280.
[58] F. Ochoa-Loza, Physico-chemical factors affecting rhamnolipid biosurfactant application for removal of metal contaminants from soil, Ph.D. dissertation, University of Arizona, Tucson, 1998.
[59] H. Tan, J.T. Champion, J.F Artiola, M.L Brusseau and R.M. Miller, Environ. Sci. Technol., 28 (1994) 2402.
[60] F.J. Ochoa-Loza, J.E. Artiola and R.M. Maier, J. Environ. Qual., 30 (2001) 479.
[61] C.N. Mulligan, On the capability of biosurfactants for the removal of heavy metals from soil and sediments, Department of Civil Engineering and Applied Mechanics, Ph.D. Thesis, McGill University, Montreal, Canada, 1998.
[62] D.A. Edwards, Z. Liu and R.G. Luthy, J. Environ. Eng., 120 (1994) 5.
[63] C.N. Mulligan, R N. Yong and B.F. Gibbs, J. Hazard. Mater., 85 (2001) 111.
[64] L. Fraser, Environ. Health Perspect., 108 (2000) A320.
[65] J.W. Neilson, J.F. Artiola and R.M. Maier, J. Environ. Qual., 32 (2003) 899.
[66] B. Dahr Azma and C.N. Mulligan, Pract. Periodical Hazard. Toxic Radioact. Waste Manage., 8 (2004) 166.
[67] H. Massara, C.N. Mulligan and J. Hadjinicolaou, Soil Sediment Contam. Int. J., 16 (2007) 1–14.
[68] C.N. Mulligan, R.N. Yong and B.F. Gibbs, Environ. Sci. Technol., 33 (1999) 3812.
[69] P. Somasundaran and S. Krishnakumar, Colloids Surf. A, 123–124 (1997) 491.
[70] D.C. Herman, R.J. Lenhard and R.M. Miller, Environ. Sci. Technol., 31 (1997) 1290.

[71] M. Garcia-Junco, C. Gomez-Lahoz, J.-L. Niqui-Arroyo and J.-J. Ortego-Calvo, Environ. Sci. Technol., 37 (2003) 2988.
[72] H.M. Poggi-Varaldo and N. Rinderknecht-Seijas, Acta Biotechnol., 2–3 (2003) 271.
[73] K.Y. Cheng, Y. Zhao and W.C. Wong, Environ. Technol., 25 (2004) 1159.
[74] B. Dahrazma and C.N. Mulligan, ASTM STP, 1482 (2006) 200.
[75] C.N. Mulligan and B.F. Gibbs, J. Chem. Technol. Biotechnol., 47 (1990) 23.

Thermodynamics, Solubility and Environmental Issues
T.M. Letcher (editor)

Chapter 16

Sorption, Lipophilicity and Partitioning Phenomena of Ionic Liquids in Environmental Systems

Piotr Stepnowski

Faculty of Chemistry, University of Gdańsk, Gdańsk, Poland

1. INTRODUCTION

Nowadays, more than one hundred thousand chemicals are known to the international trade, and every year more than two thousand new chemicals are registered in Europe and North America [1]. Since the production and utilization of these substances almost inevitably lead to their presence in the environment, the management of their risk represents an international challenge [2]. Most industrially developed nations have established agencies that assess the risks posed by chemicals to human health and the environment. In the case of substances known to come into direct contact with humans or living organisms in general (in the form of food additives, drugs, cosmetics or pesticides), risk assessment measures are introduced and harmonized in national and international legislation systems. These usually involve a number of steps: hazard identification, effect or dose–response assessment, exposure assessment and risk characterization and classification [3]. In the case of new industrial chemicals, such as novel reaction media, solvents or electrolytes, only very rough evaluations of their acute toxicity are made prior to their introduction. That is why until now "safer chemical" has usually translated into "less toxic". However, in order to be able to predict the behavior of a chemical substance in environmental and biological systems, it is important to know their physical chemical properties, for example, their various partition and sorption coefficients, solubility in water, Henry's law constant, melting/boiling temperature, vapor pressure, bioconcentration factors and diffusion properties. Environmental chemistry is therefore an important aspect in the design of chemicals, since those that persist in the environment remain available to exert toxic effects for longer periods of time than those which do not, and they may bioaccumulate [4]. The incorporation of prospective environmental chemical assessments in the design of new industrial chemicals is thus of the utmost importance. This is very well put in the 10th principle of Green Chemistry, which states that "any chemicals should be designed so that at the end of their function they do not persist in the environment and break down into innocuous degradation products" [5].

2. IONIC LIQUIDS

Ionic liquids are a good example of the industrial chemicals of the future. They consist of an organic cation, such as alkylimidazolium, alkylpyridinium, alkylphosphonium or alkylammonium, and an organic or inorganic anion, such as tetrafluoroborate, hexafluorophosphate, tosylate or bis(trifluoromethylsulfonyl)imide. Table 1 presents some common examples of ionic

Table 1
Common examples of ionic liquids

Ionic liquid type	Structure	Ionic liquid type	Structure
Alkylammonium	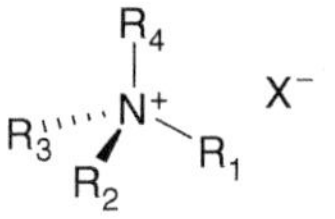	Alkylpyridinium	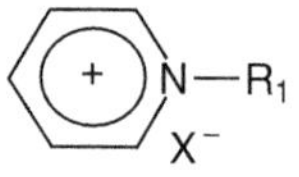
Alkylphosphonium	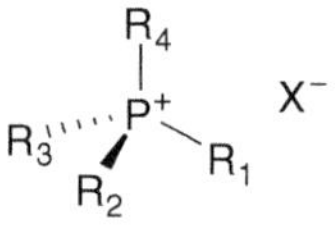	Alkylpyrolidinium	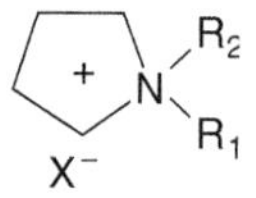
Alkylsulfonium	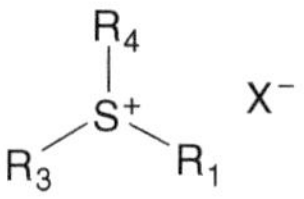	Alkyltioazolium	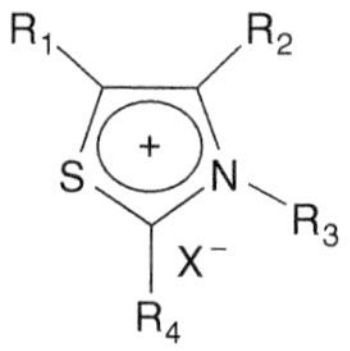
Alkylimidazolium	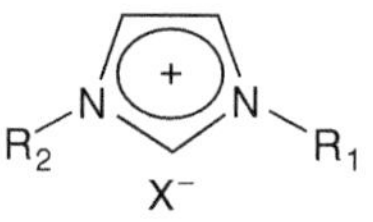	Alkyltriazolium	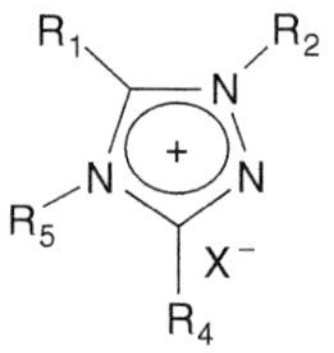

X^-: BF_4^-, PF_6^-, $AlCl_4^-$, SbF_6^-, $CF_3SO_3^-$, $(CF_3SO_2)_2N^-$, etc. R_x: From $-CH_3$ to $-C_9H_{19}$.

liquids. Their liquidity at room temperature is due mainly to the cation's high degree of asymmetry, which inhibits crystallization at low temperatures, whereas the anion is usually characterized by considerable delocalization of the electron cloud, which tends to decrease interionic interaction. Ionic liquids are thermally and water stable, liquid at room temperature (more precisely below 100°C), have a non-measurable vapor pressure and are able to solvate a variety of organic and inorganic species. Moreover, the vast numbers of possible cation and anion combinations possess widely tunable properties with regard to polarity, hydrophobicity and solvent miscibility. With the great variety of such combinations enabling the fine-tuning of their chemical properties, ionic liquids have already become recognized by the chemical industry as new, target-oriented reaction media. Nowadays, ionic liquids are emerging as alternative green solvents, in other words, as alternative reaction media for synthesis, catalysis and biocatalysis [6–10].

The "green" aspect of ionic liquids derives mainly from their practically undetectable vapor pressure, but this is obviously insufficient justification for calling a technology "cleaner". Since so many ion combinations are possible, it is of the greatest importance to outline rational guidelines for developing ionic liquids that are both technologically suitable and environmentally innocuous [11–12]. Once their large-scale implementation has begun, it will not be long before ionic liquids become a permanent component of industrial effluents. In view of their great stability, they could slip through classical treatment systems to become persistent components of the environment, where the long-term consequences of their presence are still unknown. The current review by Jastorff et al. [12] summarizes the studies of ionic liquid toxicology done so far. The "alkyl side chain length effect" with a multitude of cation and anion combinations has been compared for several biological test systems. In every case it was found that the shorter the alkyl side chain length, the lower the cytotoxicity. It is therefore the side chain length, i.e., the lipophilicity, of the cation that seems to be one of the major determinants of the overall observed biological effect.

3. SORPTION OF IONIC LIQUIDS IN THE ENVIRONMENT

The extent to which a compound is sorbed on to soil and sediment particles is extremely important, because it may dramatically affect its fate in and its impact on the environment. Transport of the compound in porous media, such as soils, sediments and aquifers, is strongly influenced by its tendency to sorb on to the various components of the solid matrix. Furthermore, since molecular transfer is a prerequisite for the uptake of organic pollutants by organisms, the bioavailability of a given compound, and thus its rate of biotransformation, is affected by sorption as well [13]. Therefore, if we are to fully understand the behavior of ionic liquids in the environment, it is of prime interest to analyze the possible interaction sites available as a result of the chemical structure of these entities. Several molecular interactions are possible between ionic liquids and the environmental compartments. Table 2 summarizes them, using the imidazolium cation as an example. The predominant sorption mechanism of ionic liquids in the soil and/or sediment matrix appears to be ion exchange. The high electron acceptor potential of delocalized aromatic systems in the cationic compartments of ionic liquids will be responsible for electrostatic attractions with polar moieties on the particle surfaces, such as oxides, oxyhydroxides, aluminosilicates or clay minerals, but also with the ionized carboxylic groups in humic and fulvic acids. It is also clear that elongating the alkyl chain of ionic liquids will result in the molecule's elevated hydrophobicity. It is therefore also possible that, with greater loadings, ionic liquids will be retained in the soil by both ion exchange and hydrophobic interactions. The latter can occur between the ionic liquid's alkyl chain and soil organic matter, as well as with other alkyl chains of ionic liquid cations previously bound to the soil surface. This mechanism, moreover, will be strongly dependent on the anions present in the aquifer, with which selected cations will form ion-pairs, thereby enabling the molecule to be partitioned.

Recently, Gorman-Lewis and Fein [15] measured the adsorption of the 1-butyl-3-methylimidazolium chloride ionic liquid on to a range of surfaces designed to represent those commonly found in the near-surface environment. Ionic liquids were also found to interact with the mineral montmorillonite; this is attributed solely to electrostatic interaction between the cation and the outer and interlayer surface of clay.

In a recent contribution of ours, the sorption of selected imidazolium ionic liquid cations with different alkyl side chain lengths (from C_3 to C_6) was determined and compared [16]. This enabled us to follow the changes in sorption behavior due only to the variation in lipophilicity of the cations under scrutiny. Three types of soils differing in their physical and chemical properties were chosen for this experiment. A batch-equilibrium technique was employed to determine the sorption of the ionic liquid cations. The experiments were performed according to OECD guidelines [17]. The determinations of the residual concentrations of ionic liquids were performed using HPLC methods developed by our group [18–21].

Table 2
Bonding types and energies of interaction between ionic liquid cations and certain environmental compartments

Kind of interaction	Energy (kJ/mol)	Possible environmental compartment	Example system
Strong ionic binding	−40	Soils and sediments (e.g., humic and fulvic acids)	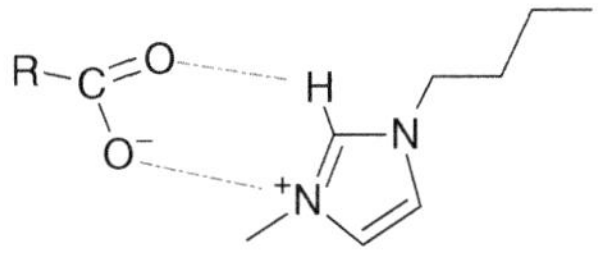
Ionic binding	−20	Saline- and freshwaters (e.g., Cl^-, HCO_3^-), mineral surfaces of suspensions and soils (e.g., oxides, silanols)	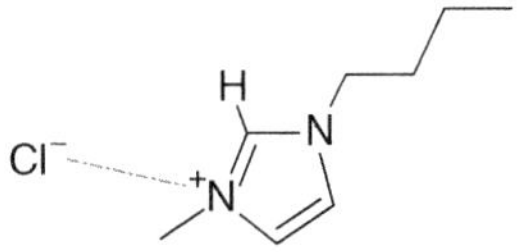
Ion–dipole interaction	−(4–17)	Reductive sediments (e.g., amines, proteins, humic acids)	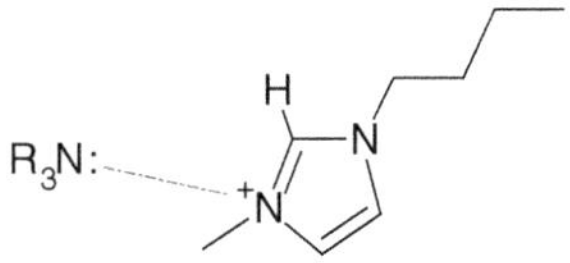
Dipole–dipole interaction	−(4–17)	Bioactive surfaces (e.g., lignins, cutins)	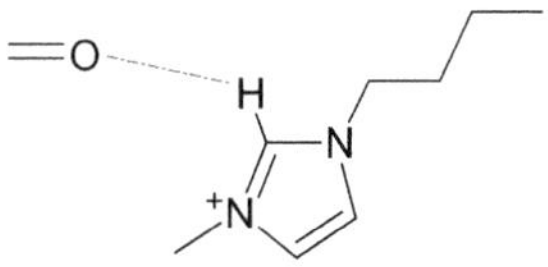
Hydrogen binding interaction	−(4–17)	Biogenic organic matter (e.g., unsaturated fatty acids, waxes)	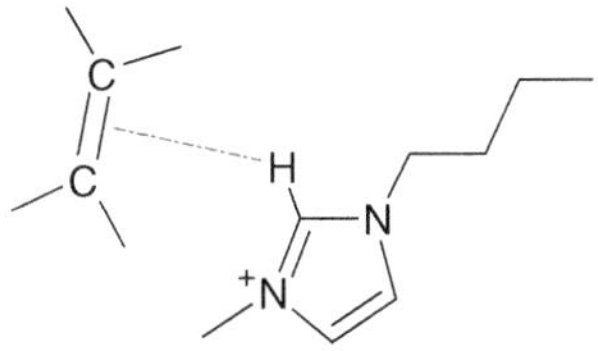
Hydrophobic interaction	−4	Incomplete degradation matter (e.g., black carbon)	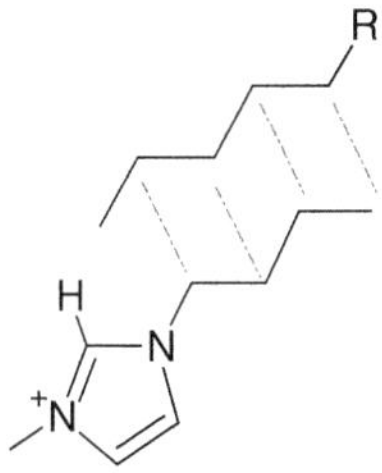

Note: Concept adapted from ref. [14].

The results showed that all ionic liquid cations were strongly bound to all the soils studied. As shown in Fig. 1, the sorption of the 1-alkyl-3-methylimidazolium species decreased in the order $C_6MIM > A\ C_5MIM > C_4MIM > C_3MIM$, i.e., with decreasing lipophilicity. The sorption coefficients of the most strongly sorbed ionic liquid – the 1-hexyl-3-methylimidazolium entity – obtained for the three soils at the initial concentration of 1 mM were 226, 81 and 24 ml g^{-1} for agricultural, clayey and peaty soils respectively. We further fitted experimental data from agricultural and clayey soils to the non-linear Freundlich isotherm (Fig. 2), expressed by the following equation:

$$S = K_f \times C^{1/n}$$

where S is the adsorbed quantity per unit mass of sorbents, K_f is known as the Freundlich coefficient, C the equilibrium concentration and $1/n$ the Freundlich exponent describing the isothermal slope. Many studies have demonstrated that this type of isotherm supports the assumption of a heterogeneous surface, which fairly accurately characterizes the sorption to the soil. Table 3 sets out the isotherm parameters obtained. In the case of the shorter alkyl chain entities (C_3–C_5) the slopes of the isotherms ($1/n$) are less than one. This means that the isotherm is concave downward, which probably indicates that added sorbates are bound by ever weaker free energies. In the case of the longest alkyl chain ionic liquid (C_6), the isothermal slope is greater than unity, which indicates that the isotherm is convex upward, from which it may be inferred that the greater sorbate presence in the sorbent enhances the free energies of further sorption. Theoretically, it is likely that ionic liquids with longer alkyl chains will substantially modify the available soil surface, enhancing the sorptive capabilities of the soil. This effect is illustrated schematically in Fig. 3. In our study, referred to above [16], we also determined the desorption of sorbed ionic liquids with 0.01 M calcium chloride solutions after a 24-h dynamic extraction. Desorption from the soils under investigation decreased with increasing alkyl chain length. In practice, therefore, compounds such as C_6MIM or C_5MIM become irreversibly bound to the soil component.

Apart from sorption to the soil component, interaction between the compound and the sediment is equally important if we are to understand the behavior and fate of ionic liquids in aquatic systems. It is likely that both the soil and the aquatic environment will be recipients of ionic liquid contamination. We ran one preliminary experiment with marine sediment collected in Puck Bay (southern Baltic Sea) [16]. The sediments from this part

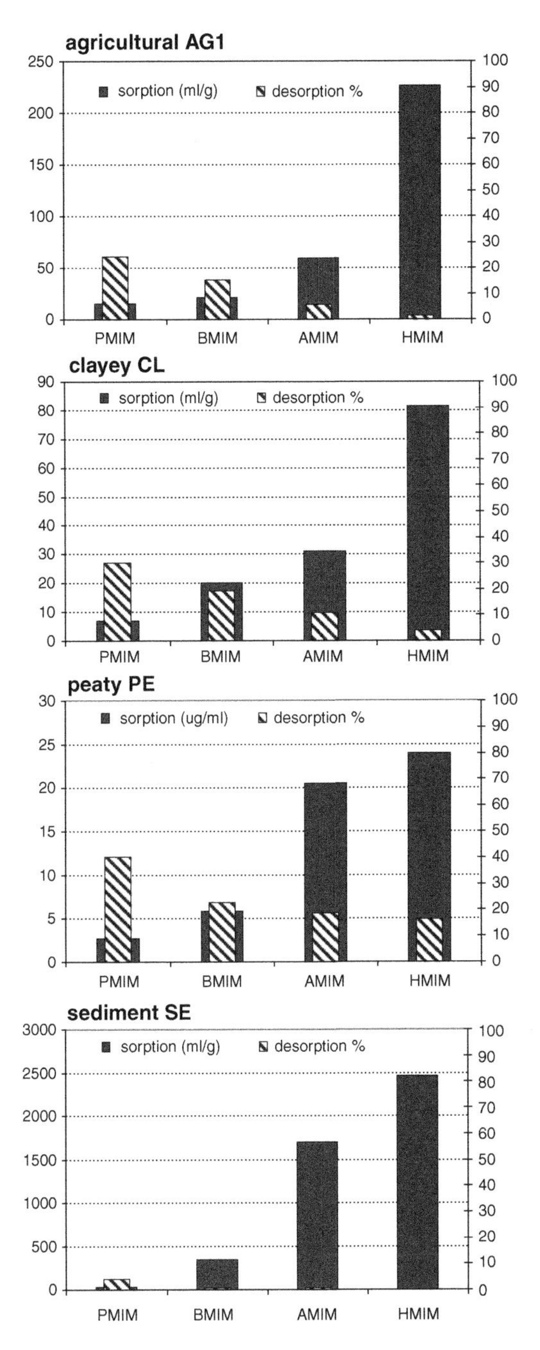

Fig. 1. Sorption coefficients at equilibrium K_d (left y-axis) obtained for three soils at an initial concentration of 1 mM and desorption characteristics (right y-axis) of ionic liquids. AG1, agricultural soil; CL, clayey soil; PE, peaty soil; SE, marine sediment. Values are the means of two determinations. Reprinted from ref. [12] with the permission of CSIRO PUBLISHING (http://www.publish.csiro.au/journals/ajc).

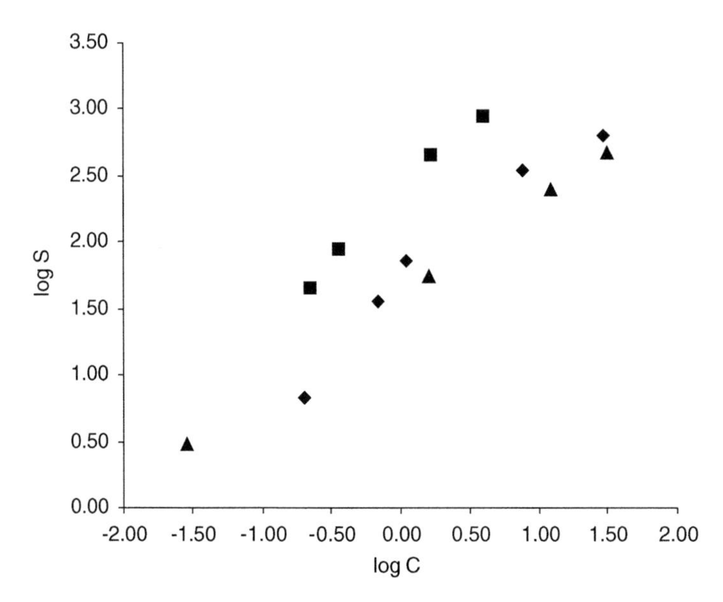

Fig. 2. Examples of Freundlich isotherms of ionic liquid sorption on agricultural soil (AG1). (◆), 1-butyl-3-methylimidazolium; (■), 1-hexyl-3-methylimidazolium and (▲), 1-butyl-3-ethylimidazolium entities. Values are the means of two determinations. Reprinted from ref. [12] with the permission of CSIRO PUBLISHING (http://www.publish.csiro.au/journals/ajc).

Table 3
Freundlich isotherm coefficients determined by linear regression of data for ionic liquid sorption on agricultural soil

Ionic liquid	Freundlich isotherm		
	K_f	$1/n$	R^2
1-propyl-3-methylimidazolium	1.59	0.72	0.99
1-butyl-3-methylimidazolium	1.63	0.89	0.96
1-amyl-3-methylimidazolium	2.16	0.56	0.98
1-hexyl-3-methylimidazolium	2.37	1.03	0.99

of the bay are a greenish-gray sapropel, a mixed mineral suite, which fits the model of "swelling clays" that can dominate the sorption of cationic compounds in sediments. We found that all the ionic liquids were strongly and practically irreversibly sorbed on to the sediment from the southern Baltic Sea, and the capacity factors obtained were one to two orders of magnitude higher than those obtained for soils.

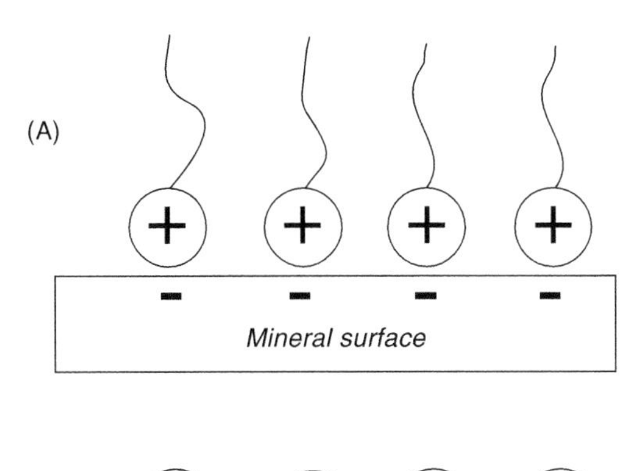

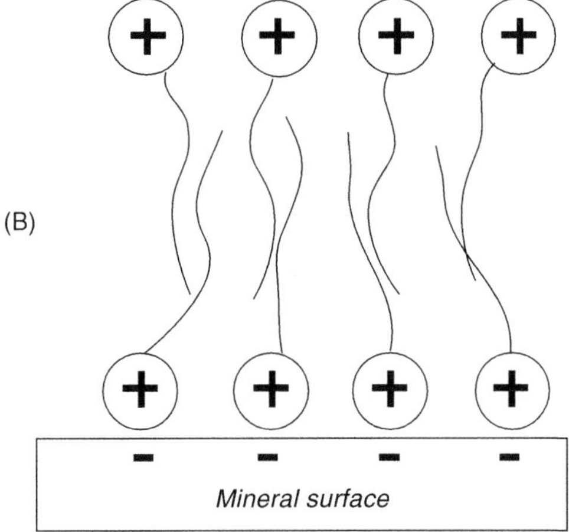

Fig. 3. Hypothetical types of sorption of ionic liquid cations in natural soils. (A) Monolayer of short chain cations; (B) admicelle of longer chain cations on a mineral surface.

Though only a preliminary study, this work outlined the general principles of ionic liquid cation sorption on soil components. Several questions are now open, which should be resolved by future research. Those requiring the most urgent attention are the effect of pH, ionic strength and temperature, and also biodegradability potential, on the sorption and desorption of these compounds. Future studies should also use a larger pool of compounds, including the very common pyridinium and phosphonium liquid derivatives.

4. LIPOPHILICITY AND THE PARTITIONING OF IONIC LIQUIDS

As outlined in the previous paragraphs, the most important factor contributing to the sorption, desorption, toxicity and bioaccumulation of ionic liquids is the length of the alkyl chain in the cationic structure. Though different in the detailed mode of action, these effects are attributed to changes in the lipophilicity of the compound. The reference lipophilicity scale is defined by the logarithm of the partition coefficient determined in the 1-octanol:water partition system (log K_{ow}). The coefficient defining the equilibrium partitioning of a chemical is nowadays a well-recognized parameter in environmental chemistry for its usefulness in predicting the tendency of a substance to interact with geochemical and biological surfaces. Therefore, any prospective risk analysis of ionic liquids should include data on lipophilicity, a key parameter in hazard assessment, which provides an understanding of their potential to cause adverse effects in all environmental compartments.

In a recent study, Domańska et al. [22] were the first to calculate the 1-octanol:water partition coefficients of 1-alkyl-3-methylimidazolium chlorides. The experimental partition coefficients (log K_{ow}) for a homologous series of four compounds differing in their alkyl chain lengths were found to be negative at room temperature and ranged from −0.31 to −0.14. In another study, Ropel et al. [23] measured the partition coefficients of 12 imidazolium-based ionic liquids at room temperature, using the slow-stirring method. For the butylmethylimidazolium cation, values of log K_{ow} ranged from −2.52 to 1.04, depending on the choice of anion. It seems, however, that the results obtained by these procedures are hard to extrapolate to the real environmental situation, where it is likely that partitioning of the cation to organic phases will occur together with the anions already present in the environment, and not with those with which it entered the environment. The possible pathway of ionic liquids entering the environment is graphically sketched in Fig. 4.

In another recent study we therefore chose the chromatographic methodology, since it yields average values corresponding to cation partition regardless of the anion present in the native structure of the ionic liquid [24]. As a result, this somewhat simplified technique minimizes uncertainty in lipophilicity measurements and also eliminates the influence of the anionic part on the overall solubility of the organic cation; the anion effect is averaged by the use of a high-concentration buffer system in the mobile phase. The relative lipophilicity of a series of 1-alkyl-3-methylimidazolium salts (alkyl chain from C_3 to C_6) was measured on a high performance liquid chromatography setup equipped with an octyl-bonded silica reversed phase (RP) and an immobilized artificial membrane (IAM) phase that is believed to closely mimic the surface of a biological membrane (Fig. 5). We also examined the lipophilicity question with the use of theoretical calculations, using a classical

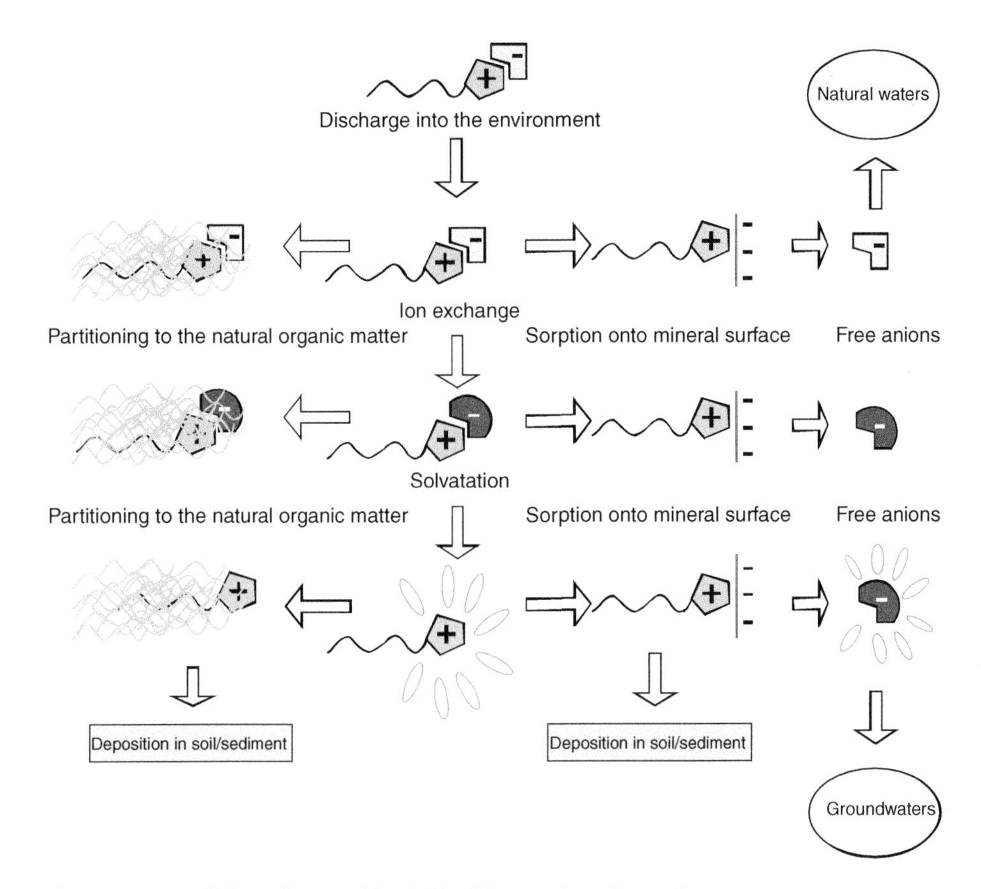

Fig. 4. The possible pathway of ionic liquids entering the environment.

fragmental approach modified by manual calculations of the geometric bond factor for quaternary ammonium and of the electronic bond factor.

The experimentally measured lipophilicity coefficients obtained for all the ionic liquids studied generally indicated a relatively low value of this parameter, which implied their preferential partitioning to the aqueous phase. Values determined on the reversed phase (log K_{RP}) ranged from -0.39 to 1.23 for 1-propyl- and 1-hexyl-3-methylimidazolium salts respectively, whereas the retention-derived lipophilicities on the IAM (log K_{IAM}) were substantially higher, varying from 0.92 to 1.7 for this set of compounds. The higher values were the net result of several interactions occurring during partitioning to this phase. It is clear that not only hydrophobic bonding but also interactions of positively charged nitrogen with the carbonyl and phosphate

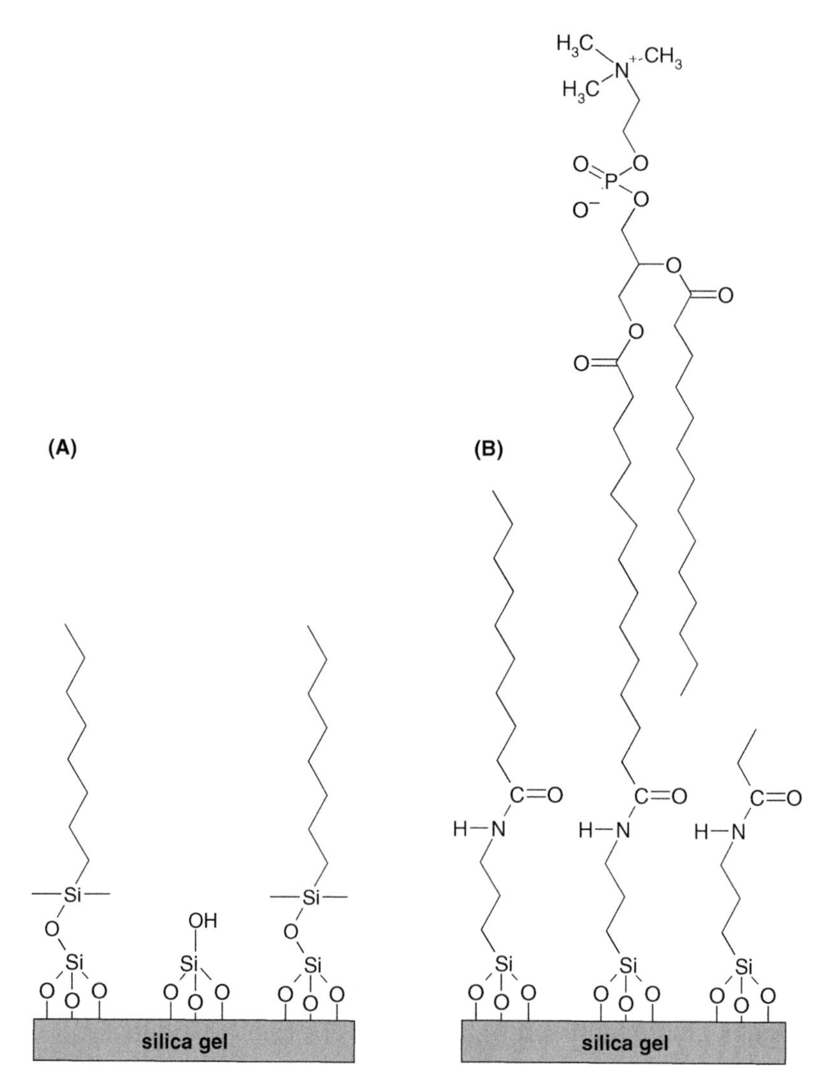

Fig. 5. Chemical structures of stationary phases used to determine relative lipophilicity. (A) Reversed (*n*-octyl) stationary phase; (B) immobilized artificial membrane stationary phase.

groups present in the IAM stationary phase occurred here. The results suggest that ionic liquids, so far considered to be highly polar entities, can also interact with biological membranes.

In the theoretical calculation of lipophilicity (log P) we used the procedure proposed by Hansch and Leo [25]. This combined the fragmentary value for all carbons, hydrogens and quaternary nitrogens, the geometric bond factor and the electronic bond factor describing electronic portion of the bond factor extending through a chain of up to five alkane carbons [24].

The square correlation coefficient of the linear relationship between the theoretically predicted log P and chromatographic log K_{RP} values is 0.996, whereas the correlation between log P and log K_{IAM} is much less significant ($R^2 = 0.936$). Since the calculation method includes only the partial hydrophobicities of the cation molecule, it is likely that in the case of reversed phase chromatography the molecular mechanisms of the ionic liquid separation are outweighed by hydrophobic interactions. In the case of IAM chromatography it is clear that the weak correlation is due to more than the hydrophobic interactions taking place when ionic liquid cations are partitioning into this phase. The result shows that when approaching the biological membrane, the ionic liquid cation may not only partition with its alkyl tail into the hydrophobic part of the phospholipid bilayer but also become permanently built into the membrane structure via additional ion exchange and ion–dipole interactions.

5. CONCLUSIONS

On entering the environment, anions and cations of ionic liquids will be substantially solvated and to a large extent separated. Furthermore, these anions and cations will interact in various ways with different environmental compartments. Ion exchange will be the preferred type of bonding for cations, though at elevated concentrations, and for compounds with longer alkyl chains, hydrophobic interactions will also be of importance. The extent to which any particular soil type adsorbs ionic liquids will therefore depend not only on the nature and properties of the soil but also on the type of cation. Sorption of ionic liquids on to soils will increase with increasing alkyl side chain length (lipophilicity) of the ionic liquid cation. Stronger interactions of the more lipophilic ionic liquids on soils are the net effect of bonding between the ionic liquid alkyl chain and soil organic matter, as well as with other alkyl chains of ionic liquid cations already bound to the soil surface. It is also understood that sorption on to fine-textured mineral sediments will be extremely strong and practically irreversible, so that any remobilization of these compounds in marine environments with such sediments is rather unlikely.

The importance of the lipophilicity parameter in the environmental distribution of ionic liquids induced us to study further the partitioning of these entities in model systems. The experimentally measured and theoretically estimated lipophilicity obtained for alkylimidazolium ionic liquids generally indicate a relatively low partitioning availability, and hence a preference for the aqueous phase. This can only be interpreted in the light of the well solvated, dissociated cation of the ionic liquid, and not the ionic liquid in its native, ion-pair form. Moreover, the data obtained on an IAM does not exclude possible interactions with biological and environmental barriers.

ACKNOWLEDGMENT

Financial support was provided by the Polish Ministry of Research and Higher Education under grants 2PO4G 083 29, 2P04G 118 29 and DS 8390-4-0141-6.

REFERENCES

[1] P. Calow, Controlling Environmental Risks from Chemicals, Principles and Practice, Wiley, New York, 1997.
[2] G. Vollmer, L. Giannoni, B. Sokull-Kluüttgen and W. Karcher (eds.), Risk Assessment – Theory and Practice, ECSC-EC, Luxemburg, 1996.
[3] C.J. van Leeuwen and J.L.M. Hermens (eds.), Risk Assessment of Chemicals, An Introduction, Kluwer, Dordrecht, 1996.
[4] R.S. Boethling and D. Mackay, Handbook of Property Estimation Methods for Chemicals. Environmental and Health Sciences, Lewis Publishers, Boca Rota, FL, 2000.
[5] P.T. Anastas and J.C. Warner, Green Chemistry: Theory and Practice, Oxford University Press, New York, 1998.
[6] M.J. Earle and K.R. Seddon, Pure Appl. Chem., 72 (2000) 1391.
[7] P. Wasserscheid and W. Keim, Angew. Chem., 112 (2000) 3926.
[8] T. Welton, Chem. Rev., 99 (1999) 2071.
[9] J.D. Holbrey, K.R. Seddon and R. Wareing, Green Chem., 3 (2001) 33.
[10] C.M. Gordon, Appl. Catal. A, 222 (2001) 101.
[11] B. Jastorff, R. Störmann, J. Ranke, K. Mölter, F. Stock, B. Oberheitmann, W. Hoffman, J. Hoffmann, M. Nüchter, B. Ondruschka and J. Filser, Green Chem., 5 (2003) 136.
[12] B. Jastorff, K. Mölter, P. Behrend, U. Bottin-Weber, J. Filser, A. Heimers, B. Ondruschka, J. Ranke, M. Schaefer, H. Schröder, A. Stark, P. Stepnowski, F. Stock, R. Störmann, S. Stolte, U. Welz-Biermann, S. Ziegert and J. Thöming, Green Chem., 7 (2005) 362.
[13] R.P. Schwarzenbach, P.M. Gschwend, D.M. Imboden, Environmental Organic Chemistry, Wiley, New York, 1993.
[14] R. Kaliszan, Structure and Retention in Chromatography, Harwood Academic Publishers, Amsterdam, 1997.
[15] D.J. Gorman-Lewis and J.B. Fein, Environ. Sci. Technol., 38 (2004) 2491.
[16] P. Stepnowski, Aust. J. Chem., 58 (2005) 170.

[17] OECD, Adsorption–Desorption Using a Batch Equilibrium Method, p. 106, OECD, Paris, 2000.
[18] Stepnowski, A. Müller, P. Behrend, J. Ranke, J. Hoffmann and B. Jastorff, J. Chromatogr. A, 993 (2003) 173.
[19] P. Stepnowski and W. Mrozik, J. Sep. Sci., 28 (2005) 149.
[20] S. Kowalska, B. Buszewski and P. Stepnowski, J. Sep. Sci., 29 (2005) 1116.
[21] P. Stepnowski., Anal. Bioanal. Chem., 381 (2005) 189.
[22] U. Domańska, E. Bogel-Lukasik and R. Bogel-Lukasik, Chem. Eur. J., 9 (2003) 3033.
[23] L. Ropel, L.S. Belveze, S.N.V.K. Aki, M.A. Stadtherr and J.F. Brennecke, Green Chem., 7 (2005) 83.
[24] P. Stepnowski and P. Storoniak, Environ. Sci. Pollut. Res., 12 (2005) 199.
[25] C. Hansch and A. Leo, Exploring QSAR – Fundamentals and Applications in Chemistry and Biology, ACS, Washington DC, 1995.

Thermodynamics, Solubility and Environmental Issues
T.M. Letcher (editor)

Chapter 17

The Solubility of Hydroxyaluminosilicates and the Biological Availability of Aluminium

Christopher Exley

Birchall Centre for Inorganic Chemistry and Materials Science, Lennard-Jones Laboratories, Keele University, Staffordshire, UK

1. WHAT ARE HYDROXYALUMINOSILICATES?

Hydroxyaluminosilicates ($HAS_{(s)}$) are formed from the reaction of neutral monomeric silicic acid ($Si(OH)_{4(aq)}$) with aluminium hydroxide ($Al(OH)_{3(s)}$) [1]. $HAS_{(s)}$ are not formed via the hydrolysis/condensation of soluble aluminosilicates such as the putative $AlOSi(OH)_{3(aq)}{}^{2+}$ [2]. The evidence to support the formation of the latter soluble complex is exclusively potentiometric and thus has relied on the precise measurement of differences in $[H^+]$ in acidic solutions [3]. There is no direct experimental evidence to support the formation and existence of $AlOSi(OH)_{3(aq)}{}^{2+}$. The identification of this soluble aluminosilicate is not accessible through ^{27}Al NMR since it is not possible to discriminate between it and $Al(H_2O)_{6(aq)}{}^{3+}$. The confirmation of its identity might be obtained if it was formed in solutions which included ^{29}Si-enriched $Si(OH)_{4(aq)}$ and ^{29}Si NMR was used to either observe it directly as a shift which was distinguishable from $Si(OH)_{4(aq)}$ or indirectly as a quantifiable depletion in the $Si(OH)_{4(aq)}$ signal. In the absence of direct evidence of its existence one must rely on the aforementioned potentiometric data which have ascribed a weak binding constant to the complex and have predicted its dissolution at solution pH above ca. 4.00. It is this binding constant which has been used in myriad modelling exercises to discount the importance of $Si(OH)_{4(aq)}$ in aluminium solubility control in soil and surface waters [4]. However, there is a burgeoning recognition that $Si(OH)_{4(aq)}$ is a significant influence on the biological availability of aluminium and it has been our contention for more than 20 years that it exerts its influence through the formation of $HAS_{(s)}$ [5]. These are neither soluble aluminosilicates nor are they derived from soluble aluminosilicates per se; they are the result of the condensation of $Si(OH)_{4(aq)}$ at solid surfaces of $Al(OH)_{3(s)}$. This reaction is arguably the only environmentally significant inorganic chemistry of $Si(OH)_{4(aq)}$ though it is not yet deemed to be sufficiently important to warrant inclusion in leading textbooks on inorganic chemistry.

We now know a great deal about the formation and structures of $HAS_{(s)}$ and all of what we know to-date continues to reinforce the contention that they are discrete solid phases which have analogues in the natural environment and are unrelated to soluble aluminosilicates such as the putative $AlOSi(OH)_{3(aq)}^{2+}$.

For example:

(i) $HAS_{(s)}$ are only formed in solutions which are saturated with respect to $Al(OH)_{3(s)amorphous}$.
(ii) $HAS_{(s)}$ are formed in solutions in which the precipitation of $Al(OH)_{3(s)}$ is approached from either acidic or basic conditions or where $Si(OH)_{4(aq)}$ is mixed with preformed $Al(OH)_{3(s)}$.
(iii) The initial step in the formation of $HAS_{(s)}$ is the 'competitive' condensation of $Si(OH)_{4(aq)}$ across adjacent hydroxyl groups on a framework of $Al(OH)_{3(s)}$.
(iv) There are two discrete and structurally distinct forms of $HAS_{(s)}$. They are not simply an amorphous 'mixture' of $Al(OH)_{3(s)}$ and $Si(OH)_{4(aq)}$.
(v) $HAS_{A(s)}$ is composed of aluminium and silicon in the ratio of 2:1 and is the predominant form of $HAS_{(s)}$ in solutions in which the initial $[Si(OH)_{4(aq)}]$ is less than or equal to $[Al_T]$.
(vi) $HAS_{B(s)}$ is composed of aluminium and silicon in the ratio of 1:1 and is the predominant form of $HAS_{(s)}$ in solutions in which the initial $[Si(OH)_{4(aq)}]$ is significantly in excess (more than twice) of $[Al_T]$.
(vii) The formation of $HAS_{A(s)}$ is a prerequisite to the formation of $HAS_{B(s)}$ and this is irrespective of the initial solution ratio of $Si(OH)_{4(aq)}$ and Al_T. Thus, $HAS_{B(s)}$ is formed via the further condensation of $Si(OH)_{4(aq)}$ across adjacent hydroxyl groups on a framework of $HAS_{A(s)}$.

2. A SUMMARY OF THE EVIDENCE

The evidence to support each of the aforementioned statements can be summarised as follows:

$HAS_{(s)}$ are only formed in solutions which are saturated with respect to $Al(OH)_{3(s)amorphous}$.

We have used both indirect and direct methods in identifying the formation of $HAS_{(s)}$ in solutions which were either under- or over-saturated with respect to $Al(OH)_{3(s)amorphous}$. We used the formation of the morin–Al complex as an estimate of $[Al_{(aq)}^{3+}]$ and showed that for all solutions where $pH \leq 4.00$ and $[Al_T] \leq 1$ mmol/L all $Al_{(aq)}^{3+}$ was bound by morin irrespective of the presence of $[Si(OH)_{4(aq)}]$ up to 2 mmol/L. Similar solutions though in the pH range >4.00 to ≤5.50, where $Al(OH)_{3(s)}$ was predicted to be formed, showed significant reductions in the binding of $Al_{(aq)}^{3+}$ by morin and in particular in the presence of $Si(OH)_{4(aq)}$ and suggested that $HAS_{(s)}$ were formed in these solutions. Following the collection by membrane filtration of precipitated materials we used SEM–EDX to confirm that the precipitates which were formed in the presence of $Si(OH)_{4(aq)}$ had Si:Al ratios in the range 0.50–1.00 [6, 7]. Recently we have shown that when the formation of $Al(OH)_{3(s)}$ was prevented, not by a too acidic pH but, for example, due to a significant excess of either fluoride (soluble aluminium fluoride complexes predominate) or phosphate (insoluble aluminium hydroxyphosphate predominates), then in both situations $HAS_{(s)}$ were not formed [8] which once again pointed towards $Al(OH)_{3(s)}$ as a prerequisite to the formation of $HAS_{(s)}$.

$HAS_{(s)}$ are formed in solutions in which the precipitation of $Al(OH)_{3(s)}$ is approached from either acidic or basic conditions or where $Si(OH)_{4(aq)}$ is mixed with preformed $Al(OH)_{3(s)}$.

We have shown that once formed $HAS_{(s)}$ are slower to aggregate towards a filterable size than the corresponding phase of $Al(OH)_{3(s)}$ [1, 9]. We have used this property to discriminate between the pH-dependent solubilities of $Al(OH)_{3(s)}$ and $HAS_{(s)}$ and to demonstrate that $HAS_{(s)}$ are formed whether minimum solubility for $Al(OH)_{3(s)}$ (ca. pH 6.2) was approached from acidic (pH 3.0) or basic (pH 9.0) solution. Titration of acidic solutions indicated the formation of $HAS_{(s)}$ from ca. pH 4.5 and showed an $HAS_{(s)}$ solubility minimum at ca. pH 5.5, whereas titration of basic solutions resulted in the formation of $HAS_{(s)}$ at ca. pH 8.5 with, again, a recognisable solubility minimum at pH ca. 5.5. When aggregates of $Al(OH)_{3(s)}$ were preformed over a range of pH and mixed with $Si(OH)_{4(aq)}$ it was shown that $Si(OH)_{4(aq)}$ reduced the aggregate size and suggested that the formation of $HAS_{(s)}$ involved the disaggregation of preformed frameworks of $Al(OH)_{3(s)}$ [1].

The initial step in the formation of $HAS_{(s)}$ is the 'competitive' condensation of $Si(OH)_{4(aq)}$ across adjacent hydroxyl groups on a framework of $Al(OH)_{3(s)}$.

The observation that $Si(OH)_{4(aq)}$ slowed the rate of growth of the solid phase formed was interpreted as a 'poisoning' of the polymerisation of

$Al(OH)_{3(s)}$ [9]. In its simplest form the reaction was described as the condensation of $Si(OH)_{4(aq)}$ across adjacent hydroxyl groups on a dimer of aluminium hydroxide. A similar reaction had been described previously for the interaction of a dimeric silicate anion with a calcium hydroxide lattice in the mechanism which describes the retardation of ordinary portland cement by sugars [10]. We shall see that the suggested mechanism of formation of $HAS_{(s)}$ is strongly supported by the known stoichiometries and structures of the $HAS_{(s)}$ which are formed. We speculated that the interaction between $Si(OH)_{4(aq)}$ and $Al(OH)_{3(s)}$ was unique in that adjacent hydroxyl groups on the aluminium hydroxide template were ideally spaced to accommodate $Si(OH)_{4(aq)}$ [5, 11] and we tested this hypothesis by studying the potential interaction of the substituted silicic acids, dimethylsilane-diol (DMSD; $Si(OH)_2(CH_3)_2$) and tetramethyldisilane-diol (TMDS; $Si_2O(OH)_2(CH_3)_4$), with $Al(OH)_{3(s)}$. If our speculation was sound then neither of these substituted silicic acids should co-precipitate with $Al(OH)_{3(s)}$ since in DMSD the silanol groups are separated by methyl groups and in TMDS the silanol groups are on either silicon. We used solution, ^{27}Al, ^{29}Si and ^{13}C, and solid state, ^{27}Al and ^{29}Si, NMR to demonstrate that neither of these substituted silicic acids either reacted with $Al_{(aq)}^{3+}$ or were co-precipitated with $Al(OH)_{3(s)}$. SEM–EDX of collected solid phases confirmed the total absence of Si in these solids [12]. Thus, the unique geometries/orientation of $Si(OH)_{4(aq)}$ and the hydroxyl groups on aluminium hydroxide templates are critical to the formation of $HAS_{(s)}$; however, the reaction is also competitive with the autocondensation of $Al(OH)_{3(s)}$. For example, as will be explained later, a significant excess of $Si(OH)_{4(aq)}$ (i.e. with respect to the stoichiometry of the solid phase) is required before the formation of $HAS_{(s)}$ is preferred over $Al(OH)_{3(s)}$ [13]. The reaction is truly a competitive condensation reaction!

There are two discrete and structurally distinct forms of $HAS_{(s)}$. They are not simply an amorphous 'mixture' of $Al(OH)_{3(s)}$ and $Si(OH)_{4(aq)}$.

Our earlier attempts to identify both the formation and the stoichiometries of $HAS_{(s)}$ involved their retention on cation exchange resins and subsequent elution using dilute acid [1, 9]. The technique relied on the assumptions that $HAS_{(s)}$ were small enough to enter the matrix of the resin (<100 nm), that they were positively charged and that their retention by the resin functional groups would not alter the form in which they had entered the resin. We successfully applied this method in dilute mildly acidic solutions to identify $HAS_{(s)}$ with Si:Al ratios between ca. 0.2 and 0.7 and, at that time, we interpreted these results as confirmation of the existence of one general form of $HAS_{(s)}$ with an ideal Si:Al ratio of 0.5. It was not until we had developed a method to collect significant quantities (hundreds of milligrams) of precipitated $HAS_{(s)}$ that we were able to use a battery of structural and compositional techniques to positively identify and characterise two distinct forms of $HAS_{(s)}$ [6].

$HAS_{A(s)}$ is composed of aluminium and silicon in the ratio of 2:1 and is the predominant form of $HAS_{(s)}$ in solutions in which the initial $[Si(OH)_{4(aq)}]$ is less than or equal to $[Al_T]$.

When 10 L volumes of acidic (pH 3.0) solutions which included $[Al_T] \geq [Si(OH)_{4(aq)}]$ ($[Si(OH)_{4(aq)}] \leq 2.0$ mmol/L) were titrated up to pH 6.2 and subsequently aged in the dark at a constant temperature for 3 months it was, thereafter, possible to collect up to 500 mg of solid phase by filtration using 0.2 μm membrane filters. Larger pore size filters collected substantially less solid phase, suggesting that the precipitate was extremely slow to aggregate towards an easily filterable size. The precipitates were air-dried to a constant weight and ground to a fine powder in an agate mortar and pestle. Analysis of these precipitates by SEM–EDX revealed the 'expected' Si:Al ratio of ca. 0.5 while solid state ^{27}Al and ^{29}Si NMR showed that the structure of these $HAS_{(s)}$ was dominated by Si coordinated through three Si–O–Al linkages ($Q^3(3Al)$) to Al in an octahedral geometry (Al^{VI}). Contact AFM images of $HAS_{(s)}$ precipitated under the same conditions revealed flat (1–2 nm thick) rectangular (up to 170 nm long) particles [6, 7]. We called this solid phase $HAS_{A(s)}$ and we speculated that it was similar to the naturally occurring secondary mineral phase known as protoimogolite [14].

$HAS_{B(s)}$ is composed of aluminium and silicon in the ratio of 1:1 and is the predominant form of $HAS_{(s)}$ in solutions in which the initial $[Si(OH)_{4(aq)}]$ is significantly in excess (more than twice) of $[Al_T]$.

In the same series of experiments as described above solid phases were rapidly and visibly precipitated from solutions where $[Si(OH)_{4(aq)}] >> [Al_T]$ (again, $[Si(OH)_{4(aq)}] \leq 2.0$ mmol/L). Following ageing for 3 months these precipitates were easily collected by filtration using a 2.0 μm membrane filter. Clearly under the identical conditions of pH and ionic strength these solid phases aggregated very much more quickly to a filterable size and this immediately suggested a difference to those solid phases collected from solution in which $[Al_T] \geq [Si(OH)_{4(aq)}]$. Analysis of air-dried, finely ground precipitates by SEM–EDX revealed an Si:Al ratio of ca. 1.0 while solid state ^{27}Al and ^{29}Si NMR showed the totally unexpected finding of up to 50% of the constituent Al in tetrahedral geometry (Al^{IV}). This shift from octahedral to tetrahedral geometry was supported by changes in the organisation of Si from $Q^3(3Al)$, the signature for $HAS_{A(s)}$, to predominantly $Q^3(1–2Al)$. When these $HAS_{(s)}$ were imaged using AFM they were found to be flat (1–2 nm in height), discoid (up to ca. 40 nm in diameter) structures which were sometimes found to be in chain-like assemblies [6, 7]. We called this solid phase $HAS_{B(s)}$ and again likened it to a naturally occurring secondary mineral phase known as protoimogolite allophane [14].

The formation of $HAS_{A(s)}$ is a prerequisite to the formation of $HAS_{B(s)}$ and this is irrespective of the initial solution ratio of $Si(OH)_{4(aq)}$ and Al_T. Thus, $HAS_{B(s)}$ is formed via the further condensation of $Si(OH)_{4(aq)}$ across adjacent hydroxyl groups on a framework of $HAS_{A(s)}$.

The unexpected structural identity of $HAS_{B(s)}$ required additional thinking on the mechanism of formation of $HAS_{(s)}$ and we speculated that $HAS_{B(s)}$ would be formed via the further reaction of $Si(OH)_{4(aq)}$ with $HAS_{A(s)}$ [6]. Recent research has suggested that there may, additionally, be an intermediate between the two forms, which we have called $HAS_{AB(s)}$, in which $Si(OH)_{4(aq)}$ has reacted with $HAS_{A(s)}$ but has not, yet, fuelled the dehydroxylation to $HAS_{B(s)}$ whereupon up to 50% of the total Al^{VI} is converted to Al^{IV} [8]. We have tested the hypothesis that the formation of $HAS_{A(s)}$ is a prerequisite to $HAS_{B(s)}$ by initiating the time-dependent precipitation of solid phases from solutions in which $HAS_{B(s)}$ was predicted to predominate and by applying a battery of analytical techniques to determine the form of $HAS_{(s)}$ which precipitated at each time point [13]. The earliest time point at which sufficient solid phase could be precipitated and collected by membrane filtration was ca. 30 min. Electron probe microanalysis of this phase revealed an Si:Al ratio of 0.6 while solid phases collected under the same conditions after longer periods of ageing showed increased content of Si up to an Si:Al ratio of 0.92 for solid phases collected after ageing for 336 h. We were able to use thermogravimetric analysis to discount the possibility that any of the solid phases included $Al(OH)_{3(s)}$ [13, 15] and so we concluded that these precipitates included both $HAS_{A(s)}$ and $HAS_{B(s)}$ with proportionately larger amounts of the latter with increased ageing time. These conclusions were supported further by solid state ^{27}Al and ^{29}Si NMR, in particular, the former which indicated an increased content of Al^{IV} over Al^{VI} with increased ageing time [13].

3. WHAT NEXT FOR HYDROXYALUMINOSILICATES?

We are continuing to apply proven analytical methods and develop new analytical tools to understand further the structure and properties of $HAS_{(s)}$. For example, preliminary investigations with X-ray photoelectron spectroscopy (XPS) have been very useful in discriminating between $HAS_{A(s)}$ and $HAS_{B(s)}$ in only a few milligrams of solid phase while a novel method of particle sizing has been used to identify different populations of particles of $HAS_{A(s)}$ and $HAS_{B(s)}$ under a range of solution conditions. We are also investigating small angle X-ray scattering (SAXS) using synchrotron radiation to identify the earliest stages of HAS formation and in particular the nature of the solids (nanoparticles) which are formed.

4. HYDROXYALUMINOSILICATES AND THE BIOLOGICAL AVAILABILITY OF ALUMINIUM

The overriding aim of our research on $HAS_{(s)}$ is to elucidate their role in the biogeochemical cycle of aluminium [16] and specifically to understand how they keep aluminium out of biota [17]. This has led us to try to define a solubility expression for $HAS_{B(s)}$ in order that such might be used quantitatively in predictive models of aluminium solubility control in soil and surface waters [18]. The solubility constant which we derived from our unconventional solubility expression:

$$K^{*}HAS_{B} = [Al_{r}][Si(OH)_{4}]^{2}[OH^{-}]^{4}$$
$$K^{*}HAS_{B} = 10^{-40.6\pm0.15} \quad (n=17)(20°C/I=0.1\,mol/L)$$

explained the lower solubility of $HAS_{B(s)}$ compared with $Al(OH)_{3(s)}$ and has been successfully applied to the solubility of aluminium and silicon in natural waters [19]. However, it is informative to consider how a constant which is used to define thermodynamic equilibrium (if only arbitrarily defined in the case of $HAS_{B(s)}$) might also inform about biological availability. This immediately raises issues of definition, for example, what in the current context do we mean by 'biological availability' and 'thermodynamic equilibrium'? There are properties of the latter, which necessarily render it an inaccurate assessment of the former. For example, in biological systems thermodynamic equilibrium is rarely achieved and though it may act in driving a chemical reaction in much the same way as gravity acts on the Earth, thermodynamic equilibrium is not approached asymptotically but through 'waves' of concentrations of free ions which involve periods of both under- and over-saturation. In addition, any concentration which is defined by thermodynamic equilibrium, in this case for $Al_{(aq)}^{3+}$, is a completely 'static' concept, it has no kinetic component. This means that it defines the concentration of $Al_{(aq)}^{3+}$ which is immediately (instantaneously) available to be bound by a 'biological' ligand but it does not offer any insight into how quickly or slowly the equilibrium concentration is maintained during such binding events by the subsequent equilibrium processes. The biologically available fraction of Al_T is defined as a biological response to a critical proportion of target ligands binding $Al_{(aq)}^{3+}$ [20] and will be determined by kinetic constraints as much as by thermodynamic equilibrium. Thus, the $[Al_{(aq)}^{3+}]$ which is predicted by our solubility constant for $HAS_{B(s)}$ is sufficiently small to ensure that, strictly speaking, aluminium is not biologically available. Biological ligands will bind a proportion of the $[Al_{(aq)}^{3+}]$ which is in equilibrium with the solid phase but neither the equilibrium $[Al_{(aq)}^{3+}]$ nor the rate at which it might be replaced by equilibrium-fuelled dissolution of the solid phase will be sufficient to elicit a biological response.

$HAS_{(s)}$ are effective in limiting the biological availability of aluminium because of their kinetic inertia. While there are not any stability or equilibrium constants to describe them quantitatively (they are, after all, not soluble complexes) their formation is dependent on the prior formation of $Al(OH)_{3(s)}$ and thus the existence of these ultimately highly insoluble secondary minerals is controlled by a more soluble (though insoluble) precursor. We can use this to explain how the formation of $HAS_{(s)}$ has been shown to both increase and decrease the biological availability of aluminium [21, 22]. For example (see Fig. 1), in acidic media one can postulate that the simplest form of $HAS_{(s)}$ will be the reaction of a single molecule of $Si(OH)_{4(aq)}$ with a dimeric unit of $Al(OH)_{3(s)}$. In solutions in which $[Al_T] \geq [Si(OH)_{4(aq)}]$ the formation of this simplest form of $HAS_{A(s)}$ will be competitive with the formation of $Al(OH)_{3(s)}$ and the stabilities of both of these forms will depend on how rapidly they self-aggregate towards critically stable nuclei (nanoparticles). Before they reach these critical sizes they are prone to equilibrium-fuelled disaggregation and, therefore, they remain sources of relatively rapidly available $Al_{(aq)}^{3+}$. Under certain conditions $HAS_{(s)}$ have been shown to aggregate much more slowly towards their critical size than $Al(OH)_{3(s)}$ and it is under such conditions that $HAS_{(s)}$ are less stable towards dissolution than $Al(OH)_{3(s)}$ and, consequently, more potent as sources of biologically available aluminium. We are, only now, beginning to understand the driving forces which underlie the formation and aggregation of $HAS_{(s)}$ and how these impact on the biological availability of aluminium. For example, the simplest theoretical form of $HAS_{B(s)}$ would involve the reaction of a further single molecule of $Si(OH)_{4(aq)}$ with the dimeric form of $HAS_{A(s)}$. One significant difference between the properties of the simplest forms of $HAS_{A(s)}$ and $HAS_{B(s)}$ is that in acidic media the

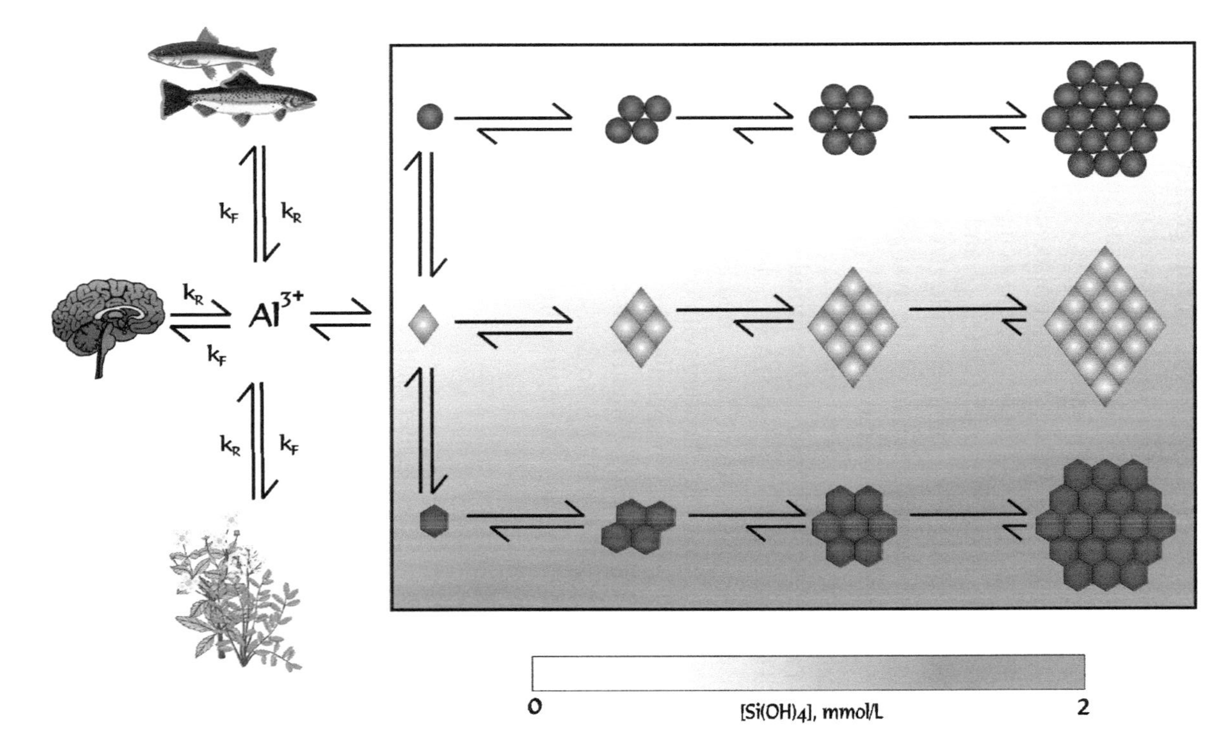

Fig. 1. The formation of hydroxyaluminosilicates and the biological availability of aluminium. The rate at which the solid phase (blue circle – $Al(OH)_{3(s)}$; orange diamond – $HAS_{A(s)}$; red hexagon – $HAS_{B(s)}$) self-aggregates towards critically sized stable nuclei (nanoparticles) will determine the availability of $Al_{(aq)}^{3+}$ to biological ligands. The reversible arrows give qualitative estimates of the relative stabilities of the different aggregates and illustrate how the formation of $HAS_{(s)}$ could both increase and decrease the biological availability of $Al_{(aq)}^{3+}$ relative to $Al(OH)_{3(s)}$. Where the rate of the forward reaction, K_F, is very much greater than the rate of the reverse reaction, K_R, equilibrium-fuelled dissolution of unstable solid phases will promote the biological availability (and possibly, toxicity) of aluminium.

latter self-aggregate at a much greater rate than $HAS_{A(s)}$ such that they achieve a critical size more rapidly and hence, may become a less potent source of $Al_{(aq)}^{3+}$ more rapidly than $HAS_{A(s)}$. We believe that one factor which influences the rates of self-aggregation of $HAS_{A(s)}$ and $HAS_{B(s)}$ is the greater number of silanol (Si–OH) groups which 'surround' $HAS_{B(s)}$. These appear to confer a greater degree of hydrolytic stability such that in mildly acidic milieu $HAS_{B(s)}$ are uncharged while $HAS_{A(s)}$ carry a net positive charge.

In summary the kinetic inertia of $HAS_{(s)}$ is maximised under conditions in which they form critically sized nuclei (nanoparticles) as rapidly as possible (see Fig. 1). A major driving force is the $[Si(OH)_{4(aq)}]$ and specifically an excess of $Si(OH)_{4(aq)}$ relative to Al_T. For example, while $HAS_{A(s)}$ is composed of Si and Al in the ratio 1:2 it is unusual for all of the available aluminium to be precipitated as $HAS_{A(s)}$ unless the initial ratio of $Si(OH)_{4(aq)}$ to Al_T is at least 1:1. Similarly, $HAS_{B(s)}$, which is composed of Si and Al in the ratio 1:1, is only the predominant precipitated phase when the initial ratio of $Si(OH)_{4(aq)}$ to Al_T is at least 2:1. Thus, both $HAS_{A(s)}$ and $HAS_{B(s)}$ depend on an excess of $Si(OH)_{4(aq)}$ to stabilise them relative to $Al(OH)_{3(s)}$ and $HAS_{A(s)}$, respectively [13]. The excess of $Si(OH)_{4(aq)}$ provides the longevity or timeframe required for either $HAS_{(s)}$ to self-aggregate to their respective ultra-stable critical nuclei. However, the $[Si(OH)_{4(aq)}]$ is not the only critical factor as in dilute solutions (i.e. low $[Al_T]$) the time required for $HAS_{(s)}$ to self-aggregate and achieve stable nuclei may be extended beyond a biological limit!

ACKNOWLEDGEMENTS

Many thanks to F.J. Doucet, C. Schneider and S. Strekopytov for their invaluable contributions to this research. Thanks to O. Exley for her help with Fig. 1. The majority of this research was funded by EPSRC, NERC, The Royal Society of London and Dow Corning Ltd.

REFERENCES

[1] C. Exley and J.D. Birchall, Polyhedron, 12 (1993) 1007.
[2] C. Exley and J.D. Birchall, Geochim. Cosmochim. Acta, 59 (1995) 1017.
[3] V.C. Farmer and D.G. Lumsdon, Geochim. Cosmochim. Acta, 58 (1994) 3331.
[4] P. van Hees, U. Lundström, R. Danielsson and L. Nyberg, Chemosphere, 45 (2001) 1091.
[5] C. Exley, C. Schneider and F.J. Doucet, Coord. Chem. Rev., 228 (2002) 127.
[6] F.J. Doucet, C. Schneider, S.J. Bones, A. Kretchmer, I. Moss, P. Tekely and C. Exley, Geochim. Cosmochim. Acta, 65 (2001) 2461.

[7] F.J. Doucet, M. Rotov and C. Exley, J. Inorg. Biochem., 87 (2001) 71.
[8] S. Strekopytov and C. Exley, Polyhedron, 24 (2005) 1585.
[9] C. Exley and J.D. Birchall, Polyhedron, 11 (1992) 1901.
[10] J.D. Birchall and N.L. Thomas, Br. Ceram. Proc., 35 (1984) 305.
[11] C. Exley, J. Inorg. Biochem., 69 (1998) 139.
[12] C. Schneider, Ph.D. Thesis, Keele University, UK, 2003.
[13] S. Strekopytov, E. Jarry and C. Exley, Polyhedron, 25 (2006) 3399.
[14] P. Cradwick, V.C. Farmer, J. Russell, C. Masson, K. Wada and N. Yoshinaga, Nat. Phys. Sci., 240 (1972) 187.
[15] S. Strekopytov and C. Exley, Polyhedron, 25 (2006) 1707.
[16] C. Exley, J. Inorg. Biochem., 97 (2003) 1.
[17] J.D. Birchall, C. Exley, J.S. Chappell and M.J. Phillips, Nature, 338 (1989) 146.
[18] C. Schneider, F. Doucet, S. Strekopytov and C. Exley, Polyhedron, 23 (2004) 3185.
[19] D. Dobrzyński, Acta Geol. Pol., 55 (2005) 445.
[20] C. Exley and J.D. Birchall, J. Theor. Biol., 159 (1992) 83.
[21] C. Exley, N.C. Price and J.D. Birchall, J. Inorg. Biochem., 54 (1994) 297.
[22] C. Exley, J.K. Pinnegar and H. Taylor, J. Theor. Biol., 189 (1997) 133.

Thermodynamics, Solubility and Environmental Issues
T.M. Letcher (editor)

Chapter 18

Apatite Group Minerals: Solubility and Environmental Remediation

M. Clara F. Magalhães[a] and Peter A. Williams[b]

[a]*Department of Chemistry and CICECO, University of Aveiro, P-3810-193 Aveiro, Portugal*
[b]*School of Natural Sciences, University of Western Sydney, Locked Bag 1797, Penrith South DC NSW 1797, Australia*

1. INTRODUCTION

Soils, wastes and waters contaminated by hazardous geochemical trace elements represent actual or potential threats to living organisms. Depending on the degree of environmental threat or the level of risk, contaminated systems may be subjected to remediation with the primary aim of reduction of the threat to an acceptable level. In this connection it is noted that the acceptable maximum contaminant level of a chemical element or substance is defined on a country-by-country basis, and may involve national or international standards. Several remediation techniques are available to remove, reduce or immobilize hazardous elements in wastes, soils and waters. Remediation processes based on chemical treatments aim to destroy contaminants or convert them into less environmentally hazardous forms [1]. Environmental problems that result from the contaminant's high solubility can sometimes be solved by in situ immobilization by addition of appropriate reactants. Here, attention is drawn to the *apatite* group of minerals, and especially to the arsenate and phosphate members that contain calcium and lead. These have recognised applications, and limitations, in stabilization processes used for soils, wastes and waters contaminated by certain hazardous geochemical trace elements.

2. *APATITE* GROUP MINERALS

The *apatite* group minerals are hexagonal or pseudohexagonal monoclinic arsenates, phosphates and vanadates of general formula $A_5(XO_4)_3Z$, where

A = Ba, Ca, Ce, K, Na, Pb, Sr, Y, X = As, P, Si, V and Z = F, Cl, O, OH, H_2O. Solid-solution between end-members is extensive, but complete only in certain cases. Carbonate ion may partially replace the XO_4 group, with appropriate charge compensation [2, 3]. Fluorapatite ($Ca_5(PO_4)_3F$) is the most common member of the group and the major constituent of phosphorites, which are the main raw materials for the manufacture of phosphoric acid derivatives including fertilizers, foods, pharmaceuticals and other chemicals. Fluorapatite in phosphorites is usually partially carbonated and hydroxylated. Hydroxyapatite ($Ca_5(PO_4)_3OH$) is the primary mineral constituent of bones and teeth and hosts a variety of chemical substituents in its structure. These materials are stable for billions of years, even during tectonic events, and over a wide range of solution pH and geological conditions, as can be inferred from the existence of ancient sedimentary phosphorites [4].

Solids belonging to the *apatite* group have been investigated as host materials for long-term immobilization of a large number of stable and radioactive elements including cadmium, copper, lead, nickel, uranium, zinc, iodide and bromide [4–28]. This body of work demonstrates conclusively that members of the *apatite* group have great potential in remediation techniques, and field experiments are now being implemented [28]. In situ soil remediation by addition of calcium phosphate rock reduces cadmium, lead and zinc solubility, and decreases human bioavailability associated with accidental ingestion of soil, and phytoavailability [27, 28]. Permeable reactive barriers containing hydroxyapatite are a promising technology for removing cadmium, lead and zinc from highly contaminated mining drainage waters with a substantial reduction of costs [4, 29].

The *apatite* group of minerals displays a large solubility range, with calcium-bearing species being more soluble than their lead analogues, as can be seen from the data of Table 1. Many reports of solubility properties of these compounds have appeared in the literature in the past. The data in Table 1 represent a critical compilation that may be used for modelling purposes. Earlier data have been summarised elsewhere [30]. Nevertheless, solubility relations are complex and subtle solid-solution phenomena serve to modify them to considerable extent. For example, the presence of trace foreign metal ions in the hydroxyapatite structure tends to stabilize it, decreasing its solubility and increasing its kinetic inertness [44]. Aside from this, the solubilities of the end-member $A_5(XO_4)_3Z$ species display an extraordinary dependence on the nature of the Z group at 298.15 K, and this relationship is quite different for Ca versus Pb analogues. For $Ca_5(PO_4)_3Z$ the magnitude of log K varies in the order $F << OH << Cl$, whereas for

Table 1
Apatite group Ca and Pb phosphates and arsenates

Mineral	Chemical composition	log *K*	References
Chlorapatite	$Ca_5(PO_4)_3Cl$	−53.08	[31]
Fluorapatite	$Ca_5(PO_4)_3F$	−59.99	[31–34]
Hydroxyapatite	$Ca_5(PO_4)_3OH$	−56	[35]
Johnbaumite	$Ca_5(AsO_4)_3OH$	−38.04	[36]
Fermorite	$Ca_5(AsO_4)_3OH$		
Svabite	$Ca_5(AsO_4)_3F$		
Turneaureite	$Ca_5(AsO_4)_3Cl$	−51.7[a]	[37]
Pyromorphite	$Pb_5(PO_4)_3Cl$	−83.61	[38–41]
Fluorpyromorphite	$Pb_5(PO_4)_3F$	−71.56	[39, 41]
Hydroxypyromorphite	$Pb_5(PO_4)_3OH$	−76.71	[39, 41, 42]
Mimetite	$Pb_5(AsO_4)_3Cl$	−83.37	[38]
Clinomimetite	$Pb_5(AsO_4)_3Cl$		
Hydroxymimetite	$Pb_5(AsO_4)_3OH$	$-73 < \log K < -70$	[42, 43]
Hedyphane	$Ca_2Pb_3(AsO_4)_3Cl$		

Note: Solubility products at 298.15 K are listed where known. Members of the group in italics are not yet recognised as discrete mineral species. Solid solution phenomena are important features of the group and only end-member compositions are listed.
[a]At 310.15 K.

$Pb_5(PO_4)_3Z$ the order is $Cl << OH << F$. This situation is further complicated in the case of the arsenate-bearing phases. For both $Pb_5(AsO_4)_3Z$ and $Ca_5(AsO_4)_3Z$, the above order is $Cl <<< OH$, a situation that reflects a very subtle combination of lattice and hydration energies. It is further noted that the Pb members of the group are much less soluble than their Ca congeners. Nevertheless, these relationships have been used to considerable effect in remediation studies and pilot projects.

3. LEAD PHOSPHATE MINERALS

Lead is widespread in nature and has been widely mined and used since ancient Roman times. Lead is not an essential element for living organisms and is quite toxic. The main anthropic sources of lead in the environment come from agricultural activities such as application of lead-containing biocides (lead arsenates) and sewage sludge, industrial activities such as mining and ore processing, disposal of lead batteries and other lead-containing materials, and urban activities such as the use of leaded gasoline, which has, however, been phased out in most countries.

Soils with a high lead content are potentially toxic to humans and other living beings. Soils contaminated with lead are found in both rural and urban settings [45, 46]. In situ lead immobilization in soils, wastes and waters by phosphates is now recognized as an appropriate technology for the reclamation of lead-contaminated systems [12–17, 20, 21, 25, 28, 42, 47–51]. Procedures have involved the addition of solid phosphates with the *apatite* structure, including synthetic hydroxyapatite [5–9, 12, 29, 49, 52–55], biogenic *apatite* derived from bones [4, 22], rock phosphate [12, 15, 28, 45, 54, 55], phosphate fertilizers [45, 54, 55], or aqueous phosphate ions, to soil water solutions containing dissolved lead [12, 41, 45, 49, 51].

Lead immobilization by phosphates arises from the low solubility of pyromorphite ($Pb_5(PO_4)_3Cl$). As stated, lead members of the *apatite* group minerals are much less soluble than their calcium congeners. Nriagu [39], as early as 1974, proposed the removal of lead from wastewaters and the stabilization of lead in contaminated soils and sediments by reaction with phosphate ions to precipitate pyromorphite.

These techniques are based not only on the principle that lead-containing phosphates with the *apatite* structure are highly insoluble, but also that rapid reactions occur with *apatite* and lead ions at the solid/aqueous solution interface [12, 13, 15, 20, 29, 48, 53, 56]. Removal of lead from aqueous solutions using synthetic hydroxyapatite gives aqueous lead concentrations below the maximum contamination level after 1 h [12, 53]. Other workers [9] observed the formation of calcium–lead *apatite* solid-solutions after 3 mins contact between synthetic hydroxyapatite and aqueous solutions containing lead, and no lead was detected in the aqueous solution after 24 h contact. However, the efficiency of lead removal depends on the characteristics of the phosphate rock employed [15]. It has been shown that the composition and crystallinity of the phosphate influence the speed of the surface reactions [4, 44]. More highly crystalline solids have lower solubilities and dissolution rates, making the *apatite* less reactive [4]. The presence of fluoride in the hydroxyapatite structure decreases its solubility and dissolution rate, while the presence of carbonate decreases structural stability, and increases solubility and the dissolution rate [4, 35].

Several authors [11–16, 18, 19, 41, 42] have suggested that the precipitation of the lead-bearing *apatites* pyromorphite and hydroxypyromorphite results from prior dissolution of hydroxyapatite, which is much more soluble than the lead phases. Continuous dissolution of hydroxyapatite was observed as the result of the formation of less soluble species [12, 15, 29, 53]. Some workers have studied the effect of different phosphate amendments (synthetic hydroxyapatite, phosphate fertilizers, phosphoric acid, calcium hydrogen-phosphate, phosphate rock) on in situ lead immobilization in contaminated soils [45, 54, 55], the bioaccessibility of lead in soils for humans by in vitro tests with simulated gastric and intestinal solutions [54] and the bioaccessibility for plants by a sequential extraction technique [27, 54, 55]. The formation of hydroxypyromorphite and pyromorphite has also been observed with an increase in the original, residual lead in soil using a sequential extraction method [45]. These authors, as well as many others [12–15], concluded that hydroxyapatite is the best amendment to remediate lead-contaminated soils in that it leads to the lowest levels of bioaccessibility of the heavy element. This is explained by dissolution of hydroxyapatite to provide phosphate for formation of insoluble lead phosphate phases as well as calcium ions to replace lead ions at exchangeable sites in the soil [12]. The lower effectiveness of phosphate rock compared with hydroxyapatite in relation to lead bioaccessibility is a consequence of the lower solubility of the phosphate rock minerals, limiting the amounts of phosphate available in solution. Soluble phosphate fertilizers performed best in minimizing lead bioaccessibility in simulated gastric solutions [54]. In this connection, the solubility of phosphate rock (considered as mixtures of fluor- and hydroxyapatites) increases with an increase in phosphate substitution by carbonate [5, 35]. The higher the solubility of the phosphate rock the more effective it is in reducing lead concentrations [12, 15]. Synthetic hydroxyapatite is much more effective in reducing lead concentrations from aqueous solutions than phosphate rock, as it is able to remove more lead per mass unit of the solid used [15, 55].

The decrease of available lead in aqueous solutions in contact with media containing phosphate ions both in solution and adsorbed on goethite surfaces has been studied by several authors [49, 51]. They observed the formation of pyromorphite in the bulk solution by homogeneous nucleation from the ions present as well as heterogeneous nucleation induced by goethite that specifically adsorbs phosphate ions. It was shown that phosphate ions adsorbed on goethite react directly with dissolved lead in aqueous solution to give pyromorphite crystallized directly on the goethite surface. It was further concluded that pyromorphite precipitation is faster than phosphate desorption from the goethite surface.

Pyromorphite, hydroxypyromorphite and fluorpyromorphite form rapidly when lead and phosphate ions are available [12, 15, 29, 39, 45], depending on relative prevailing concentrations of chloride, hydroxide and fluoride ions [13, 15, 45]. Several researchers [12, 29, 45, 53] reported the formation of hydroxypyromorphite on reacting synthetic hydroxyapatite with aqueous lead (II) solutions. Members of the solid-solution between hydroxyapatite and hydroxypyromorphite are formed as intermediate phases during the continuous dissolution of hydroxyapatite crystals in the presence of aqueous ions, ending with the formation of hydroxypyromorphite [29, 53]. Phosphate rocks containing mixtures of carbonated hydroxyl- and fluorapatites give rise to calcium-bearing pyromorphite solid-solutions with a similar total composition of the initial phosphate rock as a result of partial or total calcium substitution by lead [15].

The natural occurrence of pyromorphite has been observed in road-side soils [17, 24, 39, 46, 57], contaminated soils and wastes and mine wastes [17, 24, 57]. Cotter-Howells et al. [57] reported the presence of calcium-rich pyromorphite in mine waste-contaminated soils from the South Pennine orefield in the United Kingdom. Pyromorphite is the main lead component of garden soils of a village in a related area having highly elevated lead levels from mining activities over an extended period [58]. Normal blood lead contents in the population of the village were explained by the presence of the insoluble lead phosphate limiting bioavailable lead. Several other authors [24, 28, 54] have similarly concluded that pyromorphite can limit human lead bioavailability.

4. ARSENATE MINERALS

The environmental impact of arsenic is one of the most important challenges in environmental geochemistry. Millions of people from countries in all continents experience severe health risk through the ingestion of arsenic-contaminated waters. In India and Bangladesh alone, nearly 100 million people are affected by arsenic-contaminated drinking water and food crops [59], and will have no access to arsenic-free drinking water in the near future [59, 60]. Arsenic concentrations in ground waters of the West Bengal and Bangladesh regions can reach values of up to 2 mg/L [59, 60] and after treatment in columns with different fillings the total arsenic content in the treated water is very often >0.05 mg/L. The World Health Organization (WHO) established 0.010 mg/L arsenic as a provisional guideline for the maximum contaminant level for total arsenic in potable water. This has been adopted by many countries but not in all the arsenic-affected developing countries, where the maximum contaminant level is set at 0.050 mg/L As [59, 61].

Mobilization of arsenic in the environment arises either by the natural oxidation of arsenic-bearing sulphosalt minerals or from anthropic activities related to mining and ore processing, metallurgy, agriculture, wood preservation and sewage treatment. Major sources for soil arsenic contamination are irrigation with arsenic-contaminated water and the use of arsenic-based herbicides and pesticides.

Precipitation of arsenic-bearing solid phases under oxidizing conditions by calcium and iron(III) ions (as low solubility arsenates) has been a common technique to remove arsenic from contaminated waters or metallurgical effluents [62, 63]. Distribution of arsenic-free water is seen as the optimal solution for the problem, but such an outcome is not foreseen for the near future in many of the affected regions [59, 60]. Meanwhile, remediation technologies are available mainly for industrial scale arsenic removal [61]. Chemical processes for removing arsenic from waters must include redox reactions to convert As(III) into As(V) to achieve more effective removal [64], followed by procedures that transfer the dissolved contaminant from solution to a solid phase (precipitation, co-precipitation, adsorption and ion exchange) [61]. Water treatment by chemical methods comprises the addition of coagulants (the two main coagulants are aluminium sulphate and iron(III) chloride) and lime softening; adsorption of arsenic species on activated alumina, activated carbon and iron oxide coated sand are the most used methods. New treatment technologies use iron filings, granular ferric oxides, kimberlite tailings, etc. [61]. Laboratory experiments show that these techniques are efficient for arsenic removal from waters with arsenic concentrations below the maximum contaminant level. An evaluation of arsenic treatment plants in West Bengal showed that 46, 38 and 16% of them have to deal with waters with arsenic concentrations greater than 0.010, 0.050 and 0.300 mg/L, respectively [65].

4.1. Calcium Arsenates

Svabite ($Ca_5(AsO_4)_3F$) and other calcium arsenates have been considered for safe arsenic disposal, in treating industry and mineral processing wastes with lime, and it has been assumed that these solids are extremely insoluble and stable [23, 36, 66–68]. Currently this technique has been abandoned as there is evidence that calcium arsenate compounds slowly decompose either by change of pH of the media or by contact with atmospheric carbon dioxide that under alkaline conditions gives rise to the formation of calcium carbonate with slow release of arsenate ions back into the environment [69].

At the beginning of the 20th century considerable attention was given to the preparation and properties of calcium arsenates because of their use as insecticides for certain crops [70]. These materials were used widely in the United States until the middle of the second half of the 20th century [71]. Several different calcium arsenates were employed for the purpose. Bothe and Brown [36] studied the stability of several of them at 23°C, with pH ranging between 6.5 and 12.5, and presented refined values for their solubility constants; among them was a solid with a chemical composition corresponding to that of johnbaumite ($Ca_5(AsO_4)_3OH$). The solids/aqueous solution system for this species was continued for a period of 4 years. They observed that this phase only crystallized when the calcium source was very pure, pH was in the range 9.5–10 and Ca/As ratios lay between 1.67 and 1.90. They also noted that it was not possible to form this solid in the presence of magnesium ions. For higher pH and the Ca/As between 2.0 and 2.5, a solid with the composition $Ca_2(AsO_4)OH$, sometimes mixed with johnbaumite, was obtained. Other solids, including $Ca_3(AsO_4)_2{\cdot}xH_2O$, $Ca_5H_2(AsO_4)_4{\cdot}9H_2O$ and $CaHAsO_4{\cdot}H_2O$ (haidingerite) were obtained under different experimental conditions. Fig. 1 is constructed using the published values [36] and shows relationships between the most stable calcium arsenates. From the figure it can be concluded that johnbaumite is a stable phase only under alkaline conditions.

The true value for the solubility constant of johnbaumite must be closer to the Bothe and Brown [36] value than those presented by other authors. Values of p$K \approx 45$ obtained by Mahapatra et al. [40] and Narasaraju et al. [37]

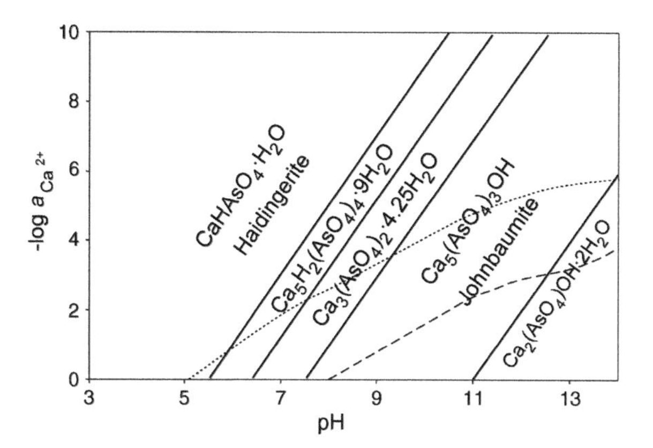

Fig. 1. Equilibrium relations at 298.15 K for selected calcium arsenates in terms of pH and calcium ion activities. The dotted curve (··········) corresponds to equilibrium conditions between solid phases and aqueous solutions of total arsenate activity 10^{-3} and the dashed curve (– – – –) to a total aqueous arsenate activity of 10^{-7}.

for the solubility constant of johnbaumite are much lower than that of Bothe and Brown [36]. Narasaraju et al. [37] also determined the solubility constant for a solid of composition $Ca_5(AsO_4)_3Cl$, and obtained the value listed in Table 1. If johnbaumite behaves in a similar fashion to hydroxyapatite, the solid/aqueous solution system will only attain equilibrium after long periods of time, probably more than 1 year. The lower values [37, 40] were obtained under acid conditions and with much shorter equilibration times, and most probably refer to undersaturated solutions. Fig. 1 shows that the data of Bothe and Brown [36] gives a coherent phase diagram, which can be used to explain reported observations on crystallization conditions for each solid phase. It shows that johnbaumite has a broad region of stability that contradicts the observations reported by Nishimura and Robins [72]. These latter authors re-evaluated the solubility and stability regions of various calcium arsenates and concluded that johnbaumite must be of high temperature origin. They conducted experiments at 25°C and did not obtain any calcium arsenate with the *apatite* structure. However, they noted that johnbaumite had been synthesized by several researchers at temperatures above 90°C and with pH ranging between 8 and 11. The difficulty in obtaining johnbaumite at room temperature seems to be kinetic rather than thermodynamic in origin. Johnbaumite has a narrower stability field than hydroxyapatite [35] but many stable calcium arsenates exist under different solution compositions. The inhibitory effect of magnesium on hydroxyapatite crystallization and dissolution is well known. While it is possible to crystallize hydroxyapatite containing traces of magnesium, laboratory experiments show that synthetic johnbaumite crystallizes only in the absence of magnesium under alkaline conditions [23, 37, 40, 70, 72, 73].

As can be seen from the data in Table 1, johnbaumite, in spite of belonging to the *apatite* group, is much more soluble than any other of the related solid phases. Turneaureite is also more soluble than chlorapatite and svabite is probably more soluble than fluorapatite. Minute amounts of arsenic-rich species from the *apatite* group, mainly hedyphane and members of the solid-solution between turneaurite and johnbaumite, were found in rocks recovered from the Långban mine dumps located at Värmland, Sweden [74]. Calcium-rich phases enriched in phosphate and depleted in lead, arsenate and chloride, were classified as phosphatian johnbaumites, while lead-rich phases had trace amounts of phosphate with a composition approaching that of pure hedyphane [74]. Thus the johnbaumite structure must be stabilized by the presence of phosphate while hedyphane must be a rather stable calcium containing arsenate. Indeed, Bothe and Brown [23] observed that small amounts of phosphate facilitated the synthesis of johnbaumite even in the presence of magnesium.

4.2. Lead Arsenates

Synthetic lead arsenates have been used from the beginning of the 20th century as insecticides, causing some soils to be contaminated with both lead and arsenic. Nevertheless, some *apatite* group lead phosphates display the same solubility characteristics as those of pyromorphite. The crystallization of mimetite gives rise to very low concentrations of dissolved arsenic and lead, removing them from solution even under conditions where other lead containing phases normally crystallize [75]. On the other hand, Rao [43] published results for the solubility of hydroxypyromorphite and its solid solution with hydroxymimetite that show that hydroxymimetite is much more soluble than hydroxypyromorphite. Unfortunately, it is not possible to calculate solubility constants for these phases from the data presented, since only pH and total lead concentrations were reported.

Of course, there can be no suggestion of remediation of arsenic-contaminated environments by addition of lead and chloride to cause the precipitation of mimetite; that would be quite counterproductive. It is noted, however, that when pyromorphite crstallization is promoted by addition of phosphate to environments rich in lead, the gratuitous scavenging of arsenate by the pyromorphite–mimetite solid-solution [38] may be a useful subsidiary outcome.

5. CONCLUSIONS

In situ metal immobilization employing phosphate is a cost-effective and environmental friendly technique of less disruptive nature [24, 28, 47, 76] than other remediation techniques such as soil removal, washing or leaching. This technique was suggested by the United States Environmental Protection Agency in 1996 as an alternative to soil removal for the amendment of urban lead-contaminated soils [27]. The treatment of waters with calcium oxide or hydroxide can precipitate dissolved arsenate. To attain a total dissolved arsenic level lower than the maximum permissible concentration for total arsenic in potable water (0.010 mg/L arsenic) it is necessary to have a pH higher than 12.5, which poses other environmental problems. Under these conditions calcium carbonate will be the thermodynamically stable solid phase under ambient surface conditions. Transformation of calcium arsenates into calcium carbonate will re-release arsenic into the environment. For $pH > 12$

the stable phase is a solid of chemical composition $Ca_2(AsO_4)(OH) \cdot 2H_2O$ [23, 36]. From Fig. 1 it can be seen that johnbaumite is stable under alkaline conditions, and is moderately soluble when compared with hydroxyapatite. The use of hydroxyapatite to decrease the arsenate content of contaminated waters is not effective as johnbaumite is much more soluble than the phosphate analogue and substitution of phosphate by arsenate in the *apatite* lattice can occur only for very high concentrations of arsenate. McNeill and Edwards [77], in a study of arsenic removal during water softening, concluded that solid *apatite* was not effective for removing arsenate from waters, and they could not crystallize solids with the *apatite* structure from solutions supersaturated in relation to hydroxyapatite. The crystallization of hydroxyapatite from dilute supersaturated solutions is a very slow process, and is thus not suitable for water treatment. No calcium arsenate is suitable for lowering dissolved arsenic in aqueous solutions unless the pH is high and large amounts of calcium are present. This situation is far removed from the usual values of calcium concentrations and pH in natural waters, which are mostly governed by geochemical processes involving calcium carbonate under alkaline conditions.

The approaches outlined in this brief review are however of great application with respect to decontamination of lead and arsenic streams and the results of field trials [78] and related environmental issues [79] continue to be aired in the literature. It is anticipated that further thermodynamic data for some of the phases that can not at present be modelled satisfactorily will be forthcoming as a result of the importance of the problems that have been mentioned. These are of particular importance with respect to long-term environmental modelling and will have to be viewed in conjunction with related kinetic studies, areas which have received much less attention that equilibrium approaches [80].

REFERENCES

[1] P.A. Wood, in "Contaminated Land and its Reclamation" (R.E. Hester and R.M. Harrison, eds.), p. 47, Vol. 7, The Royal Society of Chemistry, Cambridge, 1997.

[2] J.W. Anthony, R.A. Bideaux, K.W. Bladh and M.C. Nichols, Handbook of Mineralogy, Vol. IV. Arsenates, Phosphates, Vanadates, Mineral Data Publishing, Tucson, 2000.

[3] M. Fleischer and J.A. Mandarino, Glossary of Mineral Species, The Mineralogical Record, Inc., Tucson, 1995.

[4] J.L. Conca and J. Wright, Appl. Geochem., 21 (2006) 1288.

[5] J.O. Nriagu, Inorg. Chem., 11 (1972) 2499.

[6] T. Suzuki, T. Hatsushika and Y. Hayakawa, J. Chem. Soc. Faraday Trans. I, 77 (1981) 1059.

[7] T. Suzuki, T. Hatsushika and M. Miyake, J. Chem. Soc. Faraday Trans. I, 78 (1982) 3605.
[8] T. Suzuki, K. Ishigaki and M. Miyake, J. Chem. Soc. Faraday Trans. I, 80 (1984) 3157.
[9] Y. Takeuchi and H. Arai, J. Chem. Eng. Japan, 23 (1990) 75.
[10] R. Gauglitz, M. Holterdorf, W. Franke and G. Marx, Radiochimica Acta, 58/59 (1992) 253.
[11] L.Q. Ma, J. Environ. Qual., 25 (1996) 1420.
[12] Q.Y. Ma, S.J. Traina, T.J. Logan and J.A. Ryan, Environ. Sci. Technol., 27 (1993) 1803.
[13] Q.Y. Ma, T.J. Logan, S.J. Traina and J.A. Ryan, Environ. Sci. Technol., 28 (1994) 408.
[14] Q.Y. Ma, S.J. Traina, T.J. Logan and J.A. Ryan, Environ. Sci. Technol., 28 (1994) 1219.
[15] Q.Y. Ma, T.J. Logan and S.J. Traina, Environ. Sci. Technol., 29 (1995) 1118.
[16] Y. Xu and F.W. Schwartz, J. Contam. Hydrol., 15 (1994) 187.
[17] M.V. Ruby, A. Davis and A. Nicholson, Environ. Sci. Technol., 28 (1994) 646.
[18] V. Laperche, S.J. Traina, P. Gaddam and T.J. Logan, Environ. Sci. Technol., 30 (1996) 3321.
[19] V. Laperche, T.J. Logan, P. Gaddam and S.J. Traina, Environ. Sci Technol., 31 (1997) 2745.
[20] X.-B. Chen, J.V. Wright, J.L. Conca and L.M. Peurrung, Environ. Sci. Technol., 31 (1997) 624.
[21] T.T. Eighmy, B.S. Crannell, L.G. Butler, F.K. Cartledge, E.F. Emery, D. Oblas, J.E. Krzanowski, J.D. Eusden, Jr., E.L. Shaw and C.A. Francis, Environ. Sci. Technol., 31 (1997) 3330.
[22] W.D. Bostick, R.J. Jarabek, D.A. Bostick and J. Conca, Adv. Environ. Res., 3 (1999) 488.
[23] J.V. Bothe, Jr. and P.W. Brown, Environ. Sci. Technol., 33 (1999) 3806.
[24] P. Zhang and J.A. Ryan, Environ. Sci. Technol., 33 (1999) 625.
[25] M. Manecki, P.A. Maurice and S.J. Traina, Soil Sci., 165 (2000) 920.
[26] C.C. Fuller, J.R. Bargar, J.A. Davis and M.J. Piana, Environ. Sci. Technol., 36 (2002) 158.
[27] J.C. Zwonitzer, G.M. Pierzynski and G.M. Hettiarachchi, Water Air Soil Pollut., 143 (2003) 193.
[28] S.B. Chen, Y.G. Zhu and Y.B. Ma, J. Hazard. Mater., 134 (2006) 74.
[29] E. Mavropoulos, N.C.C. Rocha, J.C. Moreira, A.M. Rossi and G.A. Soares, Mater. Charact., 53 (2004) 71.
[30] P. Viellard and Y. Tardy, in "Phosphate Minerals" (J.O. Nriagu and P.B. Moore, eds.), p. 171, Springer-Verlag, Berlin, 1984.
[31] V.M. Valyashko, L.N. Kogarko and I.L. Kodakovsky, Geochem. Int., 5 (1968) 21.
[32] E.P. Egan, Z.T. Wakefield and K.E. Elmore, J. Am. Chem. Soc., 73 (1951) 5581.
[33] H.G. McCann, Arch. Oral Biol., 13 (1968) 987.
[34] Z. Amjad, P.G. Koutsoukos and G.H. Nancollas, J. Colloid Interface Sci., 82 (1981) 394.
[35] M.C.F. Magalhães, P.A.A.P. Marques and R.N. Correia, in "Biomineralization – Medical Aspects of Solubility" (E. Königsberger and L.-C. Königsberger, eds.), p. 71, Wiley, Chichester, 2006.

[36] J.V. Bothe, Jr. and P.W. Brown, J. Hazard. Mater., B69 (1999) 197.
[37] T.S.B. Narasaraju, P. Lahiri, P.R. Yadav and U.S. Rai, Polyhedron, 4 (1985) 53.
[38] A.I. Inegbenebor, J.H. Thomas and P.A. Williams, Miner. Mag., 53 (1989) 363.
[39] J.O. Nriagu, Geochim. Cosmochim. Acta, 38 (1974) 887.
[40] P.P. Mahapatra, L.M. Mahapatra and B. Mishra, Polyhedron, 6 (1987) 1049.
[41] M. Manecki, P.A. Maurice and S.J. Traina, Am. Miner., 85 (2000) 932.
[42] S.K. Lower, P.A. Maurice and S.J. Traina, Geochim. Cosmochim. Acta, 62 (1998) 1773 (data from [5, 39]).
[43] S.V.C. Rao, J. Indian Chem. Soc., 53 (1976) 587.
[44] P.A.A.P. Marques, Reacções de Superfície de Cerâmicos de Fosfato de Cálcio em plasma simulado, University of Aveiro, Unpublished Ph.D. Thesis, 2003.
[45] M. Chen, L.Q. Ma, S.P. Singh. R.X. Cao and R. Melamed, Adv. Environ. Res., 8 (2003) 93.
[46] J. Cotter-Howells, Environ. Pollut., 93 (1996) 9.
[47] J.D. Cotter-Howells and S. Caporn, Appl. Geochem., 11 (1996) 335.
[48] X.-B. Chen, J.V. Wright, J.L. Conca and L.M. Peurrung, Water Air Soil Pollut., 98 (1997) 57.
[49] P. Zhang, J.A. Ryan and L.T. Bryndzia, Environ. Sci. Technol., 31 (1997) 2673.
[50] P. Zhang, J. Ryan and J. Yang, Environ. Sci. Technol., 32 (1998) 2763.
[51] M. Manecki, A. Bogucka, T. Bajda and O. Borkiewicz, Environ. Chem. Lett., 3 (2006) 178.
[52] T. Suzuki and I. Kyoichi, Chem. Eng. Commun., 34 (1985) 143.
[53] S. Bailliez, A. Nzihou, E. Beche and G. Flamant, Process Safety Environ. Protect., 82 (2004) 175.
[54] X.-Y. Tang, Y.-G. Zhu, S.-B. Chen, L.-L. Tang and X.-P. Chen, Environ. Int., 30 (2004) 531.
[55] Y.G. Zhu, S.B. Chen and J.C. Yang, Environ. Int., 30 (2004) 351.
[56] D. Koeppenkastrop and E.J. De Carlo, Chem. Geol., 95 (1990) 251.
[57] J.D. Cotter-Howells, P.E. Champness, J.M. Charnock and R.A.D. Pattrick, Eur. J. Soil Sci., 45 (1994) 393.
[58] J.D. Cotter-Howells and I. Thornton, Environ. Geochem. Health, 13 (1991) 127.
[59] M.F. Hossain, Agric. Ecosyst. Environ., 113 (2006) 1.
[60] S. Sarkar, A. Gupta, R.K. Biswas, A.K. Deb, J.E. Greenleaf and A.K. SenGupta, Water Res., 39 (2005) 2196.
[61] H. Garelick, A. Dybowska, E. Valsami-Jones and N.D. Priest, J. Soils Sediments, 5 (2005) 182.
[62] D.H. Moon, D. Dermatas and N. Menounou, Sci. Total Environ., 330 (2004) 171.
[63] P. Navarro, C. Vargas, E. Araya, I. Martin and F.J. Alguacil, Rev. Metalurgia, 40 (2004) 409.
[64] R.W. Fuessle and M.A. Taylor, J. Environ. Eng. ASCE, 130 (2004) 1063.
[65] M.A. Hossain, M.K. SenGupta, S. Ahamed, M.M. Rahman, D. Mondal, D. Lodh, B. Das, B. Nayak, B.K. Roy, A. Mukherjee and D. Chakraborti, Environ. Sci. Technol., 39 (2005) 4300.
[66] T. Nishimura, C.T. Ito, K. Tozawa and R.G. Robins, Impurity, Control and Disposal, Proceedings of the 15th Annual Hydrometallurgical Meeting, Vancouver, p. 2.1, 1985.

[67] J.V. Bothe, Jr. and P.W. Brown, J. Am. Ceram. Soc., 85 (2002) 221.
[68] M. Kucharski, W. Mroz, J. Kowalczyk, B. Szafirska and M. Gluzinska, Arch. Metall., 47 (2002) 119.
[69] P.A. Riveros, J.E. Dutrizac and P. Spencer, Canad. Metall. Q., 40 (2001) 395.
[70] H.V. Tartar, L. Wood and E. Hiner, J. Am. Chem. Soc., 46 (1924) 809.
[71] E.A. Murphy and M. Aucott, Sci. Total Environ., 218 (1998) 89.
[72] T. Nishimura and R.G. Robins, Miner. Process. Extr. Metal. Rev., 18 (1998) 283.
[73] R. Stahl-Brasse, N. Ariguib-Kbir and H. Guérin, Bull. Soc. Chim. France, (8) (1971) 2828.
[74] A.G. Christy and K. Gatedal, Miner. Mag., 69 (2005) 995.
[75] M.C.F. Magalhães and M.C.M. Silva, Monatsh. Chem., 134 (2003) 735.
[76] X.D. Cao, L.Q. Ma, D.R. Rhue and C.S. Appel-Environ. Pollut., 131 (2004) 435.
[77] L.S. McNeill and M. Edwards, J. Environ. Eng.-ASCE, 123 (1997) 453.
[78] R. Melamed, X. Cao, M. Chen and L.Q. Ma, Sci. Total Environ., 305 (2003) 117.
[79] D.J. Vaughan, Elements, 2 (2006) 71; P. O'Day, ibid 77; G. Morin and G. Calas, ibid 97.
[80] P.A. Williams, Pure Appl. Chem., 77 (2005) 643.

PART V

POLYMER RELATED ISSUES

Thermodynamics, Solubility and Environmental Issues
T.M. Letcher (editor)

Chapter 19

Solubility of Gases and Vapors in Polylactide Polymers

Rafael A. Auras

140, School of Packaging, Michigan State University, East Lansing, MI 48824-1223, USA

1. INTRODUCTION

Polylactide (PLA) polymers are environment-friendly biodegradable materials that have garnered growing attention in the last few years as food packaging materials since packages made from PLA have the advantage of being produced from renewable resources, provide significant energy savings, and can be recyclable and compostable [1–3]. Economic and life cycle analysis studies have shown that PLA is an economically feasible material to be used as a packaging polymer [4–6]. Moreover, medical studies have shown that the level of lactic acid (LA) that migrates to food from packaging containers is much lower than the amount of LA used in common food ingredients [7]. Therefore, polymers derived from LA are good candidates for packaging applications.

Currently, PLA is being used as a food packaging polymer for short shelf life products with common applications such as containers, drinking cups, sundae and salad cups, overwrap and lamination films, and blister packages [8, 9]. It is also being used for bottled water. PLA's optical, physical, and mechanical properties have been compared to those of polystyrene (PS) and polyethylene terephthalate (PET) [10] although studies comparing the actual performance of PLA packages versus other plastics are scarce. PLA plastics are still in the early developmental stage for packaging applications. At this time, some of the basic properties such as optical, physical, and mechanical properties have been determined as a function of lactide content and environmental parameters such as temperature and humidity [1, 3]. Barrier properties of these polymers (i.e., solubility, diffusion, and permeation) have barely been studied. This chapter summarizes the state of the knowledge regarding the interaction and solubility of gases, vapors, and organic compounds in

PLA polymers. First, it describes the main factors that influence polymer/ chemical interactions. Second, it details the basic properties of PLA polymers, and finally it describes the interaction of PLA with gases, vapors, and organic compounds.

2. POLYMER/CHEMICAL INTERACTIONS

Unlike glass, metals, and ceramics, polymeric materials are relatively permeable to small molecules such as gases, water and organic vapors, and liquids. Chemical permeability through polymeric membranes may result in continuing change of the polymer structure. For example, in packaging applications, the loss of specific chemical compounds, aroma or flavor constituents or the gain of off-odors due to mass transfer processes can lead to a reduction of product quality, resulting in a reduction in shelf life of products. Mass transfer processes between the polymer, a permeant, and/or the environment include sorption, permeation, and migration processes. Sorption is the uptake of product components, such as aroma, flavor, odor, or coloring compounds by the polymer material. The permeation process is a diffusion-controlled process where a permeant is sorbed by, diffused through, and then desorbed from a polymeric membrane. Migration is the transfer of substances originally present in the plastic material into a packaged product [11].

For polymer application, safety, and optimization, it is important to know how a permeant transfers both into and through the polymer. Neglecting obvious troubles such as leakage through cracks or pinholes in the material, to analyze and predict these mass transfer processes one needs to (a) estimate the solubility of a chemical in a specified material and (b) characterize the diffusion coefficient.

3. THEORETICAL CONSIDERATIONS

3.1. Solubility

The thermodynamic criterion of solubility is based on the free energy of mixing ΔG_M, which states that two substances are mutually soluble if ΔG_M is zero or negative. The free energy of mixing for a solution process between a solvent and a polymer is given by the relation

$$\Delta G_M = \Delta H_M - T\,\Delta S_M \tag{1}$$

where ΔH_M is the enthalpy of mixing per unit volume, T the absolute temperature, and ΔS_M the entropy of mixing per unit volume. As ΔS_M is generally small and positive, there is a certain limiting positive value of ΔH_{M} below which dissolution is possible.

Hildebrand and Scott [12] first correlated the enthalpy of mixing with the cohesive properties of the permeant, and for a binary mixture of a polymer and a solvent, the Hildebrand–Scatchard equation is expressed as:

$$\Delta h_{\mathrm{M}} = \Phi_{\mathrm{P}}\Phi_{\mathrm{S}}(X_{\mathrm{P}}V_{\mathrm{P}} + X_{\mathrm{S}}V_{\mathrm{S}})(\delta_{\mathrm{P}} - \delta_{\mathrm{S}})^2 \tag{2}$$

$$\Phi_{\mathrm{P}} = \frac{X_{\mathrm{P}}V_{\mathrm{P}}}{X_{\mathrm{P}}V_{\mathrm{P}} + X_{\mathrm{S}}V_{\mathrm{S}}} \tag{3}$$

$$\Phi_{\mathrm{S}} = \frac{X_{\mathrm{S}}V_{\mathrm{S}}}{X_{\mathrm{P}}V_{\mathrm{P}} + X_{\mathrm{S}}V_{\mathrm{S}}} \tag{4}$$

where Φ_{P} and Φ_{S} are the volume fractions, X_{P} and X_{S} the mole fractions, V_{P} and V_{S} the molar volumes, and δ_{P} and δ_{S} the solubility parameters of the polymer and the organic compound. The term "solubility parameter" was also first introduced by Hildebrand and Scott [12]. The solubility parameter, δ, is defined as the square root of the cohesive energy density (CED):

$$\delta = (\mathrm{CED})^{1/2} = \left(\frac{\Delta E_{\mathrm{v}}}{V_{\mathrm{l}}}\right)^{1/2} \tag{5}$$

where ΔE_{v} is the molar energy of vaporization and V_{l} the molar volume of the liquid.

To derive the enthalpy of mixing ΔH_{M} per mole of organic compound, Δh_{M} is differentiated by the mole fraction X_{S} of organic compound.

$$\left(\frac{\delta \Delta h_{\mathrm{M}}}{\delta X_{\mathrm{S}}}\right)_{XP} = \Delta H_{\mathrm{M}} = V_{\mathrm{S}}\Phi_{\mathrm{P}}^2(\delta_{\mathrm{P}} - \delta_{\mathrm{S}})^2 \tag{6}$$

taking into account that the change in volume of mixing is negligible, $\Phi_{\mathrm{P}}^2 = 1$. So,

$$\Delta H_{\mathrm{M}} = V_{\mathrm{S}}(\delta_{\mathrm{P}} - \delta_{\mathrm{S}})^2 \tag{7}$$

which can also be correlated with the Flory–Huggins parameter (χ) [13].

$$\chi = \left(\frac{V_S}{RT}\right)(\delta_P - \delta_S)^2 \tag{8}$$

A basic assumption of solubility parameter theory is that a correlation exists between the CED of pure substances and their mutual solubility. Therefore, Eq.(6) predicts that $\Delta H_M = 0$ if $\delta_P = \delta_S$; hence, two substances with equal solubility parameters should be mutually soluble. This is in accordance with the general rule that chemical and structural similarities favor solubility. Therefore, the relative affinity of a polymer and solvent can be assessed using solubility parameters.

Hansen and Skaarup [14] proposed an extension of the solubility parameter approach to polar and hydrogen-bonding systems. Their approach divides the contributions to the overall solubility parameter into three components representing non-polar (δ_D), polar (δ_P), and hydrogen-bonding (δ_H) contributions to the CED. Hansen's three-dimensional solubility parameters, δ_D, δ_P, and δ_H, are among the most extensively used. However, to apply this theory some restrictions are applied. Only energetic contributions to the mixing process are involved and entropic effects are not considered. Moreover, the heat of mixing is estimated from properties of the pure substances. Despite these shortcomings, the solubility parameter approach is convenient to use and helpful as a means of estimating the expected compatibility between the polymer phase and the solvents. Thus, the Hansen solubility parameters (δ_D, δ_P, δ_H) have been combined as follows:

$$\delta_T^2 = \delta_D^2 + \delta_P^2 + \delta_H^2 \tag{9}$$

The total solubility parameter, δ_T, corresponds to the overall Hildebrand solubility parameter, and the Hansen solubility parameters, δ_D, δ_P, and δ_H, are contributions from non-polar interaction, polar interaction, and hydrogen bonding, respectively [15]. The end point of the radius vector represents the solubility parameter. This means that each solvent and each polymer can be located in a three-dimensional space.

Using an Arrhenius-type relationship [16], sorption coefficient of a permeant or chemical compound in an amorphous polymer can be represented as

$$S_a = S_0 \exp\left(\frac{-\Delta H_S}{RT}\right) \tag{10}$$

where ΔH_S is the enthalpy of sorption per unit mole that is a function of the enthalpy of condensation (ΔH_C) and the enthalpy of mixing (ΔH_M)

$$\Delta H_S = \Delta H_C + \Delta H_M \tag{11}$$

The enthalpy of condensation of a penetrant is related to its enthalpy of vaporization: $\Delta H_C = -\Delta H_V$. Experimentally, there are strong correlations between ΔH_V and any of the scales of gas condensability such as critical temperature.

Now, if we replace Eq. (7) in Eq. (10) the following expression is obtained:

$$S_a = S_0 \exp\left(\frac{-V_S(\delta_P - \delta_S)^2}{RT}\right) \tag{12}$$

Paik and Tigani [17] found that the value $\delta_P - \delta_S$ does not lead to good correlation due to incomplete treatment of mutual molecular interaction forces. Matsui et al. [13] proposed the use of Chen's theory [18] where polymer-aroma compound compatibility (δ_C) is calculated from the energy of mixing caused by dispersion and dipole forces. Although the use of Chen's theory showed a good correlation between the solubility parameters and δ_C [13, 19], it does not take into account the importance of hydrogen bonding to solubility. By using Hansen's solubility parameters, which take into account the different interaction forces between a polymer and a permeant to calculate the enthalpy of mixing, Eq. (7) becomes

$$\Delta H_M = V_S \, [(\delta_{DP} - \delta_{DS})^2 + (\delta_{PP} - \delta_{PS})^2 + (\delta_{HP} - \delta_{HS})^2] \tag{13}$$

The three components, δ_D, δ_P, and δ_H, lie as vectors along orthogonal axes. The distance $\Delta\delta$ between the end points of the vectors representing the polymer and solvent is given as

$$\Delta\delta = [(\delta_{DP} - \delta_{DS})^2 + (\delta_{PP} - \delta_{PS})^2 + (\delta_{HP} - \delta_{HS})^2]^{1/2} \tag{14}$$

Good solubility is typically predicted when $\Delta\delta$ is smaller than 2.5 $MPa^{1/2}$ or 5 $(J/cm^3)^{1/2}$ [20]. Extensive tabulations of three-dimensional solubility parameters of different solvents are available in the literature and a partial list of values of δ_D, δ_P, and δ_H for a number of solvents can be found in Hansen and Barton [15, 21]. By introducing the concept from Eq. (14), and replacing $\delta_P - \delta_S$ in Eq. (12) with the value $\Delta\delta$, the solubility coefficient can be computed as:

$$S_a = S_0 \exp\gamma\left(\frac{V_S\,\Delta\delta^2}{RT}\right) \tag{15}$$

where γ is a constant characteristic of polymer property.

3.2. Diffusion

The diffusion process for the steady state is described by the general Fick's law, where the left-hand side is the divergence of F and the right-hand side is the Laplacian of the chemical potential.

$$F = D \times \nabla\mu \tag{16}$$

The unsteady-state process is described by Fick's second law

$$\frac{\partial C}{\partial t} = \mathrm{div}(D \times \mathrm{grad}\mu) \tag{17}$$

where t is time. In the case of a membrane, and if the diffusion coefficient D is independent of the concentration and the chemical potential can be expressed by the concentration, the one-dimensional form of Eq. (17) is equal to

$$\frac{\partial c}{\partial t} = D \times \frac{\delta^2 c}{\delta x^2} \tag{18}$$

Regarding Eq. (18), values of D are normally assumed to be independent of both penetrant concentration and polymer relaxations at low concentration. This is especially true for gases such as oxygen and carbon dioxide at atmospheric pressure, and some organic compounds. Many theories have been proposed and many models have been developed to describe diffusion in polymers; a detailed description of these models can be found elsewhere [22]. The diffusion processes through the membrane are generally unidirectional and perpendicular to the flat surface, and solutions to the diffusion equations are obtained from the boundary conditions where Henry's or Langmuir–Henry's law are applied.

When the diffusion coefficient (D) is independent of concentration (c), the one-dimensional solution of Eq. (18) can be expressed as Eq. (19), which describes the process mentioned above with boundary conditions according to Fick's second law [23] for a film sheet tested under isostatic conditions [24].

$$\frac{F_t}{F_\infty} = \left(\frac{4}{\sqrt{\pi}}\right)\left(\sqrt{\frac{l^2}{4Dt}}\right)\sum_{n=1,3,5,\ldots}^{\infty} \exp\left(\frac{-n^2 l^2}{4Dt}\right) \quad (19)$$

where F_t is the flow rate of the compound permeating the film in the transient state at time t, F_∞ the transmission rate in the steady state, l the film thickness, and D the diffusion coefficient. From the transient flow rate profile, the diffusion coefficient, D, can be estimated from Eq. (20) [23].

$$D = \frac{l^2}{7.199 \times t_{0.5}} \quad (20)$$

3.3. Permeability

Using the value of permeant flow at steady state F_∞, the permeability coefficient P is determined by Eq. (21)

$$P = \frac{F_\infty l}{A \Delta p} \quad (21)$$

where A is the area of the film and Δp the partial pressure gradient across the polymer film. Then, if Henry's law of solubility is applied, such as in the case of oxygen diffusion and carbon dioxide in films at low pressure (i.e., up to 1 atm), the solubility coefficient, S, can be calculated from

$$P = S \times D \quad (22)$$

4. FACTORS AFFECTING MASS TRANSFER IN POLYMERS

Sorption of gases, vapors, or organic permeants into semi-crystalline polymers is mainly described and modeled as a function of different factors or parameters

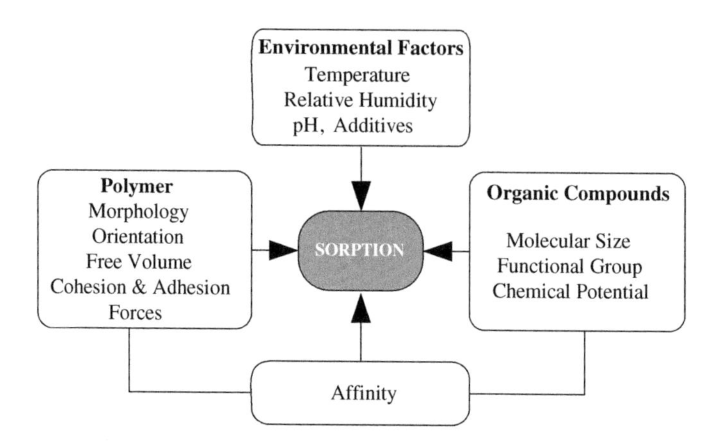

Fig. 1. Factors affecting sorption, adapted from Osajima and Matsui [30].

such as the (i) polymer-organic molecule chemistry; (ii) free volume in the amorphous phase; (iii) crystallinity; (iv) molecular orientation; and (v) physical aging and polymer thermal history [25–28]. Factors affecting the sorption of organic compounds in polymeric materials are summarized in Fig. 1. In this chapter, only the effects of the temperature, crystallinity, and free volume are addressed. More information about these effects and other factors that affect polymer permeant interactions can be found elsewhere [29].

4.1. Temperature

The temperature dependence of *P*, *S*, and *D* can be described by Van't Hoff's–Arrhenius equations [31].

$$S(T) = S_0 \exp\left(\frac{-\Delta H_S}{RT}\right) \tag{23}$$

$$D(T) = D_0 \exp\left(\frac{-E_D}{RT}\right) \tag{24}$$

$$P(T) = P_0 \exp\left(\frac{-E_P}{RT}\right) \tag{25}$$

where ΔH_S is the molar heat of sorption, E_D the activation energy of diffusion, and E_P the apparent activation energy of permeation. It is easy to see that

$$P_0 = S_0 D_0 \tag{26}$$

$$E_P = \Delta H_S + E_D \tag{27}$$

where ΔH_S is expressed by Eq. (11). As can be observed, P_0, S_0, D_0, E_P, ΔH_S, and E_D can be derived from the isostatic permeation experiment.

4.2. Free Volume

Free volume is an intrinsic property of the polymer matrix and is created by the gaps left between entangled polymer chains. Under the free volume model the absorption and diffusion of molecules in polymers depend greatly on the available free volume. For instance, many polymers show a sorption increase as the amount of free volume increases. One of the earliest models describing this behavior was developed by Fujita et al. [32]. Since then many researchers have worked with different models based on the free volume concept to describe sorption and diffusion in the glassy state [33–36].

The simplest and most used way to calculate the specific free volume in a polymer is as follows:

$$V_f = V_{SP} - 1.3V_W \tag{28}$$

where V_{SP} (cm^3/g) is the specific volume of a polymer and is equal to $1/\rho$, where ρ (g/cm^3) is the polymer density, $1.3V_W$ the occupied volume, and V_W is the sum of the increments of individual Van der Waals volumes (tabulated in Van Krevelen [20]). In general, the fractional free volume (FFV = V_f/V_{SP}) is used to express transport correlations within a family of polymers.

4.3. Crystallinity

Since crystalline regions are highly ordered as compared to amorphous regions, the free volume is lower or not existent in these regions. The crystalline regions act as excluded phases for the sorption process and as impermeable barriers for diffusion and sorption. Furthermore, crystalline regions present a constraint on the polymer chains in the amorphous region. This chain restriction influences the sorption process in the amorphous phase by limiting the effective path length of diffusion and reducing the polymer chain mobility, which creates a higher activation energy for diffusion. It should be noted that for many polymer/gas or vapor systems, the solubility coefficients are directly proportional to the free volume fraction of the amorphous phase [37–39], and for some semi-crystalline polymers it has been demonstrated [20] that Eq. (10) can be rewritten as follows:

$$S_{sc} \approx S_a \cdot (1-\chi_c) \tag{29}$$

where S_{sc} is the solubility of a semi-crystalline polymer, χ_c the degree of crystallinity of the polymer, and S_a the solubility of the amorphous structure as defined by Eq. (15). So, Eq. (15) will be expressed as:

$$S_{sc} \approx S_0' \exp\gamma\left[\frac{V_S \Delta\delta^2}{RT}\right](1-\chi_c) \tag{30}$$

Finally, as a first approximation and since some permeants such as oxygen and carbon dioxide are insoluble in the crystalline part of polymer films [40, 41], the diffusion, solubility, and permeation coefficients for semi-crystalline polymers can be estimated as:

$$D_{sc} \approx D_a \cdot (1-\chi_c) \tag{31}$$

$$S_{sc} \approx S_a \cdot (1-\chi_c) \tag{32}$$

$$P = D_{sc} \cdot S_{sc} \approx D_a \cdot S_a (1-\chi_c)^2 \tag{33}$$

where χ_c is the degree of crystallinity, S_a and D_a the diffusion and solubility coefficients in the amorphous part of the polymer as expressed by Eqs. (15 and 20), respectively, and D_{sc} the diffusion of the semi-crystalline polymer.

5. POLYLACTIDES

In the case of PLA as well as other polymers, the physical properties are determined by the constitutional unit, especially the non-functional structural groups, which are the real "building blocks" of the polymer chain, the functional structural groups or end groups such as –OH and –H, the molecular architecture (i.e., stereochemistry and arrangement, Scheme 1), and the molecular mass distribution.

The configurational and conformational differences between *meso*- and L-lactide result in different physical properties, such as melting point,

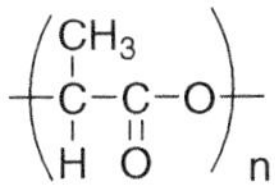

$$\left(\begin{array}{c} CH_3 \\ | \\ -C-C-O- \\ | \quad \| \\ H \quad O \end{array} \right)_n$$

Scheme 1. PLA constitutional unit.

solubility, and relative volatility. Depending on the preparation conditions, poly (L-lactide) crystallizes in three forms (α-, β-, and γ-) [1]. The stable α-form exhibits a well-defined diffraction pattern with a space group orthorhombic $P2_12_12_1$, with a unit cell containing two antiparallel chains. The chain conformation is a twofold (15 × 2/7) helix distorted periodical form of the regular s(3 × 10/7 helix). The lattice parameters are a = 10.66 Å, b = 6.16 Å, and c (chain axis) = 28.88 Å, with a crystal density ~1.26 g/cm^3. The β-form of PLA is generally prepared at a high draw ratio and a high drawing temperature. The chain conformation is a left-handed threefold helix. The β-form of PLA has an orthorhombic unit cell (containing six chains) containing a 3 1 (3 Å rise/1 monomeric unit) polymeric helix. The lattice parameters of the unit cell are: a = 10.31 Å, b = 18.21 Å, and c = 9.0 Å. The α-structure is more stable than the β-structure, with a melting point of 185°C compared to 175°C for the β-structure. Chain packing in the β-form has recently been termed a frustrate structure, where the crystal structure rests on a frustrated packing of three threefold helices in a trigonal unit cell of parameters $a = b$ = 10.52 Å and c = 8.8 Å, space group $P3_2$ [42]. The γ-form is formed by epitaxial crystallization and contains two antiparallel s(3/2) helices in the pseudoorthorhombic unit cell a = 9.95 Å, b = 6.25 Å, and c = 8.8 Å, and it assumes the known threefold helix of PLA [1].

High molecular weight PLAs are either amorphous or semi-crystalline at room temperature, depending on the amounts of L-, D-, and *meso*-lactide that are in the structure. For amorphous PLA, the glass transition temperature (T_g) is one of the most important parameters since dramatic changes in the motion of polymer chains take place at and above T_g. In the case of semi-crystalline PLA, both the T_g and melting temperature (T_m) are important physical parameters for predicting PLA behavior. The glass–rubber transition temperature is not a real thermodynamic transition point, but it is an indication of the transition between the glassy and the rubbery states of polymers. In the case of PLA, the glass transition temperature is also determined by the proportion of different lactides. Table 1 shows the glass and melting temperatures of two PLA polymers compared to those of two other standard polymers.

Table 1
Physical properties of 98% L-lactide, 94% L-lactide, PS, and PET

	98% L-lactide	94% L-lactide	PS (atactic)[a]	PET
T_g (°C)	71.4	66.1	100	80
Relaxation enthalpy (J/g)	1.4	2.9	N/A	N/A
T_m (°C)	163.4	140.8	N/A	245
Enthalpy of fusion (J/g)	37.5	21.9	N/A	47.7
Crystallinity (%)	40	25	N/A	38

[a]PS used in packaging is atactic, so it does not crystallize. Since it is an amorphous polymer, it does not have a defined melting point, but gradually softens through a wide range of temperatures.

6. POLYLACTIDE BARRIER PROPERTIES

6.1. Gases

Permeability is an intrinsic property of a gas–polymer membrane system. Correlations that relate diffusion, solubility, and permeability coefficients of diverse gases in polymers are available. Models and group contribution theories have been developed to predict permeability of gases in polymers. However, general rules or universal correlations are usually not as good at predicting permeability as rules for a particular set of polymers. A detailed description of these correlations can be found elsewhere [29]. This section presents the main barrier properties of PLA to O_2, CO_2 and N_2.

6.1.1. Oxygen

Fig. 2 presents the oxygen permeability coefficient for PLA 98% L-lactide (98L-PLA) as a function of water activity. A significant increase of the oxygen permeability coefficient as temperature increases can be observed for 98L-PLA. 98L-PLA shows an increase of the oxygen permeability coefficients from 3.5×10^{-18} kg m/(m^2 s Pa) at 5°C to 11×10^{-18} kg m/(m^2 s Pa) at 40°C when moisture is not present. At a higher water activity level ($A_w = 0.9$), the value of the oxygen permeability coefficient increases only to 8.5×10^{-18} kg m/(m^2 s Pa) at 40°C. This reduction of the oxygen permeability coefficient as water activity increases at a constant temperature is observed in Fig. 2 for the three temperatures tested. The reduction is more pronounced at 40°C.

The difference between the values of oxygen permeability measured by Auras et al. [43] and the values measured in previous studies by Lehermeier

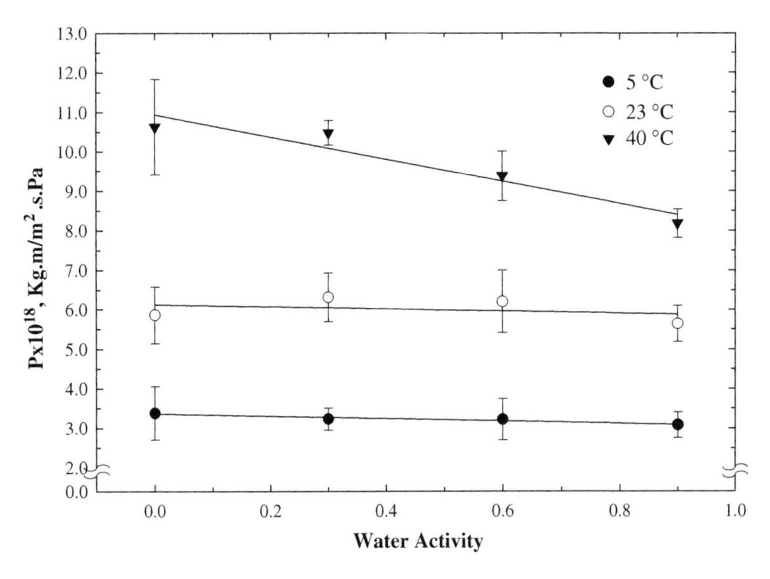

Fig. 2. Oxygen permeability coefficient as a function of water activity for 98L-PLA.

et al. [44] of 3.53×10^{-17} kg m/(m^2 s Pa) at 30°C and $A_w = 0$ for biaxially oriented PLA films can be explained by differences in the crystallinity and final processing conditions of the films.

The oxygen diffusion coefficient as a function of water activity for 98L-PLA is presented in Fig. 3a. An exponential increase of the oxygen diffusion coefficients from 2×10^{-14} to 9×10^{-14} m^2/s at 23°C can be observed as water activity increased from 0 to 0.9. The observed increase of the oxygen diffusion coefficient in the range of $A_w = 0$–0.9 can be attributed to the plasticization effect on the amorphous phase by the water molecules. Plasticization effects tend to increase the mobility of the oxygen molecules in the polymer matrix. Plasticization effects are evidenced by a reduction of the polymer's T_g [43].

Fig. 3b presents the values of the oxygen solubility coefficient in 98L-PLA films. The oxygen solubility coefficient increased as the temperature increased. A linear reduction of the solubility coefficient from 5 to 1.5×10^{-4} kg/(m^3 Pa) at 40°C can be observed as water activity increased from 0 to 0.9. This reduction can be explained, as in the case of other polymers such as PET, by the occupancy of the free volume by water molecules at higher water content, so a reduction of the oxygen solubility as water content increased is observed. PET and PLA are both hydrophobic films that absorb very low amounts of water, and show similar barrier property behavior.

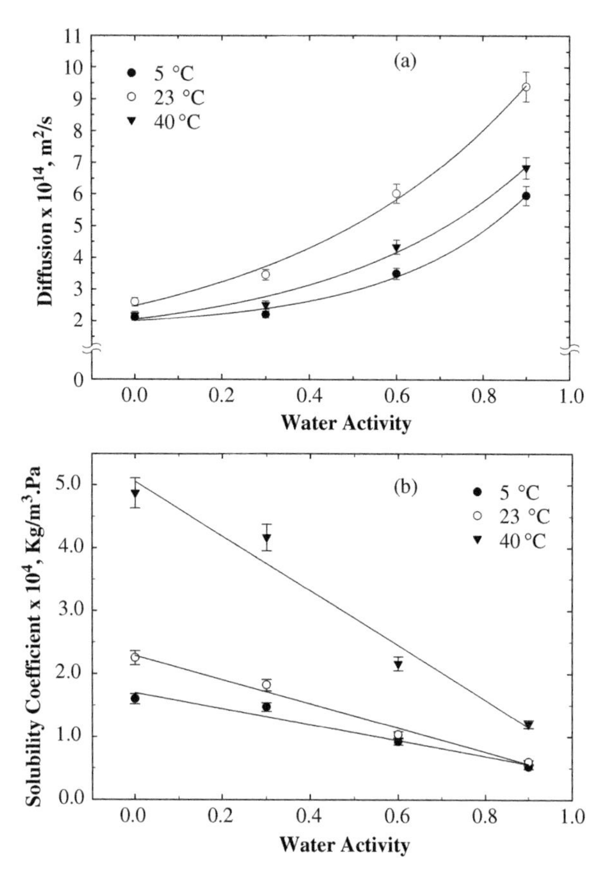

Fig. 3. (a) Oxygen diffusion coefficient of 98L-PLA films as a function of water activity; (b) solubility coefficient of 98L-PLA films as a function of water activity.

Oliveira et al. [45] determined the solubility of oxygen in PLA at 293, 303, and 313 K, and pressures between 0.11 and 0.995 bar. They reported a solubility parameter determined at 293 K of 2.24×10^{-5} kg/(m^3 Pa). The values measured by Oliveira et al. are an order of magnitude lower than the previous reported values. This difference could be attributed to the different methodology to measure oxygen permeability (i.e., quartz crystal microbalance). They also reported a Flory–Huggins parameter for O_2 in PLA between 293 and 313 K of $\chi_{12} = 1.40–2.10$.

6.1.2. Carbon Dioxide

Table 2 presents the CO_2 permeability coefficient values as a function of temperature in the range of 25–45°C at 0% RH. The activation energy was calculated according to the Arrhenius equation (25).

Table 2
CO_2 permeability coefficients and activation energy for PLA resin ($P \times 10^{17}$ kg m/(m^2 s Pa))

Resin	Temperature (°C)					E_P (kJ/mol)
	25	30	35	40	45	
98L-PLA	2.77 ± 0.05	3.12 ± 0.05	3.42 ± 0.06	3.78 ± 0.06	4.18 ± 0.15	15.65 ± 0.63

The CO_2 permeability coefficients reported here for PLA films are lower than the reported value for crystal PS at 25°C and 0% RH (1.55×10^{-16} kg m/(m^2 s Pa)) [46], but higher than those for PET (1.73 and 3.17×10^{-18} kg m/(m^2 s Pa) at 0% RH at 25 and 45°C, respectively). Carbon dioxide permeability values for PET are concomitant with the PLA values.

Oliveira et al. [45] also measured the solubility of carbon dioxide in 80% L-lactide–20% D-lactide (80L–20D) film at 293, 303, and 313 K and 0% RH. At 293 K they reported a solubility or carbon dioxide in PLA of 6.82×10^{-5} kg/(m^3 Pa), and a Flory–Huggins parameter for CO_2 in PLA at the same temperature of $\chi_{12} = 0.06$.

6.1.3. Nitrogen

The nitrogen permeability for plastic materials is generally 1/4–1/5 times smaller with respect to oxygen permeability. Therefore, if the permeability value of one of the three types of gases considered is known, it is always possible to estimate the permeability values of the other gases. Solubility of nitrogen in PLA has been studied by Oliveira et al. [45] and correlated with Flory–Huggins model. They reported nitrogen solubility coefficients of $\sim 1.59 \times 10^{-5}$ kg/(m^3 Pa) at 293 K and Flory–Huggins parameter for N_2 in PLA at 293 K of $\chi_{12} = 2.11$.

6.2. Vapors

Barrier properties to organic vapors in PLA have barely been investigated. Permeability to water and few organic compounds is reported in the literature. This section will present the barrier of PLA to water, and those organic compounds already studied in the literatures which are ethyl acetate and D-limonene. As PLA polymers become more commonly used, many other permeants are expected to be assessed.

6.2.1. Water

Fig. 4 shows the water vapor permeability coefficient (WVPC) of 98L-PLA determined at 10, 20, 30, and 37.8°C in the range of 40–90% RH. This

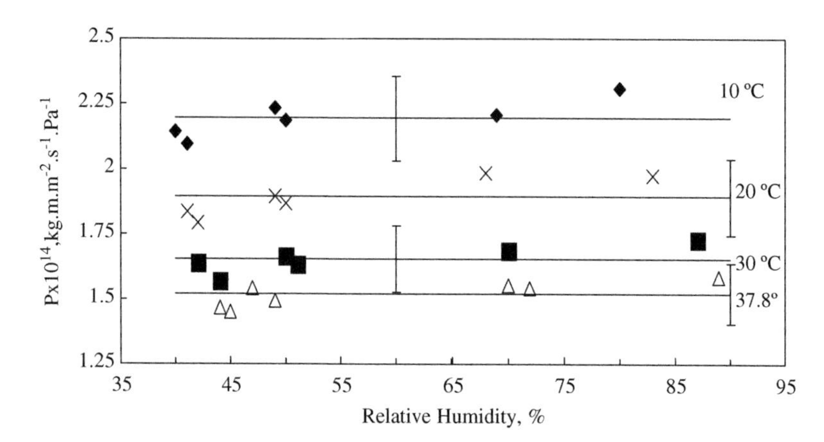

Fig. 4. Water vapor permeability coefficient of 98L-PLA as a function of percent relative humidity (% RH). Error bars represent cumulative imprecision in measurements.

figure indicates that, surprisingly, the value of permeability for 98L-PLA is practically constant over the range studied, despite PLA being a polar polymer.

The WVPC increases slightly as temperature decreases, as seen in Fig. 4. Regression analysis showed the difference in values to be statistically significant, with $p < 0.05$ in all cases. Further characteristics of the analysis can be found elsewhere [8]. The WVPC of PLA film is higher than that of PET. The average WVPC for PET is $1.1 \pm 0.1 \times 10^{-15}$ kg m/(m^2 s Pa) at 25°C in the 40–90% RH range. The reported WVPC of PS is 6.7×10^{-15} kg m/(m^2 s Pa) at 25°C [46].

The activation energy of water vapor permeation in PLA films is negative and equal to -9.73 ± 0.27 kJ/mol. Very few polymers show negative E_P values for water, and usually their magnitudes are smaller. Fig. 5 compares the activation energy of four hydrophobic polymers: PET (a hydrophobic synthetic polyester film), PS (a hydrophobic synthetic film), LDPE (a hydrophobic synthetic polyolefin film), and 98L-PLA (a hydrophobic polyester biodegradable film). PLA and PS polymers present negative water activation energies.

According to Eq. (11), the heat of sorption is a function of the heat of condensation and the heat of mixing. The heat of condensation is the energy released by a gas when it changes from the gas to liquid state. For water, the heat of condensation ΔH_C is negative and equal to -42 kJ/mol [31] while ΔH_M is positive for hydrophobic polymers and near zero or negative for polar polymers. The enthalpies of formation of water hydrogen bonds have

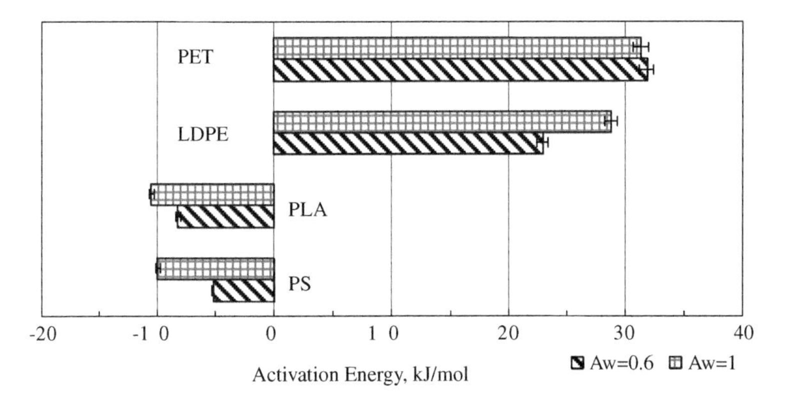

Fig. 5. Activation energy in kilojoules per mole of PET, LDPE, PLA, and PS at $A_w = 0.6$ and 1.

been measured between 14 and 27 kJ/mol. Therefore, the activation energy could be negative for polymers with low positive energy of diffusion and near zero or negative values of enthalpy of mixing ΔH_M. Further details can be found elsewhere [43].

Water solubility coefficients in PLA polymers (80L:20D) have been reported by Oliveira et al. [45]. They reported water solubility coefficients between 6.51 and 5.83×10^{-3} kg/(m^3 Pa) at 293 K. They fitted their experiment using a Flory–Huggins model with values of χ_{12} between 3.52 at 293 K and 3.67 at 313 K. Finally, Tsuji et al. [47] reported that for PLA films with molecular number between 9×10^4 and 5×10^5 g/mol and D-lactide content between 0 and 50% the WVTR changes significantly, mainly due to the fact that the change in lactide (i.e., lower D-lactide) content results in higher crystallinity. Especially, a linear decrease of WVTR was found until crystallinity amount of 20%, after that (i.e., crystallinity values >20%), the WVTR values level off.

6.2.2. *Ethyl Acetate and D-Limonene*

Ethyl acetate is a solvent found in nature and is environmentally safe and acceptable for food applications. For instance, it is used in the packaging industry because it provides high printing resolution on plastics and metals. D-Limonene is a major compound found in oil extracted from citrus and is present in orange and fruit juices. It is also used in cleaning products as a solvent or a water dilutable product. It is a very versatile chemical and therefore used in a wide variety of applications. Scalping of D-limonene from citrus juices can lead to loss of citrus flavor. Determination of the barrier properties of plastics to these organic compounds is very important in industrial applications.

Solubility and permeability coefficient values of ethyl acetate obtained from gravimetric experiments are presented in Fig. 6a and b. At 30°C, equilibrium sorption took more than 30 days, which is an indication of the low sorption of ethyl acetate in PLA. Fig. 6a presents the values of ethyl acetate sorption in PLA, PET, PP, and LDPE polymers. PS solubility coefficients are not reported in the literature, and according to Wang [48] PS dissolves in ethyl acetate. Fig. 6a shows that sorption of ethyl acetate in PLA is higher

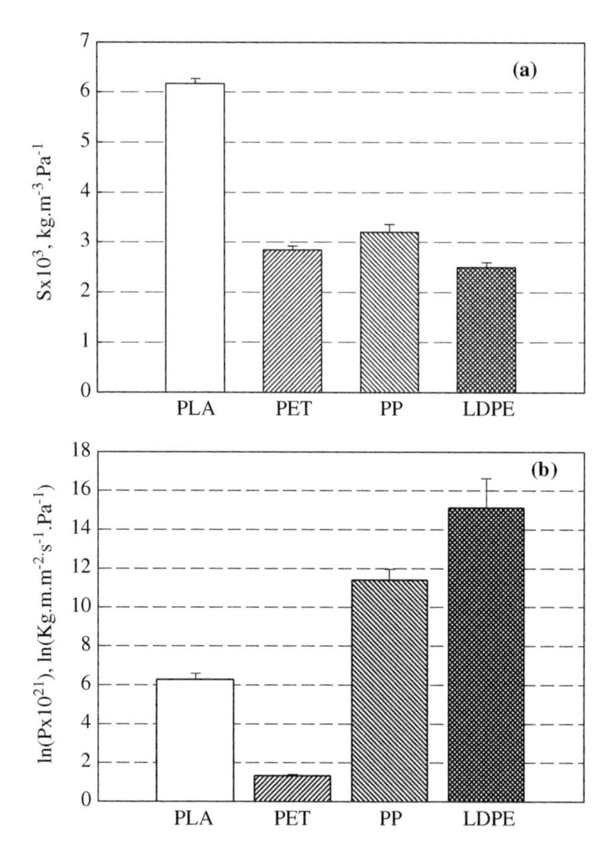

Fig. 6. (a) Ethyl acetate sorption coefficients for PLA at 30°C, p = 3030 Pa (this work); PET at 30°C, p = 9435 Pa [51]; PP at 22°C, p = 6088 Pa [50]; and LDPE at 22°C and p = 4348 Pa [49]. (b) Ethyl acetate permeability coefficients for PLA at 30°C, p = 7560 Pa (this work); PET at 30°C, p = 9435 Pa [51]; LDPE at 22°C, p = 4348 Pa [49]; and PP at 22°C, p = 6088 Pa [50].

than that in PET, PP, and LDPE. According to the Regular Solution Theory (RST) prediction, ethyl acetate sorption in PET ($\Delta\delta = 5.51$ $(J/cm^3)^{1/2}$) should be higher than that in PLA ($\Delta\delta = 7.80$ $(J/cm^3)^{1/2}$). However, as Fig. 7a shows, ethyl acetate sorption in PLA is higher than that in PET. PS, with a $\Delta\delta = 5.79$ $(J/cm^3)^{1/2}$, is dissolved in ethyl acetate. RST is useful to estimate the expected solubility coefficients but cannot exactly predict the kind of interaction (i.e., swelling, dissolution, or immiscibility) between PS or PET and ethyl acetate.

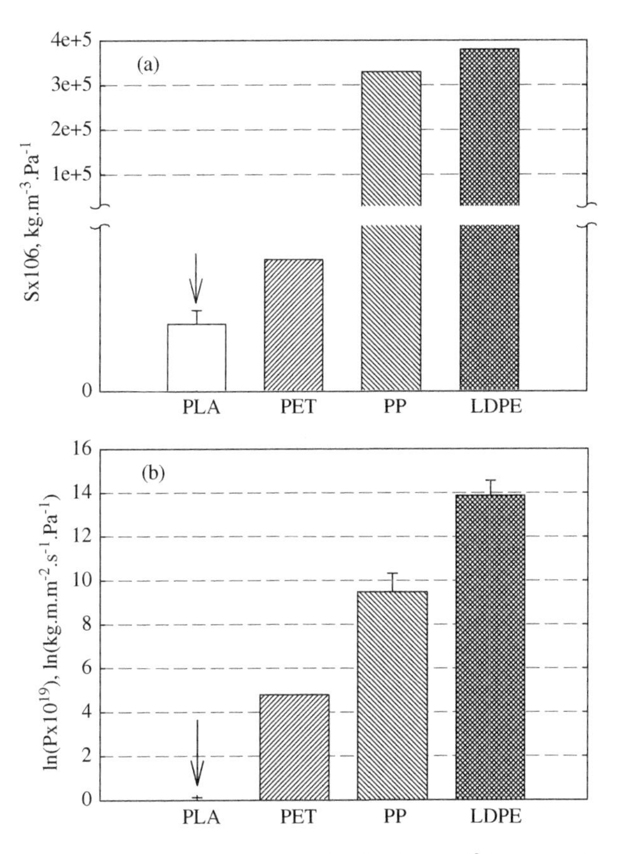

Fig. 7. (a) D-Limonene sorption coefficients for PLA at 45°C, $p = 258$ Pa (this work); PET at 23°C, $p = 40$ Pa [55]; PP at 22°C, $p = 62$ Pa [50]; and LDPE at 22°C, $p = 13$ Pa [55] (the arrow indicates that the values of PLA solubility coefficients are lower than $S \leq 4.88 \times 10^{-6}$ kg/(m^3 Pa)). (b) D-Limonene permeability coefficients for PLA at 45°C, $p = 258$ Pa (this work); PET at 23°C, $p = 40$ Pa [55]; PP at 22°C, $p = 62$ Pa [50]; and LDPE at 22°C, $p = 13$ Pa [55] (the arrow indicates that the values of PLA permeability coefficients are lower than $P \leq 9.96 \times 10^{-21}$ kg m/(m^2 s Pa)).

Solubility coefficient values are of great use in predicting permeant–polymer compatibility; however, for packaging applications permeability coefficient values better describe the quality of the barrier of a material, and are more useful in calculating the shelf life of a product. Fig. 6b compares the ethyl acetate permeability coefficient values of PLA, LDPE [49], PP [50], and PET [51]. Ethyl acetate permeability coefficients in PLA are lower than those for PP and LDPE and slightly higher than those for PET. RST cannot be used to predict permeability coefficients since it helps only to obtain the relation between solubility and $\Delta\delta$, which is not enough to predict permeability and also depends on the diffusion coefficient.

RST calculations predicted lower sorption values of D-limonene in PLA compared to those in PET, PP, LDPE, and PS [52]. The D-limonene solubility parameter at 45°C at a partial pressure of 258 Pa was estimated by Eq. (23) as $S \leq 4.88 \times 10^{-6}$ kg/(m^3 Pa). On the other hand, assuming a conservative case scenario that the sample reaches steady state at day 1, the diffusion coefficient can be calculated using Eq. (20) ($D \leq 2.04 \times 10^{-15}$ m^2/s), and the permeability coefficient values can be calculated from Eq. (22). Therefore, for D-limonene, permeability coefficients lower than 9.96×10^{-21} kg m/(m^2 s Pa) can be expected. Fig. 7a shows a comparison of the D-limonene sorption coefficients of PLA, PET, PP, and LDPE. Values for PS were not found in the literature.

Haugaard et al. [53] compared the impact of PLA, HDPE, and PS on quality changes in fresh and unpasteurized orange juice stored at 4°C for 14 days. They found that PLA is better suited for storage of orange juice than HDPE and PS due to an effective prevention of color changes, ascorbic acid degradation, and D-limonene scalping. For packaging applications permeability values are more meaningful since they can be used to calculate the shelf life of products. Fig. 7b shows the estimated D-limonene permeability coefficients of PLA, and the values reported in the literature for PET, LDPE, and PP. It is observed in Fig. 7b that PLA is a very good barrier to D-limonene. Similar findings have been reported by other researchers [53, 54]. PLA has much lower D-limonene permeability coefficients than PET, PP, and LDPE.

7. REGULAR SOLUTION THEORY: SOLUBILITY PARAMETER PREDICTIONS

Experimental determinations of polymer/chemical interactions are time consuming and expensive, requiring slow and tedious testing. Therefore, methods to predict affinity and interaction in these systems can reduce testing

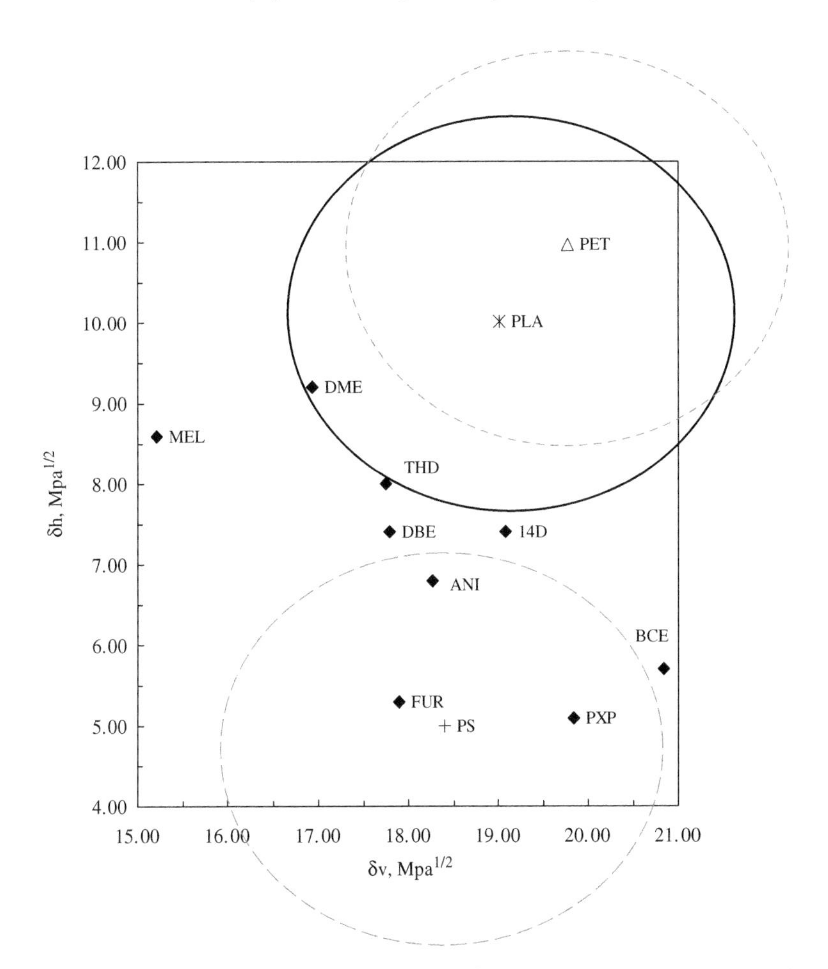

Fig. 8. Volume-dependent cohesion parameter (δ_v) versus Hansen hydrogen-bonding parameter for *polylactide* and *ethers*. Values indicated for solvents with $\Delta\delta < 5\,\mathrm{MPa}^{1/2}$ (FUR, furan; EPH, epichlorohydrin; THD, tetrahydrofuran; 14D, 1,4-dioxane; MEL, methylal (dimethoxymethane); BCE, bis(2-chloroethyl) ether; ANI, anisole (methoxybenzene); DME, di-(2-methoxyethyl) ether; DBE, dibenzyl ether; PXP, bis-(*m*-phenoxyphenol) ether).

requirements by permitting experimental evaluations to concentrate on the specific chemical compounds identified as likely to be of concern because of interaction and affinity.

A modern set of methods can be used to determine interaction or affinity between chemical compounds and polymers. These methods were in general developed to allow chemical engineers to estimate activity coefficients in

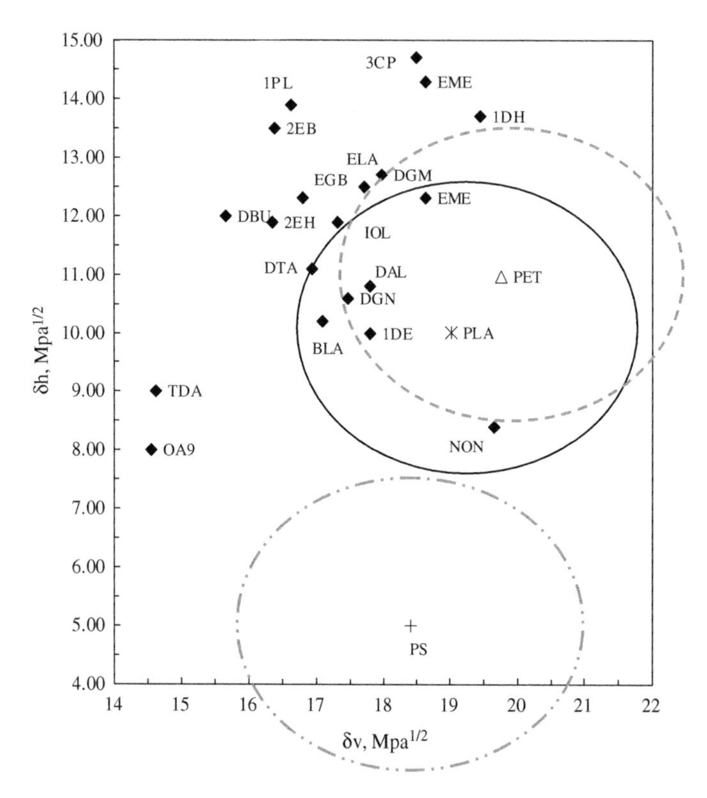

Fig. 9. Volume-dependent cohesion parameter (δ_v) versus Hansen hydrogen-bonding parameter for *polylactide* and *alcohol*. Values indicated for solvents with $\Delta\delta < 5\,MPa^{1/2}$ (3CP, 3-chloropropanol; BEA, benzyl alcohol; CHL, cyclohexanol; 1PL, 1-pentanol; 2EB, 2-ethyl-1-butanol; DAL, diacetone alcohol; DBU, 1,3-dimethyl-1-butanol; ELA, ethyl lactate; BLA, *n*-butyl lactate; EME, ethylene glycol monoethyl ether; DGM, diethylene glycol monoethyl ethermethyl; DGE, diethylene glycol monoethyl ether; EGB, ethylene glycol mono-*n*-butyl ether; 2EH, 2-ethyl-1-hexanol; IOL, 1-octanol; 2OL, 2-octanol; DGN, diethylene glycol mono *n*-butyl ether; 1DE, 1-decanol; TDA, 1-tridecanol; NON, nonyl; OA9, oleyl alcohol).

gas–liquid polymeric systems. Of these methods, three are worth mentioning: (a) RST [15, 21]; (b) UNIFAP (UNIFAC for polymers) [16]; and (c) Retention Indices Method (RIM) [56]. The first method is easy to use although sometimes lacks quantitativeness. The second method, UNIFAP, requires a large amount of effort and computing power, and the third one, RIM, is based on the potential recognized in gas chromatography that the partition of a substance between a gas and a polymer liquid can be estimated based on its structural increments. The last two methods, although simple, are more complex to apply than RST. In the next section, the RST methodology is used to estimate the interaction between PLA and some chemical compound groups such as ethers and alcohols. These interactions can be used as an indication of the kinds of interactions which should be expected between these permeants and PLA.

Figs. 8 and 9 show the affinity between a number of different permeants (ethers and alcohols) and PLA, as well as PET and PS. The solubility regions of PLA, PET, and PS are approximated by a circle with a radius of $\sim 2.5\,\delta$ units with centers at $\delta_v = 19.01$; $\delta_H = 10.01$ for PLA, $\delta_v = 19.77$; $\delta_H = 10.97$ for PET, $\delta_v = 15.90$; $\delta_H = 5.00$ for PS, which are the corresponding solubility parameters. δ_D, δ_P, and δ_H for PLA, PET, and PS were calculated by the average of the Hoftyzer–Van Krevelen and Hoy group contribution methods and confirmed by experimental values obtained from the literature [31, 57]. Note that the distortions from perfect circles in the figures are due to the differing scales of the *x* and *y* axes.

REFERENCES

[1] R. Auras, B. Harte and S. Selke, An overview of polylactides as packaging materials, Macromol. Biosci., 4 (2004) 835–864.

[2] D. Garlotta, A literature review of poly (lactic acid), J. Polym. Environ., 9 (2001) 63–84.

[3] H. Tsuji, in "Polylactides" (Y. Doi and A. Steinbuchel, eds.), pp. 129–177, Biopolymers. Polyesters III. Applications and Commercial Products, Vol. 4, Wiley-VCH Verlag GmbH, Weinheim, 2002.

[4] R. Dattaa, S.P. Tsaia, P. Bonsignorea, S.-H. Moona and J.R. Frank, Technological and economic potential of poly (lactic acid) and lactic acid derivatives, FEMS Microbiol. Rev., 16 (1995) 221–231.

[5] E.T.H. Vink, K.R. Rábago, D.A. Glassner and P. Gruber, Applications of life cycle assessment to NatureWorks™ polylactide (PLA) production, Polym. Degrad. Stab., 80 (2003) 403–419.

[6] G.M. Bohlmann, Biodegradable packaging life-cycle assessment, Environ. Prog., 23 (2004) 342–346.

[7] R.E. Conn, J.J. Kolstad, J.F. Borzelleca, et al., Safety assessment of polylactide (PLA) for use as a food-contact polymer, Fd. Chem. Toxic., 33 (1995) 273–283.

[8] R. Auras, B. Harte, S. Selke and R. Hernandez, Mechanical, physical, and barrier properties of poly (lactide) films, J. Plast. Film Sheet., 19 (2003) 123–135.

[9] R. Auras, S.P. Singh and J.J. Singh, Evaluation of oriented poly (lactide) polymers vs. existing PET and oriented PS for fresh food service containers, Packag. Technol. Sci., 18 (2005) 207–216.

[10] R. Auras, J.J. Singh and S.P. Singh, Performance evaluation of PLA existing PET and PS containers, J. Test. Eval., 34 (6) (2006) 530–536.

[11] R.J. Hernandez and J.R. Giacin, in "Factors Affecting Permeation, Sorption, and Migration Processes in Package-Product Systems" (I.A. Taub and R.P. Singh, eds.), Food Storage Stability, CRC Press LLC, Boca Raton, FL, 1997.
[12] J. Hildebrand and R.L. Scott, The Solubility of Nonelectrolytes, Reinhold, New York, 1950.
[13] T. Matsui, M. Fukamachi, M. Shimoda and Y. Osajima, Derivation of thermodynamic sorption equation of flavors with packaging films: 1, J. Agric. Food Chem., 42 (1994) 2889–2892.
[14] C.M. Hansen and K. Skaarup, Three dimensional solubility parameter – key to paint component affinities: III. Independent calculation of the parameter components, J. Paint Technol., 39 (1967) 511–514.
[15] C.M. Hansen, Hansen Solubility Parameters: A User's Handbook, CRC Press LLC, New York, 1999.
[16] J.M. Prausnitz, R.N. Lichtenthaler and E. Gomes de Azevedo, Molecular Thermodynamics of Fluid-Phase Equilibria, Prentice-Hall, Inc., Upper Saddle River, NJ, 1999.
[17] J.S. Paik and M.A. Tigani, Application of regular solution theory in predicting equilibrium sorption of flavor compounds by packaging polymers, J. Agric. Food Chem., 41 (1993) 806–808.
[18] S.-A. Chen, Polymer miscibility in organic solvents and in plasticizers – a two-dimensional approach, J. Appl. Polym. Sci., 15 (1971) 1247–1266.
[19] M. Fukamachi, T. Matsui, M. Shimoda and Y. Osajima, Derivation of thermodynamic sorption equation of flavors with packaging films: 2, J. Agric. Food Chem., 42 (1994) 2893–2895.
[20] D.W. Van Krevelen, Properties of Polymers, Ch 18. Properties Determining Mass Transfer in Polymeric Systems. p. 861, Elsevier Science B.V., Amsterdam, The Netherlands, 1997.
[21] A.F.M. Barton, Handbook of Solubility Parameters and Other Cohesion Parameters, CRC Press LLC, Boca Raton, FL, 1999.
[22] P. Neogi, Diffusion in Polymers, Marcel Dekker, Inc., New York, NY, 1996.
[23] J. Crank, in "The Mathematics of Diffusion", Publications OS, Oxford University Press, Great Britain, 1975.
[24] R. Gavara and R.J. Hernandez, Consistency test for continuous flow permeability experimental data, J. Plast. Film Sheet., 9 (1993) 126–138.
[25] M. Fujita, T. Sawayanagi, T. Tanaka, et al., Synchrotron SAXS and WAXS studies on changes in structural and thermal properties of poly[(R)-3-hydroxybutyrate] single crystals during heating, Macromol. Rapid Commun., 26 (2005) 678–683.
[26] J.F. Mano, Y. Wang, J.C. Viana, Z. Denchev and M.J. Oliveira, Cold crystallization of PLLA studied by simultaneous SAXS and WAXS, Macromol. Mater. Eng., 289 (2004) 910–915.
[27] M.E. Cagiao, D.R. Rueda, R.K. Bayer and F.J. Baltá Calleja, Structural changes of injection molded starch during heat treatment in water atmosphere: simultaneous wide and small-angle X-ray scattering study, J. Appl. Polym. Sci., 93 (2004) 301–309.
[28] R.A. Phillips, Macro-morphology of polypropylene homo-polymer tacticity mixtures, J. Polym. Sci. Polym. Phys. Ed., 38 (2000) 1947.

[29] Y. Yampolskii, I. Pinnau and B.D. Freeman, Materials Science of Membranes for Gas and Vapor Separation, John Wiley & Sons Ltd, Chichester, West Sussex, England, 2006.
[30] Y. Osajima and T. Matsui, Dynamic approach to the sorption of flavors into a food packaging film, Anal. Sci., 9 (1993) 753–765.
[31] D.W. Van Krevelen, Properties of Polymers, Elsevier Science B.V., Amsterdam, The Netherlands, 1997.
[32] H. Fujita, A. Kishimoto and K. Matsumoto, Concentration and temperature dependence of diffusion coefficients for systems polymethyl acrylate and *n*-alkyl acetates, Trans. Faraday Soc., 56 (1960) 424–437.
[33] J.S. Vrentas and J.L. Duda, Diffusion in polymer–solvent systems: I. Reexamination of the free-volume theory, J. Polym. Sci. Polym. Phys. Ed., 15 (1977) 405–416.
[34] J.E. Palamara, K.A. Mulcahy, A.T. Jones, R.P. Danner and L.J. Duda, Solubility and diffusivity of propylene and ethylene in atactic polypropylene by the static sorption technique, Ind. Eng. Chem. Res., 44 (2006) 9943–9950.
[35] M.J. Misovich and E.A. Grulke, Prediction of solvent activities in polymer solutions using an empirical free volume correction, Ind. Eng. Chem. Res., 27 (1988) 1033–1041.
[36] S.H. Jacobson, Molecular modeling studies of polymeric gas separation and barrier materials: structure and transport mechanisms, Polym. Adv. Technol., 5 (1994) 724–732.
[37] W.J. Koros, D.R. Paul and G.S. Huvard, Energetics of gas sorption in glassy polymers, Polymer, 20 (1979) 956–960.
[38] E.S. Sanders, W.J. Koros, H.B. Hopfenberg and V.T. Stannett, Pure and mixed gas sorption of carbon dioxide and ethylene in poly (methyl methacrylate), J. Membr. Sci., 18 (1984) 53–74.
[39] S. Kanehashi and K. Nagai, Analysis of dual-mode model parameters for gas sorption in glassy polymers, J. Membr. Sci., 253 (2005) 117–138.
[40] A.S. Michaels, W.R. Vieth and J.A. Barrie, Diffusion of gases in polyethylene terephthalate, J. Appl. Phys., 34 (1963) 13–20.
[41] A.S. Michaels, W.R. Vieth and J.A. Barrie, Solution of gases in polyethylene terephthalate, J. Appl. Phys., 34 (1963) 1–12.
[42] J. Puiggali, Y. Ikada, H. Tsuji, L. Cartier, T. Okihara and B. Lotz, The frustrated structure of poly (L-lactide), Polymer, 41 (2000) 8921–8930.
[43] R. Auras, B. Harte and S. Selke, Effect of water on the oxygen barrier properties of polyethylene terephthalate and poly (lactide) films, J. Appl. Polym. Sci., 92 (2004) 1790–1803.
[44] H.J. Lehermeier, J.R. Dorgan and D.J. Way, Gas permeation properties of poly (lactic acid), J. Membr. Sci., 190 (2001) 243–251.
[45] N.S. Oliveira, J. Oliveira, T. Gomes, A. Ferreira, J.R. Dorgan and I.M. Marrucho, Gas sorption in poly (lactic acid) and packaging materials, Fluid Phase Equilib., 222–223 (2004) 317–324.
[46] S. Pauly, in "Permeability and Diffusion Data" (J.I. Brandup and E.H. Grulke, eds.), p. 547, Polymer Handbook, Wiley, New York, 1999.
[47] H. Tsuji, R. Okino, H. Daimon and K. Fujie, Water vapor permeability of poly (lactide)s: effects of molecular characteristics and crystallinity, J. Appl. Polym. Sci., 99 (2006) 2245–2252.

[48] T.C. Wang, Pouch method for measuring permeability of organic vapors through plastic films, American Chemical Society, Division of Organic and Plastic Chemistry, Coat. Plast. Preprints, 31 (1975) 442–447.
[49] C.D. Barr, J.R. Giacin and R.J. Hernandez, A determination of solubility coefficient values determined by gravimetric and isostatic permeability techniques, Packag. Technol. Sci., 13 (2000) 157–167.
[50] T.J. Nielsen and J.R. Giacin, The sorption of limonene/ethyl acetate binary vapor mixtures by a biaxially oriented polypropylene film, Packag. Technol. Sci., 7 (1994) 247–258.
[51] R.J. Hernandez, J.R. Giacin, A. Shirakura and K. Jayaraman, Diffusion and Sorption of Organic Vapors through Oriented Poly (Ethylene Terephthalate) Films of Varying Thermomechanical History, Antec'90, Dallas, TX, May 7–11, 1990.
[52] R. Auras, Investigation of Polylactides as Packaging Materials, p. 268, The School of Packaging, Michigan State University, East Lansing, 2004.
[53] B.K. Haugaard, C.J. Weber, B. Danielsen and G. Bertelsen, Quality changes in orange juice packed in materials based on polylactate, Eur. Food Res. Technol., 214 (2002) 423–428.
[54] N. Whiteman, P. DeLassus and J. Gunderson, New Flavor and Aroma Barrier Thermoplastic, Polylactide, Polyolefins 2002, International Conference on Polyolefins, Houston, TX, USA, 2002.
[55] S.C. Fayoux, A.M. Seuvre and A. Voilley, Aroma transfers in and through plastic packagings: orange juice and D-limonene. A review. Part I. Orange juice aroma sorption, Packag. Technol. Sci., 10 (1997) 69–82.
[56] A.L. Baner, in "Partition Coefficients" (O. Piringer and A. Baner, eds.), pp. 79–123, Plastic Packaging Materials for Food, Vol. 1, Wiley-VCH Verlag GmbH, Weinheim, Germany, 2000.
[57] H. Tsuji and K. Sumida, Poly (L-lactide): V. Effects of storage in swelling solvents on physical properties and structure of poly (L-lactide), J. Appl. Polym. Sci., 79 (2001) 1582–1589.

Thermodynamics, Solubility and Environmental Issues
T.M. Letcher (editor)

Chapter 20

Biodegradable Material Obtained from Renewable Resource: Plasticized Sodium Caseinate Films

Jean-Luc Audic, Florence Fourcade and Bernard Chaufer

Laboratoire Chimie et Ingénierie des Procédés (CIP), Sciences Chimiques de Rennes (UMR CNRS 6226), Université de Rennes 1/ENSCR, Avenue du Général Leclerc, 35700 Rennes, France

1. INTRODUCTION

During the past 20 years research interest in the use of natural biopolymers for the manufacture of "green" biodegradable materials such as films and coatings has greatly increased. The reason for this is the drive to solve environmental problems associated with the use of synthetic petroleum polymers but also to find new markets for the excess production of Western agricultural products. Natural polymers available for forming films and coatings generally fall into three main categories; proteins, polysaccharides and lipids/fats, as seen in Table 1. Among the natural polymers cited in Table 1 that could be used in packaging applications, proteins can most easily replace conventional synthetic polymers. Proteins can be considered as thermoplastic heteropolymers based on 20 amino acid monomers that have different side group attached to the central carbon. These side groups can be chemically or enzymatically modified to improve film properties. Protein films also generally exhibit good mechanical properties and also good barrier properties to gases such as O_2 or CO_2. Among these proteins, sodium caseinate (NaCAS) was selected in this work, for the manufacture of films and coatings. NaCAS is a commercially available protein obtained by acid precipitation of casein, the main protein in cow's milk. This protein is obtained with good purity (up to 90%) and its thermoplastic and film forming properties are due to its random coil nature and its ability to form weak intermolecular interactions, i.e. hydrogen, electrostatic and hydrophobic bonds [1, 2]. This makes caseinate an interesting raw material for packaging applications in substitution of

Table 1
Natural biopolymers used in films and coating formulations

Natural polymer	Origin	Examples
Proteins	Animal	Casein, whey, egg white, collagen/gelatine, etc.
	Vegetable	Corn zein, pea protein, soy protein, wheat gluten, cottonseed...
Polysaccharides	Animal	Chitine/chitosane
	Microbial	Xanthan, pullulan, dextran, gellan
	Vegetable	Starch, cellulose, pectins...
Lipids/fats	Animal	Bees wax, fatty acids...
	Vegetable	Fatty acids, milk fat...

petroleum polymers. Considering their transparency, biodegradability and good technical properties (high barrier for gases like O_2) [3], caseinate-based films can find applications, not only in packaging [4, 5], but in edible or protective films and coatings [6, 7] or in mulching films [8]. Such films are easily obtained from casting aqueous solutions of NaCAS.

The objective in this work, was to achieve caseinate-based films with improved properties as close as possible to available packaging films based on synthetic polymers like polyethylene [6] or plasticized PVC [9]. Such films have good elongation at break point (from 150 to ~400%) and a rather low tensile strength ranging from 20 to 30 MPa. However, compared with synthetic films, casein-based films (like most protein films) have two major drawbacks [10–12]:

(i) poor mechanical properties (lower tensile strength and elongation at break point);
(ii) a high water sensitivity, i.e. high water solubility and water vapour permeability.

The aim of the present work was focused on the improvement of the mechanical properties as well as the water resistance of caseinate-based films. The first part of this study deals with the control of the mechanical properties through addition of a plasticizer. A plasticizer is defined as a low volatile organic compound which causes a decrease in polymer glass transition temperature (T_g) and an increase in flexibility and extensibility. By decreasing intermolecular forces between polymer coils, plasticizers cause an increase in material flexibility and conversely a decrease in the barrier properties due to the augmentation of the free volume [13, 14]. To summarize, an initially hard and brittle material becomes soft and flexible when sufficiently plasticized. Plasticizer efficiency is mostly governed by its molecular weight and polarity. The most compatible plasticizers are also generally the most efficient. Polyols are typical plasticizers [15, 16] for protein-based materials due to their ability to reduce intermolecular hydrogen bonding while increasing intermolecular spacing. Among different polyols tested in a previous paper [17] (different polyethylene glycols, glycerol, triethanolamine [TEA], etc.), TEA was found to be the most efficient plasticizer for NaCAS. Since water also behaves as a plasticizer [18], the plasticizing effect of TEA was measured at 53% relative humidity (RH).

In addition, caseinate-based films were crosslinked with formaldehyde (HCHO). The occurrence of covalent bridges between protein chains allows water-insoluble three-dimensional network to be achieved. The crosslinking was combined with plasticizing to produce caseinate films with improved mechanical properties and water resistance which are able to replace synthetic films and to overcome protein film deficiencies. Nevertheless, in spite of crosslinked protein material being water insoluble, it was also shown that during water immersion, plasticizer exuded out of the film, even for highly crosslinked samples.

2. EXPERIMENTAL SECTION

2.1. Materials and Reagents

NaCAS was purchased from Eurial (France). Its composition according to manufacturer was: proteins 90.2%, water 5.7%, minerals 3.5%, fat <1%. TEA and trinitrobenzene sulphonic acid were of analytical grade (purity 99+%, Acros Organics). All other reagents including magnesium nitrate (99+%), sodium azide (99%), sodium bicarbonate (99%) and HCHO (37 wt% solution in water stabilized with 10–15% methanol) were obtained from Acros Organics and used as received.

2.2. Film Preparation

An aqueous mixture of protein (5% w/v) and TEA (TEA/protein ratios = 25%, 50% and 100% w/w, respectively) was magnetically stirred at 800 rpm for ~12 h at room temperature in order to obtain a homogeneous solution. Caseinate films were then obtained by the casting method: the film forming solution was spread onto a polystyrene plate and after the excess water was evaporated the film was peeled off. All samples were kept for 7 days at 20 ± 2°C in a closed tank containing a saturated solution of $Mg(NO_3)_2$ to maintain 53% RH.

2.3. Chemical Treatment

2.3.1. Crosslinking of Plasticized Films

Protein modification was performed with increasing aliquots of HCHO (37% w/v in water) in solution of NaCAS (5% w/w) and the pH was controlled by TEA to 8.6 without the use of any buffer. The reaction with HCHO was performed with HCHO/NaCAS w/w ratios of 1%, 2%, 5% and 10%, respectively, corresponding to HCHO/Lys molar ratios of 0.67, 1.35, 3.37 and 6.75. Molar ratios were calculated in consideration of the 12.4 mol of potentially reactive amino acid residue, i.e. lysine (Lys), contained in 1 mol of NaCAS [19]. The ε-amino group of Lys was considered the primary reactive site between proteins and aldehydes [20]. The protein/HCHO dispersion was left to react with gentle stirring at room temperature or at 40°C for 6 h.

2.3.2. Determination of the Degree of Substitution

The amount of free ε-amino groups was determined by the modified 2,4,6-trinitrobenzenesulfonic acid (TNBS) procedure adapted for poorly soluble material [21]. An aliquot of 1 mL of NaCAS ($\sim$2 mg mL^{-1}) dissolved in 4% $NaHCO_3$ (buffer solution pH = 8.6) and 1 mL of 0.5% TNBS were placed in a screw cap test tube before heating at 40°C for 4 h. Any insoluble material was further hydrolyzed and dissolved by the addition of 3 mL of 6 N HCl at 120°C and 15–17 psi for 1 h. The hydrolysate was then diluted with 5 mL of H_2O and extracted with 3×20 mL portions of diethyl ether to remove both excess unreacted TNBS and trinitrophenyl-α-amino groups. A 5 mL aliquot of the aqueous phase was taken and heated for 15 min in a hot water bath to evaporate residual ether. The aliquot was diluted with 5 mL of H_2O and the absorbance was measured at 346 nm (Thermospectronic Helios γ spectrophotometer, Fisher-Bioblock, Illkirch, France). A caseinate blank was prepared according to the same procedure as the samples, except that HCl was added before TNBS. To determine the efficiency of the chemical modification, Lys residues were assumed as the more reactive amino acid. As mentioned above, Lys content in pure NaCAS was considered to be 8% [19]. Amount of Lys per moles of NaCAS was obtained by the following equation [21] (1):

$$\frac{\text{mol Lys}}{\text{mol protein}} = \frac{2(0.01)(\overline{M_w})(A_{346})}{(14600)(b)(x)} \tag{1}$$

where $\overline{M_w}$ is the average protein molecular weight ($\overline{M_w} = 22600$ g mol^{-1}), 14600 L mol^{-1} cm^{-1} the molar absorptivity of TNP–Lys, b the cell path length (in centimetres) and x the protein weight (in grams). Prior to experiments with caseinate, it was checked that bovine serum albumin free ε-amino groups reacted according to this modified TNBS procedure.

2.4. Water Solubility

2.4.1. Protein Specific Solubility

The specific solubility [22] of NaCAS was determined by assaying for protein using the 280 nm absorbance method [22]. The NaCAS films (20 × 20 mm; ~100 mg) were immersed in distilled water (75 mL) at 20°C and magnetically stirred at 250 rpm. From the absorbance at 280 nm of the supernatant, the remaining soluble protein was determined. The specific solubility is expressed in weight percentage of initial amount of dry casein in the film.

Absorbance at 280 nm of NaCAS solutions in distilled water was first measured. It was also checked that TEA and NaN_3 absorbance was negligible at 280 nm. The calibration straight line ($y = 1.457x + 0.016$) of the absorbance of NaCAS solutions versus NaCAS concentration was obtained with a correlation coefficient $r = 0.9981$. The calculated molar extinction coefficient was $\varepsilon = 32920\,L\,mol^{-1}\,cm^{-1}$. NaCAS content in films can be determined with an accuracy of better than 5% [17]. The solubility of TEA plasticized films crosslinked with HCHO was determined after 2, 5, 10, 20, 40 min and 24 h from the 280 nm absorbance of the supernatant.

2.4.2. TEA Exudation Ratio

TEA is known to have the ability to chelate certain metallic ions in highly alkaline medium (pH > 12), such as the ferric ion. In this work TEA concentration was obtained using a voltammetric technique based on the measurement of the concentration of TEA–Fe^{3+} complex [23, 24]. The NaCAS films (20 × 20 mm; ~100 mg) were immersed in distilled water (75 mL) at 20°C and magnetically stirred at 250 rpm for 2 min, 5 min, 10 min, 20 min, 40 min, and 24 h respectively. A 100 μL of the supernatant was then transferred to the voltammetric cell and diluted with 10 mL 1 M NaOH solution as supporting electrolyte. A 15–290 μL of 0.5 g L^{-1} solution $NH_4Fe_{III}(SO_4)_2 \cdot 12H_2O$ was added into the voltammetric cell (see Table 2). The solutions were de-aerated with nitrogen gas for 10 min and an inert atmosphere (N_2) was maintained over the solutions during measurements. Electrochemical experiments were performed using three electrodes. The voltammograms of these solutions were obtained with a micro autolab polarographic analyser Metrohm piloted by the GPES software. The working electrode was a hanging mercury-drop

Table 2
Samples preparation for voltammetric analysis

Immersion time in distilled water	1 M NaOH solution (mL)	$0.5\,g\,L^{-1}$ $NH_4Fe_{III}(SO_4)_2 \cdot 12H_2O$ solution (μL)
2 min	10	40
5 min	10	40
10 min	10	40
20 min	10	50
40 min	10	50
24 h	10	70

electrode (HMDE), the reference electrode was $Ag/AgCl/Cl^-$ and the counter electrode was a Pt electrode. The setting parameters were as follows: pulse height 50 mV, step potential 0.0012 V, potential −1.2 to −0.7 V versus $Ag/AgCl/Cl^-$. It was assumed that the TEA was fully converted to the TEA–Fe^{3+} complex. The amount of TEA was determined from the TEA–Fe^{3+} complex peak intensity. The calibration straight line ($y = 4E\text{-}8x + 2E\text{-}9$) of the TEA–$Fe^{3+}$ peak intensity versus TEA concentration was obtained with a correlation coefficient $r = 0.9897$.

2.5. Stress–Strain Testing

Tensile strength, elongation at break point and Young's modulus were obtained using a model 1011 testing system (Instron, Engineering Corp., Canton, MA). Uniaxial stress–strain and ultimate properties were performed on 25 mm × 20 mm × 180 μm samples employing rubber jaws and a cross-head speed of 10 or 50 mm min^{-1}. The data obtained were each calculated from an average of five samples. Stress–strain curves plotted in Figs. 1 and 3 were obtained from experimental data as close as possible to the mean mechanical properties.

3. RESULTS AND DISCUSSION

3.1. Plasticized Caseinate Films: Mechanical Testing

Among different polyols tested in a previous paper [17], TEA was selected as the best candidate for plasticizing NaCAS. Both polyol plasticizers and water molecules contain hydroxyl groups that interact with protein through hydrogen bonding. Equilibrium moisture content affects the mechanical strength of protein films through water plasticization: water

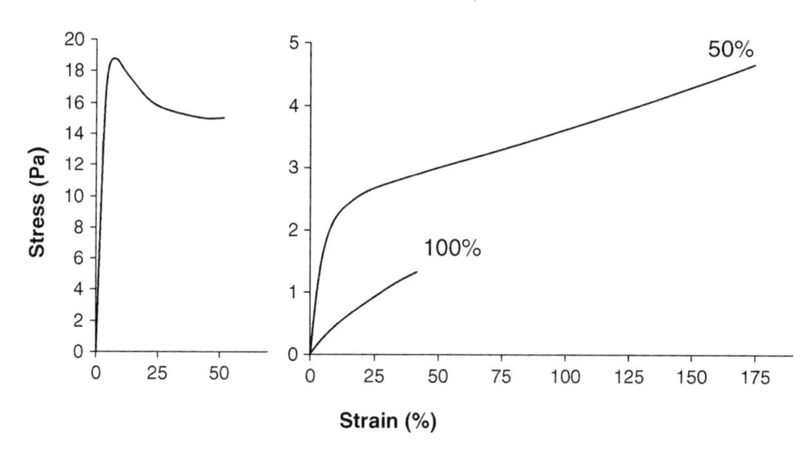

Fig. 1. Stress–strain curves of TEA plasticized samples at 53% RH. The plasticizer content (w/w) is indicated on each curve.

molecules soften the structure of films [18] giving the materials a high molecular lubricity. For the present study, all the tests were performed at constant RH of 53%.

Mechanical performances of NaCAS films versus the content of TEA are given in Fig. 1. The film containing only 25% TEA exhibited poor ultimate elongations (~55%) for tensile strength up to 19 MPa. The stress increased continuously until a maximum, corresponding to the yield point and then decreased until the sample broke. With 50 and 100% plasticizer, the load increased continuously with an increase of strain until the film broke. The maximum tensile stress decreased significantly as plasticizer concentration increased. Elongation at the breakdown of NaCAS films was also dependent on plasticizer content: tensile strain increased from ~50% for samples containing 25% TEA to over 100% for samples containing 50% TEA. Nevertheless, further increase in plasticizer content decreased the strain to 30%. This behaviour is in close agreement with the excess content of plasticizers (TEA + H_2O) already described. The simultaneous action of water and excess plasticizer resulted in a weak film with low cohesion which was difficult to handle.

These results are in agreement with the data dealing with plasticization of protein and NaCAS material. The plasticizer decreases the interactions between protein chains and this is closely connected to a decrease in cohesive tensile strength and to the glass transition temperature.

At 53% RH, the best tensile properties were obtained for 50% plasticized TEA–NaCAS films which showed ~175% elongation for a tensile strength of 4.5 MPa. The general trend of the tensile curve of 50% plasticized film with a continuous increase of strain with stress and without any yield point, is typical of rubber-like behaviour. When plasticized enough, NaCAS films exhibited ultimate properties and elastic modulus very close to those of commercial PE or PVC-based packaging films. With respect to these results the formulations containing 50% TEA were selected for their ability to produce the best mechanical properties for NaCAS films.

3.2. Crosslinking of Plasticized Caseinate Films

As a general rule for synthetic as well as natural polymers, an increase in plasticizer content lead to higher elongation and lower tensile strength. Nevertheless, plasticized protein films presents two main drawbacks: on one hand, extensible films from NaCAS exhibit rather low stresses at break point, inferior to that of synthetic polymers. On the other hand, such plasticized NaCAS films are water-soluble which greatly limit their applications. As a result, in order to improve both tensile properties and reduce their water sensibility, NaCAS-based films have been reacted with HCHO. This reaction between HCHO and protein is a two-step process: the first step corresponds to the formation of the methylol derivative and the second one corresponds to the formation of methylene bridges, i.e. crosslinks between protein chains. Crosslinking was chemically determined and its influence on mechanical properties and water solubility was studied.

3.3. Determination of Crosslinking Efficiency

The reaction of HCHO with NaCAS was followed by measuring the amount of residual ε-amino groups in the protein. Commercial HCHO contained methanol and was used without further treatment. Data given below are only representative of the occurrence of crosslinking. Fig. 2 shows the amount of unreacted amino groups in modified NaCAS as a function of the molar ratio of HCHO to the initial amount of ε-amino groups. By increasing the HCHO content, the amount of free ε-NH_2 decreased rapidly from ~50 mmoles -NH_2/100 g NaCAS to less than 25 mmoles -NH_2/100 g NaCAS. For HCHO/-NH_2 ratios upper than 3 a plateau below 25 was reached at room temperature. When the reaction was performed at 40°C, the crosslinking was more efficient, and 66% of the ε-amino groups were involved in the crosslinking procedure for HCHO/ε-NH_2 ratios higher than 3. The crosslinking of NaCAS by HCHO can accordingly be adjusted by varying the concentration of the reagent and temperature.

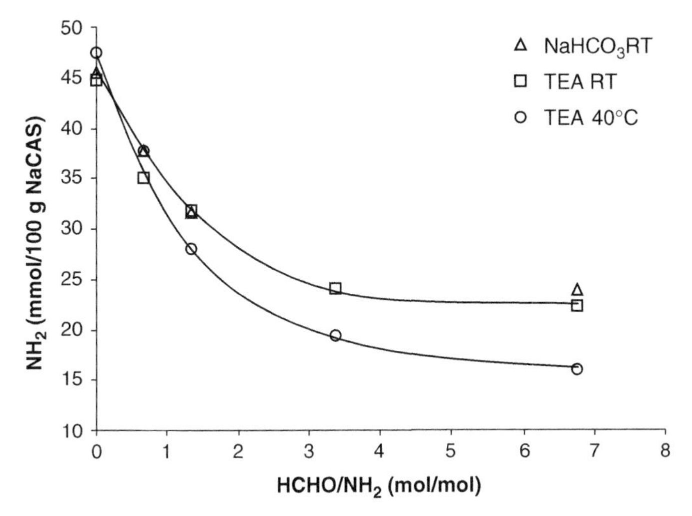

Fig. 2. Amount of residual amino groups in modified NaCAS versus molar ratio of formaldehyde to the initial amount of amino groups (RT = room temperature).

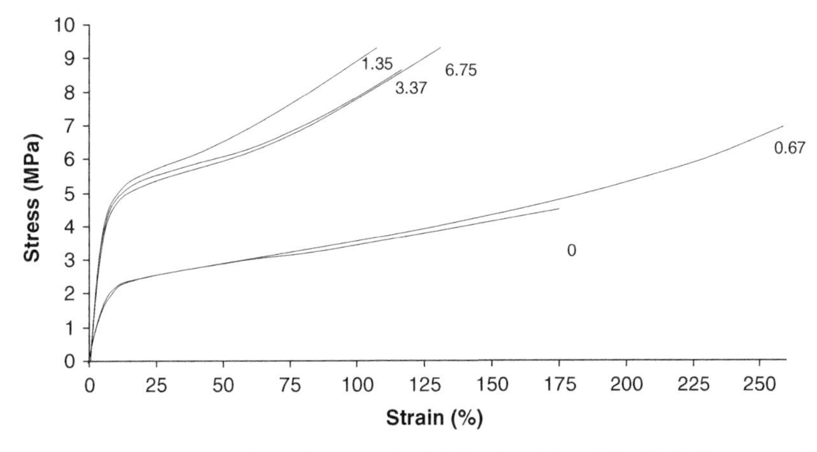

Fig. 3. Stress–strain properties of 50% w/w TEA plasticized NaCAS films crosslinked with formaldehyde (HCHO/ε-NH_2 molar ratios reported on each curve) at 53% RH.

3.4. Effect of Crosslinking on Mechanical Properties

Fig. 3 deals with the stress–strain curves of 50% TEA plasticized NaCAS films containing 0–10% w/w HCHO. At low HCHO concentration (1% w/w; HCHO/-NH_2 = 0.67), the ultimate properties of NaCAS films were improved: the stress and elongation at break point increased from 4.5 to 6.9 MPa and

Table 3
Elastic modulus of 50% TEA plasticized NaCAS films crosslinked with different HCHO/ε-NH_2 ratios

HCHO/ε-NH_2 (mol/mol)	Elastic modulus (MPa)	Standard deviation
0	46.0	2.1
0.67	46.6	2.33
1.35	109.7	13
3.37	97.5	6.4
6.75	103.3	15.2

from 175 to more than 260%, respectively. The general trend of the stress–strain curve of TEA plasticized samples crosslinked by 1% HCHO only is close to the shape of the non-crosslinked samples. For TEA plasticized films crosslinked with 2–10% w/w HCHO (HCHO/ε-NH_2 ratios from 1.35 to 6.75), the stresses at break point, at ~8–9 MPa, were more than twice the values of TEA/NaCAS films without HCHO and ultimate strain values were up to 110–125% (Fig. 3). Table 3 gives the Young's modulus of TEA plasticized films containing 0–10% w/w HCHO. Compared with only plasticized samples, small amounts (1% w/w; HCHO/ε-NH_2 = 0.67) of HCHO did not cause major changes in elastic modulus (~46 MPa in both cases). For HCHO/ε-NH_2 ratios ranging from 1.35 to 6.75 the Young's modulus increased to ~100–110 MPa which is more than twice the value of TEA/NaCAS films without HCHO. The mechanical properties of crosslinked plasticized NaCAS films increased with the concentration of crosslinking agent. HCHO/ε-NH_2 ratio of 1.35 seemed to be a threshold for enhancing mechanical properties of plasticized caseinate films. Mechanical properties are dependent on the ratio of the crosslinker to protein in a limited range; further crosslinking seems not involved in film performance from a mechanical point of view.

3.5. Effect of Crosslinking on Water Solubility

3.5.1. Protein Solubility

TEA plasticized NaCAS films with various ratios of crosslinker were immersed into distilled water. The specific water solubility of NaCAS (percentage of initial dry weight of NaCAS in films) versus immersion time is shown in Fig. 4. For the unreacted NaCAS film, the NaCAS specific solubility increased rapidly and after 10 min the film was completely dissolved. However, for the crosslinked films, the solubility remained less than 6% after 40 min of immersion in water, whatever the HCHO/NaCAS ratio, indicating

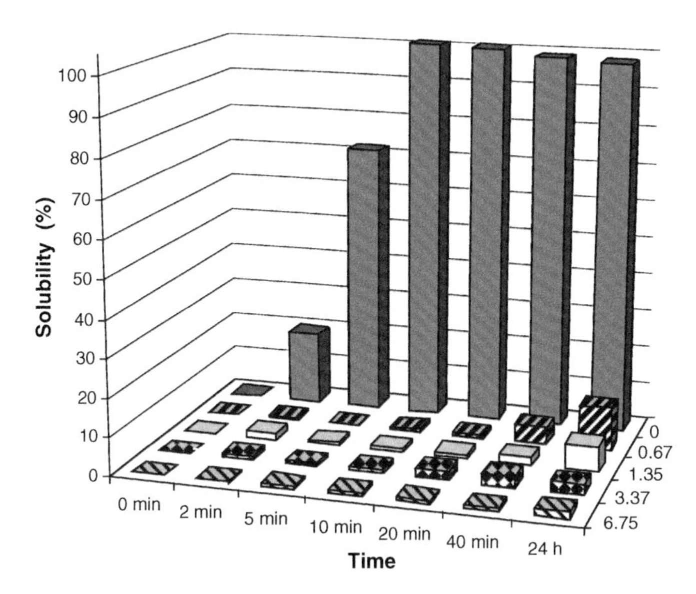

Fig. 4. NaCAS specific solubility of 50% TEA plasticized films from 280 nm absorbance method versus immersion time in water for HCHO/ε-NH_2 molar ratios reported on the right hand side.

the water resistance of crosslinked samples. Nevertheless, after 24 h immersion in water release of protein was checked versus crosslinker ratio. Protein specific solubility ranged from 11.4% for the lowest HCHO/ε-NH_2 molar ratio (0.67) to 1.8% for the highest ratio (6.75). Fig. 4 clearly points out that the higher the HCHO/ε-NH_2 ratio the lower is the solubility, contrary to tensile stress which did not increase further for HCHOH/ε-NH_2 molar ratio greater than 1.35 (Fig. 3). Specific solubility of NaCAS determined from the 280 nm absorbance method is in good agreement from a qualitative point of view with the decrease of residual amino groups versus HCHO/ε-NH_2 ratio (Fig. 2).

3.5.2. Triethanolamine Exudation

Fifty percent of TEA plasticized NaCAS films with various ratios of crosslinker were immersed into distilled water. The exudation ratio of TEA (percentage of initial dry weight of TEA initially present in films) versus immersion time is shown in Fig. 5. As suspected, the water-soluble plasticizer exudes out of the film and after 10 min immersion in distilled water ~80% of the TEA migrated out of the film, whatever the crosslinking. For all the NaCAS films, the whole TEA exuded out of the film after 24 h in water. Nevertheless, for short time of contact with water (2 min), the plasticizer

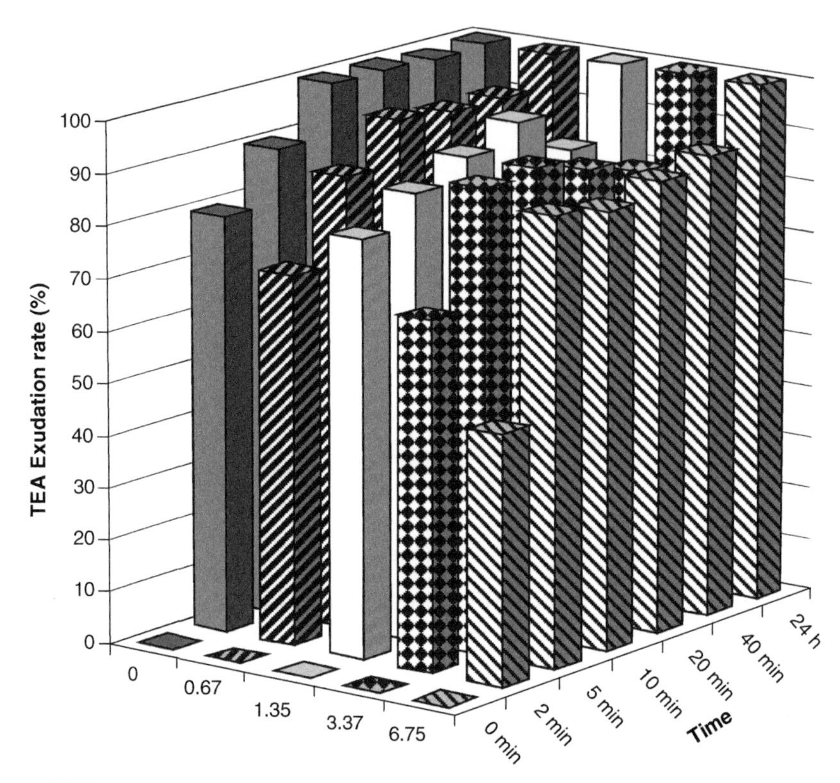

Fig. 5. TEA exudation rate for 50% TEA plasticized films from voltammetric analysis versus water immersion time and HCHO/ε-NH_2 molar ratios reported on the front side.

migration decreased with increased crosslinker content, indicating a steric hindrance to TEA exudation. In fact, TEA is a water-soluble plasticizer that migrates out of the film: it is the reason why solubility is sometimes expressed without taking into account the plasticizer content in the dry matter of the film [25], considering that the water soluble plasticizer is automatically exuded from the film during the water solubility test. However, these data needs further investigation for a better understanding of the role of crosslinking on plasticizer exudation in order to avoid pollution of the surrounding medium by plasticizers but also to confer protein films permanent mechanical properties.

4. CONCLUSION

Biodegradable protein films with improved mechanical properties close to synthetic polymer films can be achieved from NaCAS. Enhanced tensile properties were obtained by using TEA as plasticizer. Chemical crosslinking of NaCAS with HCHO is an efficient way to increase tensile properties and to overcome water sensitivity of the protein film. Crosslinking with HCHO/ε-NH_2 ratio of 1.35 at least improved mechanical strengths to 8–9 MPa with a strain at break point of ~110–125% and a Young's modulus greater than 100 MPa. The specific solubility of NaCAS obtained from a UV absorbance method shows the improvement of NaCAS film water resistance is closely related to crosslinker content. Nevertheless, it was also shown that during water immersion TEA plasticizer extensively migrate out of the films to the surrounding solution, whatever the ratio of crosslinker to reactive groups. In future work a longer and more branched molecule should be used as plasticizer in order to resist to exudation and water extraction.

REFERENCES

[1] I. Arvanitoyannis and C.G. Biliaderis, Food Chem., 62 (1998) 333.
[2] T.E. Creighton, Proteins: Structure and Molecular Principles, W.H. Freeman Publisher, New York, 1984.
[3] J. Chick and Z. Ustunol, J. Food Sci., 63 (1998) 1024.
[4] J.E. Kinsella, D.M. Whitehead, J. Brady and N.A. Bringe, Milk Proteins: Possible Relationships of Structure and Function, Elsevier Applied Science, London, 1989.
[5] J.M. Krochta, A.E. Pavlath and N. Goodman, in "Engineering and Food – Vol. 2", Preservation Processed Fruits and Vegetables, pp. 329–340, Elsevier Applied Science, New York, 1990.
[6] H. Chen, J. Dairy Sci., 78 (1995) 2563.
[7] R. Banerjee, H. Chen, G. Hendricks and J.E. Levis, J. Dairy Sci., 77 (1994) 7, 24.
[8] L.A. De Graaf and P. Kolster, Macromol. Symp., 127 (1998) 51.
[9] J.L. Audic, F. Poncin-Epaillard, D. Reyx and J.C. Brosse, J. Appl. Polym. Sci., 79 (2001) 1384.
[10] J. Grevellec, C. Marquié, L. Ferry, A. Crespy and V. Vialettes, Biomacromolecules, 2 (2001) 1104.
[11] L.J. Mauer, D.E. Smith and T.P. Labuza, Int. Dairy J., 10 (2000) 353.
[12] H. Chen and W. Wang, J. Dairy Sci., 77 (1994) 8.
[13] J. Borek and W. Osoba, J. Polym. Sci.: Part B: Polym. Phys., 34 (1996) 1903.
[14] R. Gätcher and H. Müller, Plastics Additives Handbook, 3rd edn., Hanser Publishers, New York, 1990.
[15] A.C. Sanchez, Y. Popineau, C. Mangavel, C. Larré and J. Gueguen, J. Agric. Food Chem., 46 (1998) 4539.
[16] J. Gueguen, G. Viroben, P. Noireaux and M. Subirade, Ind. Crops Prod., 7 (1998) 149.
[17] J.L. Audic and B. Chaufer, Eur. Polym. J., 41 (2005) 1934.
[18] M.T. Kalichevsky, J. Blanshard and P. Tokarczuk, Int. J. Food Sci. Technol., 28 (1993) 139.

[19] C. Dinnella, M.T. Gargaro, R. Rossano and E. Monteleone, Food Chem., 78 (2002) 363.
[20] J.W. Rhim, A. Gennadios, C.L. Weller, C. Cezeirat and M.A. Hanna, Ind. Crops Prod., 8 (1998) 195.
[21] W.A. Bubnis and C.M. Ofner III, Anal. Biochem., 207 (1992) 129.
[22] R.P. Konstance and E.D. Strange, J. Food Sci., 56 (1991) 556.
[23] N. Menek and Z. Heren, Cem. Concr. Res., 29 (1999) 777.
[24] N. Menek and Z. Heren, Cem. Concr. Res., 30 (2000) 1615.
[25] C. Marquié, C. Aymard, B. Cuq and S. Guilbert, J. Agric. Food Chem., 43 (1995) 2762.

Thermodynamics, Solubility and Environmental Issues
T.M. Letcher (editor)

Chapter 21

Supercritical Carbon Dioxide as a Green Solvent for Polymer Synthesis

Colin D. Wood, Bien Tan, Haifei Zhang and Andrew I. Cooper

Department of Chemistry, Donnan and Robert Robinson Laboratories, University of Liverpool, Liverpool L69 3BX, UK

1. INTRODUCTION

Supercritical carbon dioxide (scCO_2) is an inexpensive, non-toxic, and non-flammable solvent for materials synthesis and processing [1–3]. As such, it has emerged as the most extensively studied environmentally benign medium for organic transformations and polymerization reactions. This stems from a list of advantages ranging from solvent properties to practical environmental and economic considerations. Aside from the gains provided by CO_2 as a reaction medium in general, it finds particular advantages in the synthesis and processing of porous materials and this will be discussed here [1, 4, 5]. Although CO_2 is attractive in a range of different applications a significant technical barrier remains in that it is a relatively weak solvent: important classes of materials which tend to exhibit low solubility in scCO_2 include polar biomolecules, pharmaceutical actives, and high-molecular weight polymers [1, 2, 4, 6–9]. Until recently, the only polymers found to have significant solubility in CO_2 under moderate conditions (<100°C, <400 bar) were amorphous fluoropolymers [2] and, to a lesser extent, polysiloxanes [8]. Therefore, the discovery of inexpensive CO_2-soluble materials or "CO_2-philes" has been an important challenge [10–12]. We will focus on the application of carbon dioxide in the synthesis of porous polymers and recently developed CO_2-soluble hydrocarbon polymers. The general properties of supercritical fluids (SCFs) in relation to chemical synthesis [9] and extraction [13] have been reviewed previously and will not be emphasized here.

2. POROUS MATERIALS AND SUPERCRITICAL FLUIDS

Porous materials are used in a wide range of applications, including catalysis, chemical separations, and tissue engineering [4]. However, the synthesis of these materials is often solvent intensive. scCO_2 as an alternative solvent for the synthesis of functional porous materials can circumvent this issue as well as affording a number of specific physical, chemical, and toxicological advantages. For example, energy-intensive drying steps are required in order to dry porous materials, whereas the transient "dry" nature of CO_2 overcomes these issues. Pore collapse can occur in certain materials when removing conventional liquid solvents; this can be avoided using SCFs because they do not possess a liquid–vapour interface. Porous structures are important in biomedical applications (e.g., tissue engineering) where the low toxicity of CO_2 offers specific advantages in terms of minimizing the use of toxic organic solvents. In addition, the wetting properties and low viscosity of CO_2 offers specific benefits in terms of surface modification.

A number of new approaches have been developed in the past few years for the preparation of porous materials [4]. Current routes include foaming [14–18], CO_2-induced phase separation [19, 20], reactive [21–23], and non-reactive [24, 25] gelation of CO_2 solutions, nanoscale casting using scCO_2 [26, 27], and CO_2-in-water (C/W) emulsion templating [28, 29]. Each of these methods uses a different mechanism to generate the porosity in the material. In the following sections we will discuss some of the methods that we have developed.

3. CO_2 AS A PRESSURE-ADJUSTABLE TEMPLATE/POROGEN

Cooper et al. [21, 30] demonstrated the formation of permanently porous crosslinked poly(acrylate) and poly(methacrylates) monoliths using scCO_2 as the porogenic solvent. Materials of this type [31, 32] are useful in applications such as high-performance liquid chromatography, high-performance membrane chromatography, capillary electrochromatography, microfluidics [33], and high-throughput bioreactors [34]. In this process, no organic solvents are used in either the synthesis or purification. It is possible to synthesize the monoliths in a variety of containment vessels, including chromatography columns and narrow-bore capillary tubing. Moreover, the variable density associated with SCF solvents was exploited in order to "fine-tune" the polymer morphology. The apparent Brunnauer–Emmet–Teller (BET) surface area and the average pore size of the materials varied substantially for a series of crosslinked monoliths synthesized using $scCO_2$ as the porogen over a range of reaction pressures [23]. This approach was also extended to synthesize well-defined porous, crosslinked beads by suspension polymerization, again without using any organic solvents [22]. The surface area of the beads could be tuned over a broad range (5–500 m^2/g) simply by varying the CO_2 density. The ability to fine-tune the polymer morphology in materials of this type can be rationalized by considering the variation in solvent quality as a function of CO_2 density and the resulting influence on the mechanism of nucleation, phase separation, aggregation, monomer partitioning, and pore formation [35].

Recently, an entirely new approach to prepare porous materials was developed by templating the structure of *solid* CO_2 by directional freezing [36]. In this process, a liquid CO_2 solution was frozen in liquid nitrogen unidirectionally. The solid CO_2 was subsequently removed by direct sublimation to yield a porous, solvent-free structure with no additional purification steps (Fig. 1). Other CO_2-soluble actives could be incorporated into the porous structure uniformly. This was demonstrated by dispersing oil red uniformly in the aligned porous sugar acetates [36]. This method differs fundamentally from the other CO_2-based techniques [14–29], and offers the unique advantage

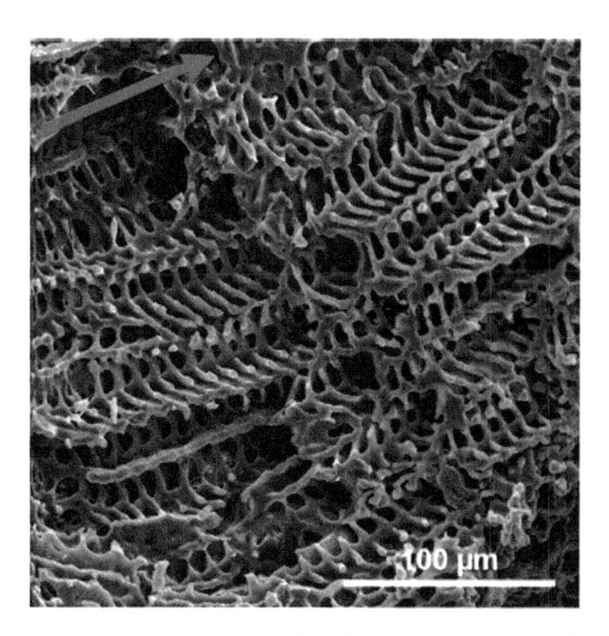

Fig. 1. Aligned porous sugar acetate produced by directional freezing of liquid CO_2 solution. The red arrow represents the approximate direction of freezing. Reproduced with permission: copyright 2005 American Chemical Society [36].

of generating materials with aligned pore structures. Materials with aligned microstructures and nanostructures are of interest in a wide range of applications such as organic electronics [37], microfluidics [38], molecular filtration [39], nanowires [40], and tissue engineering [41].

In addition to producing materials with aligned porosity, there are a number of additional advantages associated with this new technique. The method avoids the use of any organic solvent, thus eliminating toxic residues in the resulting material. The CO_2 can be removed by simple sublimation, unlike aqueous-based processes where the water must be removed by freeze-drying [42–44]. Moreover, the method can be applied to relatively non-polar, water-insoluble materials. These aligned porous structures may find numerous applications, for example, as biomaterials. Aligned porous materials with micrometre-sized pores are of importance in tissue engineering where modification with biological cells is required. We are particularly interested in the use of such porous materials as scaffolds for aligned nerve cell growth. The latter application will be greatly facilitated by the recent development of biodegradable CO_2-soluble hydrocarbon polymers as potential scaffold materials [45].

4. TEMPLATING OF SUPERCRITICAL FLUID EMULSIONS

Emulsion templating is useful for the synthesis of highly porous inorganic [46–49], and organic materials [50–52]. In principle, it is possible to access a wide range of porous hydrophilic materials by reaction-induced phase separation of concentrated oil-in-water (O/W) emulsions. A significant drawback to this process is that large quantities of a water-immiscible oil or organic solvent are required as the internal phase (usually >80 vol. %). In addition, it is often difficult to remove this oil phase after the reaction. Inspired by studies on SCF emulsion stability and formation [53], we have developed methods for templating high internal phase C/W emulsions (HIPE) to generate highly porous materials in the absence of any organic solvent – only water and CO_2 are used [28]. If the emulsions are stable one can generate low-density materials ($\sim 0.1\ g/cm^3$) with relatively large pore volumes (up to $6\ cm^3/g$) from water-soluble vinyl monomers such as acrylamide and hydroxyethyl acrylate. Fig. 2 shows a crosslinked poly(acrylamide) material synthesized from an HIPE, as characterized by SEM and confocal microscopy (scale $= 230\ \mu m \times 230\ \mu m$). Comparison of the two images illustrates quite clearly how the porous structure shown in the SEM image is templated from the C/W emulsion (as represented by the confocal microscopy image of the

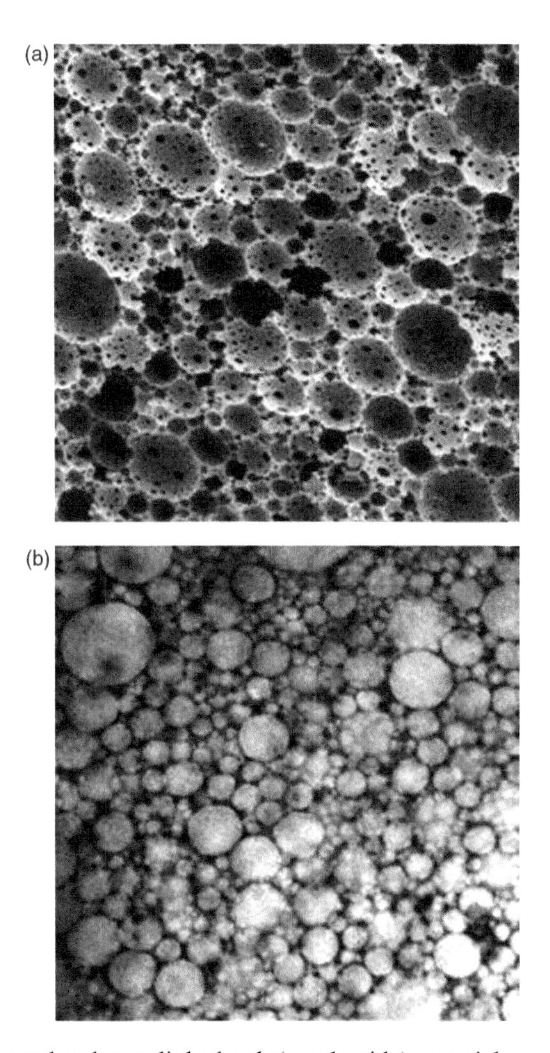

Fig. 2. Emulsion-templated crosslinked poly(acrylamide) materials synthesized by polymerization of a high internal phase CO_2-in-water emulsion (C/W HIPE). (a) SEM image of sectioned material. (b) Confocal image of same material, obtained by filling the pore structure with a solution of fluorescent dye. As such (a) shows the "walls" of the material while (b) show the "holes" formed by templating the $scCO_2$ emulsion droplets. Both images = 230 μm × 230 μm. Ratio of CO_2/aqueous phase = 80:20 (v/v). Pore volume = 3.9 cm^3/g. Average pore diameter = 3.9 μm. Reprinted with permission: copyright 2001 WILEY VCH Verlag GmbH & Co. [28].

pores). In general, the confocal image gives a better measure of the CO_2 emulsion droplet size and size distribution immediately before gelation of the aqueous phase. Initially we used low molecular weight ($M_w \sim 550$ g/mol) perfluoropolyether ammonium carboxylate surfactants to stabilize the C/W emulsions [28] but there are some practical disadvantages of using these surfactants in this particular process such as cost and the fact that the surfactant is non-degradable. It was subsequently shown that it is possible to use inexpensive hydrocarbon surfactants to stabilize C/W emulsions and that these emulsions can also be templated to yield low-density porous materials [29]. In this study it was shown that all of the problems associated with the initial approach could be overcome and it was possible to synthesize C/W emulsion-templated polymers at relatively modest pressures (60–70 bar) and low temperatures (20°C) using inexpensive and readily available hydrocarbon surfactants. Moreover, we demonstrated that this technique can, in principle, be extended to the synthesis of emulsion-templated HEA and HEMA hydrogels that may be useful, for example, in biomedical applications [54–56].

5. POLYMER SOLUBILITY IN CO_2

One of the fundamental issues that one must consider when implementing CO_2 for polymer synthesis or processing is polymer solubility. As mentioned, CO_2 is a weak solvent and there has been considerable research effort focused on discovering inexpensive biodegradable CO_2-soluble polymers from which inexpensive CO_2-soluble surfactants, ligands, and phase transfer agents could be developed. However, it is very difficult to predict which polymer structures would be CO_2-soluble, despite recent attempts to rationalize specific solvent–solute interactions by using ab initio calculations [57]. Only a few examples of CO_2-soluble polymers currently exist and, as such, there are a limited number of "design motifs" to draw upon. Moreover, it is clear that polymer solubility in CO_2 is influenced by a large number of interrelated factors [8] such as specific solvent–solute interactions [11, 57–59], backbone flexibility [11, 58, 60], topology [60], and the nature of the end-groups [60]. Given the current limits of predictive understanding, the discovery of new CO_2-soluble polymers might be accelerated using parallel or 'high-throughput' methodology. The synthetic approaches for such a strategy are already well in place; for example, a growing number of methods exist whereby one may synthesize and characterize polymer libraries [61]. By contrast, there are no examples of techniques for the rapid, parallel determination of solubility for libraries of materials in $scCO_2$ or other SCFs. The conventional method for evaluating polymer solubility in SCFs is cloud point measurement [8, 11, 58, 60], which involves the use of a variable-volume view cell. This technique is not suitable for rapid solubility measurement and would be impractical for large libraries of materials. Previously, we developed a simple method involving a 10-well reaction vessel whereby 10 solubilities could be measured in parallel using compressed fluid solvents such as liquid R134a [62]. This method was used to discover a new, inexpensive R134a-soluble stabilizer for dispersion polymerization, but suffers from a number of limitations; for example, the pressure limit is low (50 bar) and only compressed *liquid* solvents (as opposed to SCFs) can be used with this particular equipment [62].

In order to try and develop more rapid methods for elucidating polymer solubility in any solvent an understanding of the characteristics of the solvent in question must be fully understood. A brief discussion concerning the solubility of polymers in CO_2 will be presented followed by a novel high-throughput solubility screening method.

A number of research groups have synthesized 'CO_2-philic' fluoropolymers or silicone-based materials for use as steric stabilizers in dispersion polymerization [3, 63–66], phase transfer agents for liquid–liquid extraction [67], supports for homogeneous catalysis [68, 69], and surfactants for the formation of water/CO_2 emulsions and microemulsions [70, 71]. Unfortunately, the high cost of fluorinated polymers may prohibit their use on an industrial scale for some applications. Fluoropolymers also tend to have poor environmental degradability, and this could negate the environmental advantages associated with the use of $scCO_2$. The lack of inexpensive CO_2-soluble polymers and surfactants is a significant barrier to the future implementation of this solvent technology [72].

Beckman and coworkers [11, 58] reported that inexpensive poly(ether carbonate) (PEC) copolymers have been reported to be soluble in CO_2 under moderate conditions and could function as building blocks for inexpensive surfactants, but numerous practical difficulties remain. These hydrocarbon systems involve PECs synthesized by aluminium-catalyzed copolymerization of cyclic ethers with CO_2 (i.e., M1 = ethylene oxide, propylene oxide, cyclohexene oxide; $M_2 = CO_2$). These copolymers were found to be soluble in liquid CO_2 at concentrations of 0.2–1.5% (w/v) at ambient temperatures and pressures in the range 120–160 bar – that is, significantly above the liquid–vapour pressure for CO_2. These statistical copolymers were generated from very inexpensive feed-stocks and are thus appealing as 'building blocks' for cheap surfactants. The enhanced solubility of these copolymers with respect to poly(propylene oxide) is speculated to arise, at least in part, from specific

Lewis acid–base interactions that exist between CO_2 and the carbonyl groups of the carbonate moieties [11, 58, 73].

Step-growth polymerization is a method that can be used to synthesize polymers with well-defined chemical compositions; indeed, unlike non-ideal statistical chain growth polymerizations, the composition of step-growth polymers mirrors exactly the composition of the monomer feed. We recently developed step-growth polymerization routes to synthesize PEC and also poly(ether ester) (PEE) materials as potential inexpensive hydrocarbon CO_2-philes [74]. These polymers are soluble in CO_2, but only up to moderate molecular weights (<10000 g/mol).

Similarly, sugar acetates are highly soluble and have been proposed as renewable CO_2-philes [57]. Such materials could, in principle, function as CO_2-philic building blocks for inexpensive ligands and surfactants, but this potential has not yet been realized and numerous practical difficulties remain. For example, CO_2 solubility does not in itself guarantee performance in the various applications of interest. Effective surfactants, in particular, tend to require specific asymmetric topologies such as diblock copolymers [65, 70]. This in turn necessitates a flexible and robust synthetic methodology to produce well-defined architectures for specific applications.

6. HIGH-THROUGHPUT SOLUBILITY MEASUREMENTS IN CO_2

We reported a new method which allows for the rapid parallel solubility measurements for libraries of materials in SCFs [75]. The technique was used to evaluate the solubility of a mixed library of 100 synthetic polymers including polyesters, polycarbonates, and vinyl polymers. It was found that poly(vinyl acetate) (PVAc) showed the highest solubility in CO_2; the anomalously high solubility of PVAc has been shown previously [72]. This method is at least 50 times faster than other techniques in terms of the rate of useful information that is obtained and has broad utility in the discovery of novel SCF-soluble ligands, catalysts, biomolecules, dyes, or pharmaceuticals for a wide range of materials applications.

7. INEXPENSIVE AND BIODEGRADABLE CO_2-PHILES

PVAc is an inexpensive, high-tonnage bulk commodity polymer which, unlike most vinyl polymers, is moderately biodegradable and has been used in pharmaceutical excipient formulations. PVAc has also been shown to exhibit anomalously high solubility in CO_2 with respect to other vinyl hydrocarbon polymers [72], although the polymer is soluble only at relatively low molecular weights under conditions of practical relevance ($P < 300$ bar, $T < 100°C$).

We recently presented a simple and generic method for producing inexpensive and biodegradable polymer surfactants for $scCO_2$ for solubilization, emulsification, and related applications [45]. In this method, the terminal hydroxyl group of a PVAc oligomer is transformed into an imidazole ester by reaction with carbonyl diimidazole (CDI). This route has a number of advantages. First, the oligomer vinyl acetate (OVAc) imidazolide intermediate can be isolated, purified, and then coupled with a wide range of alcohols (or amines) to produce a variety of structures. Second, the route introduces a carbonate linkage that may further enhance CO_2 solubility [58, 74], and could also improve the biodegradability of the resulting materials. To illustrate the use of OVAc as a solubilizing group, an organic dye, Disperse Red 19 (DR19), was functionalized with OVAc (M_n 1070 g/mol, M_w 1430 g/mol) to produce **1** (Fig. 3). The stoichiometry of the reaction was controlled such that one OVAc chain was attached to each DR19 molecule, as confirmed by GPC and ^1H NMR. DR19 itself had negligible solubility in CO_2 up to pressures of 300 bar/25°C (no colour was observed in the CO_2 phase). By contrast, the functionalized dye, **1**, was found to be soluble in CO_2 (100–200 bar) at least up to concentrations of ~1 wt% (Fig. 3). This suggests that fractionated OVAc has potential as a less expensive and more biodegradable replacement for the highly fluorinated materials used previously to solubilize species such as dyes, catalysts, proteins, and nanoparticles in CO_2 [2, 3, 66–68, 71, 76, 77]. As mentioned previously, another important area in $scCO_2$ technology is the formation of water-in-CO_2 (W/C) and

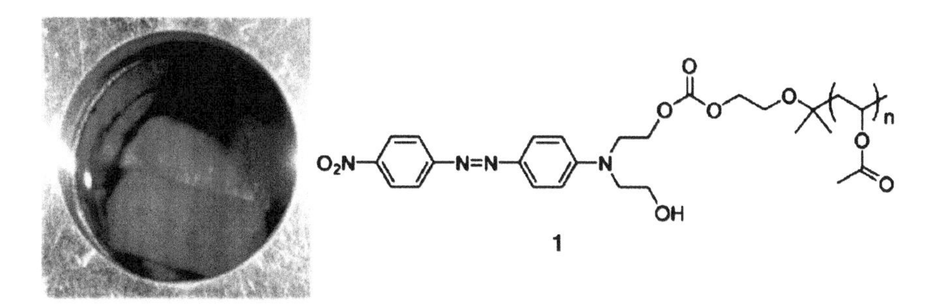

Fig. 3. Photograph showing the dissolution of an OVAc-functionalized dye, **1**, in CO_2 (200 bar, 20°C, 0.77 wt%). Reproduced with permission: copyright 2005 American Chemical Society [45].

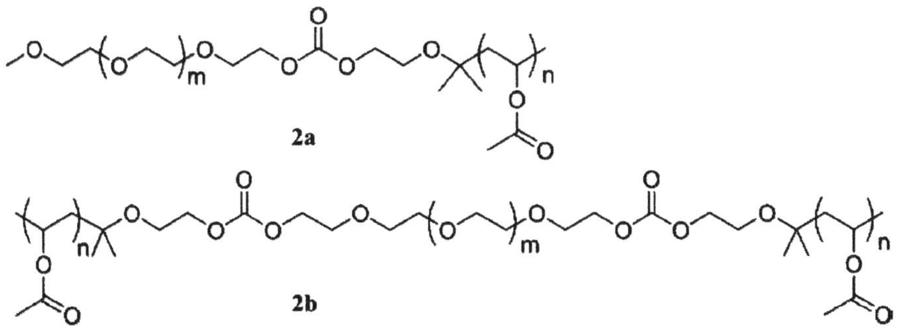

Fig. 4. Structures of CO_2-philic surfactants for C/W emulsion formation: OVAc-*b*-PEG diblock polymer (**2a**) and OVAc-*b*-PEG-*b*-OVAc triblock polymer (**2b**) [45]. Reproduced with permission: copyright 2005 American Chemical Society.

C/W emulsions and microemulsions [71, 78–81]. The same CDI route was used to couple OVAc with poly(ethylene glycol) monomethyl ethers (HO-PEG-OMe) and poly(ethylene glycol) diols (PEG) to produce diblock (Fig. 4, **2a**) and triblock (**2b**) copolymers, respectively.

These polymers were found to be useful surfactants. For example, an OVAc-*b*-PEG-OVAC triblock surfactant was found to emulsify up to 97% (v/v) C/W emulsion which was stable for at lease 48 h. An OVAc-*b*-PEG diblock copolymer was used to form a 90% (v/v) C/W emulsion. The polymerization of the continuous aqueous phase led to the formation of highly porous crosslinked poly(acrylamide) (Fig. 5), which was comparable to that prepared with perfluoropolyether ammonium carboxylate surfactants [28].

8. CONCLUSIONS

In general, CO_2 is an attractive solvent alternative for the synthesis of polymers because it is 'environment friendly', non-toxic, non-flammable, inexpensive, and readily available in high purity from a number of sources. Product isolation is straightforward because CO_2 is a gas under ambient conditions, removing the need for energy-intensive drying steps. CO_2 affords a number of specific advantages for the synthesis and modification of porous materials; moreover, it offers the potential of reducing organic solvent usage in the production of materials. We have shown a simple and generic method for producing inexpensive, functional hydrocarbon CO_2-philes for solubilization, emulsification, and related applications. We have also shown that these structures can outperform perfluorinated analogues in specific applications.

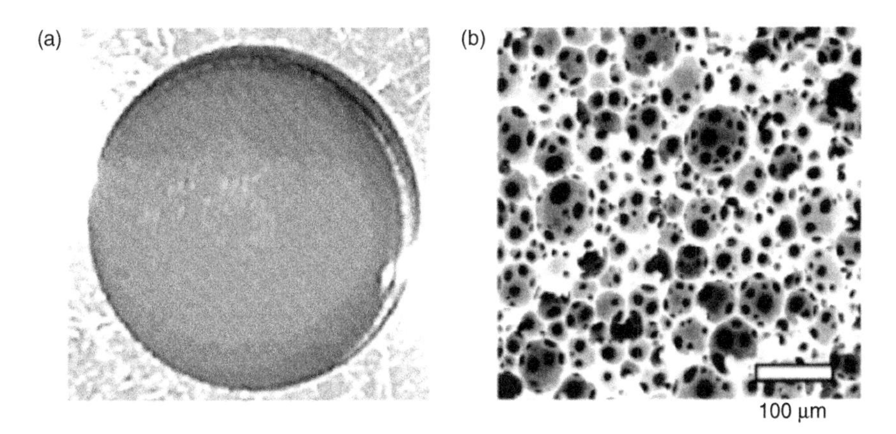

Fig. 5. (a) Photograph of highly concentrated CO_2-in-water (C/W) emulsion (97%, v/v, CO_2) stabilized by a triblock copolymer surfactant. A water-soluble dye, methyl orange, was included in the aqueous phase. (b) Emulsion-templated poly(acrylamide) material synthesized by polymerization of a concentrated C/W emulsion (90%, v/v, CO_2) stabilized with a diblock surfactant [45]. Reproduced with permission: copyright 2005 American Chemical Society.

This synthetic route should allow the design of a wide range of related CO_2-philic materials.

ACKNOWLEDGEMENTS

We thank EPSRC for financial support (EP/C511794/1) and the Royal Society for a Royal Society University Research Fellowship (to A.I.C.).

REFERENCES

[1] A.I. Cooper, J. Mater. Chem., 10 (2000), 207–234.
[2] J.M. DeSimone, Z. Guan and C.S. Elsbernd, Science, 257 (1992), 945–947.
[3] J.M. DeSimone, E.E. Maury, Y.Z. Menceloglu, J.B. McClain, T.J. Romack and J.R. Combes, Science, 265 (1994), 356–359.
[4] A.I. Cooper, Adv. Mater., 15 (2003) 1049.
[5] C.D. Wood, A.I. Cooper and J.M. DeSimone, Curr. Opin. Solid State Mater. Sci., 8 (2004) 325.
[6] J.L. Kendall, D.A. Canelas, J.L. Young and J.M. DeSimone, Chem. Rev., 99 (1999) 543.
[7] H.M. Woods, M. Silva, C. Nouvel, K.M. Shakesheff and S.M. Howdle, J. Mater. Chem., 14 (2004) 1663.
[8] C.F. Kirby and M.A. McHugh, Chem. Rev., 99 (1999) 565.

[9] P.G. Jessop and W. Leitner, Chemical Synthesis Using Supercritical Fluids, Wiley-VCH, Weinheim, 1999.
[10] E.J. Beckman, Chem. Commun., (2004) 1885–1888.
[11] T. Sarbu, T.J. Styranec and E.J. Beckman, Ind. Eng. Chem. Res., 39 (2004) 678.
[12] J. Eastoe, A. Paul, S. Nave, D.C. Steytler, B.H. Robinson, E. Rumsey, M. Thorpe and R.K. Heenan, J. Am. Chem. Soc., 123 (2001) 988.
[13] M.A. McHugh and V. Krukonis, Supercritical Fluid Extraction, 2nd edn., Butterworth Heinemann, Boston, 1994.
[14] S.K. Goel and E.J. Beckman, Polymer, 34 (1993) 1410.
[15] S.M. Howdle, M.S. Watson, M.J. Whitaker, V.K. Popov, M.C. Davies, F.S. Mandel, J.D. Wang and K.M. Shakesheff, Chem. Commun., 109 (2001).
[16] S. Siripurapu, Y.J. Gay, J.R. Royer, J.M. DeSimone, R.J. Spontak and S.A. Khan, Polymer, 43 (2002) 5511.
[17] B. Krause, G.H. Koops, N.F.A. van der Vegt, M. Wessling, M. Wubbenhorst and J. van Turnhout, Adv. Mater., 14 (2002) 1041.
[18] S. Siripurapu, J.M. DeSimone, S.A. Khan and R.J. Spontak, Adv. Mater., 16 (2004) 989.
[19] H. Matsuyama, H. Yano, T. Maki, M. Teramoto, K. Mishima and K.J. Matsuyama, Membr. Sci., 194 (2001) 157.
[20] H. Matsuyama, A. Yamamoto, H. Yano, T. Maki, M. Teramoto, K. Mishima and K.J. Matsuyama, Membr. Sci., 204 (2002) 81.
[21] A.I. Cooper and A.B. Holmes, Adv. Mater., 11 (1999) 1270.
[22] C.D. Wood and A.I. Cooper, Macromolecules, 34 (2001) 5.
[23] A.K. Hebb, K. Senoo, R. Bhat and A.I. Cooper, Chem. Mater., 15 (2003) 2061.
[24] C. Shi, Z. Huang, S. Kilic, J. Xu, R.M. Enick, E.J. Beckman, A.J. Carr, R.E. Melendez and A.D. Hamilton, Science, 286 (1999) 1540.
[25] F. Placin, J.P. Desvergne and F.J. Cansell, Mater. Chem., 10 (2000) 2147.
[26] H. Wakayama, H. Itahara, N. Tatsuda, S. Inagaki and Y. Fukushima, Chem. Mater., 13 (2001) 2392.
[27] Y. Fukushima and H. Wakayama, J. Phys. Chem. B, 103 (1999) 3062.
[28] R. Butler, C.M. Davies and A.I. Cooper, Adv. Mater., 13 (2001) 1459.
[29] R. Butler, I. Hopkinson and A.I. Cooper, J. Am. Chem. Soc., 125 (2003) 14473.
[30] A.I. Cooper, C.D. Wood and A.B. Holmes, Ind. Eng. Chem. Res., 39 (2000) 4741.
[31] F. Svec and J.M.J. Fréchet, Science, 273 (1996) 205.
[32] F. Svec and J.M.J. Fréchet, Ind. Eng. Chem. Res., 38 (1998) 34.
[33] C. Yu, M.C. Xu, F. Svec and J.M.J. Fréchet, J. Polym. Sci. A, Polym. Chem., 40 (2002) 755.
[34] M. Petro, F. Svec and J.M.J. Fréchet, Biotechnol. Bioeng., 49 (1996) 355.
[35] D.C. Sherrington, Chem. Commun., 21 (1998) 2275.
[36] H. Zhang, J. Long and A.I. Cooper, J. Am. Chem. Soc., 127 (2005) 13482.
[37] H. Gu, R. Zheng, X. Zhang and B. Xu, Adv. Mater., 16 (2004) 1356.
[38] S.R. Quake and A. Scherer, Science, 290 (2000) 1536.
[39] A. Yamaguchi, F. Uejo, T. Yoda, T. Uchida, Y. Tanamura, T. Yamashita and N. Teramae, Nat. Mater., 3 (2004) 337.
[40] R. Adelung, O.C. Aktas, J. Franc, A. Biswas, R. Kunz, M. Elbahri, J. Kanzow, U. Schuramann and F. Faupel, Nat. Mater., 3 (2004) 375.

[41] C.Y. Xu, R. Inai, M. Kotaki and S. Ramakrishna, Biomaterials, 25 (2004) 877.
[42] W. Mahler and M.F. Bechtold, Nature, 285 (1980) 27.
[43] S.R. Mukai, H. Nishihara and H. Tamon, Chem. Commun., (2004) 874.
[44] H. Nishihara, S.R. Mukai, D. Yamashita and H. Tamon, Chem. Mater., 17 (2005) 683.
[45] B. Tan and A.I. Cooper, J. Am. Chem. Soc., 127 (2005) 8938.
[46] A. Imhof and D.J. Pine, Adv. Mater., 10 (1998) 697.
[47] P. Schmidt-Winkel, W.W. Lukens, P.D. Yang, D.I. Margolese, J.S. Letlow, J.Y. Ying and G.D. Stucky, Chem. Mater., 12 (2000) 686.
[48] V.N. Manoharan, A. Imhof, J.D. Thorne and D.J. Pine, Adv. Mater., 13 (2001) 447.
[49] H. Zhang, G.C. Hardy, M.J. Rosseinsky and A.I. Cooper, Adv. Mater., 15 (2003) 78.
[50] N.R. Cameron and D.C. Sherrington, Adv. Polym. Sci., 126 (1996) 163.
[51] W. Busby, N.R. Cameron and C.A.B. Jahoda, Biomacromolecules, 2 (2001) 154.
[52] H. Zhang and A.I. Cooper, Chem. Mater., 14 (2002) 4017.
[53] C.T. Lee, P.A. Psathas, K.P. Johnston, J. deGrazia and T.W. Randolph, Langmuir, 15 (1999) 6781.
[54] J. Song, E. Saiz and C.R. Bertozzi, J. Am. Chem. Soc., 125 (2003) 1236.
[55] Y. Luo, P.D. Dalton and M.S. Shoichet, Chem. Mater., 13 (2001) 4087.
[56] F. Chiellini, R. Bizzari, C.K. Ober, D. Schmaljohann, T.Y. Yu, R. Solaro and E. Chiellini, Macromol. Rapid. Commun., 22 (2001) 1284.
[57] P. Raveendran and S.L. Wallen, J. Am. Chem. Soc., 124 (2002) 12590.
[58] T. Sarbu, T. Styranec and E.J. Beckman, Nature, 405 (2000) 165.
[59] S.G. Kazarian, M.F. Vincent, F.V. Bright, C.L. Liotta and C.A. Eckert, J. Am. Chem. Soc., 118 (1996) 1729.
[60] C. Drohmann and E.J. Beckman, J. Supercrit. Fluids, 22 (2002) 103.
[61] R. Hoogenboom, M.A.R. Meier and U.S. Schubert, Macromol. Rapid Commun., 24 (2003) 15.
[62] C.D. Wood and A.I. Cooper, Macromolecules, 36 (2003) 7534.
[63] K.A. Shaffer, T.A. Jones, D.A. Canelas, J.M. DeSimone and S.P. Wilkinson, Macromolecules, 29 (1996) 2704.
[64] C. Lepilleur and E.J. Beckman, Macromolecules, 30 (1997) 745.
[65] A.I. Cooper, W.P. Hems and A.B. Holmes, Macromolecules, 32 (1999) 2156.
[66] P. Christian, S.M. Howdle and D.J. Irvine, Macromolecules, 33 (2000) 237.
[67] A.I. Cooper, J.D. Londono, G. Wignall, J.B. McClain, E.T. Samulski, J.S. Lin, A. Dobrynin, M. Rubinstein, A.L.C. Burke, J.M.J. Fréchet and J.M. DeSimone, Nature, 389 (1997) 368.
[68] W.P. Chen, L.J. Xu, Y.L. Hu, A.M.B. Osuna and J.L Xiao, Tetrahedron, 58 (2002) 3889.
[69] I. Kani, M.A. Omary, M.A. Rawashdeh-Omary, Z.K. Lopez-Castillo, R. Flores, A. Akgerman and J.P. Fackler, Tetrahedron, 58 (2002) 3923.
[70] K.P. Johnston, Curr. Opin. Colloid Interface Sci., 5 (2000) 351.
[71] K.P. Johnston, K.L. Harrison, M.J. Clarke, S.M. Howdle, M.P. Heitz, F.V. Bright, C. Carlier and T.W. Randolph, Science, 271 (1996) 624.
[72] Z. Shen, M.A. McHugh, J. Xu, J. Belardi, S. Kilic, A. Mesiano, S. Bane, C. Karnikas, E.J. Beckman and R. Enick, Polymer, 44 (2003) 1491.
[73] T. Tsukahara, Y. Kayaki, T. Ikariya and Y. Ikeda, Angew Chem. Int. Ed., 43 (2004) 3719.

[74] B. Tan, H.M. Woods, P. Licence, S.M. Howdle and A.I. Cooper, Macromolecules, 38 (2005) 1691.
[75] C.L. Bray, B. Tan, C.D. Wood and A.I. Cooper, J. Mater. Chem., 15 (2005) 456.
[76] M.A. Carroll and A.B. Holmes, Chem. Commun., (1998) 1395.
[77] P.G. Shah, J.D. Holmes, R.C. Doty, K.P. Johnston and B.A. Korgel, J. Am. Chem. Soc., 122 (2000) 4245.
[78] S.R.P. da Rocha, P.A. Psathas, E. Klein and K.P. Johnston, J. Colloid Interface Sci., 239 (2001) 241.
[79] C.T. Lee, P.A. Psathas, K.P. Johnston, J. deGrazia and T.W. Randolph, Langmuir, 15 (1999) 6781.
[80] J.L. Dickson, B.P. Binks and K.P. Johnston, Langmuir, 20 (2004) 7976.
[81] J.L. Dickson, P.G. Smith, V.V. Dhanuka, V. Srinivasan, M.T. Stone, P.J. Rossky, J.A. Behles, J.S. Keiper, B. Xu, C. Johnson, J.M. DeSimone and K.P. Johnston, Ind. Eng. Chem. Res., 44 (2005) 1370.

Thermodynamics, Solubility and Environmental Issues
T.M. Letcher (editor)

Chapter 22

Solubility of Plasticizers, Polymers and Environmental Pollution

Ewa Białecka-Florjańczyk[a] and Zbigniew Florjańczyk[b]

[a]*Agricultural University, Institute of Chemistry, ul. Nowoursynowska 159c, 02-787 Warsaw, Poland*
[b]*Department of Chemistry, Warsaw University of Technology, ul. Noakowskiego 3, 00-664 Warsaw, Poland*

1. INTRODUCTION

The plasticizing effect of various organic substances on brittle natural or synthetic polymers is well recognized and has been applied for a long time to improve the processibility, flexibility and stretchability of many polymeric materials. The most important reason for using a plasticizer is to decrease the glass transition temperature of the polymer that is closely related to the lowest temperature limit of retention of elastic properties. The extent to which a plasticizer reduces the glass transition temperature of a polymer is often treated as a simple measure of plasticizer efficiency.

$$T_g = T_{gp} - k w_{pl} \quad (1)$$

where T_g is the glass transition temperature of polymer plasticizer mixture; T_{gp} the glass transition temperature of unplasticized polymer; k the plastcizer efficiency parameter; and w_{pl} the weight fraction of plasticizer.

Plasticizers can also affect other properties of polymers such as the degree of crystallinity, optical clarity, electric conductivity, fire behaviour, resistance to biological degradation, etc.

In the last decade the worldwide production of plasticizers was around 5 million tons per year [1]. These were applied to ~60 polymers and more than 30 groups of products. Poly(vinyl chloride) processing is by far the most important use of plasticizers, consuming over 80% of plasticizer production. Considerable amount of plasticizers is also used in combination with poly(vinylidene chloride), polyamides, polyacrylates, poly(vinyl acetate), poly(vinyl butyral), cellulose derivatives and various kind of elastomers. They are also present in paints, printing inks, lubricants and some cosmetic products.

The current database of commercially manufactured plasticizers contains more than 1200 items [2], however, only ~100 products have achieved noticeable market significance. Most of them are based on carboxylic acid esters with linear or branched aliphatic alcohols of moderate chain lengths (predominantly C_6–C_{11}). Esters of phthalic acid constitute more than 85% of the total plasticizer consumption and di-(2-ethylhexyl) phthalate (DEHP) appears to be the most widely used in Europe and North America. The other important groups of plasticizers include esters of aliphatic dicarboxylic acids, citrates, benzoates, trimellitates, epoxidized vegetable oils, phopsphates, ethylene glycols, ethylene glycols esters and polyesters. Typical chemical structures of common plasticizers are given in Table 1.

The great advantage of external plasticizing lies in the simple preparation of various formulations which are used to regulate the final properties of polymeric materials over a wide range and thus allowing one to tailor-make the product to a given application. In practical formulations the plasticizer concentration usually varied from 5 to 40 wt%. Plasticizers play also an important role as processing aids. They act as lubricants, reduce adhesion to metal surface, lower processing temperature, reduce melt viscosity, improve ductility and lower film forming temperature in dispersions. Furthermore, they are often used to prepare concentrates of several groups of additives including adhesion promoters, antistatic, biocides, curing agents, UV stabilizers to obtain a product which is easy to disperse in polymeric matrix.

The principle disadvantage of external plasticizing is the migration of plasticizers from the plasticized material to other contacting media. This process includes diffusion of plasticizers from the material bulk towards the surface, interface phenomena and sorption or evaporation into the surrounding medium. One of the consequences of the plasticizer loss is the gradual alterations to the properties of plasticized products. However, the most important issue is that plasticizers are released into the environment during their manufacture, polymer processing, and in the use of the finished product by the consumer and finally in their waste disposal stage. More than 1% of applied DEHP and ~5% of dibutyl phthalate plasticizers are estimated to be dispersed into the environment [1c]. Approximately 15% of DEHP used in paints, is known to evaporate to the atmosphere [3]. Although plasticizers have high boiling points many of them have been detected in air samples collected in

Table 1
Chemical structures and boiling points of selected plasticizers

		MW	BP[a] (°C)
Phthalates	$C_6H_4(COOR)_2$ (1,2-)		
Dibutyl phthalate (DBP)	$R = (CH_2)_3CH_3$	278	340
Di(2-ethylhexyl) phthalate (DEHP)	$R = CH_2CH(CH_2CH_3)(CH_2)_3CH_3$	390	361
Dioctyl phthalate (DOP)	$R = (CH_2)_7CH_3$	390	384
Esters of alkane-dicarboxylic acids	$RO{-}C(=O)(CH_2)_nC(=O){-}OR$		
Dibutyl sebacate (DBS)	$R = (CH_2)_3CH_3$, $n = 8$	314	178/3 mmHg
Di(2-ethylhexyl) adipate (DOA)	$R = CH_2CH(CH_2CH_3)(CH_2)_3CH_3$, $n = 4$	370	214/5 mmHg
Glycols and glycols esters	$RO[CH_2(CHR')O]_nR$		
Tetraethylene glycol	$R = H$, $R' = H$, $n = 4$	194	314
Tripropylene glycol	$R = H$, $R' = CH_3$, $n = 3$	192	265
Triethylene glycol dibenzoate	$R = C_6H_5{-}C(=O){-}$, $R' = H$, $n = 3$	314	235–7/7 mmHg
Citrates	$R'O{-}C(COOR)(CH_2COOR)_2$		
Acetyltributyl citrate (ATBC)	$R' = CH_3{-}C(=O){-}$, $R = C_4H_9$	402	173/1 mmHg
Phosphates	$RO{-}P(=O)(OR){-}OR$		
Tricresylphosphate (TCP)	$R = CH_3C_6H_4{-}$	368	265/10 mmHg

[a]According to Aldrich.

buildings, automobiles as well as in rural and remote location. The plasticizers have also been found in drinking water, food, soil and living organisms.

2. SOLUBILITY PARAMETERS AS A GUIDE FOR PLASTICIZER SELECTION

The compatibility of a plasticizer with a given polymer is very important. The maximum amount of plasticizer incorporated into a polymer at the processing temperature and retained by it without exudation during the storage, is commonly accepted as the practical limit of compatibility. From a theoretical consideration, compatibility is regarded as the miscibility on a molecular scale. As a result the classical thermodynamic theory of binary polymer–solvent system have been used to describe plasticized materials.

The Gibbs free energy of the polymer and plasticizer mixing can be described by the following equation:

$$\Delta G_{\mathrm{mix}} = RT(x_{\mathrm{p}} \ln \Phi_{\mathrm{p}} + x_{\mathrm{pl}} \ln \Phi_{\mathrm{pl}} + \chi \Phi_{\mathrm{p}} \Phi_{\mathrm{pl}}) \quad (2)$$

where x_{p} and x_{pl} are the mole fractions of polymer and plasticizer, Φ_{p} and Φ_{pl} the volume fractions of components and χ the interaction parameter which provides a measure of the thermodynamic affinity of plasticizer to polymer [4]. The smaller χ is, the more stable the solution is, relative to pure components, and the more likely that the system is miscible over a wide range of concentration. The interaction parameter is empirical and can be determined experimentally, but it can also be calculated using the Flory–Huggins and Hildebrand–Scatchard theories in terms of solubility parameters [5a].

$$\chi = (\delta_{\mathrm{p}} - \delta_{\mathrm{pl}})^2 V_{\mathrm{r}} (RT)^{-1} \quad (3)$$

where V_{r} is an average volume of plasticizer molecule and polymer repeated units and δ_{p} and δ_{pl} are solubility parameters for polymer and plasticizer. Therefore, one of the most common approaches is to use the δ values of both components for the initial estimation of compatibility. The Hansen solubility parameters are usually employed because the classical Hildebrand approach neglects the existence of polar and specific interactions which are crucial for many plasticized systems. The values of Hansen parameters for selected polymers and plasticizers are included in Tables 2 and 3, respectively. According to the Huggins criterion the value of interaction parameter must be less than 0.5 for the complete compatibility of plasticizer

Table 2
Hansen solubility parameters for selected polymers according to [6]

Polymer	Solubility parameter δ (MPa)$^{0.5}$
Poly(acrylonitrile)	25.2
Poly(butadiene)	17.0
Poly(methyl methacrylate)	22.7
Poly(styrene)	22.5
Poly(vinyl butyral)	23.1
Poly(vinyl chloride)	21.4

with a linear polymer of high molecular weight. It means that the difference between solubility parameters of compatible polymer and plasticizer should be less than approximately 4 MPa$^{0.5}$. For example, PVC having a δ value of 21.4 MPa$^{0.5}$ is compatible with DBP, DEHP and ditridecyl phthalate over the whole range of concentration. The χ values for the system (calculated according to Eq. (3)) with the most polar butyl ester are in the range 0.02–0.03 and increase up to 0.2–0.3 for the plasticizers containing longer aliphatic chains. On the other hand polyacrylontrile with a δ value of 25.2 MPa$^{0.5}$ is hardly soluble in phthalates and only butyl ester can be considered as a primary plasticizer (χ values are in the range 0.4–0.5).

The conclusions drawn from the analysis of interaction parameters must always be verified by further solubility and gelling investigation over a broad range of concentration and performance condition. The gelling behaviour of plasticizers in different polymers is monitored in practice by a rheological analysis, using laboratory compounders or cone–plate and plate–plate rheometers. The data are collected as a function of time either at constant temperature or constant rate of temperature increase. The other methods involving visual evaluation of the plasticizers solvating capabilities, the observation of morphological changes by means of scanning electron microscopy and glass transition measurements are also used to determine conditions at which the plasticized system becomes homogeneous. Plasticizers are classified into several groups, based on their behaviour. The classification of organic substances into primary and secondary plasticizers is the most common one. Primary plasticizers are considered as good solvents of high compatibility. They can be used alone to form stable homogeneous gels with the polymer at the usual processing temperature. Secondary plasticizers have limited compatibility and thus they are usually applied in combination with a primary plasticizer.

Table 3
Physicochemical properties and environmental fate data for selected plasticizers[a]

	Solubility parameter, δ (MPa)$^{0.5}$	H_2O solubility (mg/dm^3)	K_H – estimated Henry's law constants (atm m^3/mole)	K_{OC} – organic carbon–water partition coefficients (dm^3/kg)	Biodegradation half-lives	log K_{OW} octanol–water partition coefficients
Dimethyl phthalate	21.4	4000	2.0×10^{-7}, 1.1×10^{-7}	190–1590	1–7 days	1.60
Diethyl phthalate	20.6	1080	6.1×10^{-7}	69–1726	3 days–8 weeks	1.4–3.3
Dibutyl phthalate	19.6	11.2	4.6×10^{-7}	1386	2–23 days	4.45–4.90
Di(2-ethylhexyl) phthalate	18.4	0.3	1.1×10^{-5}, 1.3×10^{-7}	482000, 82429	5–23, 21–54 days	4.2–7.6
Dioctyl phthalate	18.6	3.0	4.5×10^{-7}	610000	4 days–4 weeks	8.10
Ditridecyl phthalate	18.1	0.34	2.2×10^{-4}	1200000	–[b]	>8
Benzylbutyl phthalate	20.9	2.7	4.8×10^{-6}, 1.3×10^{-6}	17000	1–7days	4.6–4.9
Dibutyl sebacate	17.8	40	4.8×10^{-8}	575	–	–
Di(2-ethylhexyl) sebacate	17.4	0.20	8.5×10^{-5}	560000	–	10.1
Triethylene glycol dibenzoate[c]	21.2	30.4	1.8×10^{-9}	1580	6–28 h	3.2
Acetyltributyl citrate	19.2	5.0	3.8×10^{-10}	5100	–	4.30
Tricresyl phosphate	23.1	0.36	8.1×10^{-7}, $1.1–2.8 \times 10^{-7}$	79000	–	5.1

[a]The data according to Ref. [1] except of solubility parameters, which were calculated according to Ref. [5b].
[b]Lack of data.
[c]All the values for triethylene glycol dibenzoate, which were taken from Ref. [17].

Plasticizers that have a good compatibility with the polymer are much less prone to migrate when the polymeric material comes into contact with liquids or other polymeric materials. There were several attempts to correlate resistance to migration with solubility parameters of contacting media. It turned out, however, that the solubility parameter method, while helpful in initial assessment, is limited for reliably predicting a plasticizer's extractability from a polymeric matrix because common cleaners, lubricants and other fluids or solid materials that come in contact with plasticized materials, are often made of mixtures of proprietary chemical formulations. In addition, standardized chemical compatibility data neglect the effect of internal stress in final products and interaction with other additives which may strongly affect the plasticizer solubility and migration resistance. Therefore, for precise description of plasticizers extraction, several testing procedures have been developed and normalized.

3. ENVIRONMENTAL AND HEALTH ISSUES

Plasticizers are key additives in many plastics and are therefore important constituents of a variety of products – not only electrical insulation, wall coverings, lubricants and floor carpets but also such items as food packaging, bottled water containers, toys, cosmetics and medical materials. Due to the high volume of the production of phthalates many reviews focus on their implications on human exposure and public health. We will discuss here the properties of some other representative plasticizers besides phthalic esters. Some of their relevant physicochemical parameters are collected in Table 3.

3.1. Outdoor Environment

Plasticizers are released into the environment during their synthesis and the preparation of plastic products as well as during the use of the products by consumers. Most of them are relatively non-volatile (see BP in Table 1), however, they have been detected at nanogram per cubic metre (ng/m^3) scale even in rural locations (in the case of phthalates 1–5 ng/m^3 [7]). In urban air, especially in office buildings the reported concentrations of phthalates are considerably greater (even up to 1–2 $\mu g/m^3$ [8]). In the atmosphere the volatile plasticizers can be subjected to degradation via the reaction with hydroxyl radicals. The half-life for this reaction has been estimated to vary from several days (diallylphthalate) to over 100 days (DMP) [7].

The solubility in water is one of the most important properties of plasticizers in the natural environment, because it determines the extent of leaching of plasticizers molecules. It arises from hydrophilic–hydrophobic properties of the molecules. Plasticizers, which are mainly the esters of relatively long alcohols, have a solubility in water of less than several milligrams per cubic decimetre (mg/dm^3) (see Table 3). Moreover, the solubility in water can affect the equilibrium between concentrations of plasticizer in vapour phase and aqueous solution – the appropriate Henry's law constants K_H are given in Table 3 – but because of their limited volatility the volatilization from water is a slow process.

The environmental fate of plasticizers results from their chemical constitution. Most of them contain the ester group and thus the hydrolysis of ester bond is the main reaction in aqueous medium. According to Wolfe et al. [9] it appears that this reaction at pH 7 may be too slow and is negligible (the half-life time exceeds 100 days).

The biodegradation processes are of greater importance in surface as well as in groundwater (see Table 2). The degradation process is sometimes difficult to analyse, as it appears to depend on the environmental conditions (e.g., mixing, temperature, aerobic or anaerobic conditions) and the results are rather incomparable. Nevertheless, it appears, that biodegradation of butylbenzyl and dibutyl phthalates proceeds more easily. Some aerobic biodegradation of phthalates or glycols dibenzoates were studied in the presence of *Rhodotorula rubra* [10a] or *Rhodococcus rhodococcus* [10b] and *Pseudomonas species* [10c].

The other way of removing plasticizers from water is their adsorption at liquid–solid interface (i.e., soil or sediments) [11]. This process depends on hydrophobicity of the substance, i.e. is proportional to organic carbon–water partition coefficient K_{OC} – plasticizers with K_{OC} value greater than 1000 dm^3/kg are regarded as having relatively significant affinity for sediments and soil. It means that it is of greater importance in the case of 2-ethylhexyl (hydrophobic alkyl chain) than in the case of dimethyl phthalate (see Table 3).

Another factor influencing the occurrence of phthalates in the environment is their potential for persisting and accumulating in organic matrices. The extent of the biocummulation process, resulting from lipophilicity, can be correlated with octanol–water partition coefficient K_{OW} (see Table 3), because octanol is believed to imitate the properties of fatty organic structures and aqueous phases of plants and animal tissues. In the case of strongly hydrophobic molecules, K_{OW} exceeds 10^4. A comparison of bio-concentration factors (BCF = concentration in organism/mean concentration in water) for various organisms and plasticizers suggests the importance of this process for hydrophobic molecules (dinonyl phthalate or di(2-ethylhexyl)sebacate

[12]), but it appears that such a bio-concentration of most of the plasticizers by aquatic organisms such as fishes may not be significant [13]. Phthalates, unlike other hydrophobic chemicals, e.g. polychlorinated biphenyls, appear to be more readily metabolized, particularly by enzymes.

3.2. Indoor Environment and Toxicological Effects

The reports on plasticizers known to be present in indoor environments, involve mainly phthalates and organophosphates. Phthalates are ubiquitous industrial chemicals used as plasticizers for PVC in most applications – wire and cables, films and sheeting, spread coatings, paints. They are also present in textiles, carpeted floor and electronic goods. Organophosphates are used as plasticizers and also as fire-retardant agent in PVC formulations. They are also used as pigment dispersants and additives in adhesives, lacquer coatings and wood preservatives.

Indoor contamination with plasticizers results mainly from leaching, depending on the characteristics of particular polymeric materials, temperature and the volatility of compound. Normally the concentrations of plasticizers in indoor air are in nanogram per cubic metre (ng/m^3) range, but plasticizers having lower boiling points and higher vapour pressure (e.g., DMP, DEP, DBP, TBP) usually show higher concentrations. Their adsorption by dust particles is responsible for the widespread distribution of phthalates in office buildings [8].

The exposure of humans to plasticizer contamination, indoors, is via inhalation of house dust, dermal penetration and non-dietary ingestion. In this context the exposure to consumer products especially food, medical and pharmaceutical materials or toys is of particular interest. The entire toxicological effect of plasticizers depends not only on their biological activity but also on their migration, solubility and leaching rates [14]. These are affected by the compatibility of plasticizer and matrix polymer [15] – molecules of plasticizer are not chemically bounded to the product and may therefore leach into surrounding medium. On the other hand, the rate of leaching depends on extracting solvent. For example saliva extracts plasticizers from denture soft-lining materials more than 20 times faster than water [16]). Plasticizers involved, other than phthalates include glycols and their esters [17], sorbitol, polyadipates [18] and acetyltributyl citrate (ATBC – its intake increases because of usage of vinylidene chloride copolymer films in microwave ovens). Recently, a new class of more environmentally benign plasticizers has been derived from vegetable oils, e.g. the epoxidized soybean oil (ESO), and has widespread acceptance in food contact applications [19].

In medical applications the migration of a plasticizer is an important problem, and concerns the direct contact of a plastic material with open wounds. Contact with plastics should not generate any changes in plasma proteins, enzymes, cellular elements of blood and they should not cause thrombosis or toxic or allergic reactions. Plasticized PVC is currently used in many medical devices, ranging from intravenous fluid containers and blood bags to medical tubings [20]. Plasticizers such as phthalates (DEHP, DBP), adipates, tri(2-ethylhexyl) trimellitate (TOM) and citrates [21] in appropriate concentrations, are used. From a pharmaceutical point of view, plasticizers can regulate the release rate of active compounds. Furthermore by selecting the type of plasticizer and its concentration [22], improvements to film forming properties or a decrease in the glass transition temperature (in some applications the low processing temperature is necessary) can be made.

The most frequently used material in the production of flexible toys is plasticized PVC. The evidences indicating the health risk associated with PVC are so far unknown, but various methods have been developed to investigate the migration and toxicity of phthalates from teething rings or other flexible PVC toys which can be chewed by children. The results were scattered over a wide range and in 1999 the European Union banned the utilization of phthalates in toys for children under 3 years old. Orally ingested phthalates are hydrolysed to appropriate monoesters by endogenous esterases of many tissues, e.g. small intestine, which are thought to be hepatocarcinogenic [23]. Instead of phthalates, other plasticizers such as ATBC, benzoates or some adipates, sebacates or azelates can be used.

Inhalation exposure to DEHP, which is the major indoor plasticizer, may increase the risk of asthma, but the acute toxicity of phthalates is low and depends on the structure (chain length) of the alcohol part of molecule [24]. Dimethyl and diethyl phthalates (used extensively in insect repellents and in cosmetics, exhibit no significant toxicological effect despite their relatively high volatility. The higher chain phthalates do show some toxic effects documented mainly on rats and mice at high doses (like liver tumours or teratogenicity) [25]. In general, the health effects of phthalic esters are well studied in animals but there is the paucity of data in humans by any route of exposure. Quantitative estimates of risk to humans at various stages of life or health, or of save levels of exposure cannot be established with confidence at this time. Regarding carcinogenic properties, the International Agency for Research on Cancer classify only BBP and DEHP as a result of some experiments on animals.

Besides these phthalates belong to so-called xenoestrogenes, i.e. have endocrine disrupting properties, although their estrogenic potency is rather small compared with the reference compound, 17-βestradiol [26].

4. CONCLUDING REMARKS

At present we do not have a sufficient number of experimental data to explicitly determine the threats towards the environment and health shown by plasticizers present in polymeric materials. There are various speculations regarding potential replacements for di(2-ethylhexyl) phthalate, but these predications may be far from future reality because various research studies still produce conflicting results. There are, however, no doubts that demands made on ecological and toxicological performance will become increasingly stringent. Because of this, low volatile plasticizers, preferentially new families of oligomeric esters which are also difficult to extract, will become more important in all areas of applications, especially in plastics which come into contact with food, body fluids, lubricating oils or fuels. The development and implementation of new plasticizers that can be used in combination with environment-friendly polymers based on polylactide is becoming a new and important challenge. The annual production of these materials is ~200 ktons, but a number of predictions indicate that during the nearest decade it will rapidly increase and that they will be widely used in packing applications [27]. One of the common expectation from a plasticizer is that it will be biologically degradable and it will increase the biodegradation rate of a polymeric material. Although many suggestions can be found in the literature of such polymers, including citrate plasticizers, triacetine, oligomeric esteramides, glycerine and fatty acid derivatives, this area is still in its infancy and needs new synthetic strategies and new data on their fundamental physicochemical and biochemical properties to enable the design and production of desirable polymers and compatible plasticizers.

RERERENCES

[1] (a) A.H. Tullo, Chem. Eng. News, 83 (2005) 29; (b) G. Wypych (ed.), "Handbook of Plasticizer", ChemTech Publishing, Toronto, New York, 2004, Chapter 2; (c) W. Butte, ibid. Chapter 17.

[2] A. Wypych, Plasticizers Database, ChemTech Publishing, Toronto, 2004 (CD-version).

[3] T.J. Wams, Sci. Total Environ., 78 (1987) 1.

[4] (a) P.J. Flory, J. Chem. Phys., 9 (1941) 660; (b) M.L. Huggins, J. Chem. Phys., 9 (1941) 440; (c) P.J. Flory, J. Chem. Phys., 10 (1942) 51.
[5] (a) C.F. Hammer, in "Polymer Blends" (D.R. Paul and S. Newman, eds.), Academic Press, New York, 1978; (b) A.M.F. Barton, in "Expanded Cohesion Parameter Formulation," Handbook of Solubility Parameters and other Cohesion Parameters, pp. 37–60, CRC Press, Boca Raton, FL, 1983.
[6] J. Brandrup and E.H. Immergut, Polymer Handbook, 3rd edn., Wiley, New York, 1989.
[7] Hazardous Substances Data Bank, National Library of Medicine [Online].
[8] C.J. Wechsler, J. Environ. Sci. Technol., 18 (1984) 648.
[9] N.L. Wolfe, W.C. Steen and L.A. Burns, Chemosphere, 9 (1980) 403.
[10] (a) J. Gartshore, D.G. Cooper and J.A. Nicell, Environ. Toxicol. Chem., 22 (2003) 1244; (b) S. Nalli, D.G. Cooper and J.A. Nicell, Biodegradation, 13 (2002) 343; (c) S. Bhaumik, C. Christodoulatos and B.W. Brodman, DTIC Conference Paper, May 1996.
[11] K. Po-Hsu, L. Fang-Yin and H. Zeng-Yei, J. Environ. Sci. Health, Part A, 40 (2005) 103.
[12] C.A. Staples, D.R. Peterson, T.F. Parkerton and W.J. Adams, Chemosphere, 35 (1997) 667.
[13] C.E. Mackintosh, J. Maldonado, J. Hongwu, N. Hoover, A. Chong, M.G. Ikonomou and F.A. Gobas, Environ. Sci. Technol., 38 (2004) 2011.
[14] C.O. Hemmerlin and Q.T. Pham, Polymer, 41 (2000) 4401.
[15] M. Hamdani and A. Feigenbaum, Food Addit. Contam., 13 (1996) 717.
[16] E.C. Munksgaard, Eur. J. Oral Sci., 113 (2005) 166.
[17] B. Stanhope and N. Netzel, Polimery, 48 (2003) 421.
[18] M. Biedermann and K. Grab, Packag. Technol. Sci., 19 (2006) 159.
[19] J. Murphy, Plast. Addit. Compound., 2 (2000) 18.
[20] B. Demoré, J. Vigneron, A. Perrin, M.A. Hoffman and M. Hoffman, J. Clin. Pharm. Ther., 27 (2002) 139.
[21] H.Q. Yin, X.B. Zhao, J.M. Courtney, C.R. Blass, R.H. West and G.D.O. Loewe, J. Mater. Sci. Mater. Med., 10 (1999) 527.
[22] M.A. Frohoff-Hulsmann, B.C. Lippold and J.W. McGinity, Eur. J. Pharm. Biopharm., 48 (1999) 67.
[23] T. Niino, T. Ishibasi, T. Itho, S. Sakai, H. Ishiwata, T. Yamada and S. Ondera, J. Health Sci., 47 (2001) 318.
[24] P.M. Lorz, F.K. Towae, W. Enke, R. Jäckh and N. Bhargawa, Phthalic Acid and Derivatives in Ullmann's Encyclopedia of Industrial Chemistry, Wiley-VCH, Weinheim, Germany, 2000.
[25] R. Kavlock, K. Boekelheide, R. Chapin, et al., Reprod. Toxicol., 16 (2002) 721 and references cited therein.
[26] C.A. Harris, P. Henttu, M.G. Parker and J.P. Sumpter, Environ. Health Perspect., 105 (1997) 802.
[27] R.E. Drumright, P.R. Gruber and D.E. Henton, Adv. Mater., 12 (2000) 1841.

PART VI

PESTICIDES AND POLLUTION EXPOSURE IN HUMANS

Thermodynamics, Solubility and Environmental Issues
T.M. Letcher (editor)

Chapter 23

Solubility Issues in Environmental Pollution

Alberto Arce and Ana Soto

Department of Chemical Engineering, University of Santiago de Compostela, E-15782 Santiago de Compostela, Spain

1. SOLUBILITY ISSUES IN ENVIRONMENTAL POLLUTION

The forms adopted by chemical pollutants, and their distribution among the various phases of the environment, are determined by their thermodynamic properties. In particular, thermodynamics is fundamental in the spread of pollutants from their natural or anthropogenic sources, which requires their dispersion in water, air or soil (Fig. 1).

Most atmospheric pollution is due to means of transport: road vehicles, trains, ships and aircraft. The main pollutants emitted by these sources are CO_2, organic volatiles (VOCs), nitrogen oxides (NO_x) and particles of heavy metals such as lead or mercury. The main immobile sources of atmospheric pollution are certain kinds of industrial plant, and power stations or heaters that burn fossil fuels. The combustion of fossil fuels releases sulphur compounds (H_2S, SO_2, etc.) as well as NO_x, CO_2, VOCs and particulate matter.

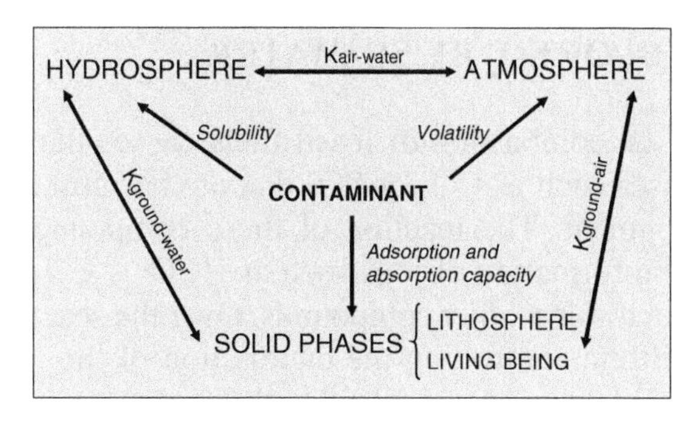

Fig. 1. Dispersion of pollutants in the environment.

Most of these pollutants both affect human health and cause environmental problems such as acid rain and global warming.

The most common pollutants of surface waters and groundwater on land are bacteria and other organic matter, hydrocarbons, industrial waste, pesticides and other agrochemicals, and household products. Efforts to limit water pollution centre on the treatment of urban and industrial wastewaters, and unchannelled inflows, such as those of agricultural origin, are more difficult to control. Mention must also be made of marine pollution, oil spills in particular.

Soil is often polluted by seepage from solid industrial and urban waste dumps and from mine spoil benches. Though spatially limited (except where seepage reaches groundwater), such pollution is frequently highly persistent.

Many of the types of pollution mentioned above involve the dissemination of the pollutant following its dissolution in water or diffusion into air. Thermodynamic analysis not only throws light onto the nature of these processes, but can also play an important role in the development of solutions to pollution problems. Thermodynamically or kinetically controlled separation operations can often both remedy existing pollution and prevent new incidents. In the remainder of this chapter we sketch three cases in point. As an example of the absorption of gaseous pollutants by a liquid, which is currently one of the most common approaches to purification of waste gases, we describe the use of seawater to absorb sulphur dioxide produced by the combustion of fossil fuels. Secondly, we review the antiknock petrol additives that have replaced methyl *tert*-butyl ether (MTBE) since the latter was found to give rise to significant pollution of surface and ground waters. Finally, with a view to the possibly not-too-distant future, we consider the desulphurization of oil refinery streams by ionic liquids.

2. ABSORPTION OF SO_2 BY SEAWATER

As noted above, the combustion of fossil fuels causes the release into the atmosphere of gases such as CO_2 or SO_2 that are harmful to human health and/or the environment. The reaction of these compounds with airborne water results in the formation of ions such as HCO_3^-, CO_3^{2-}, SO_3^- or H^+, which in turn react to form other compounds. Over the sea, which is slightly alkaline, the equilibria controlling the distribution of the various forms of CO_2 and SO_2 between air and water lead to the sea acting as a sink for these pollutants, which tends to limit both the greenhouse effect and the effects of acid rain on land. The carbonates and sulphates into which CO_2 and SO_2 are

Table 1
Composition of seawater with a salinity of ~35 g/kg

Ion	g/kg	Ion	g/kg
Cl^-	19.35	K^+	0.40
Na^+	10.77	SO_4^{2-}	2.71
Mg^{2+}	1.29	HCO_3^-	0.14
Ca^{2+}	0.40	Br^-	0.07

converted are natural components of seawater, and seawater is sufficiently abundant for its composition (Table 1) [1] to remain practically unaffected by these inputs.

The absorption by the sea of SO_2 released into the atmosphere from chimneys can be viewed as a particular case of flue gas desulphurization (FGD). In what follows we consider two examples of FGD processes using seawater: the inertion of engine exhaust on oil tankers and the reduction of SO_2 emission on coastal industrial plants. Because of its alkalinity, seawater can absorb approximately three times as much SO_2 as fresh water.

Current marine transport legislation establishes that the tanks in which crude oil and its liquid derivatives are transported must have an inert atmosphere in order to eliminate the risk of explosion and fire. In particular, the oxygen content of the atmosphere should be less than 5% and its SO_2 content less than 300 ppm (elimination of SO_2 is necessary both for immediate safety reasons and to prevent corrosion of tank structure). In oil tankers this inert atmosphere is provided by the exhaust gases from their engines, which burn fuel oil that, by weight, is typically 80–90% carbon, 12–15% hydrogen, 1.5–7.0% sulphur and up to 0.5% ash. Burning this fuel using an ~20% excess of air in order to ensure complete combustion produces a flue gas, the average composition of which is 75–80% nitrogen, 13% CO_2, 6–7% water, 4–5% oxygen and 0.5–1.0% SO_2, together with small quantities of NO_x and ash. To reduce the SO_2 content to the required level, the flue gas is scrubbed with seawater in a counterflow packed tower.

The design of efficient scrubbers for flue gas inertion requires knowledge of the solubility of relevant gases in seawater and an understanding of the chemical equilibria in which they are involved in this medium. Although rigorous studies of these equilibria have been carried out (for SO_2, for example, see Refs. 2, 3), in practice it is common to ignore minority gases such as NO_x, HCl or Cl_2 and to use simple models such as that of Abdulsattar et al. [4], which considers the only dissolved gases to be SO_2 and CO_2 (Table 2).

Table 2
Fundamental equilibria involved in the model of Abdulsattar et al. [4]

$CO_2\ (aq.) + H_2O \leftrightarrow H_2CO_3$
$H_2CO_3 \leftrightarrow HCO_3^- + H^+$
$HCO_3^- \leftrightarrow CO_3^{2-} + H^+$
$SO_2\ (aq.) + H_2O \leftrightarrow HSO_3^- + H^+$
$HSO_3^- \leftrightarrow SO_3^{2-} + H^+$
$HSO_4^- \leftrightarrow SO_4^{2-} + H^+$
Proton concentration of seawater of pH 8
$[H^+] = 10^{-8.00}/\gamma_{H^+} + [HSO_3^-] + 2[SO_3^{2-}] - [H_2CO_3^*] - [HSO_4^-]$
With $[H_2CO_3^*] = [H_2CO_3] + [CO_2\ (aq.)]$

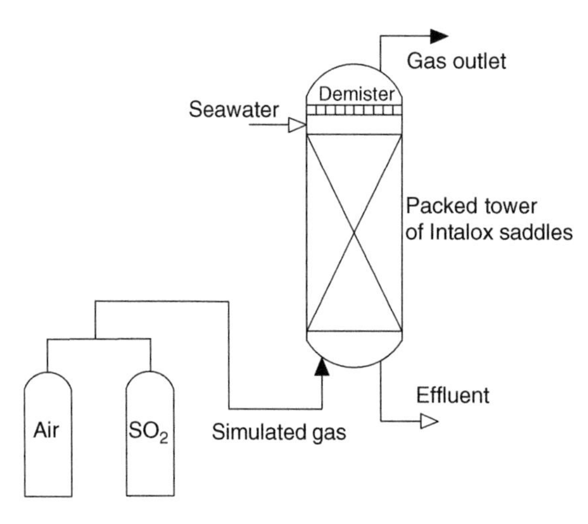

Fig. 2. Counterflow packed tower for absorption of flue gas SO_2 by seawater.

Using an experimental counterflow scrubber packed with Intalox ceramic saddles (Fig. 2) to scrub simulated flue gas at 14.5°C and atmospheric pressure, Baaliña et al. [5, 6] verified the accuracy of the Abdulsattar model under a range of FGD operating conditions (Figs. 3 and 4). Under conditions resulting in effluent acidities typical of on-board scrubbers (pH 2–3), over 97% of SO_2 was removed from the flue gas. Table 3 shows the average composition of inert gas produced by seawater FGD. Note that because of its usually high acidity, discharge of the scrubber effluent in near-shore waters is banned by industrialized countries.

Seawater FGD can be employed not only for atmosphere inertion on oil tankers, but also to reduce SO_2 emission by on-land industrial plants located

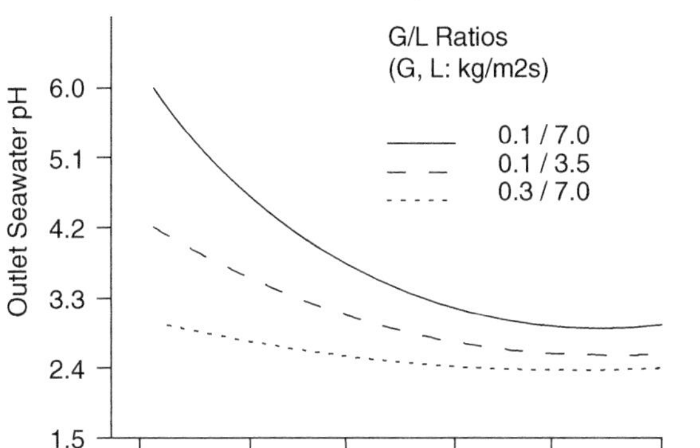

Fig. 3. Influence of input SO_2 concentration on effluent pH, for various gas/liquid flow rate ratios in the tower of Fig. 2.

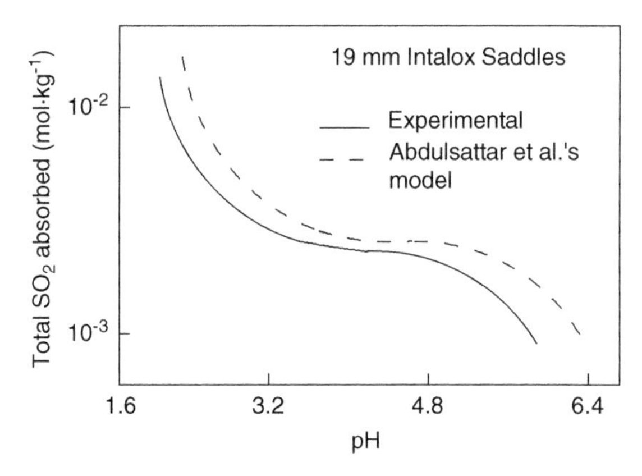

Fig. 4. SO_2 absorption by the tower of Fig. 2 plotted against effluent pH, together with the predictions of Abdulsattar et al. [4].

on the coast. In particular, for fossil-fuel-burning power stations it constitutes an attractive alternative to wet FGD systems based on the use of lime, limestone or magnesium hydroxide, because seawater is already used for gas cooling purposes [7, 8]. It is also employed by the canning industry to remove SO_2 from gases produced by anaerobic fermentation of waste, which allows the methane content of these gases to be used as fuel. The pH of the effluent produced in these applications is generally high enough to allow direct

Table 3
Average composition of inert gas produced by seawater FGD

Component	Concentration (wt%)	Component	Concentration (wt%)
O_2	4–5	N_2 + rare gases	82–83
CO_2	13	NO_x	<100 ppm
SO_2	300 ppm	Soot and ash	7–9 mg/m^3

discharge into the sea, but can in any case easily be brought to an acceptable value by means of a simple neutralization treatment before discharge.

3. REPLACEMENT OF MTBE BY OTHER TERTIARY ETHERS

Since 1970s, when the use of tetraethyl lead as an antiknock petrol additive began to be phased out because of its toxicity, its place has been occupied by alcohols and ethers. These additives increase octane rating and reduce both carbon monoxide emission and, by decreasing the reactivity of emitted VOCs, the increase in tropospheric ozone. Ethers are generally preferable to alcohols as regards compliance with oxygen content and vapour pressure specifications, although ethanol was widely employed in the 1990s, and *tert*-butyl alcohol has also sometimes been used.

Oxygenated petrol additives, like the hydrocarbon components of petrol, can enter the environment at all stages of the production and use of this fuel [9]. In refineries, for example, they can be released into the atmosphere through evaporation, and into waterbodies via wastewater. However, most losses into the environment arise during the storage, distribution and use of the additives or additive-treated petrol, in particular because of leaks or spills from underground tanks, pipelines or fuelling points. Depending on their solubility, additives contained in petrol polluting waterbodies will be transferred to the water, in some cases to a significant extent.

The most widely employed ether petrol additive in the 1990s was MTBE, but MTBE can be harmful to health [10–12], and even in the 1980s had begun to pose environmental problems, leaky underground petrol tanks and other sources having caused significant pollution of aquifers, surface waters and even water supplies [10, 13]. MTBE pollution of air has also been detected in the neighbourhood of refineries, petrol stations, car parks, garages and motorways. Although recent studies [10, 14] have shown that conventional preventive and remedial methods are capable of dealing with MTBE, as well as pointing towards future improvements in this area, it was

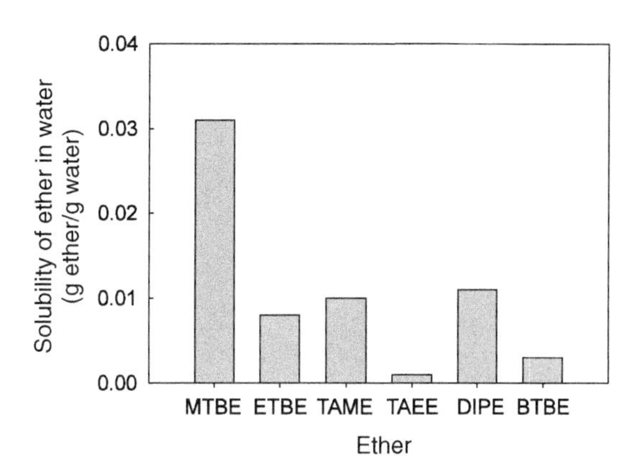

Fig. 5. Solubility of various ethers in water at 25°C [19–24]. *Abbreviations*: MTBE, methyl *tert*-butyl ether; ETBE, ethyl *tert*-butyl ether; TAME, *tert*-amyl methyl ether; TAEE, *tert*-amyl ethyl ether; DIPE, diisopropyl ether; BTBE, butyl *tert*-butyl ether.

Table 4
Rankings of five ethers by suitability for use as antiknock additives, according to five criteria (abbreviations as in Fig. 5)

	Vapour pressure	Organic carbon sorption	Blending octane	Oxygen content
MTBE	5°	1°	1°	1°
ETBE	3°	4°	1°	2°
TAME	2°	2°	3°	2°
TAEE	1°	–	4°	3°
DIPE	4°	3°	2°	2°

natural for attention to turn to alternative ethers [15–18] that, because of their greater molecular mass, are less soluble in water than is MTBE (Fig. 5) [19–24]. Most of the compounds featured in Fig. 5 endow petrol with slightly lower octane ratings than does MTBE, but they also give it a lower vapour pressure, which reduces loss of unburnt fuel. Table 4 orders these ethers by their suitability for use as antiknock additives with respect to some of the properties considered by Diaz and Drogos [10]. Unfortunately all these compounds are very resistant to biodegradation [9, 10], unlike alcohols.

It should be borne in mind that the solubility of all organic antiknock additives in water depends on the presence or absence of other solutes. For example, Fig. 6 shows how the solubilities of the ethers of Fig. 5 are affected by the presence of ethanol [19–24]. Analysis of their transfer to water is

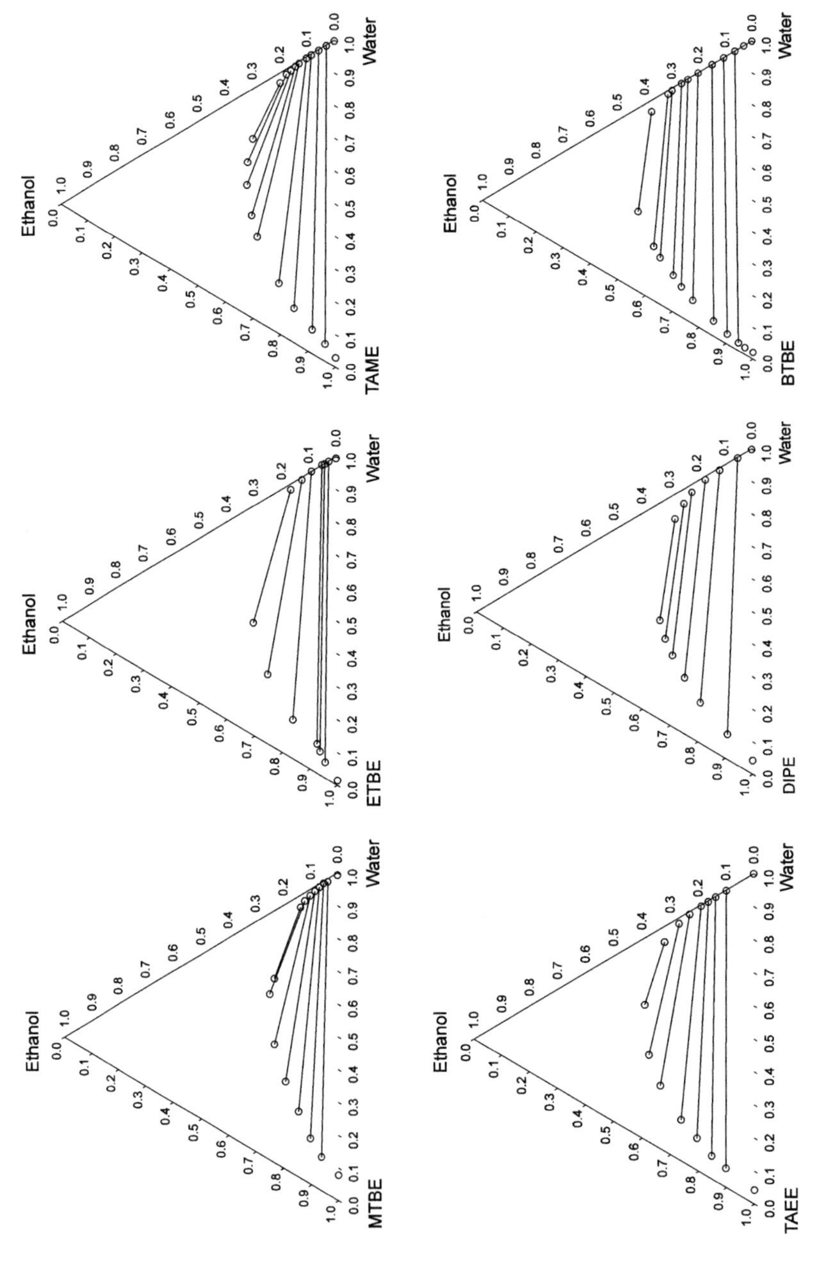

Fig. 6. Liquid–liquid equilibria of ether+ethanol+water systems (abbreviations as in Fig. 5).

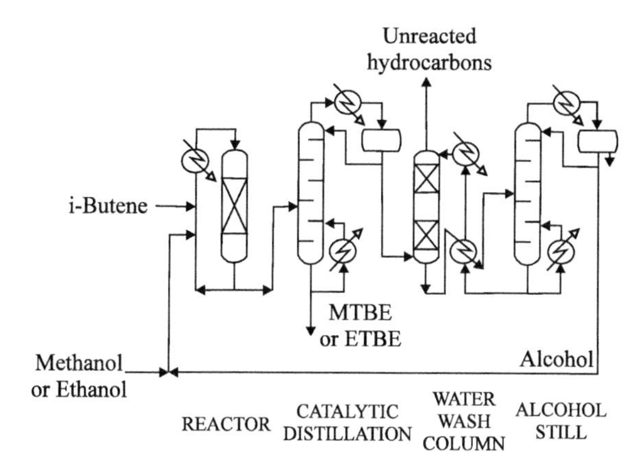

Fig. 7. Manufacture of MTBE or ETBE by catalytic etherification (abbreviations as in Fig. 5).

therefore a complex problem, commercial petrol being a mixture of more than 30 different components. Most of the literature on the pollution of water by petrol only focuses on measured concentrations of ethers [9, 25–27] and benzene, toluene, ethylbenzene and xylenes (BTEX) fraction components [9] in aquifers, lakes and water supplies.

In practice, since MTBE began to be phased out it has mainly been replaced by TAME or ETBE. TAME was already the second most widely used ether antiknock additive, and MTBE-producing facilities were easily converted to ETBE production because both ethers are produced by reaction of isobutene with an alcohol (ethanol or methanol) using a cationic ion-exchange resin as catalyst (Fig. 7); the reaction with ethanol is somewhat less efficient, but this drawback does not outweigh other factors. Apart from the plant conversion issue, the chief advantages of ETBE over TAME are its lower solubility in water, its higher octane rating (very similar to that of MTBE) and the possibility of producing it from bioethanol, which in Europe brings tax advantages [28].

Isobutene, from which both MTBE and ETBE are obtained, is the olefin of lowest molecular mass from which an ether with antiknock capacity can be obtained. Currently, attention is being turned to ethers of heavier olefins in the range C_5–C_7 [15–18, 29, 30]. As well as those mentioned above (TAEE, DIPE and BTBE), these heavier ethers include *tert*-hexyl ethyl ether (THxEE) and *tert*-heptyl ethyl ether (THpEE). TAEE, THxEE and THpEE are all obtained by reaction of the corresponding olefin with ethanol, BTBE by reaction of isobutene with *n*-butanol and DIPE as a by-product of the production of isopropyl alcohol by reaction of isopropylene with water. All these ethers raise the octane rating of petrol without raising its Reid vapour pressure, and their poor solubility in water to a large extent precludes the environmental pollution problems associated with MTBE.

4. DESULPHURIZATION OF FUEL OILS WITH IONIC LIQUIDS

Within the framework of the Kyoto Protocol on atmospheric pollution, the European Union has imposed a maximum sulphur content of 50 ppm for fossil fuels, to be lowered to 10 ppm by 2009 (Fig. 8). The U.S. Environmental Protection Agency (EPA) has imposed limits of 30 ppm for petrol and 15 ppm for diesel fuel. With current catalytic hydrodesulphurization technology it is very difficult to satisfy these limits. One of the most promising alternatives is the use of room-temperature ionic liquids (RTILs). Although these substances were once prohibitively expensive, their prices have already begun to fall and should continue to do so.

RTILs are salts with melting points below 100°C that are composed of a large organic cation and an inorganic anion. Research into RTILs has blossomed in recent years, and is regarded by the EPA as a field to be prioritized in the 21st century. RTILs are widely expected to be the solvents of the future [31, 32]. They dissolve most organic and inorganic compounds, whether polar or non-polar; are thermally stable and non-inflammable; and their negligible vapour pressure means reduced losses from evaporation and reduced risk for those who use them. Furthermore, they are highly flexible in that the physical and chemical properties of an ionic solvent can often be custom-designed by modification of its anion and/or cation.

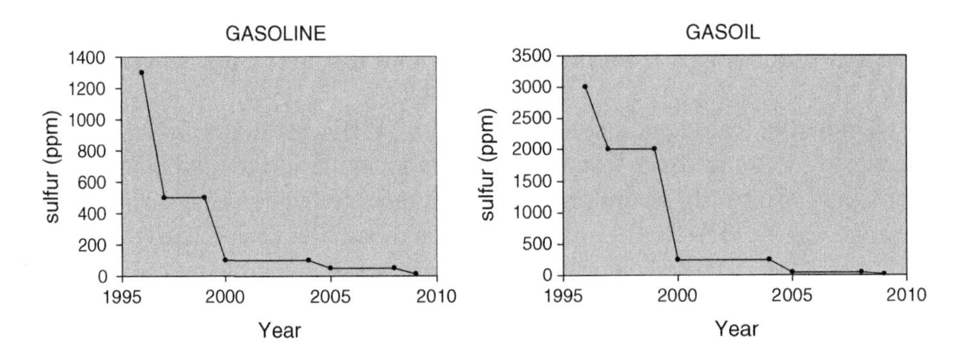

Fig. 8. Evolution of the sulphur contents of European motor fuels.

According to recent studies [33–48], RTILs are ideal, readily recoverable extractants for desulphurization of petrol and diesel oil, although the desired low residual contents mean that even with the most sulphur-avid RTIL it is necessary to use a multi-stage separation process involving, for example, a tray tower, a rotating disc contactor or an Oldshue-Rushton extractor (Fig. 9). In designing such processes using the standard MESH equations the heat equation can probably be ignored if, as is usual, the extraction is isothermal, but accurate knowledge of the relevant liquid–liquid equilibria is essential.

The main sulphur-bearing impurities of fuel oils are generally mercaptans, sulphides, disulphides, thiophenes and benzothiophenes. Because of differences between petrol and diesel oil compositions, they will have different optimal desulphurization RTILs. In both cases, the central task in designing an RTIL-based separation system is to find an RTIL that dissolves the sulphur-bearing compounds without dissolving the hydrocarbons. Because of the great solvent power of most RTILs, this task is not easy. For example, [Omim][BF_4] dissolves thiophene very well, and is practically immiscible with cyclohexane

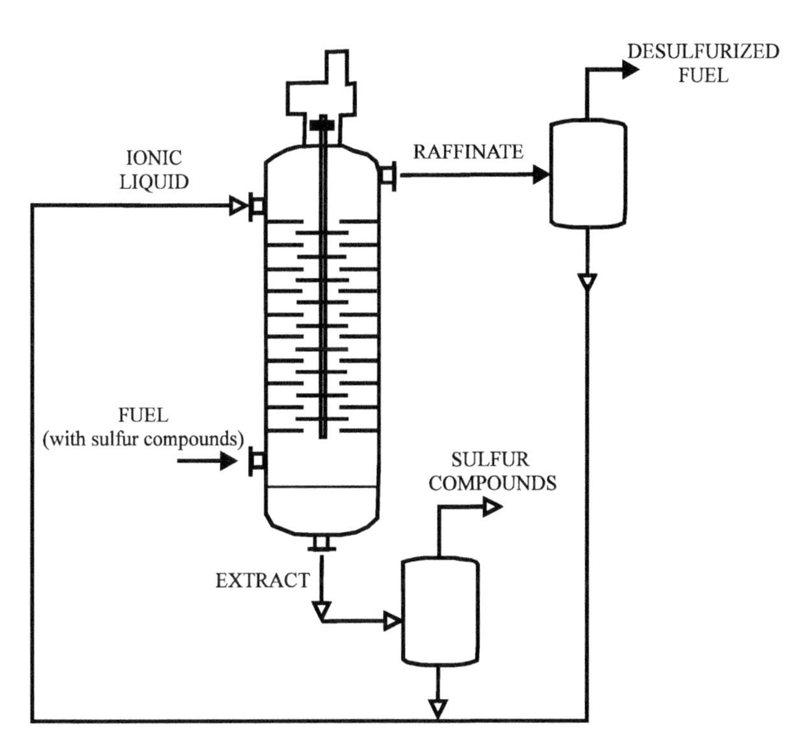

Fig. 9. An extraction unit for desulphurization of petrol or diesel oil using an ionic liquid extractant.

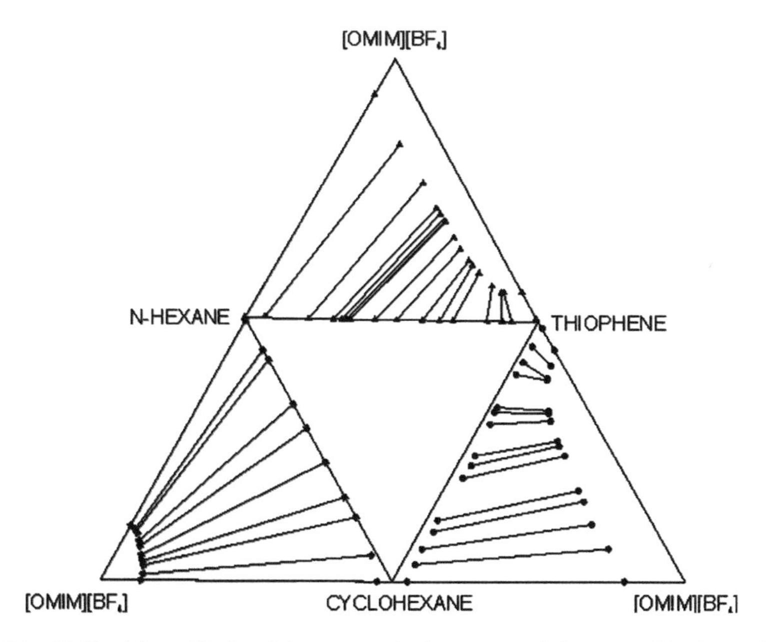

Fig. 10. Liquid–liquid equilibria of the system {*n*-hexane, cyclohexane, thiophene, [Omim][BF_4]}.

and *n*-hexane (Fig. 10), but it also dissolves aromatics that are good for octane rating, such as toluene (Table 5). Most studies that have identified RTILs capable of desulphurizing fuels (some of which simultaneously remove nitrogenated impurities) have ignored the effects on the hydrocarbon composition of the fuel. Only through quantitative liquid–liquid equilibrium studies will it be possible to identify RTILs providing an optimal balance between extraction of sulphur compounds and extraction of hydrocarbons.

Properties other than liquid–liquid equilibria that should be considered in choosing an RTIL as a desulphurizing extractant include thermal stability, toxicity (virtually no ecotoxicological data are yet available for RTILs), corrosivity and immediately process-related variables such as density, viscosity and surface tension – the extremely high viscosity of some halogenated RTILs makes them very difficult to handle. Then there is the question of how to recover the extractant: the negligible vapour pressure of RTILs suggests their recovery by distillation, but the sulphur compounds from which they are to be separated also have high boiling points; one solution might be to work under vacuum. Other options are extraction with water (for hydrosoluble RTILs) or extraction with supercritical CO_2,

Table 5
Solubility of thiophene and various hydrocarbons in various ionic liquids (unpublished data obtained in our laboratory)

	Hexane	Dodecane	Hexadecane	Cyclohexane	Toluene	Tiofene
$[C_5mim][NTf_2]$	I	I	I	I	P	P
$[C_6mim][NTf_2]$	I	I	I	I	P	P
$[C_8mim][NTf_2]$	I	I	I	I	P	P
$[C_9mim][NTf_2]$	I	I	I	I	P	P
$[P_{6,6,6,14}][NTf_2]$	P	P	I	S	S	P
$[Choline][NTf_2]$	I	I	I	I	I	I
$[C_2mim][OTf]$	I	I	I	I	I	P
$[C_4mim][OTf]$	I	I	I	I	P	P
$[P_{6,6,6,14}][OTf]$	P	I	I	S	S	S
$[C_2mim][EtSO_4]$	I	I	I	I	I	P
$[C_2mim][OMs]$	I	I	S	I	I	P
$[P_{6,6,6,14}][(CH_3)_2NCS_2]$	P	P	I	S	S	S
$[P_{6,6,6,14}][Cl]$	P	I	I	S	S	S
$[C_8mim][Cl]$	I	I	I	I	I	S
$[C_4mim][BF_4]$	I	I	I	I	I	P
$[C_8mim][BF_4]$	I	I	I	I	P	P
$[C_6mim][P(C_2F_5)_3F_3]$	I	I	I	I	P	P

Note: S, totally or highly soluble; P, partially soluble; I, insoluble or practically insoluble.

although the initial results with the latter technique are not especially encouraging [48].

The engineering aspects of RTIL desulphurization have received particular attention from Jess and coworkers [33–37], who regard extraction with RTILs as best deployed as a final desulphurization stage following conventional catalytic hydrodesulphurization. Like us, these authors point to the need for accurate thermodynamic data on which to base the design of RTIL desulphurization processes, and to the critical importance of the RTIL recovery stage. Among the chloroaluminates and other ionic liquids they examined, they found $[BMIM][OcSO_4]$ and $[EMIM][EtSO_4]$ to be particularly promising because they are halogen-free and relatively inexpensive. Table 6 highlights the most important work and conclusions established at the moment about desulphurization of fuel oils with ionic liquids. High solubility of sulphur compounds on RTILs could be the solution to an environmental pollution problem.

Table 6
Summary of work done on desulphurization of fuel oils with ionic liquids

Reference	Tested ionic liquids	Conclusions
[34]	$[BMIM]Cl/AlCl_3$; $[EMIM]Cl/AlCl_3$; $[HN(C_6H_{11})Et_2][CH_3SO_3]/[HNBu_3][CH_3SO_3]$	Ionic liquid's extraction power for DBT is not uniquely based on chemical interactions involving the acid proton
	[BMIM]X	Desulphurization is hardly affected by the chemical nature of the anion. The size of the anion is rather important for the extraction effect
	$[C_nMIM]BF_4$	Pronounced effect of the cation's size on the extraction power of the ILs. A possible explanation for this behaviour may be that the physical solubility of DBT in the IL is dependant on stearic factors
[37]	$[BMIM]Cl/AlCl_3$; $[BMIM][BF_4]$; $[BMIM][PF_6]$; $[BMIM][OcSO_4]$; $[EMIM][EtSO_4]$; $[MMIM][Me_2PO_4]$	The use of halogen-free ionic liquids avoids stability and corrosion problems. Extraction is only to a small extent influenced by the degree of alkylation of the DBT derivatives, which is a major advantage compared to HDS
[40]	$[BMIM]Cl/AlCl_3$; $[BMIM][BF_4]$; $[BMIM]Cu_2Cl_3$	Cu–Cl ionic liquids exhibit remarkable extraction desulphurization ability from gasoline and their most considerable advantage is no polymerization reaction of the olefins in the gasoline
[41]	$[BMIM][PF_6]$; $[BMIM][BF_4]$	Chemical oxidation in conjunction with ionic liquid extraction improves the yield of desulphurization of light oil
[42]	$[EMIM][BF_4]$; $[BMIM][PF_6]$	Considerable structural sensitivity towards selective absorption of S-containing aromatics from gasoline (thiophene >> methylthiophene > toluene >> trimetyl-benzene > isobutylthiol > hexene > 2-methylpentane, methylcyclopentane)

[43]	[EMIM][BF_4]; [BMIM][PF_6]; [BMIM][BF_4]	The ionic liquids can be used in multiple cycles for the removal of S-containing compounds from fuels. The absorption capacity of ionic liquids is sensitive to both the anion and the cation compositions
[45]	[EMIM][BF_4]; [BMIM][PF_6]; [BMIM][BF_4]	High selectivity, particularly towards aromatic sulphur and nitrogen compounds, for extractive desulphurization and denitrogenation from transportation fuels. The absorbed aromatic S-containing compounds were quantitatively recovered. The Lewis acidic $AlCl_3$-TMAC ionic liquids were found to have remarkably high absorption capacities for aromatics, particularly S-containing aromatic compounds, but their regeneration is problematic
[46]	[C_nMIM] NTf_2, BF_4, PF_6, Cl	Use of supercritical fluid extraction (SFE), for a more effective separation of products from ILs
[47]	[HMIM][NTf_2]	Continuous bicyclic process of extraction of sulphur-containing aromatic compounds (SAs) from diesel fuel and gasoline with ionic liquids and reextraction of SAs from ILs with supercritical carbon dioxide ($scCO_2$)

REFERENCES

[1] W. Stumm and J.J. Morgan, Aquatic Chemistry: Chemical Equilibria and Rates in Natural Waters, 3rd edn., Wiley, New York, 1996.
[2] G. Al-Enezi, H. Ettouney, H. El-Dessouky and N. Fawzi, Ind. Eng. Chem. Res., 40 (2001) 1434.
[3] J. Rodriguez-Sevilla, M. Alvarez, M.C. Diaz and M.C. Marrero, J. Chem. Eng. Data, 49 (2004) 1710.
[4] A.H. Abdulsattar, S. Sridhar and L.A. Bromley, AIChE J., 23 (1977) 62.
[5] A. Baaliña, E. Rodriguez, J.A. Santaballa and A. Arce, Environ. Technol., 17 (1996) 331.
[6] A. Baaliña, E. Rodriguez, J.A. Santaballa and A. Arce, Environ. Technol., 18 (1997) 545.
[7] R.K. Srivastava and W. Jozewicz, J. Air Waste Manag. Assoc., 51 (2001) 1676.
[8] K. Oikawa, C. Yongsiri, K. Takeda and T. Harimoto, Environ. Prog., 22 (2003) 67.
[9] Office of Science and Technology Policy, Interagency Assessment of Oxygenated Fuels, 1997.
[10] A.F. Diaz and D.L. Drogos (eds.), Oxygenates in Gasoline: Environmental Aspects. ACS Symposium Series 799, American Chemical Society, Washington, 2002.
[11] J.M. Davis and W.H. Farland, Toxicol. Sci., 61 (2001) 211.
[12] A. Hirose, A. Nishikawa, M. Ema, H. Kurebayashi, M. Yamada and R. Hasegawa, Mizu Kankyo Gakkaishi, 25 (2002) 491.
[13] O.C. Braids, Environ. Forensics, 2 (2001) 189.
[14] R.A. Deeb, K.H. Chu, T. Shih, S. Linder, I. Suffet, M.C. Kavanaugh and L. Alvarez-Cohen, Environ. Eng. Sci., 20 (2003) 433.
[15] J. Kivi, A. Rautiola, J. Kokko, J. Pentikainen, P. Aakko, M. Kyto and M. Lappi, Soc. Automot. Eng. Spec. Publ., SP-1277 (1977) 81.
[16] J. Ignatius, H. Jaervelin and P. Lindqvist, Hydrocarbon Process. Int. Ed., 74 (1995) 51.
[17] V. Macho, J. Mikulec and M. Kralik, Petroleum Coal, 41 (1999) 155.
[18] E.J. Chang Preprints Pap. – Am. Chem. Soc., Div. Fuel Chem., 39 (1994) 330.
[19] A. Arce, M. Blanco, R. Riveiro and I. Vidal, Can. J. Chem. Eng., 74 (1996) 419.
[20] A. Arce, J.M. Martínez-Ageitos, O. Rodríguez and A. Soto, J. Chem. Thermodyn., 33 (2001) 139.
[21] A. Arce, A. Machiaro, O. Rodríguez and A. Soto, J. Chem. Eng. Data, 47 (2002) 529.
[22] A. Arce, O. Rodríguez and A. Soto, Fluid Phase Equilib., 224 (2004) 185.
[23] T.M. Letcher, S. Ravindran and S. Radloff, Fluid Phase Equilib., 71 (1992) 177.
[24] M.S.H. Fandary, A.S. Aljimaz, J.A. Al-Kandary and M.A. Fahim, J. Chem. Eng. Data, 44 (1999) 1129.
[25] R. Johnson, J. Pankow, D. Bender, C. Price and J. Zogorski, Environ. Sci. Technol., 34 (2000) 210A.
[26] Y. Rong and W. Tong, Environ. Forensics, 6 (2005) 355.
[27] R.W. Gullick and M.W. LeChevallier, J. Am. Water Works Assoc., 92 (2000) 100.
[28] R. Koenen and W. Puettmann, Grundwasser, 10 (2005) 227.
[29] E. Pescarollo, R. Trotta and P.R. Sarathy, Hydrocarbon Process. Int. Ed., 72 (1993) 53.

[30] T. Zhang, K. Jensen, P. Kitchaiya, C. Phillips and R. Datta, Ind. Eng. Chem. Res., 36 (1997) 4586.
[31] M.J. Earle and K.R. Seddon, Pure Appl. Chem., 72 (2000) 1391.
[32] M.C. Uzagare and M. Salunkhe, Asian Chem. Lett., 8 (2004) 269.
[33] A. Jess and J. Eβer, in "Ionic Liquids IIIB: Fundamentals, Progress, Challenges, and Opportunities" (R.D. Rogers and K.R. Seddon, eds.), p. 83, ACS Symposium Series 902, American Chemical Society, Washington, 2005.
[34] A. Bösmann, L. Datsevich, A. Jess, A. Lauter, C. Schmitz and P. Wasserscheid, Chem. Commun., 23 (2001) 2494.
[35] J. Eβer, A. Jess and P. Wasserscheid, Chem. Ing. Tech., 75 (2003) 1149.
[36] A. Jess, P. Wasserscheid and J. Eβer, Chem. Ing. Tech., 76 (2004) 1407.
[37] J. Eβer, P. Wasserscheid and A. Jess, Green Chem., 6 (2004) 316.
[38] L.A. Aslanov and A.V. Anisimov, Neftekhimiya, 44 (2004) 83.
[39] J. Feng, C. Li, M. Hong and Z. Wang, Shiyou Huagong, 35 (2006) 272.
[40] C. Huang, B. Chen, J. Zhang, Z. Liu and Y. Li, Energy Fuels, 18 (2004) 1862.
[41] W.H. Lo, H.Y. Yang and G.T. Wei, Green Chem., 5 (2003) 639.
[42] S. Zhang and Z.C. Zhang, Green Chem., 4 (2002) 376.
[43] S. Zhang and Z.C. Zhang, Preprints Symp. – Am. Chem. Soc., Div. Fuel Chem., 47 (2002) 449.
[44] C.Z. Zhang, C.P. Huang, J.W. Li and C.Z. Qiao, Huaxue Yanjiu, 16 (2005) 23.
[45] S. Zhang, Q. Zhang and Z.C. Zhang, Ind. Eng. Chem. Res., 43 (2004) 614.
[46] H. Zhao, S. Xia and P. Ma, J. Chem. Technol. Biotechnol., 80 (2005) 1089.
[47] H.C. Zhou, N. Chen, F. Shi and Y.Q. Deng, Fenzi Cuihua, 19 (2005) 94.
[48] J. Planeta, P. Karasek and M. Roth, Green Chem., 8 (2006) 70.

Thermodynamics, Solubility and Environmental Issues
T.M. Letcher (editor)

Chapter 24

Hazard Identification and Human Exposure to Pesticides

A. Garrido Frenich, F.J. Egea González, A. Marín Juan and J.L. Martínez Vidal

Department of Analytical Chemistry, University of Almería, 04120 Almería, Spain

1. INTRODUCTION

Each day humans come into contact with thousands of chemicals presented in a multitude of consumer products and some of these products contain pesticides. Exposure takes place through occupational activity, through contact in the environment and by ingestion of food and water. This exposure is of great concern to governments who wish to estimate health risks to the general public and to workers.

Risk assessment procedures are largely concerned with hazard identification and exposure estimation. It is the basis of regulatory evaluation of pesticides and is defined as the evaluation of the potential for adverse health effects in humans from exposure to chemicals [1]. On a conceptual level, risk assessment is the bringing together of the probabilities of hazards and exposures, to produce an understanding of the likelihood of some type of adverse outcome. The process involves three components: data gathering to estimate the hazard and exposure; characterisation (combining the factors) to produce an estimate of the magnitude and estimating the probability of the anticipated adverse effect. The United States Environmental Protection Agency (EPA) [2] divides the risk assessment process into four steps: hazard identification, dose–response assessment, exposure assessment and risk characterisation.

Hazard identification refers to the potential health effects that may occur from different types of pesticide exposure. It is strongly related to the extent and the type of a pesticide's toxic properties. This phase, usually involves the gathering of data on whether exposure to a pesticide causes an adverse effect, for example, cancer or endocrine disruption. This is usually determined by a battery of studies on several species of laboratory animals. It is important to know whether the adverse effects observed in one species will occur in other species. Hazard identification may also involve characterising the behaviour of a pesticide or metabolite within the human body and chemical interactions within organs, cells, or even parts of cells.

The dose–response assessment considers the magnitude or incidence of effects that occur, or are predicted to occur, at a given dose level, having taken into account the amount of a substance a person is exposed to. Dose–response assessment considers the dose levels at which adverse effects are observed in test animals, and using these dose levels to calculate a corresponding dose for humans.

The third EPA step, exposure assessment, estimates how often a person comes in contact with a particular pesticide. In some cases a given pesticide may be highly labile and will not persist in the environment while other products may have a longer lifespan, resulting in greater probability of people coming in contact with it. It is necessary to identify the exposed population, the routes of exposure (inhalation exposure, dermal exposure and oral exposure) and the different exposure sources. Distinguishing between non-occupational exposure and worker exposure, we have the following exposure sources: (i) food; (ii) residential uses; (iii) drinking water; (iv) bystander exposure and finally (v) worker exposure (e.g., pesticide manufacturers, agricultural workers and pesticide applicators).

Risk characterisation integrates data from hazard identification, dose–response and exposure assessments to describe the overall risk from a pesticide. It develops a qualitative or quantitative estimate of the likelihood that any of the hazards associated with the pesticide will occur in exposed people. It also involves the assumptions used in assessing exposure as well as the uncertainties that are built into the dose–response assessment [3].

The European Union (EU) follows a similar scheme [4], but it combines hazard identification and dose response effect into a unique step called, "assessment of effects".

Once the risk assessment process is completed for a given pesticide, regulators begin the risk management step; to decide how much exposure will be allowed and, if necessary, establish risk reduction options to ensure that with reasonable certainty, a pesticide will not be harmful to humans. These risk assessments support two major types of regulatory decisions [5]: the approval and registration of pesticides and the setting of standards for acceptable exposure levels in air, water and food. Using the conclusions of the risk assessment, regulators can approve a pesticide as proposed, or adopt protective measures to limit its exposure (occupational or non-occupational).

Decisions taken by regulators have important economical consequences for the industry that manufactures pesticides, the farmers, the workers and the consumers. Such consequences, justify the need for regulators to have well supported data related to the four steps of the risk assessment process. Risk assessment requires a robust analytical basis; studies addressed either to hazard identification or exposure estimation. Both have to be conducted under good laboratory practice (GLP) certification [6]. This paper deals with the key points of hazard identification and exposure assessment, both aspects highly influenced by the solubility and physical-chemical properties of pesticides.

2. PRIORITY PROPERTIES AFFECTING THE HAZARDS OF PESTICIDES

A knowledge of the physicochemical properties of pesticides is very important in environmental risk assessment, because they influence the distribution, persistence and fate of either parent compounds or metabolites in the environment. The importance was considered by Marino [7] who evaluated different data sources of physical properties of chemicals for their consistency. The more important properties are: molecular weight, melting point, boiling point, vapour pressure, water solubility (WS), Henry's law constant (which assess the ability of a chemical to partition between air and water) and the octanol–water partition coefficient (K_{ow}). The latter property is very important in distribution processes such as sorption to soils and sediments. The technical guide issued by the EU [4], highlights the determination of the above parameters, and includes others properties which may be required, depending on the environmental compartment to be assessed, such as photodegradation rates, soil-organic carbon partition coefficients and sediment–aquatic fate data. Finizio et al. [8] compared the different methodologies used in obtaining K_{ow} for chemicals, including pesticides. He showed that often results depended on the methodology used. He also included a large set of K_{ow} data obtained from the literature. Shen and Wania compile a set of physicochemical properties of organochlorine (OC) pesticides by treating existing data in the literature, indicating the difficulties in assessing such properties as a result of insufficient experimental data [9]. Halfon et al. [10] published a set of priority data for use in the environmental risk assessment and potential transfer of pesticides from soils to the River Po. The properties selected were: the half-life of microbial transformation in soil ($T_{0.5}$), as an indication of persistence; the WS, as a measure of leaching potential; the vapour pressure to indicate pesticide losses due to volatilisation and the pesticide usage habits as an indication of the intake source.

Physicochemical properties and chemical functionality can be useful in hazard assessments relating to human health, in cases lacking in experimental data [11]. For example, the uptake and absorption by inhalation, may be calculated from melting point, boiling point and vapour pressure at ambient temperature data. In summary, it is possible that compounds with a vapour pressure greater than 20 Pa at 20°C and a boiling point below 200°C can enter the body by inhalation. Furthermore, particles of 5 mm diameter are unable to reach alveoli, whereas pulmonary re-sorption is possible for particles smaller than 2 mm. Dermal absorption is assumed to be negligible for pollutants having a log K_{ow} value of below −1 or above 6 and a molecular mass of greater than 500. Absorption from the gastrointestinal tract can be regarded as negligible when log K_{ow} is below 2 or above 7 and the molecular mass is above 1000. The presence of structural elements and functional groups may give some indication about the metabolic fate and also about probable reactivity towards biological molecules (e.g., proteins).

When a toxic compound enters the body, the following processes should be considered: absorption, distribution, metabolism and excretion [12]. The extent and the rate of absorption of a pollutant, is a function of the molecular mass, charge, physical state, solubility, stability and reactivity. Lipid-soluble chemicals such as OC pesticides can readily dissolve in the membranes and therefore can diffuse across cell walls. In contrast, ionic substances do not readily enter the lipid membrane matrix in an ionised form and therefore only un-ionised forms freely diffuses across membranes, except in the case of very low molecular mass substances, which may diffuse through aqueous pores.

Distribution is the process by which an absorbed substance and/or its bio-transformation products circulate and partition within the body. The most important chemical characteristics that affect distribution are lipophilicity, molecular size and shape and degree of ionisation. Lipophilic, small, non-ionic molecules can diffuse readily across cellular membranes. Very low molecular, hydrophilic substances may diffuse through aqueous pores. A number of tissue parameters also affect tissue distribution, such as blood flow rate, the permeability of capillary and cell membranes and the nature of the extracellular fluid matrix.

Metabolism involves the bio-transformation of chemicals. Many compounds entering the systemic circulation are lipophilic and weakly ionisable, being poorly excreted because of re-absorption from the renal tubule after filtration in the glomerulus. The bio-transformation of xenobiotics generally leads to the formation of more polar, hydrophilic metabolites that may be more readily excreted via urine. In first phase xenobiotics are oxidised, reduced or hydrolysed (through enzymes such as P450), and in a second phase they are conjugated with glucuronic acid, sulphate, acetate, amino acids or glutathione. Finally, excretion, the process involving to removal of xenobiotics from the body, involves one of three paths: urine is excreted via the kidneys; faeces via the liver and bile and exhalated air, via the lungs. The molecular mass is important for kidney excretion; glomeruli can filter out compounds of molecular masses which are below 40–60 kDa. Lipophilicity and pK_a are also important, because these properties determine whether or not a pollutant will be re-absorbed in the tubules and returned to circulation. Highly bound compounds (to proteins) will show slower elimination rates than weakly bound compounds.

3. HAZARD IDENTIFICATION AND MECHANISMS OF TOXICITY

The extent of a hazard is dependent on the mechanism of the toxicity, which involves the metabolic pathway, transport, distribution and interaction processes of the pesticide or metabolites in the different human tissues. This hazard must be assessed not only from isolated studies, but also by considering common mechanisms of toxicity that different pesticides may share.

The annex V of Directive 67/548/EEC, in the EU document, OECD guidelines for testing method for chemicals, addressing three areas of concern: (i) determination of physical-chemical properties; (ii) methods for determining effects on human health and (iii) methods for environmental effects. Annex IA of Directive 93/67/EEC is involved with the risk assessment of potential toxic effects concerning human health. The technical guidance document on risk assessment published by the European Commission [4], is involved with human health assessment. Analysis of hazard identification and dose–response assessment is based on the following key points:

1. toxic-kinetics, metabolism and distribution;
2. acute toxicity;
3. irritation;
4. corrosivity;
5. sensitisation;

6. repeated dose toxicity;
7. mutagenicity;
8. carcinogenicity;
9. toxicity for reproduction.

EPA considers the following toxicity tests for hazard identification: (i) acute testing; (ii) sub-chronic testing; (iii) chronic toxicity testing; (iv) developmental and reproductive testing; (v) mutagenicity testing: assess the potential of a pesticide to affect the genetic components of the cell and (vi) hormone disruption.

The primary purpose of the acute toxicity tests and other short-term studies are to identify hazards, provide a basis for classification and labelling, and enable the selection of exposure ranges for long-term studies. Usually, acute studies are conducted in young adult animals and involve measuring, mortality, body mass, changes at necropsy, histopathological examination of organs showing gross pathology, etc. Other studies may cover the assessment of a specific organ, such as liver, kidney and several indicators of cellular damage. Apart from these studies, acute neuro-toxicity studies can be developed to assess the motor activity and the control of the central and peripheral nervous system. Most of studies assessing hazard of pesticides aims to establish reference doses or reference concentrations to limit the intake of such pesticide in humans (hazard characterisation). The EPA sets reference dose values and reference dose concentrations referred to the intake of pesticides via oral/dermal or via inhalation respectively. Solecki et al. [13] reviewed the work developed by the Joint FAO/WHO Meeting on Pesticide Residues for acute health risk assessment of agricultural pesticides. The review addresses general considerations in the setting of acute reference doses (ARfDs) and selecting different toxicological endpoints. Haematotoxicity, immunotoxicity, neuro-toxicity, liver and kidney toxicity, endocrine effects, as well as developmental effects are taken into account as acute toxic alerts, relevant for the consideration of ARfDs for pesticides.

Sub-chronic and chronic toxicity tests provide an in-depth look at a number of organ systems, prenatal developmental toxicity and reproduction and fertility effects. A complete set of all these tests is available at the integrated risk information system (IRIS) of EPA [14].

Concerning carcinogenic effects, genotoxicity information is very useful as part of a weight-of-evidence approach to evaluate potential human carcinogenicity [5]. Cimino reviews the tiers of different regulatory agencies for conducting genetic toxicological tests [15]. In general, most of regulatory entities use the following three genotoxicity tests: a bacterial gene mutation assay, an in vitro mammalian cell assay for gene mutation and/or chromosome aberrations and an in vivo assay for chromosomal effects using either metaphase aberrations or micro-nucleus analysis. If geno-toxicity is identified, assessment of the ability of the chemical to interact with DNA in the gonad may be required. Mutagens that show a positive test in this second stage may be tested in a third stage with in vivo rodent tests to provide data for a quantitative risk assessment.

4. HUMAN EXPOSURE ASSESSMENT

An exposure assessment consists of quantifying the level of pesticides to which human populations, population subgroups and individuals, are exposed to, in terms of magnitude, duration and frequency. There are several approaches to estimate or quantifying human exposures. Direct methods involve measurements of exposure in the exposed subject, for example, personal monitoring and bio-monitoring. Indirect methods involve extrapolating exposure estimates from other measurements and existing data, for example, environmental monitoring, food monitoring or dietary and exposure models [16].

Exposure models represent important tools for indirect exposure assessments. They are typically used where direct measurements of exposure or biological monitoring data are not available or where these techniques are not appropriate for the exposure assessment situation. An exposure model is defined as a logical or empirical construct which allows the estimation of individual or population exposure parameters from available input data [17].

The exposure models can be categorised in terms of the types of exposure sources, such as: (i) environmental, (ii) dietary, (iii) consumer product, (iv) occupational, (v) aggregate and (vi) cumulative. Aggregate exposure models consider multiple exposure pathways, while cumulative models consider multiple chemicals. Fryer et al. [18] review different current approaches and assess its policy implications. The use of models is necessary for preliminary public health assessment, and most countries do have monitoring programs.

Despite the usefulness of exposure models to estimate exposure of population in different scenarios, the necessity of quantitative exposure data either for registration or for epidemiological studies, is widely recognised [19]. The quantification of the exposure is a critical issue and can be addressed from many different points of view. It is not always easy to select the correct strategy. Biomarkers for monitoring pesticides, among other chemicals, can be divided into three categories [20]: (i) biomarkers of exposure, (ii) biomarkers of susceptibility and (iii) biomarkers of effect. The first ones provide information on the dose of a pesticide that can be related to the exposure. Biomarkers of susceptibility indicate the variables that affect the response of an individual to a particular pesticide. Biomarkers of effect provide information on an event, usually in the pre-clinical stage, occurring at a target site after exposure that directly correlates to the manifestation of the disease. As the later two types of biomarkers are more related to clinical and medical sciences, in this paper we will deal with the first type of biomarkers (biomarkers of exposure), which are more related to the properties of chemicals.

4.1. Biomarkers of Exposure

Considering the multi-route and multimedia exposure of pesticides, the exposure assessment can be considered from three aspects: (i) external (or potential) dose, (ii) internal or absorbed dose and (iii) biologically active dose. The external dose measurements determine the potential exposure of individuals or population. It involves a proper monitoring of the exposed environment, including air, water, food consumptions and workplace environment, to achieve an accurate quantification of the potential exposure of individuals. This kind of monitoring does not give information about the absorption of pesticides into the body.

Biomarkers of internal dose integrate all pathways of exposure by estimating the amount of a pesticide that is absorbed into the body via measurements of the pesticide, its metabolite, or its reaction product in biological media. The most studied biological samples have been urine and blood, but also breast milk, saliva, placenta, etc. Finally the biologically effective dose is the amount of a pesticide that has interacted with a target site and altered a physiological function, for example, the inhibition of cholinesterase enzymes or the development of DNA adducts.

4.1.1. Biomarkers of Internal Dose Bio-Monitoring

In most of developed countries, programs to monitor the presence of pesticides in the environment are being carried out, dealing with air, water and soil contamination. These programs detect exposures to pesticides or other chemicals that are likely to cause disease. The goal for public health agencies is to obtain the amounts of these pesticides that are to be found in the population [21].

Bio-monitoring is the assessment of internal dose by measuring the parent chemical or its metabolite or reaction product in human blood, urine, milk, saliva, adipose or other tissue [22]. The advances of analytical techniques and extraction methods allow one to obtain a picture of the actual absorbed dose into the body at very low concentrations. Several papers deal with analytical methods for quantifying internal dose of pesticides in body fluids and human tissues [23–25].

Attending to the likely cumulative effects, pesticides with similar functional groups are supposed to have a similar behaviour in terms of body absorption and toxicological properties. In a first classification we can distinguish between persistent and non-persistent pesticides. The convention of Stockholm [26] classified OC pesticides as persistent organic pollutants, because they have high environmental half-lives, being accumulated in adipose tissues and biomagnified through the food chain. They are transported by migratory birds and through a process involving volatilisation, condensation, evaporation and deposit, known as the grasshopper effect [27]. Most other pesticides are listed in a group of non-persistent chemicals which can be classified in terms of its functional groups, such as organophosphate, carbamate, pyrethroids, triazine herbicides, phenoxyacid, chloroacetanilides, quats, dithiocarbamates, etc.

Bio-monitoring requires accurate analytical methods to comply with regulatory requirements. Method development is being carried out in order to achieve selectivity, sensitivity and accuracy in the identification of pesticides (parent compounds and metabolites) in the different body media. Such properties are being improved either in the sample processing steps or in the detection step.

Sample processing efforts focus on the minimisation of the solvent consumption, and the time reduction, by developing automated methodologies, which may involve extraction, clean-up, concentration and derivatisation, steps on-line. The most lipophilic compounds are OC pesticides ($\log K_{ow} > 3$, most of them >6 [8]). Urine has been used occasionally for analysing pollutants. However, because of the tendency for most pollutants, including OCs, many metabolites of pesticides, such as endosulfan [28], to bioaccumulation in lipid parts of the body, it makes sense to choose blood, human adipose tissues, breast milk, etc. for analysis. The extraction from such lipid matrixes are often difficult. The use of solid phase disk extraction (SPDE) [29–31] has been used to analyse OC pesticides in several human body fluids achieving a high-throughput sample processing. Effective liquid extraction, with clean-up steps based on automatised solid phase extraction (SPE) or liquid chromatography (LC) based clean-up methods, allows automatisation in the sample processing of human adipose tissues diminishing the uncertainty of the analytical results [32–34].

The more hydrophilic pesticides, such as organophosphorus (OPs), have a log K_{ow} value of less than 3, and as a result tend to be excreted by urine and as expected most of bio-monitoring on these substances are performed in urine or serum samples. Several researchers have used a solid phase micro-extraction (SPME) technique, which has the advantages of integrating sampling, sample extraction, pre-concentration and sample introduction in a single step. It has been applied to the analysis of OPs and metabolites in blood and urine (either specific metabolite of individual pesticides as common alkylphosphate derivatives). Reported methods based on SPME, either sampling the head space of samples [35, 36] or by immersion of fibres in the samples [37]. The influence of the physicochemical parameters in the extraction step is very important [38], and WS and hydrophobicity (log K_{ow}) are properties that influence the transference of pesticides from the sample to the fibre. In addition the vapour pressure and the Henry's law constant, governs the distribution between the headspace and the liquid phase, making the assessment of the transfer of matter between sample and fibre, more complex. Most of the reported extraction methods for OPs and metabolites in urine or blood (plasma, serum) are based on SPE [39–43].

LC technique has proved to be very good for analysing polar compounds, such as pesticide metabolites. It is done without derivatisation and this reduces the sample extraction time considerably [62, 63]. The compatibility with polar mobile phases and aqueous samples makes this technique one of the best for the bio-monitoring of polar compounds in human fluids.

4.1.2. Biomarkers of Active Dose

Another strategy for exposure assessment analysis is based on the measurable binding of pesticides or metabolites to specific cell receptors. In the areas of hazard assessment and exposure assessment, in vitro studies to assess mechanistic processes of toxicity, have been developed. These studies utilise new and innovative technologies such as genomics, transcriptomics and proteomics which allows the identification of in vitro clusters of genes and proteins that can be induced or silenced by pesticides or metabolites. The key point in these cases is to validate in vitro studies with in vivo exposure assessment by testing if mechanistic responses found in vitro correlates with in vivo exposure doses. This type of integrated, mechanism-driven in vitro to in vivo approach relies extensively on the use of cell assays to develop new biomarkers.

Examples of this kind of biomarker can be specific DNA adducts, DNA breaks, oxidative DNA damage or micro-nuclei cells (e.g., in blood lymphocytes and selected somatic cells), which may be indicative of an increased hazard derived from exposure. DNA damage will enhance the probability of mutations occurring in critical target genes and cells. The damage may be detected in peripheral blood lymphocytes, in any primary or cultured cell system, or by analysing the excretion in urine or plasma of reaction products indicative of genotoxic interactions [64].

One of the most used tests for genotoxicity, is the alkaline comet assay that has also been proposed as a tool for bio-monitoring [65–67]. It was first used by Östling and Johanson [68] and modified by Singh et al. [69]. Comet assay detects DNA single-strand breaks, alkali labile sites and DNA cross-linking in individual cells. It is based on the electrophoresis of cells embedded and lysed in agarose on a microscope slide, liberating the DNA. Other authors have emphasised the necessity of harmonisation with data inter-comparison purposes [70, 71]. Comet assay, in combination with micro-nucleus analysis, chromosome aberrations and sister chromatic exchanges have been used for bio-monitoring occupational exposure to atrazine, alachlor, cyanazine, 2,4-dichloro-phenoxy-acetic acid, malathion and other OC pesticides [72–75], revealing the strong probability that pesticides cause DNA damage.

From another point of view, the toxic effect of a xenobiotic compound on a biological system can be reflected at the cellular level by its impact on gene expression. Consequently, measurements of the transcription (mRNA) and translation products of gene expression (proteins) can reveal valuable information about the potential toxicity of chemicals. Furthermore it can be used as the basis of new biomarkers by developing appropriate measurement methods for detecting likely changes.

The rapid progress in DNA sequentiation (genomic), in gene expression (transcriptomic) and the development of techniques which allow one to analyse changes in proteins expressed (proteomic), together with the availability of databases and libraries, presents a great potential for hazard characterisation and for biomarker development.

The sensitive requirements for exposure assessment have been met by the recent development of mass spectrometry (MS) techniques for the characterisation of proteins expressed by a genome, tissue or cell. Of particular interest is the matrix-assisted laser-desorption/ionisation (MALDI), which makes it possible to obtain protein mass fingerprinting for a wide range of proteins by MS. This method involves selectively cutting proteins by enzymatic actions, and comparing the fragment masses with theoretical peptides available in bio-informatic databases.

5. CONCLUSIONS

Risk assessment is a multidisciplinary task related to toxicology, analytical chemistry, biochemistry, molecular biology, health disciplines, politics, etc. The four key aspects of risk assessment are; hazard identification, dose response, exposure assessment and risk characterisation. They are all driven by dynamics based on intake, absorption and effect.

Physical-chemical properties of pesticides are fundamental in such dynamics, governing the processes of environment contamination, absorption, distribution, metabolism and finally excretion from the body. The most important parameters determining the fate of pesticides in the environment and the body are; solubility, K_{ow}, lipophilicity, molecular mass, charges and volatility. The reactivity to bind to proteins also influences the bioaccumulation process.

Important advances have recently occurred in analytical methodologies that allow the determination of biomarkers of internal dose as part of national monitoring programmes. These advances are responsible for: (i) the development of high-throughput methodologies such as those described in Section 4.1.1 allows the analysis of a high number of samples per day; (ii) the development of LC techniques as well as detection strategies such as MS in tandem mode (MS/MS) allow to identify pesticides and polar metabolites on the basis of confident structural information at low levels; (iii) the development quality assurance programmes and the inclusion of quality control measures in the methodologies allowing a comparison of data on internal doses, which is essential in risk assessment. There are remaining difficulties associated with the availability of samples and the metabolite standards (either isotope as native). In this content the creation of international bio-banks, where biological samples can be stored in optimal conditions, and documented with well-planned sampling protocols, can be useful for epidemiological studies.

Biomarkers of effective dose have a great potential either in hazard characterisation or in exposure assessment. The rapid progress in genomic, transcriptomic and proteomic technologies, in combination with the ever-increasing power of bio-informatics, creates a unique opportunity to test genotoxic effects, such as DNA damage and protein modifications at low levels, which is essential in hazard identification and acute and chronic exposure assessment. The weak points are: the lack of standardisation of methodologies (mainly those referred to as comet assays) that make it difficult to compare results; and the many factors that can interfere in the analysis, such as age, smoking habits, sex, diet, etc. Despite these inconveniences, their potential as screening methods for epidemiologic studies, as well as the relevant information on the effect produced by the xenobiotics as consequence of exposure, is interesting and important. The most important aspect is the use of biomarkers to determine an effective dose and to correlate exposure with effect. Furthermore these kinds of biomarkers can provide evidence of exposure and probability of developing diseases, which is one of the main goals of risk assessment.

REFERENCES

[1] Risk Assessment, Toxicology Steering Committee (RATCS), Risk assessment approaches used by UK Government for evaluating human health effects of chemicals, Institute for Environment and Health, Leicester, UK, 1999.
[2] National Research Council (NRC), Risk Assessment in the Federal Government: Managing the Process, National Academy Press, Washington, DC, March, 1983.
[3] J.R. Fowlle and K.L. Dearfield, Risk Characterization Handbook, USEPA, Washington, DC, 2000.
[4] Technical Guidance Document on Risk Assessment (Parts I to III), European Commission Joint Research Centre, EUR 20418 EN/1, 2003.
[5] K.L. Dearfield and M.M. Moore, Environ. Mol. Mutagen., 46 (2005) 236.
[6] OCDE Principles on Good Laboratory Practices, 1997, OECD Environment Directorate, Environmental Health and Safety Division, http://www.oecd.org/ehs/
[7] D.J. Marino, Risk Anal., 26 (2006) 185.
[8] A. Finizio, M. Vighi and D. Sandroni, Chemosphere, 34 (1997) 131.
[9] L. Shen and F. Wania, J. Chem. Eng. Data, 50 (2005) 742.
[10] E. Halfon, S. Galassi, R.B. Oggemann and A. Provini, Chemosphere, 33 (1996) 1543.
[11] U. Bernauer, A. Oberemm, S. Madle and U. Gundert-Remy, Basic Clin. Pharmacol. Toxicol., 96 (2005) 176.
[12] E. Dybinga, J. Doeb, J. Grotenc, J. Kleinerd, J. O'Briene, A.G. Renwickf, J. Schlatterg, P. Steinbergh, A. Tritscheri, R. Walkerj and M. Younesk, Food Chem. Toxicol., 40 (2002) 237.
[13] R. Solecki, L. Davies, V. Dellarco, I. Dewhurst, M. van Raaij and A. Tritscher, Food Chem. Toxicol., 43 (2005) 1569.
[14] http://www.epa.gov/iris/backgr-d.htm.
[15] M.C. Cimino, Environ. Mol. Mutagen., 47 (2006) 362.
[16] M.J. Nieuwenhuijsen (ed.), Exposure Assessment in Occupational and Environmental Epidemiology, Oxford University Press, Oxford, 2003.
[17] World Health Organisation (WHO), Methods of Assessing Risk to Health from Exposure to Hazards Released from Landfills, World Health Organization, Copenhagen, 2000.
[18] M. Fryer, C.D. Collins, H. Ferrier, R.N. Colvile and M.J. Nieuwenhuijsen, Environ. Sci. Policy, 9 (2006) 261.

[19] R.E. Shore, Am. J. Public Health, 85 (1995) 474.
[20] Committee on Advances in Assessing Human Exposure to Airborne Pollutants, Human Exposure Assessment for Airborne Pollutants: Advances and Opportunities, National Academy of Sciences, Washington, DC, 1991.
[21] J.L. Pirkle, J. Osterloh, L.L. Needham and E.J. Sampson, Int. J. Hyg. Environ. Health, 208 (2005) 1.
[22] D. Paustenbach and D. Galbraith, Reg. Toxicol. Pharmacol., 44 (2006) 249.
[23] D.B. Barr and L.L. Needham, J. Chromatogr. B, 778 (2002) 5.
[24] C. Aprea, C. Colosio, T. Mammone, C. Minoia and M. Maroni, J. Chromatogr. B, 769 (2002) 191.
[25] J.L. Martinez Vidal and A. Garrido Frenich (eds.), Pesticide Protocols, Humana Press, New Jersey, 2006.
[26] United Nations Environment Program, Final Act of the Conference of Plenipotentiaries on Stockholm Convention on Persistent Organic Pollutants, May 2001.
[27] F. Wania and D. Mackay, Ambio, 22 (1993) 10.
[28] J.L. Martinez Vidal, F.J. Arrebola, A. Fernandez and M.A. Rams, J. Chromatogr. B, 719 (1998) 71.
[29] A. Covaci and P. Schepens, Chemosphere, 43 (2001) 439.
[30] A. Covaci, P. Jorens, Y. Jacquemyn and P. Schepens, Sci. Total Environ., 298 (2002) 45.
[31] A. Covaci, in "Methods in Biotechnology, Pesticide Protocols" (J.L. Martínez Vidal and A. Garrido Frenich, eds.), p. 49, Vol. 19, Humana Press, New Jersey, 2006.
[32] J.L. Martínez Vidal, A. Garrido Frenich, F.J. Egea Gonzalez and F.J. Arrebola Liebanas, in "Methods in Biotechnology, Pesticide Protocols" (J.L. Martínez Vidal and A. Garrido Frenich, eds.), p. 3, Vol 19, Humana Press, New Jersey, 2006.
[33] J.L. Martínez Vidal, M. Moreno Frías, A. Garrido Frenich, F. Olea-Serrano and N. Olea, Anal. Bioanal. Chem., 372 (2002) 766.
[34] F. Hernández, E. Pitarch, R. Serrano, J.V. Gaspar and N. Olea, J. Anal. Toxicol., 26 (2002) 94.
[35] H. Tsoukali, G. Theodoridis, N. Raikos and I. Grigoratou, J. Chromatogr. B, 822 (2005) 194.
[36] F. Hernandez, E. Pitarch, J. Beltran and F.J. Lopez, J. Chromatogr. B, 769 (2002) 65.
[37] F.J. López, E. Pitarch, S. Egea, J. Beltran and F. Hernández, Anal. Chim. Acta, 433 (2001) 217.
[38] J. Dugay, C. Miege and M.C. Hennion, J. Chromatogr. A, 795 (1998) 27.
[39] A.O. Olsson, J.V. Nguyen, M.A. Sadowski and D.B. Barr, Anal. Bioanal. Chem., 376 (2003) 808.
[40] J.V. Sancho, O. Pozo, F.J. López and F. Hernández, Rapid Commun. Mass Spectrom., 16 (2002) 639.
[41] F. Hernández, J.V. Sancho and O.J. Pozo, J. Chromatogr. B, 808 (2004) 229.
[42] J. Ueyama, I. Saito, M. Kamijima, T. Nakajima, M. Gotoh, T. Suzuki, E. Shibata, T. Kondo, K. Takagi, K. Miyamoto, J. Takamatsu, T. Hasegawa and K. Takagi, J. Chromatogr. B, 832 (2006) 58.
[43] G.K. Hemakanthi De Alwis, L.L. Needham and D.B. Barr, J. Chromatogr. B, 83 (2006) 34.

[44] S.E. Baker, A.O. Olsson and D.B. Barr, Arch. Environ. Contam. Toxicol., 46 (2004) 281.
[45] G. Leng and W. Gries, in "Methods in Biotechnology", Pesticide Protocols (J.L. Martínez Vidal and A. Garrido Frenich, eds.), p. 17, Vol. 19, Humana Press, New Jersey, 2006.
[46] G. Leng, K.H. Kuehn and H. Idel, Toxicol. Lett., 88 (1996) 215.
[47] C. Aprea, A. Stridori and G. Sciarra, J. Chromatogr. B, 695 (1997) 227.
[48] D.L. Hughes, D.J. Ritter and R.D. Wilson, J. Environ. Sci. Health, 36 (2001) 755.
[49] C. Aprea, G. Sciarra, N. Bozzi and L. Lunghini, in "Methods in Biotechnology", Pesticide Protocols (J.L. Martínez Vidal and A. Garrido Frenich, eds.), p. 91, Vol. 19, Humana Press, New Jersey, 2006.
[50] K. Wittke, H. Hajimiragha, L. Dunemann and J. Begerow, J. Chromatogr. B, 755 (2001) 215.
[51] S. Fustinoni, L. Campo, C. Colosio, S. Birindelli and V. Foa, J. Chromatogr. B, 814 (2005) 251.
[52] M. Cruz Márquez, F.J. Arrebola, F.J. Egea González, J.L. Martínez Vidal and M.L. Castro Cano, J. Chromatogr. A, 939 (2001) 79.
[53] M.R. Driss, S. Sabbah and M.L. Bouguerra, J. Chromatogr., 552 (1991) 213.
[54] E.S. DiPietro, C.R. Lapeza, W.E. Turner, V.G. Green, J.B. Gill and D.G. Patterson, Jr., Organohalogen Comp., 31 (1997) 26.
[55] J.A. van Rhijn, W.A. Traag, P.F. van de Spreng and L.G. Tuinstra, J. Chromatogr., 630 (1993) 297.
[56] J.R. Barr, V.L. Maggio, D.B. Barr, W.E. Turner, A. Sjodin, C.D. Sandau, J.L. Pirkle, L.L. Needham and D.G. Patterson, J. Chromatogr. B, 794 (2003) 137.
[57] C.D. Sandau, A. Sjodin, M.D. Davis, J.R. Barr, V.L. Maggio, A.L. Waterman, K.E. Preston, J.L. Preau, D.B. Barr, L.L. Needham and D.G. Patterson, Anal. Chem., 75 (2003) 71.
[58] D.B. Barr, R. Bravo, J.R. Barr and L.L. Needham, in "Methods in Biotechnology", Pesticide Protocols (J.L. Martínez Vidal and A. Garrido Frenich, eds.), p. 35, Vol. 19, Humana Press, New Jersey, 2006.
[59] M. Nichkova and M.P. Marco, in "Methods in Biotechnology", Pesticide Protocols (J.L. Martínez Vidal and A. Garrido Frenich, eds.), p. 133, Vol. 19, Humana Press, New Jersey, 2006.
[60] J.M. Van Emon, J. AOAC Int., 84 (2001) 125.
[61] J.C. Chuang, J.M. Van Emon, J. Durnford and K. Thomas, Talanta, 67 (2005) 658.
[62] J.M. Pozzebon, W. Vilegas and I. Jardim, J. Chromatogr. A, 987 (2003) 375.
[63] A.O. Olsson, S.E. Baker, J.V. Nguyen, L.C. Romanoff, S.O. Udunka, R.D. Walker, K.L. Flemmen and D.B. Barr, Anal. Chem., 76 (2004) 2453.
[64] G. Eisenbrand, B. Pool-Zobel, V. Baker, M. Balls, B.J. Blaauboer, A. Boobis, A. Carere, S. Kevekordes, J.C. Lhuguenot, R. Pieters and J. Kleiner, Food Chem. Toxicol., 40 (2002) 193.
[65] R.J. Albertini, D. Anderson, G.R. Douglas, L. Hagmar, K. Hemminki, F. Merlo, A.T. Natarajan, H. Norppa, D.E.G. Shu-ker, R. Tice, M.D. Waters and A. Aitio, Mutat. Res., 463 (2000) 111.
[66] E. Rojas, M.C. Lopez and M. Valverde, J. Chromatogr. B, 722 (1999) 225.

[67] R.R. Tice, E. Agurell, D. Anderson, B. Burlison, A. Hartmann, H. Kobayashi, Y. Miyamae, E. Rojas, J.C. Ryu and Y.F. Sasaki, Environ. Mol. Mutagen., 35 (2000) 206.
[68] O. Östling and K.J. Johanson, Biochem. Biophys. Res. Commun., 123 (1984) 291.
[69] N.P. Singh, M.T. McCoy, R.R. Tice and E.L. Schneider, Exp. Cell Res., 175 (1988) 184.
[70] P. Møller, Mutat. Res., 612 (2006) 84.
[71] F. Faust, F. Kassie, S. Knasmüller, R.H. Boedecker, M. Mann and V. Mersch-Sundermann, Mutat. Res., 566 (2004) 209.
[72] V. Garaj-Vrhovac and D. Zeljezic, Mutat. Res., 469 (2000) 279.
[73] D. Zeljezic and V. Garaj-Vrhovac, Mutagenesis, 16 (2001) 359.
[74] P. Grover, K. Danadevi, M. Mahboob, R. Rozati, B. Saleha and M.F. Rahman, Mutagenesis, 18 (2003) 201.
[75] L. Yañez, V.H. Borja-Aburto, E. Rojas, H. de la Fuente, R. Gonzalez-Amaro, H. Gomez, A.A. Longitud and F. Diaz-Barriga, Environ. Res., 94 (2004) 18.

Thermodynamics, Solubility and Environmental Issues
T.M. Letcher (editor)

Chapter 25

Solubility and Body Fluids

Erich Königsberger and Lan-Chi Königsberger

School of Chemical and Mathematical Sciences, Murdoch University, Murdoch, WA 6150, Australia

1. INTRODUCTION

Solubility phenomena (i.e. dissolution and precipitation reactions) are the physicochemical basis of numerous biological processes. These include, for instance:

- gas solubilities in respiratory and photosynthetic processes; the solubility of volatile anaesthetics;
- the crystallisation, both in biologically controlled and pathological processes, of biogenic minerals in a variety of body fluids; the resorption of mineralised tissue;
- the accumulation, due to their higher solubility, of lipophilic substances, such as pesticides, in liquid fat contained, e.g. in adipose tissue or human milk;
- the incorporation of metal ions such as strontium (including the radioactive ^{90}Sr isotope) in bone, by co-precipitation and solid-solution formation. Since Sr stabilises bone apatite crystals (i.e. decreases solubility), it may retard the resorption of the calcified matrix and thus have therapeutic potential in the prevention and treatment of osteopenic disorders [1]. In dental enamel, a combination of strontium and fluoride was reported to be more effective in stabilising the apatetic structure than each element alone [2]. This results in an improved crystal resistance to degradation by bacterial acids and hence may be useful for the prevention of dental caries [3].

All of these solubility phenomena are governed by the laws of thermodynamics and kinetics. The human body is essentially an isothermal system (a notable exception, related to gout, has been reported in the literature – see below). Thus, the pertinent in vitro measurements have almost always been performed at 37°C. However, reactions in body fluids are complicated by the presence of organic complexing agents which affect the speciation of metal ions (i.e. their distribution among these complexes). Computer speciation modelling of biofluids, which has a long history [4], has also to be considered in the modelling of solubility equilibria. The presence of organic macromolecules such as proteins provides templates or matrices which control the crystallisation of biogenic minerals and modify their morphologies. The mineral phase in organic/inorganic composites (such as teeth or bone) may form nanosized crystals, which exhibit unusual dissolution and crystallisation behaviour [5]. A new biomineralisation mechanism invoking liquid precursors and their importance for normal and pathological biomineralisation processes has been proposed [6]. These and other aspects of normal and pathological biomineralisation processes have been reviewed recently [7].

2. BODY FLUIDS

The human body contains 60% water, which is an excellent solvent for electrolytes and plasma proteins [8]. Because all cell membranes are freely permeable to water, intra- and extracellular fluids are generally considered to be in osmotic equilibrium. Therefore, the *osmolality* (the total molality of individual ionic and neutral solute species that contribute to the osmotic pressure) of the extracellular fluid is approximately equal to that of the intracellular fluid (ICF) and hence plasma osmolality is a guide to intracellular osmolality. Normal osmolality in plasma is ~0.30 'osmoles' (kg water)$^{-1}$. This is contributed mainly by sodium, chloride, potassium, urea and glucose, and additionally by other ions and substances in the blood. Most of the body fluids are *isotonic* ("of equal osmotic pressure", when only impermeant solutes are taken into account), with the notable exceptions of urine, sweat and saliva [9].

Forty-two litres of solutions involved in the major body fluids of a 70 kg human are distributed as follows [8]:

1. *Intracellular fluid* (ICF, 28 l) is the sum of all the solutions inside the ca. 10^{14} cells of the body. Although each of them contains a separate individual solution and different cell types are chemically different, their internal solutions all share a few common features that distinguish them from extracellular fluids. For instance, intracellular solutions are high in potassium, magnesium and phosphate ions and low in

sodium and chloride, and contain high concentrations of organic acids and proteins. These bind metal ions efficiently so that crystallisation of solids does not normally occur in ICF.

2. *Extracellular fluids* have various functions that are beneficial to the organism [9]. These include heat homeostasis (through blood and sweat), transport functions (energy sources, nutrients, electrolytes, dissolved gases, hormones, excretory products, etc.), wetting (saliva), lubrication (tears, synovial fluid) and neutralisation (saliva, biliary and pancreatic juices), among others [9]. Various extracellular fluids are prone to pathological calcification and stone formation. They are commonly subdivided as follows:
   a. *Interstitial fluid* (ISF, 10.5 l) is the sum of all the thin layers of tissue fluid in the interstices between cells throughout the body and constitutes the internal environment whose stable composition is essential for homeostasis. Its major chemical features are low potassium and magnesium ion concentrations, high sodium, chloride and bicarbonate concentrations, and very low or negligible concentrations of proteins or other organic acids. A special subset of ISF is the lymphatic fluid, or lymph, which plays a part in the transport of tissue fluid until it rejoins the blood. Since tissue fluid passes into the surrounding lymph vessels and lymph can thus be sampled and analysed, measurements of lymph composition are often cited as approximations of the composition of ISF.
   b. *Blood plasma* (3 l) is the ISF of a very special (liquid) tissue, namely whole blood. It circulates rapidly throughout the body and is in effective diffusion equilibrium with the ISF for most solutes except macromolecules. Plasma therefore differs in composition from ISF mainly in its protein content.
   c. *Other fluids* amount to 0.5 l. This category includes a variety of small volumes of special, usually rather small and localised, solutions such as aqueous humour (the clear, watery fluid in the eye that fills the space between the back surface of the cornea and the front surface of the vitreous humour [10]), synovial fluid (in joints), bile, saliva and many others. These are sometimes referred to as 'transcellular' fluids. Two of these special fluids have extreme pH values: gastric and pancreatic secretion. Gastric acid can be as concentrated as $0.15\,mol\,kg^{-1}$ HCl, and the alkaline pancreatic juice contains high concentrations of sodium, calcium and hydrogen carbonate, but almost no chloride.

3. SOLUBILITY PHENOMENA IN BODY FLUIDS

3.1. Gas Solubilities

Except for blood plasma passing the pulmonary capillaries, body fluids are not even in contact with, let alone at equilibrium with, a gas phase. They nonetheless contain dissolved gases, especially CO_2, whose concentration is an important determinant of pH and therefore of acid–base behaviour in all body fluids [11]. Attempts to correlate and predict the solubilities of non-electrolytes, particularly gases and volatile anaesthetics [12], in biological tissues and fluids are more complicated compared to pure solvents, since biological systems are heterogeneous and cases of specific binding of solute molecules are common. In these situations, Henry's law is no longer obeyed. For instance, oxygen solubility in blood plasma is ca. $5\,\mathrm{ml\,l^{-1}}$ at 37°C but considerably higher (ca. $200\,\mathrm{ml\,l^{-1}}$) in whole blood due to the binding of oxygen to haemoglobin, which follows a well-known sigmoid binding curve resulting from the cooperative action of the haemoglobin alpha and beta chains. The effect of haematocrit (the fraction of blood composed of red blood cells) on the solubility of several volatile anaesthetics in blood is much less pronounced but significant (10–20%) [13].

Abraham et al. [14] have reported solubility correlations for a wide variety of systems that follow Henry's law. Gas/liquid partition coefficients for biofluids are modelled by a suitable combination of gas/water and gas/oil partition coefficients, thus allowing for the hydrophobicity of a given biological tissue or fluid. This method is furthermore able to provide a measure for the tissue/blood distribution of non-electrolytes (which may also be estimated using octanol–water partition coefficients [15]).

3.2. Normal Biogenic Mineralisation

Biomineralisation commonly refers to the complex biological process by which living organisms synthesise the inorganic materials of their hard tissues. Normal (as opposed to pathological) biomineralisation is frequently characterised by a high degree of specificity and control, which is exerted during the interaction between the mineral and the organic constituents on different hierarchical levels and directs the nucleation, growth and morphology of 'normal' biomaterials such as bone and teeth. The structure and mechanism of formation of these organic/mineral biocomposites, as well as the amazing diversity and beauty of the very elaborate and complex crystal morphologies produced by various organisms, have been comprehensively described in classical and modern textbooks, e.g. [16–18], and reviews, e.g. [6].

The authors of the latter review present a somewhat different perspective on biomineralisation, with emphasis on the role of amorphous precursor phases, which may be relevant to biologically controlled mineralisation in bones and teeth (and also to pathological mineralisation such as in kidney stones) [6].

The solubilities of the various calcium phosphates found in mineralised tissues at different states of maturation have been systematically reviewed [19]. Besides its skeletal support function, bone also provides an important metabolic function in the regulation of the ionic composition of the body fluids, serving as a reservoir for cationic and anionic species by resorption and deposition of this mineralised tissue [20]. The mineral phase in bones and teeth consists of apatite nanocrystals, which are not only crucial for the superior mechanical properties of these biocomposites, but also exhibit some unusual solubility behaviour. While these nanocrystals are usually considered to be stabilised by their intimate association with the organic matrix and thus become unstable once the collagen is removed (such as by osteoclastic secretion of enzymes enabling the biological resorption of bone) [6], studies of their dissolution kinetics have shown that they can become dynamically stabilised (i.e. they do not dissolve any further) even at considerable undersaturation. These and other aspects of peculiar solubility phenomena concerning nanosized biominerals have been discussed recently [5].

3.3. Pathogenic 'Calcifications'

Pathological intra- and extracellular calcifications are frequently initiated at the biologic membranes of mitochondria or matrix vesicles, respectively, through the interaction of phosphatase enzymes with calcium-binding phospholipids, both of which are membrane-bound [21]. Sequestration of calcium by mitochondria is a common biochemical mechanism mediating various forms of toxic cell death. Pathological cellular calcification may therefore be characterised by mineral-laden mitochondria, as described for a case of hepatocellular calcification [22]. Cellular microcalcification has also been observed in a diversity of human pathologies, such as vascular dementia, Alzheimer's disease, Parkinson's disease, astrogliomas and post-traumatic epilepsy. Rodent models of central nervous system neurodegeneration indicate that this is due to the inability of neurons to regulate intracellular calcium levels properly [23]. Hydroxyapatite crystals are formed first within the protective microenvironment of the membrane-enclosed microspace, aggregate progressively throughout the cell and at exposure to the Ca^{2+}-rich extracellular fluid, for instance, after cell death, they can serve as nuclei or templates supporting progressive, autocatalytic mineral crystal proliferation [21]. The nucleoside triphosphates have been found to have a significant inhibiting effect on the hydroxyapatite crystal growth process and thus may play a beneficial role in regulating intracellular calcification [24].

In dystrophic calcifications, the mineralisation occurs without a systemic mineral imbalance as a response to previous cell injury on the microscopic level or any soft tissue damage, including that involved in implantation of medical devices or bioprosthetic heart valves [25]. Injury and cell death can cause the release of intracellular phosphate ions and fatty acids into the extracellular environment, where sparingly soluble calcium salts are precipitated in tissues. This type of calcification is particularly common in atherosclerosis and diseases associated with chronic inflammation.

Pathological intracellular calcifications can also occur in conditions associated with hypercalcaemia [26] – excess levels of calcium in the blood, caused, e.g. by hyperparathyroidism (excess secretion of parathyroid hormone by overactive parathyroid glands which normally regulate calcium levels tightly) or malignancy [27]. Intramitochondrial calcifications often precede necrosis, and fusion or rupture of calcified mitochondria results in calcareous masses in the cytoplasm or outside of the cell. Much progress has been made in the ultramicroscopic identification and systematic classification of these calcifications [28].

Since extracellular body fluids exhibit high calcium concentrations, many of them are supersaturated with respect to calcium compounds, even in the healthy organism. Examples of supersaturated body fluids are urine (oxalates, phosphates), blood plasma and saliva (phosphates), pancreatic secretions and bile (carbonates). Body fluids may also become supersaturated with respect to intermediate or final metabolic products, e.g. urine (uric acid, xanthine, cystine), bile and blood plasma (cholesterol), and synovial fluid (sodium hydrogenurate monohydrate). Despite considerable research effort, it is presently not clear why pathological mineralisation does not occur indiscriminately in the supersaturated body fluids occurring in all humans. It seems that (i) the presence of heterogeneous nucleants and (ii) a deficit of crystallisation inhibitors play a crucial role in pathological situations [29]. For instance, magnesium has found to be an important inhibitor of many pathological calcifications [30]. Macromolecules can interact with crystals both as inhibitors and as promoters of crystallisation and the relationship between pathological and benign mineralisation is not fully understood as yet [31]. A paramount example is the multifaceted role of the glycoprotein osteopontin in bone remodelling, urolithiasis, atherosclerosis, inflammatory and immune response, and cancer progression [32]. A somewhat controversial debate concerns the role of putative *nanobacteria* in various forms of calcification (see the discussion in Ref. 5).

3.3.1. Renal Calculi

Approximately 10% of the human population (with regional differences indicating both genetic and environmental factors [33]) is affected by the formation of stones or calculi in the urinary tract. Urolithiasis is not only a painful condition, but also causes annual costs to the health system in the order of billions of dollars in the USA alone [34, 35]. Based on their composition, structure and location in the urinary tract, renal stones have been classified into 11 groups and their formation mechanisms have been discussed together with alterations in urinary parameters and metabolic risk factors for renal lithiasis [35]. Approximately 70% of these stones contain calcium oxalate monohydrate (COM) and dihydrate as major components, while other calculi are composed of ammonium magnesium phosphate (struvite), calcium phosphates (hydroxyapatite and brushite), uric acid and urates, cystine and xanthine. An accurate knowledge of the solubilities of these substances is necessary to understand the cause of renal or bladder calculi formation and find ways towards its prevention and treatment [36].

Due to its vital functions related to water balance and the excretion of metabolic end products, the composition and pH of urine can vary widely (which is in contrast to the homeostatic internal environment provided by most other body fluids). Although the solubilities of uric acid (the end product of purine metabolism in humans), xanthine (an intermediate product of purine metabolism) and L-cystine (the least soluble amino acid) strongly depend on pH, accurate measurements performed by our group (Fig. 1) have shown that they hardly depend on the nature and concentration of urinary constituents [36–41], including organic urine components such as urea and creatinine [38]. The solubility of cystine in real urine was found to be comparable to that in synthetic solutions [39]. For xanthine [40] and uric acid [42], thermodynamic consistency of solubility and calorimetric data has been demonstrated.

Calcium oxalate and phosphate solubilities, on the other hand, strongly depend on the concentration of ions that form complexes with calcium, phosphate or oxalate, particularly citrate or magnesium ions [43–45]. Due to the protonation of these anions, the concentration of these complexes depends on pH, which contributes to the pH dependence of solubility. Whereas the solubility of calcium oxalate is only slightly pH dependant in the urinary pH range [45], the solubility of phosphates decreases with pH [43, 44]. We have developed a urine model [43–45] based on the JESS Expert Speciation

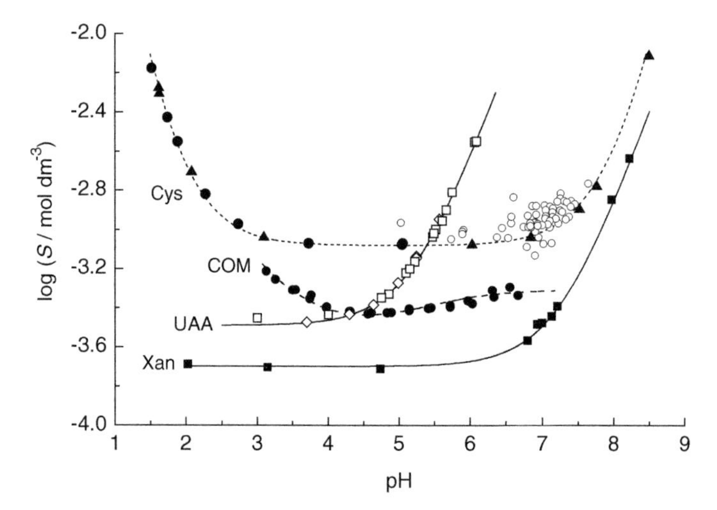

Fig. 1. Solubilities S of stone-forming substances in urine-like liquors at 37°C. (Cys) L-Cystine in standard-reference artificial urine (dots), real urine (circles) and 0.30 mol kg^{-1} NaCl (triangles) [39]. (COM) Calcium oxalate monohydrate (as $[Ca^{2+}]_{tot}$) in standard-reference artificial urine (dots) [45]. (UAA) Anhydrous uric acid in standard-reference artificial urine (diamonds) and 0.30 mol kg^{-1} NaCl + 0.30 mol kg^{-1} urea (squares) [37, 38]. (Xan) Xanthine in 0.30 mol kg^{-1} NaCl (squares) [40]. The experimental data are compared to model calculations (lines) described in the respective publications.

System [46–49] that permits reliable solubility calculations by taking all of these complexes into account. This and other urine models and their applications have been reviewed recently [5].

Due to deprotonation reactions occurring at higher pH, the solubilities of uric acid, xanthine and cystine increase with pH. For cystine and particularly for uric acid stones, therapies based on the administration of substances that increase the urinary pH have been successfully applied. However, the solubility increase of xanthine at high pH is of little therapeutical use, as is that of cystine at low pH (caused by the protonation of the amino groups). Both of these pH ranges are outside the normal urinary pH range and attempts to access them by medication may cause the precipitation of uric acid and phosphates at low and high pH, respectively. It should be noted that the solubility of the metastable uric acid dihydrate, which is also found in kidney stones and is regarded as the kinetically favoured precipitation product, is twice as high as that of anhydrous uric acid [37].

The solubility of COM in artificial urine is almost pH independent (at low pH the solubility increases because of the protonation of the oxalate ion and at high pH it increases slightly because of the deprotonation of citrate which then complexes Ca^{2+}). The solubility of calcium oxalate dihydrate, which is also found in kidney stones and whose precipitation in complex solutions is often kinetically controlled, in artificial urine is three times higher than that of the monohydrate [45].

Renal lithiasis is a multifactorial disease, with numerous factors contributing to the actual stone formation [6, 33, 35]. Although various conditions promote, or fail to inhibit, the crystallisation of stone-forming substances, supersaturation is an indispensable factor. Solubility modelling can therefore give crucial information on the risk of pathological mineralisation in general and renal lithiasis in particular. A variety of computer codes have been applied to assess the effects of diet, fluid intake and medication [5], and an application of a JESS speciation model similar to ours [43–45] has been reported [50].

3.3.2. Atherosclerosis

Cardiovascular disease due to atherosclerosis is the major cause of mortality and morbidity in industrialised countries. Atherosclerosis is characterised by hardening (and loss of elasticity) of medium or large arteries, eventually leading to narrowing of the vessel lumen. Initial steps are an immune reaction to cell damage caused by oxidation of low-density lipoprotein (LDL) in the blood vessel lining (endothelium), which involves redox-active copper and iron ions [51]. The resulting inflammation response leads to deposition of atheromatous plaques in the inner wall of the arteries (intima), followed by intracellular microcalcifications within vascular smooth muscle cells of the muscular layer surrounding the plaques. As cells die, extracellular calcium salt deposits are formed between the muscular wall and the outer portion of the atheromatous plaques [52]. The main components of human atherosclerotic lesions are calcium salts, particularly apatites, and cholesterol crystals [53, 54], which accumulate predominantly within the central core region of the lesions and probably nucleate the deposition of apatite [55].

3.3.3. Crystal Arthritis: Gout and Pseudogout

The very painful inflammations associated with gouty arthritis are induced by the deposition of needle-shaped sodium hydrogenurate monohydrate crystals in synovial fluids around joints [56]. The solubility of sodium hydrogenurate monohydrate is strongly pH dependant and is lowest at physiological pH [37]. It has been observed that only a small percentage of individuals with hyperuricaemic body fluids (which are supersaturated with respect to sodium hydrogenurate monohydrate) have ever had a gouty attack. Indeed, normal synovial fluids, serum albumin and heparin have been found to inhibit sodium hydrogenurate monohydrate crystallisation, whereas synovial fluids of gouty patients have nucleated this substance [57]. It has also been established in our solubility study [37] that even in 0.15 mol kg^{-1} NaCl at higher pH, saturated uric acid solutions can become highly supersaturated with respect to sodium hydrogenurate monohydrate without any crystallisation occurring. The solubility of sodium hydrogenurate monohydrate is markedly temperature dependant (Fig. 2) [37] and it has been speculated that the preferred occurrence of gouty attacks in the joints of the extremities is due to the lower temperature of these parts of the body [58].

Pseudogout refers to the clinical syndrome associated with the deposition of calcium diphosphate dihydrate crystals in the hyaline articular cartilage or fibrocartilage. The shedding of crystals in the joint space after rupture of a calcium diphosphate dihydrate deposit produces an acute inflammatory synovitis, which resembles a classic gouty attack and is thus often misdiagnosed as gout. Calcium diphosphate dihydrate crystal deposition is often the result of an underlying disease (such as haemochromatosis or hyperparathyroidism) and cannot be reversed. These and other aspects of pseudogout and the solubilities of the substances involved have been reviewed recently [19].

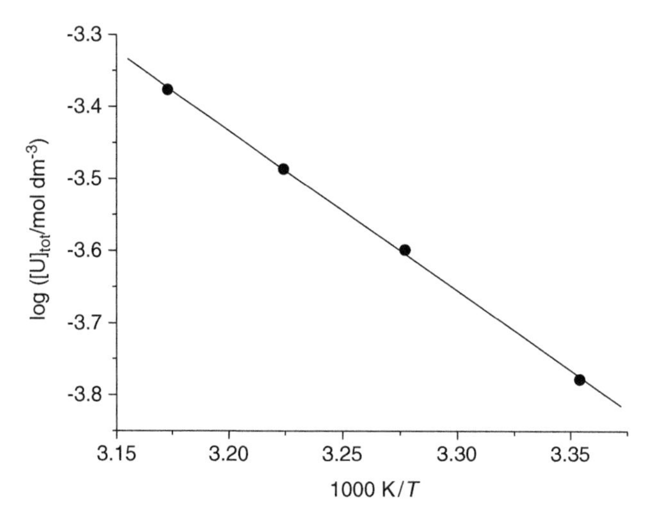

Fig. 2. Solubility of sodium hydrogenurate monohydrate in 0.15 mol kg^{-1} NaCl at physiological pH at 25, 32, 37 and 42°C [37].

3.3.4. Gallstones

More than 80% of gallstones in the Western world are cholesterol gallstones [59]. Other gallstones contain bilirubin as their main constituent, but inorganic or organic calcium salts (carbonate, phosphate, bilirubinate, palmitate) are almost always present [19].

Crystallisation of biliary cholesterol monohydrate is a multiphase process not yet fully understood [60]. Bile is normally supersaturated with respect to cholesterol [61] which is solubilised by bile salts (the soluble end product of cholesterol metabolism, such as sodium glycocholate and sodium taurocholate [9]) within micelles, whose solubilising capacity is considerably increased by the incorporation of phospholipid molecules such as lecithin [62]. Biliary vesicles contain virtually no bile salts but may accumulate cholesterol up to a cholesterol/phospholipid ratio of 2:1 (by phospholipid transfer to micelles) [63]. These thermodynamically unstable (but kinetically stabilised) vesicles then aggregate and nucleate cholesterol crystals [64, 65]. The mechanism of this crucial micelle-to-vesicle transition has been the subject of various physicochemical studies, including, e.g. calorimetric, turbidimetric, dynamic light and neutron scattering methods [66–69].

The solubility of cholesterol has been studied as a function of several physiologically important variables, such as the concentrations of various bile salts and phospholipids, e.g. [70, 71]. Carey [72] has generated extensive cholesterol solubility tables for native bile by identifying two key physicochemical variables, the bile salt-to-lecithin ratio and the total lipid concentration (bile salts + lecithin + cholesterol), that determine the solubility of cholesterol in bile [70]. These tables permit calculation of the lithogenic index or percent cholesterol saturation of native bile.

3.3.5. Pancreatic Stones

Calcium carbonate is a major constituent of pancreatic stones (consisting of ca. 95% calcite) and is found occasionally in salivary stones and many pigment gallstones, since these three gastrointestinal secretions have high pH values and contain high hydrogen carbonate concentrations. The normal function of pancreatic secretion is to contribute to the neutralisation of gastric acid and support digestion by a variety of enzymes. A physicochemical model of calcite solubility in pancreatic juice has been developed that takes the complexation constants of calcium ions with (hydrogen-) carbonate and proteins into account [73]. All of these ligands are important buffers for calcium ions in the juice [73] and various in vitro studies have been performed to investigate the dissolution of stones as well as the influence of additives on calcium carbonate saturation in pancreatic secretions (see Ref. 5).

3.3.6. Salivary and Dental Calculi

Williams et al. [74] have developed a thermodynamic model of saliva which is able to simulate the distribution and speciation of metal ions in this biological fluid. The model predicts that in the pH range of 5–6, saliva becomes supersaturated with respect to calcium phosphates in the order octacalcium phosphate $<$ hydroxyapatite $<$ fluoroapatite. This indicates that under normal conditions, no demineralisation of teeth should occur. However, dietary acids are detrimental and are thus normally neutralised by increased secretion of saliva.

Therefore, a thermodynamic driving force also exists for the formation of pathological calcifications that obstruct the salivary glands. These salivary calculi or sialoliths usually contain organic matter and various calcium phosphates [19], particularly carbonateapatites and hydroxyapatite. The latter exhibit macro- and microstructures that are practically identical to those found in hydroxyapatite renal calculi [35]. It has also been reported that the saliva of stone formers, when compared to that of healthy subjects, was characterised by a higher calcium phosphate supersaturation and deficit of crystallisation inhibitors (magnesium, phytate). These factors thus favoured crystallisation, particularly when combined with morphoanatomic disorders (stenosis, diverticuli) within the salivary duct [35].

Dental calculi, i.e. calcifications of the dental plaque biofilm, contain various calcium phosphates, since these inorganic ions are provided by saliva or crevicular fluids. Although the pattern of calcification of oral microorganisms, either intra- or extracellularly, is mainly a characteristic of each bacterial species or strain [75], it may be influenced by nutritional factors, such as saliva proteins, as well [76]. The interactions of saliva with dental calculi and its role in preventing dental caries by controlling the enamel de- and remineralisation processes have been reviewed [19].

3.4. Solubility of Other Metal Compounds

The deposition of other sparingly soluble metal compounds is usually associated with disturbances in homeostasis, i.e. the delicate regulation of metal ion concentrations in body fluids, due to various forms of disease. If these disorders result in metal overload of body fluids, solids may deposit in various tissues and damage them. In humans, metal overload diseases concern primarily iron and copper ions, which are moreover intimately involved in free radical reactions associated with, e.g. atherosclerosis and central nervous system diseases [77]. The toxicity due to oxidative cell damage of iron oxyhydroxide deposits has been related to their solubilities [78].

Iron overload in humans can be non-anaemic (primary haemochromatosis) or anaemic (thalassemia). While non-anaemic iron overload is effectively treated by phlebotomy (bloodletting), a chelating agent is necessary to increase iron excretion in the case of thalassemia [79]. Without treatment, excess iron deposits as haemosiderin, which may contain iron(III) oxyhydroxides that are either amorphous or based on ferrihydrite or goethite structures [78]. Recent progress on the (Mössbauer) spectroscopic characterisation of these deposits and their reactivity with chelating agents has been reviewed [78]. Copper overload in humans is mainly caused by Wilson's disease and has to be treated by life-long chelation therapy, cf. [5].

Other metals, which may precipitate in body fluids or affect the solubilities of other biominerals, are sometimes ingested with food or medications. A prominent example is aluminium, which is contained in food and various antacids in variable amounts. Although acute aluminium intoxications are very rare, subacute and chronic forms of aluminium intoxications due to oral intake have been debated [80]. In particular, a combination of aluminium-containing medications with (complexing) citrate or tartrate solutions (often administered to uraemic patients or contained in soft drinks, respectively) can substantially elevate the absorption of aluminium [81] and has been repeatedly associated with the onset of neurological illness [80]. The evidence implicating aluminium with Alzheimer's disease is controversial [82]. Aluminium hydroxide has a very low solubility at physiological (and intestinal) pH, but the metal rather complexes with phosphate in the intestine, resulting in excess phosphate excretion, inhibition of phosphate adsorption and reduced urinary and serum phosphate levels. This phosphorous depletion syndrome induces adverse effects on calcium metabolism and may cause severe calcium loss which eventually leads to the development of osteoporosis or osteomalacia [83]. Aluminium also forms complexes with intestinal fluoride, an element important for the normal bone structure [83].

4. CONCLUSION

With the exception of urine whose composition can vary over wide ranges, physiological solutions are usually not highly supersaturated with respect to biominerals. A careful control of ion balance, which must be maintained for physiological function, including avoidance of undesirable precipitates, can then be accomplished by biologically controlled resorption and deposition of mineralised tissue. Homeostatic regulation of ion concentration is true of all biological systems, and most likely lies at the evolutionary foundation of regulating biomineral deposition [6].

Considerable research has been devoted to the study of solubility phenomena in body fluids, particularly in the field of pathological mineralisation, aiming at its prevention and treatment. Although the underlying mechanisms and interactions with low-molecular-weight ligands and biological macromolecules are inherently complex, computerised modelling of solubility equilibria in body fluids has proven a valuable tool for predicting supersaturation and hence the risk of undesirable precipitation.

REFERENCES

[1] P.J. Marie, P. Ammann, G. Boivin and C. Rey, Calcif. Tissue Int., 69 (2001) 121.
[2] M.E.J. Curzon, Nutr. Res., 8 (1988) 321.
[3] J. Featherstone, C. Shields, B. Khademazad and M.D. Oldershaw, J. Dent. Res., 62 (1983) 1049.
[4] P.M. May, in "Handbook of Metal–Ligand Interactions in Biological Fluids. Bioinorganic Chemistry" (G. Berthon, ed.), p. 1184, Marcel Dekker, New York, 1995.
[5] E. Königsberger and L.-C. Königsberger (eds.), in "Biomineralization – Medical Aspects of Solubility", p. 1, Wiley, Chichester, 2006.
[6] F.F. Amos, M.J. Olszta, S.R. Khan and L.B. Gower, in "Biomineralization – Medical Aspects of Solubility" (E. Königsberger and L.-C. Königsberger, eds.), p. 125, Wiley, Chichester, 2006.
[7] E. Königsberger and L.-C. Königsberger (eds.), Biomineralization – Medical Aspects of Solubility, Wiley, Chichester, 2006.
[8] A.C. Guyton and J.E. Hall, Textbook of Medical Physiology, Elsevier, Amsterdam, 2006.
[9] A.F. Hofmann, in "Handbook of Metal–Ligand Interactions in Biological Fluids. Bioinorganic Medicine" (G. Berthon, ed.), p. 38, Marcel Dekker, New York, 1995.
[10] B. Cassin and S. Solomon, Dictionary of Eye Terminology, Triad Publishing Company, Gainesville, FL, 1990.
[11] P.A. Stewart, How to Understand Acid-Based: a Quantitative Acid–Base Primer for Biology and Medicine, Elsevier, Amsterdam, 1981.
[12] J.C. Shim, Y. Kaminoh, C. Tashiro, Y. Miyamoto and H.K. Yoo, J. Anesth., 10 (1996) 276.
[13] J. Lerman, G.A. Gregory and E.I. Eger, Anesth. Analg., 62 (1984) 911.
[14] M.H. Abraham, M.J. Kamlet, R.W. Taft, R.M. Doherty and P.K. Weathersby, J. Med. Chem., 28 (1985) 865.
[15] J. Sangster, Octanol–Water Partition Coefficients: Fundamentals and Physical Chemistry, Wiley, Chichester, 1997.

[16] H.A. Lowenstam and S. Weiner, On Biomineralization, Oxford University Press, New York, 1989.
[17] S. Mann, J. Webb and R.J.P. Williams (eds.), Biomineralization: Chemical and Biochemical Perspectives, Wiley-VCH, Weinheim, 1989.
[18] S. Mann, Biomineralization: Principles and Concepts in Bioinorganic Materials Chemistry, Oxford University Press, Oxford, 2001.
[19] M.C.F. Magalhães, P.A.A.P. Marques and R.N. Correia, in "Biomineralization – Medical Aspects of Solubility" (E. Königsberger and L.-C. Königsberger, eds.), p. 71, Wiley, Chichester, 2006.
[20] W.J. Landis, Phosphorus Sulfur Silicon Relat. Elem., 146 (1999) 185.
[21] H.C. Anderson, Rheum. Dis. Clin. North Am., 14 (1988) 303.
[22] D.J. Pounder, Pathology, 17 (1985) 115.
[23] D. Ramonet, L. de Yebra, K. Fredriksson, F. Bernal, T. Ribalta and N. Mahy, J. Neurosci. Res., 83 (2006) 147.
[24] J.L. Meyer, J.T. McCall and L.H. Smith, Calcif. Tissue Res., 15 (1974) 287.
[25] F.J. Schoen and R.J. Levy, Eur. J. Cardiothorac. Surg., 6 (2006) S91.
[26] B. Leyland-Jones, Semin. Oncol., 30 (2003) 13.
[27] L.H.J. Riley, H.A. Paschall and R.A. Robinson, J. Surg. Res., 6 (1966) 171.
[28] F.N. Ghadially, Ultrastruct. Pathol., 25 (2001) 243.
[29] F. Grases and A. Costa-Bauzá, Anticancer Res., 19 (1999) 3717.
[30] M.S. Seelig, in "Handbook of Metal–Ligand Interactions in Biological Fluids. Bioinorganic Medicine" (G. Berthon, ed.), p. 914, Marcel Dekker, New York, 1995.
[31] G.M. Parkinson, Curr. Opin. Urol., 8 (1998) 301.
[32] H. Rangaswami, A. Bulbule and G.C. Kundu, Trends Cell Biol., 16 (2006) 79.
[33] R.W.E. Watts, Q. J. Med., 98 (2005) 241.
[34] J.H. Parks and F.L. Coe, Kidney Int., 50 (1996) 1706.
[35] F. Grases and A. Costa-Bauzá, in "Biomineralization – Medical Aspects of Solubility" (E. Königsberger and L.-C. Königsberger, eds.), p. 39, Wiley, Chichester, 2006.
[36] E. Königsberger and L.-C. Königsberger, Pure Appl. Chem., 73 (2001) 785.
[37] Z. Wang and E. Königsberger, Thermochim. Acta, 310 (1998) 237.
[38] E. Königsberger and Z. Wang, Monatsh. Chem., 130 (1999) 1067.
[39] E. Königsberger, Z. Wang and L.-C. Königsberger, Monatsh. Chem., 131 (2000) 39.
[40] E. Königsberger, Z. Wang, J. Seidel and G. Wolf, J. Chem. Thermodyn., 33 (2001) 1.
[41] G. Sadovska, I. Kron and E. Königsberger, Monatsh. Chem., 134 (2003) 787.
[42] Z. Wang, J. Seidel, G. Wolf and E. Königsberger, Thermochim. Acta, 354 (2000) 7.
[43] F. Grases, A.I. Villacampa, O. Söhnel, E. Königsberger and P.M. May, Cryst. Res. Technol., 32 (1997) 707.
[44] E. Königsberger and L.-C. Tran-Ho, Curr. Top. Solut. Chem., 2 (1997) 183.
[45] J. Streit, L.-C. Tran-Ho and E. Königsberger, Monatsh. Chem., 129 (1998) 1225.
[46] P.M. May and K. Murray, Talanta, 38 (1991) 1409.
[47] P.M. May and K. Murray, Talanta, 38 (1991) 1419.
[48] P.M. May and K. Murray, Talanta, 40 (1993) 819.
[49] P.M. May and K. Murray, J. Chem. Eng. Data, 46 (2001) 1035.
[50] A. Rodgers, S. Allie-Hamdulay and G. Jackson, Nephrol. Dial. Transplant., 21 (2006) 361.

[51] J.W. Heinecke, in "Handbook of Metal–Ligand Interactions in Biological Fluids. Bioinorganic Medicine" (G. Berthon, ed.), p. 986, Marcel Dekker, New York, 1995.
[52] H.C. Stary, Z. Kardiol., 89 (2000) 28.
[53] D. Hirsch, R. Azoury and S. Sarig, J. Cryst. Growth, 104 (1990) 759.
[54] D. Hirsch, R. Azoury, S. Sarig and H.S. Kruth, Calcif. Tissue Int., 52 (1993) 94.
[55] S. Sarig, T.A. Weiss, I. Katz, F. Kahana, R. Azoury, E. Okon and H.S. Kruth, Lab. Invest., 71 (1994) 782.
[56] A. Weinberger, Curr. Opin. Rheumatol., 7 (1995) 359.
[57] R.W. Fiddis, N. Vlachos and P.D. Calvert, Ann. Rheum. Dis., 42 (1983) 12.
[58] J.N. Loeb, Arthritis Rheum., 15 (1972) 189.
[59] G. Paumgartner and T. Sauerbruch, Lancet, 338 (1991) 1117.
[60] F.M. Konikoff, A. Kaplun and T. Gilat, Scanning Microsc., 13 (1999) 381.
[61] R.T. Holzbach, M. Marsh, M. Olszewski and K. Holan, J. Clin. Invest., 52 (1973) 1467.
[62] T. Gilat and G.J. Somjen, Biochim. Biophys. Acta, 1286 (1996) 95.
[63] K.J. van Erpecum, Biol. Cell, 97 (2005) 815.
[64] B.J.M. van de Heijning, M.F.J. Stolk, K.J. van Erpecum, W. Renooij, A.K. Groen and G.P. van Berge-Henegouwen, J. Lipid Res., 35 (1994) 1002.
[65] Z. Halpern, M.A. Dudley, M.P. Lynn, J.M. Nader, A.C. Breuer and R.T. Holzbach, J. Lipid Res., 27 (1986) 295.
[66] C.H. Spink, V. Lieto, E. Mereand and C. Pruden, Biochemistry, 30 (1991) 5104.
[67] E.-O. Lee, J.-G. Kim and J.-D. Kim, J. Biochem., 112 (1992) 671.
[68] M.A. Long, E.W. Kaler and S.P. Lee, Biophys. J., 67 (1994) 1733.
[69] J. Leng, S.U. Egelhaaf and M.E. Cates, Biophys. J., 85 (2003) 1624.
[70] M.C. Carey and D.M. Small, J. Clin. Invest., 61 (1978) 998.
[71] A. Bandyopadhyay and S.P. Moulik, J. Phys. Chem., 95 (1991) 4529.
[72] M.C. Carey, J. Lipid Res., 19 (1978) 945.
[73] E.W. Moore and H.J. Verine, Am. J. Physiol., 252 (1987) G707.
[74] D.R. Williams, C.C. Coombes and L. Wu, in "Handbook of Metal–Ligand Interactions in Biological Fluids. Bioinorganic Chemistry" (G. Berthon, ed.), p. 1195, Marcel Dekker, New York, 1995.
[75] J. Ennever, J.L. Streckfuss and I. Takazoe, J. Dent. Res., 52 (1973) 305.
[76] T. Lie and K.A. Selvig, Scand. J. Dent. Res., 82 (1974) 135.
[77] G. Minotti, A. Mordente and A.F. Cavaliere, in "Handbook of Metal–Ligand Interactions in Biological Fluids. Bioinorganic Medicine" (G. Berthon, ed.), p. 962, Marcel Dekker, New York, 1995.
[78] W. Chua-anusorn and T.G. St. Pierre, in "Biomineralization – Medical Aspects of Solubility" (E. Königsberger and L.-C. Königsberger, eds.), p. 219, Wiley, Chichester, 2006.
[79] G.E. Jackson and N. Jarvis, in "Handbook of Metal–Ligand Interactions in Biological Fluids. Bioinorganic Chemistry" (G. Berthon, ed.), p. 1206, Marcel Dekker, New York, 1995.
[80] A. Lione, in "Handbook of Metal–Ligand Interactions in Biological Fluids. Bioinorganic Medicine" (G. Berthon, ed.), p. 1401, Marcel Dekker, New York, 1995.

[81] C. Orvig and G. Berthon, in "Handbook of Metal–Ligand Interactions in Biological Fluids. Bioinorganic Chemistry" (G. Berthon, ed.), p. 1266, Marcel Dekker, New York, 1995.
[82] D.R. McLachlan, W.J. Lukiw and T.P.A. Kruck, in "Handbook of Metal–Ligand Interactions in Biological Fluids. Bioinorganic Medicine" (G. Berthon, ed.), p. 935, Marcel Dekker, New York, 1995.
[83] H. Spencer, in "Handbook of Metal–Ligand Interactions in Biological Fluids. Bioinorganic Medicine" (G. Berthon, ed.), p. 954, Marcel Dekker, New York, 1995.

Index

S

Printed and bound by CPI Group (UK) Ltd, Croydon, CR0 4YY
08/05/2025
01864806-0009